Mathematical and Analytical Techniques with Applications to Engineering

The importance of mathematics in the study of problems arising from the real world, and the increasing success with which it has been used to model situations ranging from the purely deterministic to the stochastic, in all areas of today's Physical Sciences and Engineering, is well established. The progress in applicable mathematics has been brought about by the extension and development of many important analytical approaches and techniques, in areas both old and new, frequently aided by the use of computers without which the solution of realistic problems in modern Physical Sciences and Engineering would otherwise have been impossible. The purpose of the series is to make available authoritative, up to date, and self-contained accounts of some of the most important and useful of these analytical approaches and techniques. Each volume in the series will provide a detailed introduction to a specific subject area of current importance, and then will go beyond this by reviewing recent contributions, thereby serving as a valuable reference source.

More information about this series at http://www.springer.com/series/7311

Il Han Park

Design Sensitivity Analysis and Optimization of Electromagnetic Systems

Springer

Il Han Park
Information and Communication
Engineering
Sungkyunkwan University
Suwon, Gyeonggi-do, Korea (Republic of)

ISSN 1559-7458 ISSN 1559-7466 (electronic)
Mathematical and Analytical Techniques with Applications to Engineering
ISBN 978-981-13-0229-9 ISBN 978-981-13-0230-5 (eBook)
https://doi.org/10.1007/978-981-13-0230-5

Library of Congress Control Number: 2018946570

Printed on acid-free paper

This Springer imprint is published by the registered company Springer Nature Singapore Pte Ltd
The registered company address is: 152 Beach Road, #21-01/04 Gateway East, Singapore 189721, Singapore

Dedicated to
Miree, Seoyeon and Hajin

Preface

Design is the process of properly placing materials in a space to obtain a desired performance. The placement of materials sets a device's shape, which determines its performance. The performance of the electromagnetic system is also determined by its shape. But, the performance of the electromagnetic system is expressed with the electromagnetic field; its performance is indirectly related to the shape. This book presents the design sensitivity analysis for the electromagnetic system, which is on the relation between the performance and the geometric design variables. The design sensitivity, which is the variation rate of the system performance with respect to the design variables, provides information on how the design variables affect the performance.

The electromagnetic systems are diverse in type and size, ranging from micro-electronic devices to large power apparatus. For analysis of such various electromagnetic systems, the finite element method is popular among the engineers, researchers and graduate students. But, the finite element code is an analysis tool not a design tool; the design process using the finite element code needs much trial and error, which requires considerable time and effort.

In the mechanical engineering, a large number of research papers and books for the optimal structure design are found and some commercial codes with the design sensitivity analysis are available. By contrast, there are only few books on the optimal design of the electromagnetic system. This book may be the first one devoted to the sensitivity analysis for the electromagnetic system.

This book aims to cover the theory and application of the shape sensitivity analysis for the electromagnetic system in a unified manner. The focus is on the continuum sensitivity analysis, which has great advantages over the other sensitivity methods: the finite difference method and the discrete method. The continuum design sensitivity is obtained as an analytical form; thus, it makes it easy to calculate the sensitivity and provides accurate sensitivity. In addition, it can be easily implemented with existing numerical analysis codes such as finite element method and boundary element method since its sensitivity calculation does not depend on the analysis method.

The continuum shape sensitivity for the electromagnetic system is derived by taking the material derivative of the design performance and the variational state equation. In this differentiation, the Lagrange multiplier method is introduced to deal with the implicit equality constraint of the variational state equation. An adjoint variable technique is also employed to express explicitly the sensitivity in terms of the design variables. The variational identities are used to transform the sensitivity of a domain integral into a boundary integral on the design surface. This continuum shape sensitivity analysis, which is applied to four electromagnetic systems: the electrostatic system, the magnetostatic system, the eddy current system and the DC conductor system, provides the sensitivity formulas for each electromagnetic system. The sensitivity formulas so obtained are the general three-dimensional ones of an analytical form. These analytical sensitivity formulas provides not only physical insight but also great advantages in numerical implementation.

The book contains eight chapters and four appendices. In Chap. 1 a brief review of optimal design process and design steps for the electromagnetic system is presented and the geometric design variables are classified. The Maxwell's equations and the governing differential equations are introduced and the characteristics of the electromagnetic system are described for comparison with the structural system in the mechanical engineering. An overview of design sensitivity calculation method is also provided.

In Chap. 2, the four variational state equations for the electrostatic system, the magnetostatic system, the eddy current system and the DC current-carrying conductor are formulated by the variational method of the virtual work principle. The variational equations are derived from the differential equations with boundary conditions and they are used for deriving the continuum sensitivity formulas for the four electromagnetic systems in Chaps. 3–6.

In Chap. 3, the general three-dimensional continuum shape sensitivities for the electrostatic system are derived by using the material derivative and are applied to design problems. The shape sensitivity for the electrostatic system is classified into two types according the design variable. One is for the design problem of outer boundary and the other is for the design problem of interface. Each one has also two different types of objective functions: domain integral and system energy. The sensitivity for the system energy is examined in the electric-circuit point of view to show its sign dependency on the source condition and to derive the capacitance sensitivity. The general sensitivity formulas are applied to analytical and numerical design examples to be validated.

In Chap. 4, the general three-dimensional continuum shape sensitivities for the magnetostatic system are derived and are applied to design problems. Unlike in the electrostatic system, the shape sensitivity for the magnetostatic system has only one type for the design problem of interface. The interface design problem has also two different types of objective functions: domain integral and system energy. The magnetostatic system may have four different material regions: ferromagnetic material, permanent magnet, source current, air; thus, the general sensitivity is expressed as the sensitivity formulas for nine interfaces. The system energy sensitivity is derived in the electric-circuit point of view, and it is used to the inductance sensitivity. The general sensitivity formulas are applied to analytical and numerical design examples to be validated.

In Chap. 5, the three-dimensional continuum shape sensitivities for the eddy current system are derived and are applied to design problems. Like in the magnetostatic system, the shape sensitivity for the eddy current system has only one type for the design problem of interface. The interface design problem has also two different types of objective functions: domain integral and system power. The eddy current system may have four different material regions: ferromagnetic material, conductive material, source current, air; thus, the general sensitivity is expressed as the sensitivity formulas for nine interfaces. The system power sensitivity is derived in the electric-circuit point of view, and then the inductance sensitivity and the resistance sensitivity are derived. The two sensitivity formulas are applied to numerical examples to be validated.

In Chap. 6, the general three-dimensional continuum shape sensitivity for the DC conductor system is derived and are applied to design problems. The design problem of the DC conductor system is similar to that of the electrostatic system, but it has only the design variable of outer boundary. The design problem of outer boundary has also two different types of objective functions: domain integral and system loss power. The derived sensitivity formula is expressed as a boundary integral of Dirichlet boundary and Neumann boundary. The loss power sensitivity is used to derive the resistance sensitivity. The general sensitivity formulas are applied to analytical and numerical design examples to be validated.

The shape optimal design using the sensitivity requires the optimization algorithms and the successive geometry modeling for evolving shapes. For this purpose, Chap. 7 introduces the level set method. The level set method expresses the shape variation with the velocity field; thus, it matches well with the continuum shape sensitivity, whose sensitivity formulas are expressed with the velocity. The level set method and the continuum sensitivity are coupled to transform the usual iterative optimization into the solving process of the level set equation, which is the transient analysis in the time domain. The adaptive level set method and the artificial diffusion technique are also presented for solving the coupled level set equation with existing finite element codes.

In Chap. 8, the hole and the dot sensitivity analyses are presented for the topology optimization of the electrostatic and the magnetostatic systems. The hole sensitivity formulas in the dielectric and the magnetic material regions are derived by using a hole sensitivity concept and the continuum sensitivity in the electrostatic and the magnetostatic system. The dot sensitivity formulas in the dielectric and the magnetic material regions are also derived by using a dot sensitivity concept and the continuum sensitivity. The derived hole and the dot sensitivity formulas are applied to numerical examples to show its usefulness.

The four Appendices A-D provide more analytical and numerical examples for the four electromagnetic systems, most of which are ones for other coordinates and interfaces not included in the examples of the Chaps. 3–6.

Suwon, Korea (Republic of) Il Han Park

Acknowledgements

I am indebted to all the colleges who have contributed to the preparation of this book. Without their help, this book would not have been written.

My interest in the optimization of the electromagnetic system began in 1987, when Prof. Song Yop Hahn of Seoul National University introduced me to the topic. After I completed my Ph.D. thesis in 1990, Prof. J. C. Sabonnadiere and Prof. J. L. Coulomb of Grenoble Institute of Technology provided the opportunity to continue the work on the optimization.

I would like to thank my graduate students at Sungkyunkwan University, who work with me on the optimization problems. Special thanks are due to Kang Hyouk Lee, Kyung Sik Seo, and Seung Geon Hong, who prepared many design data and figures for this book. In particular, I had the pleasure to incorporate the co-work results of Dr. Joon Ho Lee, Dr. In Gu Kwak, Prof. Dong Hun Kim, and Prof. Jin Kyu Byun in this book. Thanks are also due to Prof. Hong Soon Choi, Prof. Young Sun Kim, Prof. Se-Hee Lee, and Dr. Myung Ki Baek, the research results of whom enrich the contents in this book. I have learned many interesting things about mechanical systems from my friend Wonkyu Moon at Pohang University of Science and Technology, and it is a pleasure to acknowledge his help.

Financial support for my research from the KETEP (grant No. 2016403020098) is gratefully acknowledged. I am delighted to express my thanks to Springer editors for their friendly cooperation in the publication of this book. Last but not least, I would like to thank my wife, Miree, and our children, Seoyeon and Hajin, for their love and understanding.

Contents

Chapter 1
Introduction

1.1 Optimal Design Process

Optimal design of electromagnetic system consists of procedures to improve the performance by evolving design variables. There are many kinds of performance measures such as electric/magnetic field distribution, system energy, system power, force/torque, energy loss, equivalent circuit parameters, induced voltage, material volume, etc. Moreover, the electromagnetic system has various constraints and design variables since it is composed of many different materials such as dielectrics, conductor, insulator, charge, magnetic material, current, permanent magnet and electrolet.

The structure of the electromagnetic system is usually so complex and sophisticated that its design process has been dependent on the engineer's experience and intuition. A systematic design process will enable the designer to develop an improved device with less time and cost. For this purpose, simulation-based design is efficient for development and production of the better electromagnetic devices [1].

The simulation-based design consists of modeling, system analysis, sensitivity analysis, and optimization. The optimal design process is shown in Fig. 1.1, where the system analysis and the sensitivity analysis are important procedures [2].

1.2 Design Steps of Electromagnetic System

Choosing the design variables in system modeling is an important step to a successful design. It is often difficult to identify the design variables that have substantial influence on the performance. It is mainly due to system structure's complexity. Wrong choice of the design variables, which limits the size of design space for searching the design variables, results in a wrong design.

I. H. Park, *Design Sensitivity Analysis and Optimization of Electromagnetic Systems*,
Mathematical and Analytical Techniques with Applications to Engineering,
https://doi.org/10.1007/978-981-13-0230-5_1

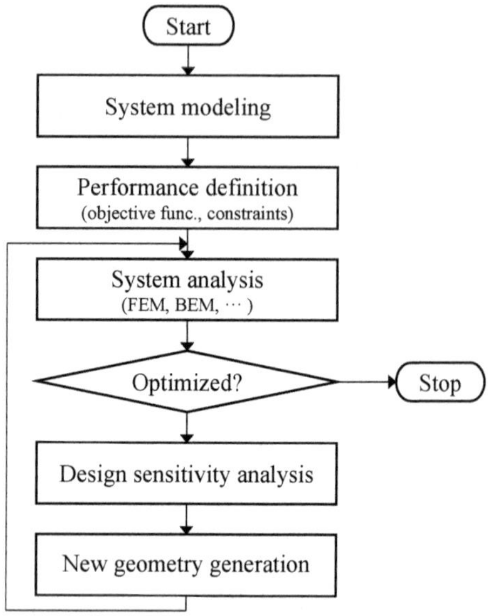

Fig. 1.1 Design process of electromagnetic system

The result of system analysis is used to evaluate the system performance. Nowadays, most of the system analyses for electromagnetic devices are carried out using numerical methods such as the finite element method, the boundary element method, etc. The finite element method, which is widely applicable to various electromagnetic systems including nonlinear system, provides reliable and accurate result; it is most frequently employed by the researchers and engineers. This book also employs the finite element method to analyze the electromagnetic system.

The objective function (performance measure, cost function), which is a criterion to ascertain whether the design is satisfactory or not, is evaluated with the results of the system analysis. Definition of the objective function, which has great influence on sensitivity evaluation and convergence, is also important to obtain a successful design. For example, the force/torque of the electromechanical system can be easily controlled with the objective function of the system energy in comparison with the objective function of force/torque, which often leads to difficulty and complexity of sensitivity evaluation. During the design optimization process, the objective function is minimized or maximized by the optimization algorithm of the mathematical programming.

The objective function for the electromagnetic system is usually nonlinear to the design variables. It is common to use the gradient-based method for the optimization algorithm. The gradient, which is called the design sensitivity of the objective function, is obtained by differentiating the objective function with respect to the design variables. The gradient information is used as the searching direction in the design space. The sensitivity analysis is the main concern of this book. The sensitivity, which means the effect of the design variables on the objective function, provides the information on how the design variables have influence on the objective function. The sensitivity information can be also used for identifying the key design variables.

The optimization problems are usually subject to constraints. If a design satisfies given constraints, it is called feasible design. If not, it is infeasible design. Whereas some design problems have simple constraints such as the upper/lower limits of the design variables and the constant volume of used material, others have complex constraints that are indirectly affected by the design variable. For example, when an electric field at a point of an electrostatic system is given as a constraint, it is usually impossible to explicitly express that with the design variables. This kind of constraint is called implicit constraint. The constraint of an explicit function of the design variables is simple, whereas the constraint of an implicit function of the design variables is complex to deal with. In this book, the electromagnetic state equation of the variational form is taken as a constraint and incorporated into the objective function by the Lagrange multiplier method.

1.3 Design Variables

The optimal design is the searching process for the better design variables providing the desired performance. Unless the design variables are well defined, the design space is limited so that a good design is not obtained no matter how accurate solution is used. For example, while a coarse geometrical modeling does not matter for an insensitive design region, a fine geometrical modeling is needed for a sensitive design region.

The material of the electromagnetic systems can be classified into two categories: active ones and passive ones. The active materials generate the source field, whereas the passive materials only react to the external field. The ferromagnetic material, the dielectrics, the electric conductor, the electric insulator, and the air belong to the passive materials. The electric charge, the electric current, the permanent magnet, and the electrolet are the active materials. For the optimal design of the electromagnetic system, the property of these materials is not taken as the design variable in this book, since it is neither controllable nor continuous in the available materials.

In this sensitivity analysis, only the geometric parameters of the material structure are taken as the design variables. The geometric design variables are classified into three categories: size, shape and topology as shown in Fig. 1.2. The size design variables such as width, height, depth, radius, angle, etc. are used for simple structures. The shape design variables, which cannot be defined with the size design variables, are used for more complex geometry. During the shape design process, its initial topology is maintained. The topology design variables are related to system layout. When a new material domain is generated outside a given material domain or an air hole is generated inside the material domain, the system topology changes. Recently, some topology design methods have been introduced to the electromagnetic system.

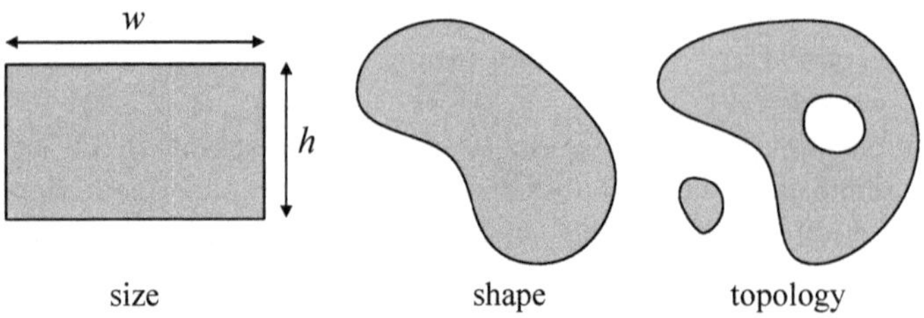

Fig. 1.2 Geometric design variables

Mathematically, the size design variables are a subset of the shape design variables, and the shape design variables are also a subset of the topology design variables. This book deals with the shape and topology design of the electromagnetic system. The optimal shape design is carried out by using the shape design sensitivity, which is derived as the analytical integral forms in the subsequent chapters. The sensitivity with respect to the size design variables are easily calculated by the design variable parametrization, which relates the size design variable to the shape design sensitivity. The topology sensitivity can be also derived with the concept of topology sensitivity and the shape design sensitivity.

1.4 Equations and Characteristics of Electromagnetic Systems

The electromagnetic systems, which are represented with the Maxwell's equations, are usually modeled by the partial differential equations for the electric and magnetic potentials. The electromagnetic systems are classified into four systems: electrostatic system, magnetostatic system, eddy current system, and wave system. These four systems are also represented with the governing partial differential equations: elliptic, parabolic and hyperbolic equations [3]. The governing equations for the electrostatic and magnetostatic systems are elliptic, and the ones for the eddy current and the wave systems are parabolic and hyperbolic, respectively. The understanding of the characteristics of the electromagnetic system is important to the development of the sensitivity analysis. In particular, recognition of differences between the electromagnetic system and the mechanical structure is helpful.

1.4.1 Maxwell's Equations and Governing Equations

The electromagnetic systems are generally represented with the Maxwell's equations and the constitutive relations [4–7]. The Maxwell equations in the differential form are:

$$\nabla \cdot \mathbf{D} = \rho \tag{1.4.1}$$

$$\nabla \times \mathbf{E} = -\frac{\partial \mathbf{B}}{\partial t} \tag{1.4.2}$$

$$\nabla \cdot \mathbf{B} = 0 \tag{1.4.3}$$

$$\nabla \times \mathbf{H} = \mathbf{J} + \frac{\partial \mathbf{D}}{\partial t} \tag{1.4.4}$$

where **D** is the electric flux density, ρ the volume charge density, **E** the electric field intensity, **H** the magnetic field intensity, **J** the volume current density and **B** the magnetic flux density. The constitutive relations are given as

$$\mathbf{D} = \varepsilon \mathbf{E} + \mathbf{P}_o \tag{1.4.5}$$

$$\mathbf{H} = \nu \mathbf{B} - \mathbf{M}_o \tag{1.4.6}$$

$$\mathbf{J} = \sigma \mathbf{E} \tag{1.4.7}$$

where ε is the electric permittivity, $\mathbf{P}_o$ the permanent polarization, ν the magnetic reluctivity, $\mathbf{M}_o$ the permanent magnetization, and σ the electric conductivity. This book deals with only the low frequency system, where the displacement current in (1.4.4) is ignored. The wave system is out of the scope of this book.

The electrostatic system is represented by two equations from Maxwell's equations and one constitutive relation;

$$\nabla \cdot \mathbf{D} = \rho \tag{1.4.8}$$

$$\nabla \times \mathbf{E} = \mathbf{0} \tag{1.4.9}$$

$$\mathbf{D} = \varepsilon \mathbf{E} + \mathbf{P}_o \tag{1.4.10}$$

With the electric scalar potential ϕ introduced, the governing partial differential equation for the electrostatic system is obtained as Poisson equation;

$$-\nabla \cdot \varepsilon \nabla \phi = \rho - \nabla \cdot \mathbf{P}_o \tag{1.4.11}$$

The magnetostatic system is represented by two equations from Maxwell's equations and one constitutive relation;

$$\nabla \times \mathbf{H} = \mathbf{J} \tag{1.4.12}$$

$$\nabla \cdot \mathbf{B} = 0 \tag{1.4.13}$$

$$\mathbf{H} = \nu\mathbf{B} - \mathbf{M}_o \tag{1.4.14}$$

With the magnetic vector potential **A** introduced, the governing partial differential equation for the magnetostatic system is obtained as

$$\nabla \times \nu(\nabla \times \mathbf{A}) = \mathbf{J} + \nabla \times \mathbf{M}_o \tag{1.4.15}$$

The eddy current system is represented by three equations from Maxwell's equations and two constitutive relations;

$$\nabla \times \mathbf{H} = \mathbf{J} + \mathbf{J}_\mathrm{e} \tag{1.4.16}$$

$$\nabla \cdot \mathbf{B} = 0 \tag{1.4.17}$$

$$\nabla \times \mathbf{E} = -\frac{\partial \mathbf{B}}{\partial t} \tag{1.4.18}$$

$$\mathbf{H} = \nu\mathbf{B} \tag{1.4.19}$$

$$\mathbf{J}_e = \sigma\mathbf{E} \tag{1.4.20}$$

where **J** is the source current density and $\mathbf{J}_e$ is the eddy current density. By introducing the magnetic vector potential **A** and the electric scalar potential ϕ, the governing partial differential equation for the eddy current system is obtained as

$$\nabla \times \nu\nabla \times \mathbf{A} = \mathbf{J} - \sigma\left(\frac{\partial \mathbf{A}}{\partial t} + \nabla\phi\right) \tag{1.4.21}$$

In the linear eddy current system without the term$\nabla\phi$, when the harmonic source is considered, the governing equation for the steady state is expressed using the complex variables as

$$\nabla \times \nu\nabla \times \mathbf{A} + j\omega\sigma\mathbf{A} = \mathbf{J} \tag{1.4.22}$$

In this book, the DC current-carrying conductor is separately described. The DC current-carrying conductor, although it has the same form of governing equation as the electrostatic system, is quite different in physics and related to the resistance of the equivalent circuit, the Joule loss, the current distribution, etc. The DC current-carrying conductor is represented by two equations from Maxwell's equations and one constitutive relation;

$$\nabla \cdot \mathbf{J} = 0 \tag{1.4.23}$$

$$\nabla \times \mathbf{E} = \mathbf{0} \tag{1.4.24}$$

$$\mathbf{J} = \sigma \mathbf{E} \tag{1.4.25}$$

where (1.4.23) is the continuity equation, which is implicit in (1.4.4) of Maxwell's equations. With the electric scalar potential ϕ introduced, the governing partial differential equation for the DC current-carrying conductor is obtained as Laplace equation;

$$-\nabla \cdot \sigma \nabla \phi = 0 \tag{1.4.26}$$

1.4.2 Characteristics of Electromagnetic Systems

The shape design sensitivity analysis has been well developed for optimal design of mechanical structures, for which a large number of research results are found in books and papers. Such a wealth of research results is very helpful for the sensitivity analysis of the electromagnetic system. There are, however, some differences between the electromagnetic system and the mechanical structure. Recognition of them helps to develop the sensitivity analysis for the electromagnetic systems.

The electromagnetic field exists even in the vacuum, whereas the mechanical fields such as stress, strain, fluidic velocity exist only where the media exist [8, 9]. In the electromagnetic system, the electric/magnetic field exists not only inside the materials but also in the air near the materials. In electromagnetics, the vacuum and the air have the material properties of dielectric constant ε_0 and magnetic permeability μ_0. Thus, the design variable of the electromagnetic system is basically the interface where two different materials meet. For example, the design problem of a magnet, of which the design objective is to produce a uniform magnetic field, is to optimize the interface shape between the ferromagnetic material, the air, and the current coil [10].

The sources of the electromagnetic field can be charge, current, permanent magnet, or electrolet, whereas the source of the mechanical field is only the force. In addition, the sources of the electromagnetic system are usually supplied by the voltage source or the current source through the circuit terminal. The permanent magnet and the electrolet are, however, treated as materials with the source. The electromagnetic system, which is connected to the external circuit, is driven or controlled by the external circuit. Thus, it is important to extract its equivalent circuit parameter. If the equivalent circuit parameter representing the electromagnetic system is incorporated into the external circuit system, the operating characteristics of the electromagnetic system can be easily obtained by analyzing the circuit system.

There are two kinds of nonlinearity in the structural system: the material nonlinearity and the geometrical nonlinearity. The geometrical nonlinearity comes from deformation of the structure geometry. But there is only the material nonlinearity in the electromagnetic system, which appears mainly in the magnetic saturation of the ferromagnetic material [11].

1.5 Design Sensitivity Analysis

The sensitivity calculation is the mathematical procedure of obtaining the derivatives of the objective function with respect to the design variables. The sensitivity calculation of state variables with respect to the design variables often costs the major computational time for optimization process. It is, therefore, crucial to have an efficient algorithm for calculating the sensitivity.

There are two approaches to obtain the design sensitivity. One is finite difference method and the other is analytic differentiation method. The analytic differentiation method is also divided into two methods: discrete one and continuum one [1, 12].

1.5.1 Finite Difference Method

The finite difference method is the simplest technique to obtain the sensitivity. When the objective function is given as a function $F(p)$ of a design variable p, its sensitivity can be approximated by comparing $F(p)$ with $F(p + \Delta p)$ perturbed by Δp in the design variable;

$$\frac{\mathrm{d}F}{\mathrm{d}p} \simeq \frac{F(p + \Delta p) - F(p)}{\Delta p} \tag{1.5.1}$$

This approximation method is so easy to implement that it is popular among engineers. This approximate sensitivity is frequently compared with the sensitivity obtained by the other methods for evaluating their efficiency and accuracy. When design variables are numerous, the finite difference method is computationally expensive. When the number of design variable is n, it requires $n + 1$ times analyses of the system matrix equation. In addition, it has a serious problem of accuracy since its accuracy is strongly dependent on the perturbation size Δp. Too-small perturbation causes numerical truncation errors, and too-large perturbation leads to inaccurate results. Thus, this method is unsuitable for the shape design problem with many design variables. The number of design variables for the shape design is the number of all nodes on the design surface.

1.5.2 Discrete Method

The discrete method of the analytical approach is based on the discretized system equation, which is obtained by numerical analysis methods such as finite element method, boundary element method. [1, 13–19]. The state equation of discretized model is expressed as an algebraic matrix equation;

$$[K(p)][\phi] = [f(p)] \tag{1.5.2}$$

where $[K(p)]$ is the $n \times n$ system matrix, $[\phi]$ the $n \times 1$ state variable vector at nodes, $[f(p)]$ the $n \times 1$ source vector, and n the number of nodes for unknown state variables. The system matrix $[K(p)]$ is determined by the system geometry and the passive material property. The source vector $[f(p)]$ is determined by the system geometry and the active material property of the source. The change in the system geometry causes the changes of $[K(p)]$ and $[f(p)]$, which result in the change of the state variable $[\phi]$. Since the state variable $[\phi]$ depends on the design variable, it can be written as $[\phi(p)]$, which is implicitly affected by the design variable in the system Eq. (1.5.2).

The objective function is usually a function of the design variables and the state variable;

$$F = F\{[p], [\phi(p)]\} \tag{1.5.3}$$

where $[p]$ is the $m \times 1$ design variable vector, $[\phi(p)]$ the $n \times 1$ state variable vector, and m the number of design variables.

The derivative of the objective function is obtained by taking the derivative of (1.5.3) with respect to the design variable vector;

$$\frac{\mathrm{d}F}{\mathrm{d}[p]} = \frac{\partial F}{\partial [p]} + \frac{\partial F}{\partial [\phi]} \frac{\mathrm{d}[\phi]}{\mathrm{d}[p]} \tag{1.5.4}$$

In this sensitivity expression, the two partial derivatives of F are easily obtained since F is an explicit function of $[p]$ and $[\phi]$. But the derivative of the state variable in the second term needs some calculations since the state variable is implicitly related to the design variable in (1.5.2). By taking the derivative of (1.5.2) with respect to the design variable vector, the derivative of the state variable is obtained as

$$\frac{\mathrm{d}[\phi]}{\mathrm{d}[p]} = [K]^{-1} \frac{\partial}{\partial [p]} \left[[f] - [K][\tilde{\phi}] \right] \tag{1.5.5}$$

where $[\tilde{\phi}]$ is the solution of (1.5.2). By inserting (1.5.5) into (1.5.4), the sensitivity is expressed as

$$\frac{\mathrm{d}F}{\mathrm{d}[p]} = \frac{\partial F}{\partial [p]} + \frac{\partial F}{\partial [\phi]} [K]^{-1} \frac{\partial}{\partial [p]} \left[[f] - [K][\tilde{\phi}] \right] \tag{1.5.6}$$

After the derivative of the state variable is calculated in (1.5.5), its values can be inserted into (1.5.5). But it requires m times analyses of the system Eq. (1.5.2). This problem is solved by introduction of an adjoint variable technique, which requires only one analysis. An adjoint variable equation is introduced;

$$[K]^{\mathrm{T}}[\lambda] = \frac{\partial F}{\partial [\phi]^{\mathrm{T}}} \tag{1.5.7}$$

where $[\lambda]$ is the $n \times 1$ adjoint variable vector, which is the nodal values like the state variable [20, 21]. By using the adjoint variable Eq. (1.5.7), the sensitivity is obtained as

$$\frac{\mathrm{d}F}{\mathrm{d}[p]} = \frac{\partial F}{\partial [p]} + [\lambda]^{\mathrm{T}} \frac{\partial}{\partial [p]} \left[[f] - [K][\tilde{\phi}] \right] \tag{1.5.8}$$

The adjoint variable vector, which is calculated in (1.5.7), is inserted into (1.5.8) to provide the sensitivity.

On the other hand, this sensitivity can be also derived using the Lagrange multiplier method. The system matrix (1.5.2), which is a kind of equality constraint, is taken a constraint subject to the objective function (1.5.3). The augmented objective function G with the Lagrange multiplier is written as

$$G = F\{[p], [\phi(p)]\} + ([f(p)] - [K(p)][\phi])[\lambda]^{\mathrm{T}} \tag{1.5.9}$$

where $[\lambda]$ is the $n \times 1$ Lagrange multiplier vector. The derivative of objective function is obtained by taking the derivative of (1.5.9) with respect to the design variable vector;

$$\frac{\mathrm{d}G}{\mathrm{d}[p]} = \frac{\partial F}{\partial [p]} + \frac{\partial F}{\partial [\phi]} \frac{\mathrm{d}[\phi]}{\mathrm{d}[p]} + \left(\frac{\partial}{\partial [p]} \left[[f] - [K][\tilde{\phi}] \right] - [K] \frac{\mathrm{d}[\phi]}{\mathrm{d}[p]} \right) [\lambda]^{\mathrm{T}} + ([f] - [K][\phi]) \frac{\mathrm{d}[\lambda]^{\mathrm{T}}}{\mathrm{d}[p]} \tag{1.5.10}$$

The last term of this equation vanishes by the system state equation (1.5.2);

$$\frac{\mathrm{d}G}{\mathrm{d}[p]} = \frac{\partial F}{\partial [p]} + \frac{\partial F}{\partial [\phi]} \frac{\mathrm{d}[\phi]}{\mathrm{d}[p]} + \left(\frac{\partial}{\partial [p]} \left[[f] - [K][\tilde{\phi}] \right] - [K] \frac{\mathrm{d}[\phi]}{\mathrm{d}[p]} \right) [\lambda]^{\mathrm{T}} \tag{1.5.11}$$

In order to avoid the calculation of $\frac{\mathrm{d}[\phi]}{\mathrm{d}[p]}$ and explicitly express this equation with the design variable, an adjoint equation is introduced:

$$[K]^{\mathrm{T}}[\lambda] = \frac{\partial F}{\partial [\phi]^{\mathrm{T}}} \tag{1.5.12}$$

where $[\lambda]$ is the adjoint variable vector, which is the Lagrange multiplier in (1.5.9). Inserting the relation (1.5.12) into (1.5.11) provides the sensitivity:

$$\frac{\mathrm{d}G}{\mathrm{d}[p]} = \frac{\partial F}{\partial [p]} + [\lambda]^{\mathrm{T}} \frac{\partial}{\partial [p]} \left[[f] - [K][\tilde{\phi}] \right] \tag{1.5.13}$$

This sensitivity is the same as the (1.5.8).

The Lagrange multiplier method is also used for the continuum method in the subsequent chapters. The discrete method is relatively simple to understand since the implicit relation between the state variable and the design variable is clearly shown. The analogy between the discrete method and the continuum method is helpful in developing the continuum sensitivity for the electromagnetic system.

The above sensitivity calculation by the discrete method is summarized as

(a) solve the state variable Eq. (1.5.2) for $[\phi]$.
(b) solve the adjoint variable Eq. (1.5.7) for $[\lambda]$.
(c) calculate the sensitivity (1.5.8) using the obtained $[\phi]$ and $[\lambda]$.

This sensitivity calculation requires only two analyses for the state and adjoint variables. In the adjoint equation, its source term in the right-hand side is easily obtained since the F is an explicit function of $[\phi]$. But the computation of $\frac{\partial [f]}{\partial [p]}$ and $\frac{\partial [K]}{\partial [p]}$ is dependent on discretization since $[K]$ and $[f]$, which are assembled with the element matrices, depend on the element such as the shape function and the mesh data. Thus, their computation requires access to the source code of the analysis program, which makes it difficult to implement the numerical program. It is unfortunate that most of the commercial programs do not provide access to the source code. It is desired to develop a sensitivity evaluation method that does not depend on discretization nor requires access to the inside of the source code.

1.5.3 Continuum Method

In the continuum method, the shape sensitivity is derived using the material derivative concept and the variational formulation for the governing equation of electromagnetic system. The continuum method is the core subject of this book. The material derivative concept of continuum mechanics is employed to relate the shape variation of electromagnetic system to the objective function [22–27]. For general application, the objective function is defined as arbitrary function of the state variables. The electromagnetic system is represented with the variational equation of the continuous model. This variational state equation for the electromagnetic system, which holds regardless of the shape variation, is taken as an equality constraint. For a systematic derivation of the continuum sensitivity, the Lagrange multiplier method is used for the equality constraint. The constraint of the variational state equation is added to the objective function to provide an augmented objective function. By taking the material derivative of this augmented objective function and using the variational identities, the continuum sensitivity formula is obtained. This shape sensitivity formula is expressed in the simple analytical form of surface integral on the design boundary. The integrand of the surface integral is written in terms of the shape variation and physical quantities such as the material properties, the state variable, and the adjoint variable.

If the exact solution for the state variable is given, the sensitivity, which is derived as an analytical form, will be exact. But the exact solution for complex electromagnetic system is not given; the sensitivity formulas are evaluated with the approximate solution by the numerical methods such as finite element method, boundary element method.

The major advantage of the continuum sensitivity is that since the variational system equation is differentiated before discretized, it does not only depend on discretization method but also provide more accurate sensitivity information than the discrete method.

In Chaps. 3–6, for deriving the shape sensitivity formulas, this continuum method is applied to the four electromagnetic systems: electrostatic system, magnetostatic system, eddy current system, and DC current-carrying conductor.

References

1. Choi, K.K., Kim, N.H.: Structural Sensitivity Analysis and Optimization 1: Linear Systems. Springer, New York (2005)
2. Arora, J.S.: Introduction to Optimum Design. Mcgraw-Hill, New York (1989)
3. Zachmanoglou, E.C., Thoe, D.W.: Introduction to Partial Differential Equations with Applications. Williams & Wilkins, Baltimore (1976)
4. Griffiths, D.J.: Introduction to Electrodynamics. Pearson, Boston (2013)
5. Stratton, J.A.: Electromagnetic Theory. McGraw-Hill, New York (1941)
6. Purcell, E.M.: Electricity and Magnetism. Education Development Center Inc., Newton (1965)
7. Reitz, J.R., Milford, F.J., Christy, R.W.: Foundations of Electromagnetic Theory. Addison-Wesley, Reading (1979)
8. Timoshenko, S., Goodier, J.N.: Theory of Elasticity. McGraw-Hill, New York (1951)
9. Sokolnikoff, I.S.: Mathematical Theory of Elasticity. McGraw-Hill, New York (1956)
10. Pironneau, O.: Optimal Shape Design for Elliptic Systems. Springer-Verlag, New York (1984)
11. Cullity, B.D.: Introduction to Magnetic Materials. Addison Wesley, Reading (1972)
12. Haftka, R.T., Grandhi, R.V.: Structural shape optimization—a survey. Comp. Methods Appl. Mech. Eng. **57**, 91–106 (1986)
13. Adelman, H.M., Haftka, R.T.: Sensitivity analysis of discrete structural systems. AIAA J. **24**, 823–832 (1986)
14. Gitosusastro, S., Coulomb, J.L., Sabonnadiere, J.C.: Performance derivative calculations and optimization. IEEE Trans. Magn. **25**, 2834–2839 (1989)
15. Park, I.H., Lee, B.T., Hahn, S.Y.: Pole shape optimization for reduction of cogging torque by sensitivity analysis. COMPEL **9**, Supplement A, 111–114 (1990)
16. Kwak, I.G., Ahn, Y.W., Hahn, S.Y., Park, I.H.: Shape optimization of electromagnetic devices using high order derivatives. IEEE Trans. Magn. **35**, 1726–1729 (1999)
17. Park, I.H., Kwak, I.G., Lee, H.B., Lee, K.S., Hahn, S.Y.: Optimal design of transient eddy current systems driven by voltage source. IEEE Trans. Magn. **33**, 1624–1629 (1997)

18. Park, I.H., Kwak, I.G., Lee, H.B., Hahn, S.Y., Lee, K.S.: Design sensitivity analysis for transient eddy current problems using finite element discretization and adjoint variable method. IEEE Trans. Magn. **32**, 1242–1245 (1996)
19. Park, I.H., Lee, B.T., Hahn, S.Y.: Design sensitivity analysis for nonlinear magnetostatic problems using finite element method. IEEE Trans. Magn. **28**, 1533–1536 (1992)
20. Dems, K., Mróz, Z.: Variational approach by means of adjoint systems to structural optimization and sensitivity analysis—I: variation of material parameters within fixed domain. Int. J. Solids Struct. **19**, 677–692 (1983)
21. Dems, K., Mróz, Z.: Variational approach by means of adjoint systems to structural optimization and sensitivity analysis—II: structure shape variation. Int. J. Solids Struct. **20**, 527–552 (1984)
22. Choi, K.K., Haug, E.J., Hou, J.W., Sohoni, V.N.: Pshenichy's linearization method for mechanical system optimization. J. Mech. Transm. Autom. Des. **105**, 97–103 (1983)
23. Choi, K.K., Haug, E.J.: Shape design sensitivity analysis of elastic structures. J. Struct. Mech. **11**, 231–269 (1983)
24. Haug, E.J., Choi, K.K., Komkov, V.: Design Sensitivity Analysis of Structural Systems. Academic Press, Orlando (1988)
25. Soares, C.A.M.: Computer Aided Optimal Design. Springer-Verlag, Berlin (1987)
26. Park, I.H.: Sensitivity analysis for shape optimization of electromagnetic devices. Ph.D. thesis, Seoul National University (1990)
27. Park, I.H., Coulomb, J.L., Hahn, S.Y.: Implementation of continuum sensitivity analysis with existing finite element code. IEEE Trans. Magn. **29**, 1787–1790 (1993)

Chapter 2
Variational Formulation of Electromagnetic Systems

In order to derive the continuum sensitivity for the electromagnetic system, the variational state equation is differentiated with respect to the design variables by using the material derivative concept in the subsequent Chaps. 3–6. In this chapter, the variational state equations for electrostatic system, magnetostatic system, eddy current system, and DC current-carrying conductor are formulated by the variational method of virtual work principle. Each variational equation is derived from its corresponding differential equation with boundary conditions. Electromagnetic systems are usually represented by a differential (point) form of Maxwell's equations that holds at all points of the field domain. Introducing the potentials such as the electric scalar potential and the magnetic vector potential, the governing differential equations are obtained as the second-order partial differential equations. Thus, the equations require continuous second-order derivatives of the potentials. The variational state equations reduce the required order of the derivatives by one so that the variational (weak) formulation provides a general solution that cannot be obtained by the differential equations. It is also the mathematical basis for the finite element method, which is widely applicable to the electromagnetic systems. Furthermore, since the variational state equation is expressed in integral form that contains the geometry information, it is more suitable to the shape design sensitivity analysis than the differential equation [1–7].

2.1 Variational Formulation of Electrostatic System

In this section, the differential state equation for electrostatic system is derived from Maxwell's equations by using the electric scalar potential ϕ, and then, its variational state equation is obtained by applying the variational formulation of the virtual work principle [8].

I. H. Park, *Design Sensitivity Analysis and Optimization of Electromagnetic Systems*, Mathematical and Analytical Techniques with Applications to Engineering,
https://doi.org/10.1007/978-981-13-0230-5_2

2.1.1 Differential State Equation

The differential state equations for electrostatic field system are derived from Maxwell's equations with the electric scalar potential. The electrostatic system is represented by two equations from Maxwell's equations;

$$\nabla \cdot \mathbf{D} = \rho \tag{2.1.1}$$

$$\nabla \times \mathbf{E} = \mathbf{0} \tag{2.1.2}$$

where $\mathbf{D}$ is the electric flux density, ρ is the volume charge density, and $\mathbf{E}$ is the electric field intensity. The electric flux density is written with the electric field intensity and the permanent polarization by the constitutive relation;

$$\mathbf{D} = \varepsilon\mathbf{E} + \mathbf{P}_0 \tag{2.1.3}$$

where ε is the electric permittivity and $\mathbf{P}_0$ is the permanent polarization, The electric permittivity ε is $\varepsilon_r\varepsilon_0$ with the relative electric permittivity ε_r and the vacuum permittivity ε_0. ε_r is assumed to be constant. The permanent polarization for electrolet materials is included for general description of the electrostatic system. With the electric scalar potential ϕ introduced from (2.1.2), the electric field intensity is written as

$$\mathbf{E} = -\nabla\phi \tag{2.1.4}$$

Inserting (2.1.3) and (2.1.4) into (2.1.1), we obtain the Poisson equation of the electrostatic system, which is the governing equation for the state variable of the electric scalar potential ϕ;

$$-\nabla \cdot \varepsilon\nabla\phi = \rho - \nabla \cdot \mathbf{P}_0 \tag{2.1.5}$$

where $-\nabla \cdot \mathbf{P}_0 = \rho_P$ and ρ_P is the bound charge density of permanent polarization. The governing differential equation of the electrostatic system (2.1.5) has a unique solution with boundary conditions. We employ most common boundary conditions;

$$\phi = C(\mathbf{x}) \quad \text{on } \Gamma^0 \text{ (Dirichlet boundary condition)} \tag{2.1.6}$$

$$\frac{\partial\phi}{\partial n} = 0 \quad \text{on } \Gamma^1 \text{ (homogeneous Neumann boundary condition)} \tag{2.1.7}$$

2.1.2 Variational State Equation

The differential state equation (2.1.5) for the electrostatic system can be reduced to a variational state equation by multiplying both sides by an arbitrary virtual potential $\bar{\phi}$ as

$$\int_{\Omega} -\nabla \cdot (\varepsilon \nabla \phi - \mathbf{P}_o)\bar{\phi} \mathrm{d}\Omega = \int_{\Omega} \rho \bar{\phi} \mathrm{d}\Omega \quad \forall \bar{\phi} \in \Phi \tag{2.1.8}$$

The arbitrary virtual potential $\bar{\phi}$ belongs to the space of admissible potential, defined as

$$\Phi = \left\{ \bar{\phi} \in H^1(\Omega) \middle| \bar{\phi} = 0 \text{ on } \quad \mathbf{x} \in \Gamma^0 \right\} \tag{2.1.9}$$

where Γ^0 is the Dirichlet essential boundary and $H^1(\Omega)$ is the Sobolev space of order one [3, 9]. $H^n(\Omega)$ is the Sobolev space of the order n, whose functions are continuously differentiable up to $n - 1$, and nth partial derivatives belong to $L_2(\Omega)$, which is the space of square integrable functions such that

$$L_2(\Omega) = \left\{ f \middle| \int_{\Omega} |f(\mathbf{x})|^2 \mathrm{d}\Omega < \infty \right\} \tag{2.1.10}$$

By the vector identity $\nabla \cdot (\nabla \psi \bar{\psi}) = (\nabla \cdot \nabla \psi)\bar{\psi} + \nabla \psi \cdot \nabla \bar{\psi}$, (2.1.8) is written as

$$\int_{\Omega} \left[(\varepsilon \nabla \phi - \mathbf{P}_0) \cdot \nabla \bar{\phi} - \nabla \cdot \left((\varepsilon \nabla \phi - \mathbf{P}_0)\bar{\phi} \right) \right] \mathrm{d}\Omega = \int_{\Omega} \rho \bar{\phi} \mathrm{d}\Omega \quad \forall \bar{\phi} \in \Phi \tag{2.1.11}$$

By the divergence theorem $\int_{\Omega} \nabla \cdot (\nabla \psi \bar{\psi}) \mathrm{d}\Omega = \int_{\Gamma} (\nabla \psi \cdot \mathbf{n}) \bar{\psi} \mathrm{d}\Gamma$, (2.1.11) is rewritten as

$$\int_{\Omega} (\varepsilon \nabla \phi \cdot \nabla \bar{\phi} - \rho \bar{\phi} - \mathbf{P}_0 \cdot \nabla \bar{\phi}) \mathrm{d}\Omega = \int_{\Gamma} (\varepsilon \nabla \phi - \mathbf{P}_0) \cdot \mathbf{n} \bar{\phi} \mathrm{d}\Gamma \quad \forall \bar{\phi} \in \Phi \tag{2.1.12}$$

Inserting the relation (2.1.3) into the right side of (2.1.12) provides the variational identity for the state equation of electrostatic system;

$$\int_{\Omega} (\varepsilon \nabla \phi \cdot \nabla \bar{\phi} - \rho \bar{\phi} - \mathbf{P}_0 \cdot \nabla \bar{\phi}) \mathrm{d}\Omega = -\iint_{\Gamma} D_n(\phi) \bar{\phi} \mathrm{d}\Gamma \quad \forall \bar{\phi} \in \Phi \tag{2.1.13}$$

where

$$D_n(\phi) = (-\varepsilon\nabla\phi + \mathbf{P}_0) \cdot \mathbf{n} = -\varepsilon\frac{\partial\phi}{\partial n} + \mathbf{P}_0 \cdot \mathbf{n} \quad (2.1.14)$$

Imposing boundary conditions on the variational identity (2.1.13) results in the variational state equation. The boundary conditions of (2.1.6) and (2.1.7) can be rewritten for the variational equation as

$$\bar{\phi} = 0 \quad \text{on } \Gamma^0 \text{ (Dirichlet boundary condition)} \quad (2.1.15)$$

$$D_n(\phi) = 0 \quad \text{on } \Gamma^1 \text{ (homogeneous Neumann boundary condition)} \quad (2.1.16)$$

In (2.1.13), $D_n(\phi)$ physically means the surface charge density on the boundary. But there is no surface charge on the homogeneous Neumann boundary, which normally comes from symmetry of a system structure. The right-hand side of (2.1.13) vanishes with the boundary conditions of (2.1.15) and (2.1.16), and the variational state equation corresponding to the differential Eq. (2.1.5) is obtained as

$$\int_\Omega \varepsilon\nabla\phi \cdot \nabla\bar{\phi}\mathrm{d}\Omega = \int_\Omega (\rho\bar{\phi} + \mathbf{P}_0 \cdot \nabla\bar{\phi})\mathrm{d}\Omega \quad \forall\, \bar{\phi} \in \Phi \quad (2.1.17)$$

We define the energy bilinear form $a(\phi, \bar{\phi})$ and the source linear form $l(\bar{\phi})$ as

$$a(\phi, \bar{\phi}) \equiv \int_\Omega \varepsilon\nabla\phi \cdot \nabla\bar{\phi}\mathrm{d}\Omega \quad (2.1.18)$$

$$l(\bar{\phi}) \equiv \int_\Omega (\rho\bar{\phi} + \mathbf{P}_0 \cdot \nabla\bar{\phi})\mathrm{d}\Omega \quad (2.1.19)$$

The variational Eq. (2.1.17) is rewritten with the energy bilinear form and the source linear form as

$$a(\phi, \bar{\phi}) = l(\bar{\phi}) \quad \forall\, \bar{\phi} \in \Phi \quad (2.1.20)$$

2.2 Variational Formulation of Magnetostatic System

Here, after the differential state equation for the magnetostatic system is derived from Maxwell's equations by using the magnetic vector potential **A**, its variational state equation is obtained by applying the variational formulation of the virtual work principle [10–12].

2.2.1 Differential State Equation

The differential state equations for the magnetostatic system is derived from Maxwell's equations with the magnetic vector potential. The magnetostatic system is represented by two equations from the Maxwell equations;

$$\nabla \times \mathbf{H} = \mathbf{J} \tag{2.2.1}$$

$$\nabla \cdot \mathbf{B} = 0 \tag{2.2.2}$$

where **H** is the magnetic field intensity, **J** is the volume current density, and **B** is the magnetic flux density. The magnetic field intensity is written with the magnetic flux density and the permanent magnetization by the constitutive relation;

$$\mathbf{H} = v\mathbf{B} - \mathbf{M}_0 \tag{2.2.3}$$

where v is magnetic reluctivity, which is the reciprocal of the magnetic permeability μ, and $\mathbf{M}_0$ is the permanent magnetization. The magnetic permeability μ is $\mu_r\mu_0$ with the relative magnetic permeability μ_r and the vacuum permeability μ_0. μ_r is constant in linear magnetic system. The permanent magnetization for permanent magnet materials is included for general description of the magnetic system. With the magnetic vector potential **A** introduced from (2.2.2), the magnetic flux density is written as

$$\mathbf{B} = \nabla \times \mathbf{A} \tag{2.2.4}$$

Inserting (2.2.3) and (2.2.4) into (2.2.1), we obtain the governing equation of the magnetostatic system for the state variable of the magnetic vector potential **A**;

$$\nabla \times v\nabla \times \mathbf{A} = \mathbf{J} + \nabla \times \mathbf{M}_o \tag{2.2.5}$$

where $\nabla \times \mathbf{M}_0 = \mathbf{J}_m$ and $\mathbf{J}_m$ is the magnetization current density of permanent magnetization. The governing differential equation of the magnetostatic system (2.2.5) has a unique solution with boundary conditions. We employ most common boundary conditions;

$$\mathbf{A} = \mathbf{0} \quad \text{on } \Gamma^0 \text{ (homogeneous Dirichlet boundary condition)} \tag{2.2.6}$$

$$\frac{\partial \mathbf{A}}{\partial n} = \mathbf{0} \quad \text{on } \Gamma^1 \text{ (homogeneous Neumann boundary condition)} \tag{2.2.7}$$

2.2.2 Variational State Equation

The differential state equation (2.2.5) for the magnetostatic system can be reduced to a variational state equation by multiplying both sides by an arbitrary virtual vector potential $\bar{\mathbf{A}}$ as

$$\int_{\Omega} [\nabla \times \nu(\nabla \times \mathbf{A})] \cdot \bar{\mathbf{A}} d\Omega = \int_{\Omega} (\mathbf{J} + \nabla \times \mathbf{M}_o) \cdot \bar{\mathbf{A}} d\Omega \quad \forall \bar{\mathbf{A}} \in \Phi \tag{2.2.8}$$

The arbitrary virtual potential $\bar{\mathbf{A}}$ belongs to the space of admissible vector potential, defined as

$$\Phi = \left\{ \bar{\mathbf{A}} \in \left[H^1(\Omega)\right]^3 | \bar{\mathbf{A}} = 0 \quad \text{on } \mathbf{x} \in \Gamma^0 \right\} \tag{2.2.9}$$

where Γ^0 is the Dirichlet essential boundary and $\mathbf{H}^1(\Omega)$ is the Sobolev space of order one [9]. For convenience of expression, we define an operator $\mathbf{B}$ for any vector function $\mathbf{S}$ as

$$\mathrm{B}(\mathbf{S}) \equiv \nabla \times \mathbf{S} \tag{2.2.10}$$

Thus, we have the relations: $\mathbf{B}(\mathbf{A}) = \nabla \times \mathbf{A}$ and $\mathbf{B}(\bar{\mathbf{A}}) = \nabla \times \bar{\mathbf{A}}$. Using this expression, (2.2.8) is written as

$$\int_{\Omega} \nabla \times (\nu \mathbf{B}(\mathbf{A}) - \mathbf{M}_o) \cdot \bar{\mathbf{A}} d\Omega = \int_{\Omega} \mathbf{J} \cdot \bar{\mathbf{A}} d\Omega \quad \forall \bar{\mathbf{A}} \in \Phi \tag{2.2.11}$$

By the vector identity $\nabla \cdot (\mathbf{a} \times \mathbf{b}) = (\nabla \times \mathbf{a}) \cdot \mathbf{b} - \mathbf{a} \cdot (\nabla \times \mathbf{b})$, the integrand of the left side in (2.2.11) is expressed as

$$\begin{aligned} \nabla \times (\nu \mathbf{B}(\mathbf{A}) - \mathbf{M}_0) \cdot \bar{\mathbf{A}} &= (\nu \mathbf{B}(\mathbf{A}) - \mathbf{M}_0) \cdot \left(\nabla \times \bar{\mathbf{A}}\right) + \nabla \cdot \left[(\nu \mathbf{B}(\mathbf{A}) - \mathbf{M}_0) \times \bar{\mathbf{A}}\right] \\ &= (\nu \mathbf{B}(\mathbf{A}) - \mathbf{M}_0) \cdot \mathbf{B}(\bar{\mathbf{A}}) + \nabla \cdot \left(\mathbf{H}(\mathbf{A}) \times \bar{\mathbf{A}}\right) \end{aligned} \tag{2.2.12}$$

This relation (2.2.12) and the divergence theorem are applied to (2.2.11):

$$\int_{\Omega} \left(\nu \mathbf{B}(\mathbf{A}) \cdot \mathbf{B}(\bar{\mathbf{A}}) - \mathbf{J} \cdot \bar{\mathbf{A}} - \mathbf{M}_o \cdot \mathbf{B}(\bar{\mathbf{A}})\right) d\Omega = -\int_{\Gamma} \left(\mathbf{H}(\mathbf{A}) \times \bar{\mathbf{A}}\right) \cdot \mathbf{n} d\Gamma \quad \forall \bar{\mathbf{A}} \in \Phi \tag{2.2.13}$$

By the vector identity $\left(\mathbf{H} \times \bar{\mathbf{A}}\right) \cdot \mathbf{n} = (\mathbf{n} \times \mathbf{H}) \cdot \bar{\mathbf{A}}$, (2.2.13) is rewritten to provide the variational identity for the state equation of the magnetostatic system;

$$\int_{\Omega} (\nu \mathbf{B}(\mathbf{A}) \cdot \mathbf{B}(\bar{\mathbf{A}}) - \mathbf{J} \cdot \bar{\mathbf{A}} - \mathbf{M}_o \cdot \mathbf{B}(\bar{\mathbf{A}}))\mathrm{d}\Omega = -\int_{\Gamma} (\mathbf{n} \times \mathbf{H}(\mathbf{A})) \cdot \bar{\mathbf{A}}\mathrm{d}\Gamma \quad \forall \bar{\mathbf{A}} \in \Phi \tag{2.2.14}$$

Imposing boundary conditions on the variational identity (2.2.14) results in the state variational equation. The boundary conditions of (2.2.6) and (2.2.7) can be rewritten for the variational equation as

$$\bar{\mathbf{A}} = \mathbf{0} \quad \text{on } \Gamma^0 \text{ (homogeneous Dirichlet boundary condition)} \tag{2.2.15}$$

$$\mathbf{n} \times \mathbf{H}(\mathbf{A}) = \mathbf{0} \quad \text{on } \Gamma^1 \text{ (homogeneous Neumann boundary condition)} \tag{2.2.16}$$

The $\mathbf{n} \times \mathbf{H}(\mathbf{A})$ in (2.2.14) physically means the surface current density on the boundary. But there is no surface current on the homogeneous Neumann boundary, which normally comes from symmetry of a system structure. Thus, the right side of (2.2.14) vanishes with the boundary conditions of (2.2.15) and (2.2.16), and the variational state equation corresponding to the differential Eq. (2.2.5) is obtained as

$$\int_{\Omega} \nu \mathbf{B}(\mathbf{A}) \cdot \mathbf{B}(\bar{\mathbf{A}})\mathrm{d}\Omega = \int_{\Omega} (\mathbf{J} \cdot \bar{\mathbf{A}} + \mathbf{M}_o \cdot \mathbf{B}(\bar{\mathbf{A}}))\mathrm{d}\Omega \quad \forall \bar{\mathbf{A}} \in \Phi \tag{2.2.17}$$

We define the energy bilinear form $a(\mathbf{A}, \bar{\mathbf{A}})$ and the source linear form $l(\bar{\mathbf{A}})$ as

$$a(\mathbf{A}, \bar{\mathbf{A}}) \equiv \int_{\Omega} \nu \mathbf{B}(\mathbf{A}) \cdot \mathbf{B}(\bar{\mathbf{A}})\mathrm{d}\Omega \tag{2.2.18}$$

$$l(\bar{\mathbf{A}}) \equiv \int_{\Omega} (\mathbf{J} \cdot \bar{\mathbf{A}} + \mathbf{M}_o \cdot \mathbf{B}(\bar{\mathbf{A}}))\mathrm{d}\Omega \tag{2.2.19}$$

The variational state equation (2.2.17) is rewritten with the energy bilinear form and the source linear form as

$$a(\mathbf{A}, \bar{\mathbf{A}}) = l(\bar{\mathbf{A}}) \quad \forall \bar{\mathbf{A}} \in \Phi \tag{2.2.20}$$

2.3 Variational Formulation of Eddy Current System

In this section, the differential state equation for the eddy current system is derived from Maxwell's equations by using the magnetic vector potential $\mathbf{A}$ and the electric scalar potential ϕ , and then it is applied to a simple model without the term $\nabla\phi$ whose state variable is the complex number for the harmonic steady state.

The governing differential equation of the complex variable is reduced to the complex variational equation by using the formulation method of the generalized variational principle [13].

2.3.1 Differential State Equation

The differential state equations for the eddy current system are derived from Maxwell's equations with the magnetic vector potential. The eddy current system is represented by three equations from Maxwell's equations;

$$\nabla \times \mathbf{H} = \mathbf{J}_t \tag{2.3.1}$$

$$\nabla \cdot \mathbf{B} = 0 \tag{2.3.2}$$

$$\nabla \times \mathbf{E} = -\frac{\partial \mathbf{B}}{\partial t} \tag{2.3.3}$$

where **H** is the magnetic field intensity, **B** is the magnetic flux density, **E** is the electric field intensity, and $\mathbf{J}_\mathrm{t}$ is the volume density of total transport current, which is the sum of the given source current density **J** and the eddy current density $\mathbf{J}_\mathrm{e}$ as

$$\mathbf{J}_\mathrm{t} = \mathbf{J} + \mathbf{J}_\mathrm{e} \tag{2.3.4}$$

The magnetic field intensity and the magnetic flux density are related by the constitutive relation;

$$\mathbf{H} = v\mathbf{B} \tag{2.3.5}$$

where v is magnetic reluctivity, which is the reciprocal of the magnetic permeability μ. The magnetic permeability μ is $\mu_\mathrm{r}\mu_0$ with the relative magnetic permeability μ_r and the vacuum permeability μ_0. In this eddy current system, μ_r is assumed to be constant and the permanent magnet materials is not included. The eddy current $\mathbf{J}_\mathrm{e}$ is written by the constitutive relation of Ohm's law as

$$\mathbf{J}_\mathrm{e} = \sigma \mathbf{E} \tag{2.3.6}$$

where σ is the electric conductivity and is assumed to be constant. With the magnetic vector potential **A** introduced from (2.3.2), the magnetic flux density is written as

$$\mathbf{B} = \nabla \times \mathbf{A} \tag{2.3.7}$$

Inserting (2.3.7) into (2.3.3) provides the relation:

$$\nabla \times \left(\mathbf{E} + \frac{\partial \mathbf{A}}{\partial t}\right) = \mathbf{0} \tag{2.3.8}$$

With the electric scalar potential ϕ introduced from (2.3.8), the electric field intensity is written as

$$\mathbf{E} = -\frac{\partial \mathbf{A}}{\partial t} - \nabla\phi \tag{2.3.9}$$

This is inserted into (2.3.6) to provide

$$\mathbf{J}_{\mathrm{e}} = \sigma\mathbf{E} = -\sigma\left(\frac{\partial \mathbf{A}}{\partial t} + \nabla\phi\right) \tag{2.3.10}$$

Using (2.3.4), (2.3.5), and (2.3.10) from (2.3.1), we obtain the governing equation of the eddy current system for the state variables of the magnetic vector potential $\mathbf{A}$ and the electric scalar potential ϕ;

$$\nabla \times \nu\nabla \times \mathbf{A} = \mathbf{J} - \sigma\left(\frac{\partial \mathbf{A}}{\partial t} + \nabla\phi\right) \tag{2.3.11}$$

where the term $\nabla\phi$ comes from the electric charge induced on the conductor surface. The term $\nabla\phi$ vanishes in the axi-symmetric system and in the two-dimensional Cartesian system with the end-connected symmetrical conductor. In this book, we deal with only the eddy current system that does not have the term. Thus, we have the governing equation of the eddy current system of

$$\nabla \times \nu\nabla \times \mathbf{A} = \mathbf{J} - \sigma\frac{\partial \mathbf{A}}{\partial t} \tag{2.3.12}$$

In this linear eddy current system, when the harmonic source is considered, the governing equation for the steady state is expressed using the complex variable method as

$$\nabla \times \nu\nabla \times \mathbf{A} + j\omega\sigma\mathbf{A} = \mathbf{J} \tag{2.3.13}$$

where $\mathbf{J}$ is the complex source current density, $\mathbf{A}$ is the complex magnetic vector potential, and ω is the given source frequency. Henceforth, all the state and adjoint variables for the eddy current system are assumed to be the complex variables. The governing differential equation of the eddy current system (2.3.13) has a unique solution with boundary conditions. We employ most common boundary conditions;

$$\mathbf{A} = \mathbf{0} \quad \text{on } \Gamma^0 \text{ (homogeneous Dirichlet boundary condition)} \tag{2.3.14}$$

$$\frac{\partial \mathbf{A}}{\partial n} = \mathbf{0} \quad \text{on } \Gamma^1 \text{ (homogeneous Neumann boundary condition)} \tag{2.3.15}$$

2.3.2 Variational State Equation

The differential state equation (2.3.13) for the eddy current system can be reduced to a variational state equation by multiplying both sides by an arbitrary virtual complex vector potential $\bar{\mathbf{A}}$ as

$$\int_{\Omega} (\nabla \times \nu \nabla \times \mathbf{A} + j\omega\sigma\mathbf{A}) \cdot \bar{\mathbf{A}} d\Omega = \int_{\Omega} \mathbf{J} \cdot \bar{\mathbf{A}} d\Omega \quad \forall \bar{\mathbf{A}} \in \Phi \tag{2.3.16}$$

The arbitrary virtual potential $\bar{\mathbf{A}}$ belongs to the space of admissible complex vector potential, defined as

$$\Phi = \left\{ \bar{\mathbf{A}} \in \left[H^1(\Omega)\right]^3 | \bar{\mathbf{A}} = 0 \quad \text{on } \mathbf{x} \in \Gamma^0 \right\} \tag{2.3.17}$$

where Γ^0 is the Dirichlet essential boundary and $\mathbf{H}^1(\Omega)$ is the complex Sobolev space of order one [9]. As in Sect. 2.2, using the vector identity and the divergence theorem, (2.3.16) is rewritten to provide the variational identity for the state equation of the eddy current system;

$$\int_{\Omega} \left(\nu\mathbf{B}(\mathbf{A}) \cdot \mathbf{B}(\bar{\mathbf{A}}) + j\omega\sigma\mathbf{A} \cdot \bar{\mathbf{A}} - \mathbf{J} \cdot \bar{\mathbf{A}}\right) d\Omega = -\int_{\Gamma} (\mathbf{n} \times \mathbf{H}(\mathbf{A})) \cdot \bar{\mathbf{A}} d\Gamma \quad \forall \bar{\mathbf{A}} \in \Phi \tag{2.3.18}$$

Imposing boundary conditions on the variational identity (2.3.18) results in the variational state equation. The boundary conditions of (2.3.14) and (2.3.15) can be rewritten for the variational equation as

$$\bar{\mathbf{A}} = \mathbf{0} \quad \text{on } \Gamma^0 \text{ (homogeneous Dirichlet boundary condition)} \tag{2.3.19}$$

$$\mathbf{n} \times \mathbf{H}(\mathbf{A}) = \mathbf{0} \quad \text{on } \Gamma^1 \text{ (homogeneous Neumann boundary condition)} \tag{2.3.20}$$

The $\mathbf{n} \times \mathbf{H}(\mathbf{A})$ in (2.3.18) physically means the surface current density on the boundary. But there is no surface current on the homogeneous Neumann boundary, which normally comes from symmetry of a system structure. Thus, the right side of

(2.3.18) vanishes with the boundary conditions of (2.3.19) and (2.3.20), and the variational state equation corresponding to the differential Eq. (2.3.13) is obtained as

$$\int_{\Omega} \left(v\mathbf{B}(\mathbf{A}) \cdot \mathbf{B}(\bar{\mathbf{A}}) + j\omega\sigma\mathbf{A} \cdot \bar{\mathbf{A}}\right)\mathrm{d}\Omega = \int_{\Omega} \mathbf{J} \cdot \bar{\mathbf{A}}\mathrm{d}\Gamma \quad \forall\, \bar{\mathbf{A}} \in \Phi \tag{2.3.21}$$

We define the bilinear form $a(\mathbf{A}, \bar{\mathbf{A}})$ and the source linear form $l(\bar{\mathbf{A}})$ as

$$a(\mathbf{A}, \bar{\mathbf{A}}) \equiv \int_{\Omega} \left(v\mathbf{B}(\mathbf{A}) \cdot \mathbf{B}(\bar{\mathbf{A}}) + j\omega\sigma\mathbf{A} \cdot \bar{\mathbf{A}}\right)\mathrm{d}\Omega \tag{2.3.22}$$

$$l(\bar{\mathbf{A}}) \equiv \int_{\Omega} \mathbf{J} \cdot \bar{\mathbf{A}}\mathrm{d}\Omega \tag{2.3.23}$$

The variational Eq. (2.3.21) is rewritten with the bilinear form and the source linear form as

$$a(\mathbf{A}, \bar{\mathbf{A}}) = l(\bar{\mathbf{A}}) \quad \forall\, \bar{\mathbf{A}} \in \Phi \tag{2.3.24}$$

2.4 Variational Formulation of DC Conductor System

Here, the differential state equation for the DC current-carrying conductor is derived from Maxwell's equations by introducing the electric scalar potential ϕ, and then, its variational state equation is obtained by applying the variational formulation of the virtual work principle.

2.4.1 Differential State Equation

The differential state equation for the DC current-carrying conductor is derived from the continuity equation in the steady-state condition and Ohm's law. The continuity equation in the steady-state condition is written as

$$\nabla \cdot \mathbf{J} = 0 \tag{2.4.1}$$

where $\mathbf{J}$ is the volume current density. Ohm's law is written as

$$\mathbf{J} = \sigma\mathbf{E} \tag{2.4.2}$$

where σ is the electric conductivity and is assumed to be constant. The electric field intensity is written with the electric scalar potential ϕ as

$$\mathbf{E} = -\nabla\phi \tag{2.4.3}$$

Inserting (2.4.2) and (2.4.3) into (2.4.1), we obtain the Laplace equation for the DC current-carrying conductor as

$$-\nabla \cdot \sigma\nabla\phi = 0 \tag{2.4.4}$$

which is the governing equation for the state variable of the electric scalar potential ϕ. This governing equation of the DC current-carrying conductor has a unique solution with the boundary conditions given as

$$\phi = C(\mathbf{x}) \quad \text{on } \Gamma^0 \text{ (Dirichlet boundary condition)} \tag{2.4.5}$$

$$\frac{\partial\phi}{\partial n} = 0 \quad \text{on } \Gamma^1 \text{ (homogeneous Neumann boundary condition)} \tag{2.4.6}$$

2.4.2 *Variational State Equation*

The differential state Eq. (2.4.4) can be reduced to a variational identity by multiplying both sides by an arbitrary virtual potential $\bar{\phi}$ and by using the vector identity and the divergence theorem;

$$\int_\Omega \sigma\nabla\phi \cdot \nabla\bar{\phi}\,\mathrm{d}\Omega = \int_\Gamma \sigma\frac{\partial\phi}{\partial n}\bar{\phi}\,\mathrm{d}\Gamma \quad \forall\,\bar{\phi} \in \Phi \tag{2.4.7}$$

The arbitrary virtual potential $\bar{\phi}$ belongs to the space of admissible potential, defined as

$$\Phi = \left\{\bar{\phi} \in H^1(\Omega) \middle| \bar{\phi} = 0 \quad \text{on } \mathbf{x} \in \Gamma^0\right\} \tag{2.4.8}$$

where Γ^0 is the Dirichlet essential boundary and $H^1(\Omega)$ is the Sobolev space of order one. In the integrand of the right-hand side in (2.4.7),

$$\sigma\frac{\partial\phi}{\partial n} = \sigma\nabla\phi \cdot \mathbf{n} = -\mathbf{J} \cdot \mathbf{n} = -J_n \tag{2.4.9}$$

where J_n is the normal component of the current density, which is always zero on the conductor surface. The boundary conditions of (2.4.5) and (2.4.6) can be rewritten for the variational equation as

$$\bar{\phi} = 0 \quad \text{on } \Gamma^0 \text{ (Dirichlet boundary condition)} \tag{2.4.10}$$

$$\frac{\partial \phi}{\partial n} = 0 \quad \text{on } \Gamma^1 \text{ (homogeneous Neumann boundary condition)} \tag{2.4.11}$$

The Dirichlet boundary condition is imposed on the electrode surfaces by the external voltage source. The homogeneous Neumann boundary condition holds on all the conductor surfaces except the electrode surfaces. The current density on the conductor surface has only the tangential component. That is, its normal component is always zero on the conductor surface [14]. So, the normal component of the electric field is zero on the conductor surface, and it leads to the homogeneous Neumann boundary condition. The Neumann boundary is the inside surface of the conductor, which belongs to the conductor. But on the outside surface normally are both the tangential and normal components of the electric field. This discontinuity of the electric field comes from the surface charges on the conductor surface. This surface charges make the electric field to be tangential on the inside surface of the conductor. Since the concerned field region is inside of the conductor in this problem, the homogeneous Neumann boundary condition is enough for analyzing this DC current-carrying conductor. Unlike other electromagnetic systems, the homogeneous Neumann boundary condition on the conductor surface is not due to the symmetry of the system structure. The right-hand side of (2.4.7) vanishes with the boundary conditions of (2.4.10) and (2.4.11), and the variational state equation corresponding to the differential Eq. (2.4.4) is obtained as

$$\int_{\Omega} \sigma \nabla \phi \cdot \nabla \bar{\phi} d\Omega = 0 \quad \forall \, \bar{\phi} \in \Phi \tag{2.4.12}$$

We define the energy bilinear form $a(\phi, \bar{\phi})$ as

$$a(\phi, \bar{\phi}) \equiv \int_{\Omega} \sigma \nabla \phi \cdot \nabla \bar{\phi} d\Omega \tag{2.4.13}$$

The variational Eq. (2.4.12) is rewritten with the energy bilinear form as

$$a(\phi, \bar{\phi}) = 0 \quad \forall \, \bar{\phi} \in \Phi \tag{2.4.14}$$

References

1. Hammond, P.: Energy Methods in Electromagnetism. Clarendon Press, Oxford (1981)
2. Hammond, P., Sykulski, J.K.: Engineering Electromagnetism: Physical Processes and Computation. Oxford University Press, Oxford (1994)

3. Becker, E.B., Carey, G.F., Oden, J.T.: Finite Elements, An Introduction, vol. 1. Prentice-Hall, New Jersey (1981)
4. Reddy, J.N.: Applied Functional Analysis and Variational Methods in Engineering. McGraw-hill, New York (1986)
5. Hughes, T.J.R.: The Finite Element Method: Linear Static and Dynamic Finite Element Analysis. Prentice-Hall, Englewood Cliffs (1987)
6. Zienkiewicz, O.C.: The Finite Element Method. McGraw-Hill, London (1977)
7. Ciarlet, P.G.: The Finite Element Method for Elliptic Problems. North-Holland, Amsterdam (1978)
8. Sabonnadière, J.C., Coulomb, J.L.: Finite Element Methods in CAD. Springer, New York (1989)
9. Adams, R.A.: Sobolev Spaces. Academic Press, New York (1975)
10. Hoole, S.R.: Computer-Aided Analysis and Design of Electromagnetic Devices. Elsevier, New York (1989)
11. Silvester, P.P., Ferrari, R.L.: Finite Elements for Electrical Engineers. Cambridge University Press, Cambridge (1983)
12. Chari, M.V.K., Silvester, P.P.: Finite Elements in Electrical and Magnetic Field Problems. Wiley, New York (1980)
13. Jin, J.: The Finite Element Method in Electromagnetics. Wiley, Hoboken (1962)
14. Hernandes, J.A., Capelas de Oliveira, E., Assis, A.K.T.: Resistive plates carrying a steady current: electric potential and surface charges close to the battery. Found. Phys. Lett. **18**, 275–289 (2005)

Chapter 3
Continuum Shape Design Sensitivity of Electrostatic System

The electrostatic system is composed of dielectric material, electric charge, and electrodes. We have examples of the design objective such as reducing electric breakdown, obtaining a desired distribution of electric field, obtaining a desired capacitance. For these objectives, the shape of the dielectric material and the electrodes is optimally designed. The shape design sensitivity for the electrostatic system, which provides the information on the shape-variation effect on the performance, helps designers to improve its performance.

The electrostatic systems include high-voltage apparatus, energy conversion machines, electric devices, microelectromechanical systems (MEMS), typical examples of which are circuit breakers, high-voltage cable, insulator, bushing, surge divider, transformer, generator, motor, actuator, sensor, capacitor, electric precipitator, piezoelectric devices, battery, dielectrophoresis devices, plasma generator.

In this chapter, the continuum shape sensitivity for the electrostatic system is derived by taking the material derivatives of the objective function and the variational state equation. The Lagrange multiplier method is introduced to deal with the implicit equality constraint of the variational state equation. An adjoint variable technique is employed to express explicitly the sensitivity in terms of the design variables, and the variational identities are used to transform the sensitivity of a domain integral into a boundary integral on the design surface. The obtained sensitivity formula, which is a general three-dimensional shape sensitivity, provides physical insight and advantages in numerical implementation.

The shape sensitivity for the electrostatic system is, according to the design variable, classified into two categories: the outer boundary design and the interface design, each of which is, according to the objective function, also divided into two kinds: the domain integral objective function and the system energy objective function. In the outer boundary design, the design variable is the Dirichlet and/or Neumann boundary where the boundary conditions are imposed. In the interface design, the design variable is the interface where two different materials meet. The domain integral objective function is the domain integral of a function of the electric potential and the electric field. The system energy objective function is the stored

I. H. Park, *Design Sensitivity Analysis and Optimization of Electromagnetic Systems*, Mathematical and Analytical Techniques with Applications to Engineering, https://doi.org/10.1007/978-981-13-0230-5_3

electric field energy. The electric field energy, although it can be thought to be a sort of domain integral, does not require solving for the adjoint variable and is used for the equivalent capacitance. For this reason, it is taken as another objective function.

In Sect. 3.1, the concept of the material derivative in continuum mechanics is introduced and the material derivative of the state variable in the electromagnetic system is derived for use of the subsequent chapters. The material derivative formula of a domain integral is also introduced. By using these, the material derivative of the domain integral of the state variable function is derived.

In Sects. 3.2 and 3.3, two general three-dimensional sensitivities of the outer boundary design are derived for two kinds of objective functions, respectively. In Sects. 3.4 and 3.5, two general three-dimensional sensitivities of the interface design are also derived for two objective functions, respectively. The interface sensitivity in the electrostatic system is only for an interface of two dielectric materials, and it is obtained as a surface integral on the outer boundary or the interface.

In Sects. 3.3 and 3.5, the two energy sensitivities for the outer boundary design and the interface design are examined in the electric-circuit point of view to show how the sign of the energy sensitivity depends on the source condition. The capacitance sensitivity is then derived using the result of the energy sensitivity.

At the ends of the Sects. 3.2–3.5, the four general sensitivity formulas are applied to design examples to be validated. The design examples are divided into two kinds. One is one-dimensional simple examples with the analytic solution. The other is two-dimensional numerical examples that do not have the analytic solution. In the former examples, the results of the analytical sensitivity are compared with the results from the sensitivity formulas in order to show the correctness of the derived sensitivity formula. In the latter examples, the sensitivity evaluated using the finite element method is applied to optimal shape design problems in order to show that the shape optimization method using the sensitivity formulas works well for the electrostatic systems.

3.1 Material Derivative and Formula

In the shape optimization, the domain shape of a material is taken as the design variable. The domain is assumed to be continuous material, and then its behaviors by the shape variation are described by using the concept of the material derivative in continuum mechanics [1, 2]. In this section, definition of the material derivative and a material derivative formula, which are used for sensitivity derivation in this chapter and the following Chaps. 4–6, are introduced.

3.1.1 Material Derivative

The time rate of change of a quantity of a material particle is known as a material derivative. The geometric change of a domain Ω is given with its boundary Γ. The domain Ω is a collection of material particles changing their positions in time [3]. The space occupied by the particles changes from Ω to a new domain Ω_t at time t as in Fig. 3.1. This deforming process is a transformation $\mathbf{T}_t$:

$$\mathbf{T}_t : \mathbf{x} \rightarrow \mathbf{x}_t, \quad \mathbf{x} \in \Omega \tag{3.1.1}$$

which describes the motion of each particle in the domain Ω:

$$\mathbf{x} = \mathbf{T}_t(\mathbf{X}) \equiv \mathbf{x}(\mathbf{X}, t), \quad \mathbf{X} \in \Omega \tag{3.1.2}$$

such that

$$\mathbf{T}_0(\Omega) = \Omega \tag{3.1.3}$$

The transformed domain is written as

$$\Omega_t = \mathbf{T}_t(\Omega) \tag{3.1.4}$$

where Ω_t is the image of Ω with respect to $\mathbf{T}_t$.

This transformation describes the deforming process of the continuum domain. $\mathbf{T}_t$ is assumed to be a one-to-one transformation of Ω onto Ω_t. The point $\mathbf{X}$ is the Lagrangian coordinates while $\mathbf{x}$ is the Eulerian coordinates of a particle. The parameter t means the amount of change of the geometry. The Eulerian velocity $\mathbf{V}(\mathbf{x}, t)$ at the point is defined as

$$\mathbf{V}(\mathbf{x}, t) = \frac{\partial \mathbf{x}_t}{\partial t} = \frac{\partial \mathbf{T}(\mathbf{x}, t)}{\partial t} \tag{3.1.5}$$

which is here called a design velocity.

If the transformation is assumed to be regular enough in the neighborhood $[0, \varepsilon]$ of t, it can be expanded by Taylor series expansion as

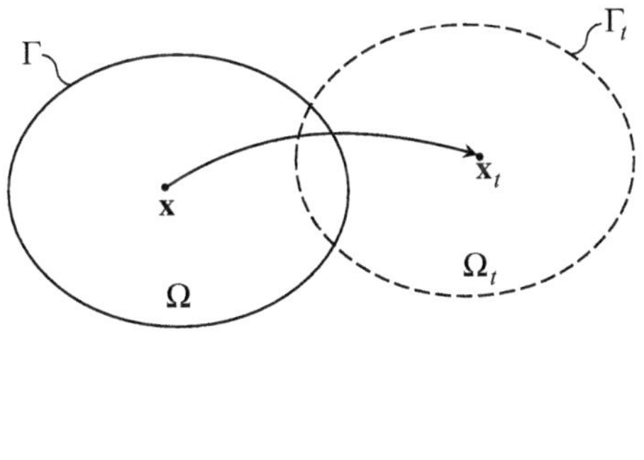

Fig. 3.1 Transformation of Ω with respect to t

$$\mathbf{T}(\mathbf{x},t) = \mathbf{T}(\mathbf{x},0) + t\frac{\partial \mathbf{T}}{\partial t}(\mathbf{x},0) + \cdots = \mathbf{x} + t\mathbf{V}(\mathbf{x},0) + \cdots \tag{3.1.6}$$

By ignoring the higher-order terms, the transformation becomes

$$\mathbf{T}(\mathbf{x},t) = \mathbf{x} + t\mathbf{V}(\mathbf{x},0) \tag{3.1.7}$$

where

$$\mathbf{V}(\mathbf{x}) = \mathbf{V}(\mathbf{x},0) \tag{3.1.8}$$

When this transformation is applied to the domain Ω, it is shown as in Fig. 3.2.

Consider an electromagnetic system that is governed by a differential equation in domain Ω;

$$L\phi = f, \quad \mathbf{x} \in \Omega \tag{3.1.9}$$

where L is a differential operator for the electromagnetic systems and $\phi(\mathbf{x})$ is the scalar state variable of the boundary-value problem in Ω. When the domain Ω is deformed in time, the state variable $\phi(\mathbf{x})$ also changes. Suppose that ϕ_t is the solution of the boundary-value problem for the deformed domain Ω_t;

$$L\phi_t = f, \quad \mathbf{x} \in \Omega_t \tag{3.1.10}$$

where $\phi_t = \phi_t(\mathbf{x}_t)$ is the solution at the moved position $\mathbf{x}_t$ but not the original position $\mathbf{x}$. When the deformation is small in (3.1.7), the moved position is written as

$$\mathbf{x}_t = \mathbf{x} + t\mathbf{V}(\mathbf{x}) \tag{3.1.11}$$

The solution $\phi_t(\mathbf{x}_t)$ at the moved position $\mathbf{x}_t$ is written as

$$\phi_t(\mathbf{x}_t) = \phi_t(\mathbf{x} + t\mathbf{V}) \tag{3.1.12}$$

which is defined in Ω and depends on t in Ω_t. $\phi_t(\mathbf{x}_t)$ is the solution to the boundary-value problem in Ω_t, which is evaluated at moving point $\mathbf{x}_t$ with t.

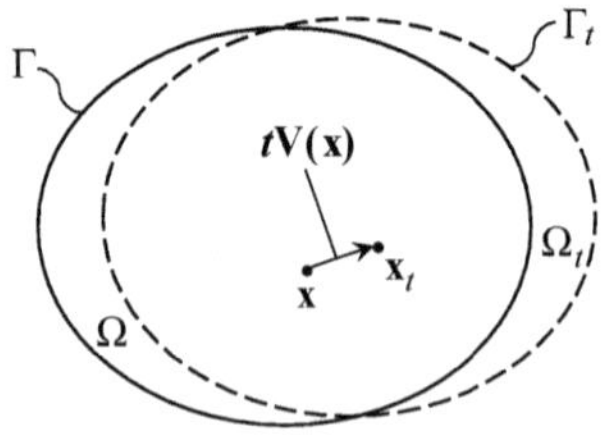

Fig. 3.2 Small variation of Ω near $t = 0$

The material derivative of ϕ is defined as

$$\dot{\phi}(\mathbf{x}) \equiv \lim_{t \to 0} \frac{\phi_t(\mathbf{x} + t\mathbf{V}(\mathbf{x})) - \phi(\mathbf{x})}{t} \tag{3.1.13}$$

where $\phi_t(\mathbf{x} + t\mathbf{V}(\mathbf{x})) - \phi(\mathbf{x})$ is the difference between ϕ at the original position and ϕ_t at the moved position of a moving particle. The material derivative of (3.1.13) can be expressed as

$$\dot{\phi}(\mathbf{x}) = \lim_{t \to 0} \frac{\phi_t(\mathbf{x}) - \phi(\mathbf{x})}{t} + \lim_{t \to 0} \frac{\phi_t(\mathbf{x} + t\mathbf{V}(\mathbf{x})) - \phi_t(\mathbf{x})}{t} \tag{3.1.14}$$

The first term of Eq. (3.1.14) is the time partial derivative of ϕ at the original position $\mathbf{x}$, and the second term is the directional (convective) derivative of ϕ at the original position. Thus, the material derivative of the state variable ϕ is expressed as

$$\dot{\phi} = \phi' + \mathbf{V} \cdot \nabla \phi \tag{3.1.15}$$

where

$$\nabla \phi = \left[\frac{\partial \phi}{\partial x}, \frac{\partial \phi}{\partial y}, \frac{\partial \phi}{\partial z} \right]^T \tag{3.1.16}$$

in the rectangular coordinates.

When the state variable of electromagnetic system is a vector $\mathbf{A}$, its material derivative is obtained by the same procedure;

$$\dot{\mathbf{A}} = \mathbf{A}' + \mathbf{V} \cdot \nabla \mathbf{A} \tag{3.1.17}$$

where the first term is the time partial derivative of $\mathbf{A}$ and the second term is

$$\mathbf{V} \cdot \nabla \mathbf{A} = V_x \frac{\partial \mathbf{A}}{\partial x} + V_y \frac{\partial \mathbf{A}}{\partial y} + V_y \frac{\partial \mathbf{A}}{\partial z} \tag{3.1.18}$$

in the rectangular coordinates.

3.1.2 Material Derivative Formula

A material derivative formula, which will be used for the sensitivity derivation in this chapter and Chaps. 4–6, is introduced. A scalar function F is given as a domain integral over Ω_t;

$$F = \int_{\Omega_t} g_t(\mathbf{x}_t)\mathrm{d}\Omega_t \tag{3.1.19}$$

where the function g_t is regular enough in Ω_t. The material derivative of F is obtained as

$$\dot{F} = \int_{\Omega} [g'(\mathbf{x}) + \nabla g(\mathbf{x}) \cdot \mathbf{V} + g(\mathbf{x})(\nabla \cdot \mathbf{V})]\mathrm{d}\Omega \tag{3.1.20}$$

By using the vector identity, this is written as

$$\dot{F} = \int_{\Omega} [g'(\mathbf{x}) + \nabla \cdot (g(\mathbf{x})\mathbf{V})]\mathrm{d}\Omega \tag{3.1.21}$$

This is rewritten by the divergence theorem;

$$\dot{F} = \int_{\Omega} g'(\mathbf{x})\mathrm{d}\Omega + \int_{\Gamma} g(\mathbf{x})V_n\mathrm{d}\Gamma \tag{3.1.22}$$

where Γ is the boundary of the domain Ω and $V_n = \mathbf{V} \cdot \mathbf{n}$ is the normal component of the velocity vector $\mathbf{V}$ on the boundary Γ. This formula was proved in the reference [2].

When the integrand g is a function $g(\mathbf{x}) = g(\phi, \nabla\phi)$ of a scalar state variable ϕ and its gradient, its partial derivative is written in the rectangular coordinate by the chain rule;

$$g'(\mathbf{x}) = \frac{\partial g}{\partial \phi}\phi' + \frac{\partial g}{\partial \nabla\phi} \cdot \nabla\phi' = g_\phi\phi' + \mathbf{g_E} \cdot \nabla\phi' \tag{3.1.23}$$

where

$$g_\phi \equiv \frac{\partial g}{\partial \phi} \quad \text{and} \quad \mathbf{g_E} \equiv \frac{\partial g}{\partial \nabla\phi} = \left[\frac{\partial g}{\partial(\nabla\phi)_x}, \frac{\partial g}{\partial(\nabla\phi)_y}, \frac{\partial g}{\partial(\nabla\phi)_z}\right]^T \tag{3.1.24}$$

By using the relation (3.1.15), (3.1.23) is rewritten as

$$g'(\mathbf{x}) = g_\phi\dot{\phi} - g_\phi(\mathbf{V} \cdot \nabla\phi) + \mathbf{g_E} \cdot \nabla\dot{\phi} - \mathbf{g_E} \cdot \nabla(\mathbf{V} \cdot \nabla\phi) \tag{3.1.25}$$

By inserting this relation into (3.1.22), the material derivative of F is expressed as

$$\dot{F} = \int_{\Omega} \left[g_{\phi}\dot{\phi} - g_{\phi}(\mathbf{V} \cdot \nabla\phi) + \mathbf{g_E} \cdot \nabla\dot{\phi} - \mathbf{g_E} \cdot \nabla(\mathbf{V} \cdot \nabla\phi) \right] \mathrm{d}\Omega + \int_{\Gamma} g(\mathbf{x}) V_n \mathrm{d}\Gamma \tag{3.1.26}$$

When the integrand g is a function $g(\mathbf{x}) = g(\mathbf{A}, \mathbf{B}(\mathbf{A}))$ of a vector state variable $\mathbf{A}$ and its curl $\mathbf{B}(\mathbf{A}) = \nabla \times \mathrm{A}$, its partial derivative is written in the rectangular coordinate by the chain rule;

$$g'(\mathbf{x}) = \frac{\partial g}{\partial \mathbf{A}} \cdot \mathbf{A}' + \frac{\partial g}{\partial \mathbf{B}} \cdot \mathbf{B}(\mathbf{A}') = \mathbf{g_A} \cdot \mathbf{A}' + \mathbf{g_B} \cdot \mathbf{B}(\mathbf{A}') \tag{3.1.27}$$

where

$$\mathbf{g_A} \equiv \frac{\partial g}{\partial \mathbf{A}} = \left[\frac{\partial g}{\partial A_x}, \frac{\partial g}{\partial A_y}, \frac{\partial g}{\partial A_z} \right]^T \quad \text{and} \quad \mathbf{g_B} \equiv \frac{\partial g}{\partial \mathbf{B}} = \left[\frac{\partial g}{\partial B_x}, \frac{\partial g}{\partial B_y}, \frac{\partial g}{\partial B_z} \right]^T \tag{3.1.28}$$

By using the relation (3.1.17), (3.1.27) is rewritten as

$$g'(\mathbf{x}) = \mathbf{g_A} \cdot \dot{\mathbf{A}} - \mathbf{g_A} \cdot (\mathbf{V} \cdot \nabla\mathrm{A}) + \mathbf{g_B} \cdot \mathbf{B}(\dot{\mathbf{A}}) - \mathbf{g_B} \cdot \mathbf{B}(\mathbf{V} \cdot \nabla\mathrm{A}) \tag{3.1.29}$$

By inserting this relation into (3.1.22), the material derivative of F is expressed as

$$\dot{F} = \int_{\Omega} \left[\mathbf{g_A} \cdot \dot{\mathbf{A}} - \mathbf{g_A} \cdot (\mathbf{V} \cdot \nabla\mathrm{A}) + \mathbf{g_B} \cdot \mathbf{B}(\dot{\mathbf{A}}) - \mathbf{g_B} \cdot \mathbf{B}(\mathbf{V} \cdot \nabla\mathrm{A}) \right] \mathrm{d}\Omega + \int_{\Gamma} g(\mathbf{x}) V_n \mathrm{d}\Gamma \tag{3.1.30}$$

3.2 Shape Sensitivity of Outer Boundary

The analysis of the electromagnetic system is basically an open boundary problem, where the electromagnetic field exists in the vacuum (or air) and extends to the infinity. Most of its design problems are, therefore, the interface design problem. But the electrostatic system has both of the interface design problem and the boundary design. When a domain is surrounded by electrodes or conductors connected to the voltage source, the electric field is limited to the inner area inside the boundary. Such a boundary is called the outer boundary, which can be also the infinite boundary but is not taken as the design variable. The shape variation of the outer boundary results in the change in the electric field in the domain. In this section, the general three-dimensional shape sensitivity for the outer boundary design is derived. First, the electrostatic system for the outer boundary design is depicted and a general objective function is defined as a domain integral. Second, the Lagrange multiplier method is introduced to handle the equality constraint of the variational state equation [4–6]. Third, the adjoint variable method is used to express explicitly the sensitivity in terms of design variable. Fourth, the variational identities are used to transform the sensitivity of domain integral into the boundary integral, which provides the general three-dimensional sensitivity formula for the outer boundary design. Finally, the obtained sensitivity formula is tested and validated with analytical and numerical examples.

3.2.1 Problem Definition and Objective Function

An electrostatic system is given as in Fig. 3.3, where the domain Ω has a distribution of ε, ρ, and $\mathbf{P}_o$. The charge density ρ is assumed to be fixed and constant in the domain. The domain Ω has the outer boundary Γ where $\mathbf{n}$ is defined as the outward normal vector. The outer boundary consists of the Dirichlet boundary Γ^0 and the Neumann boundary Γ^1. These two boundaries are taken as design variable in this shape sensitivity analysis for the outer boundary design.

Consider a general objective function of integral form;

$$F = \int_{\Omega} g(\phi, \nabla\phi) m_p \mathrm{d}\Omega \tag{3.2.1}$$

where g can be any function that is continuously differentiable with respect to the arguments of ϕ and $\nabla\phi$, and m_p is a characteristic function that is defined as

$$m_p = \begin{pmatrix} 1 & \mathbf{x} \in \Omega_p \\ 0 & \mathbf{x} \notin \Omega_p \end{pmatrix} \tag{3.2.2}$$

The region $\Omega_p \subset \Omega$, which is the integral domain for the objective function, can intersect with the outer boundary of the Dirichlet boundary Γ^0 or the Neumann boundary Γ^1 as shown schematically in Fig. 3.3. In (3.2.1), since g can be any

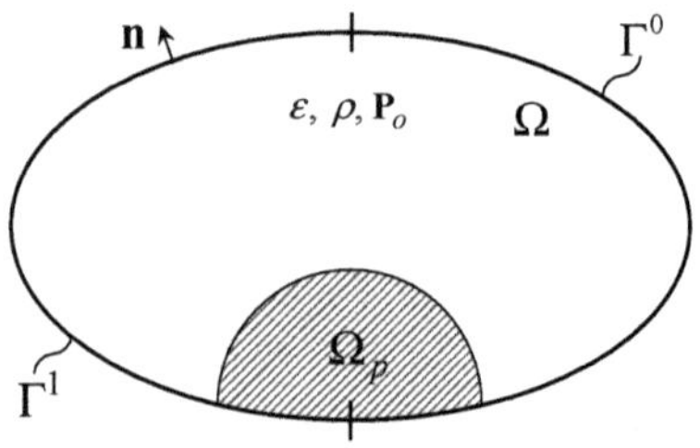

Fig. 3.3 Outer boundary design of electrostatic system

function of electric potential ϕ and electric field **E**, the objective function can represent a wide range of design problems for the electrostatic system.

3.2.2 Lagrange Multiplier Method for Sensitivity Derivation

The Lagrange multiplier method is introduced for the implicit equality constraint of the variational state equation [6]. The variation of the objective function F depends on both the integral domain Ω_p and the state variable ϕ. While its dependency on the integral domain is explicit, the dependency on the state variable ϕ is implicit. When the domain Ω is perturbed by the boundary shape, the state variable ϕ in the whole domain Ω is also perturbed through the state Eq. (2.1.20).

In derivation of the shape sensitivity, the variational state Eq. (2.1.20), which holds regardless of the change of the boundary shape, can be treated as an equality constraint. For the objective function F subject to the constraint (2.1.20), we employ the method of Lagrange multipliers, which is convenient in dealing with implicit constraints. The method of Lagrange multipliers provides an augmented objective function G;

$$G = F + l(\bar{\phi}) - a(\phi, \bar{\phi}) \quad \forall \bar{\phi} \in \Phi \tag{3.2.3}$$

where the arbitrary virtual potential $\bar{\phi}$ plays the role of Lagrange multipliers and Φ is the space of admissible potential defined in Sect. 2.1.2 as

$$\Phi = \left\{ \bar{\phi} \in H^1(\Omega) \middle| \bar{\phi} = 0 \quad \text{on} \quad \mathbf{x} \in \Gamma^0 \right\} \tag{3.2.4}$$

To develop the design sensitivity, the augmented objective function G is differentiated using the concept of material derivative;

$$\dot{G} = -\dot{a}(\phi, \bar{\phi}) + \dot{l}(\bar{\phi}) + \dot{F} \quad \forall \bar{\phi} \in \Phi \tag{3.2.5}$$

The differentiability of the bilinear form $a(\phi, \bar{\phi})$ and the source linear form $l(\bar{\phi})$, which was proved in [7], is only used here. By applying the material derivative formula (3.1.22) to the variational state Eq. (2.1.17) and the objective function (3.2.1) and using the relation (3.1.15), each term in (3.2.5) is obtained as followings:

$$
\begin{aligned}
\dot{a}(\phi,\bar{\phi}) &= \int_{\Omega} (\varepsilon\nabla\phi'\cdot\nabla\bar{\phi} + \varepsilon\nabla\phi\cdot\nabla\bar{\phi}')\mathrm{d}\Omega + \int_{\Gamma} \varepsilon\nabla\phi\cdot\nabla\bar{\phi}V_n\mathrm{d}\Gamma \\
&= \int_{\Omega} \Big(\varepsilon\nabla\dot{\phi}\cdot\nabla\bar{\phi} - \varepsilon\nabla(\mathbf{V}\cdot\nabla\phi)\cdot\nabla\bar{\phi} + \varepsilon\nabla\phi\cdot\nabla\dot{\bar{\phi}} - \varepsilon\nabla\phi\cdot\nabla(\mathbf{V}\cdot\nabla\bar{\phi})\Big)\mathrm{d}\Omega \\
&\quad + \int_{\Gamma} \varepsilon\nabla\phi\cdot\nabla\bar{\phi}V_n\mathrm{d}\Gamma
\end{aligned} \tag{3.2.6}
$$

$$
\begin{aligned}
\dot{l}(\bar{\phi}) &= \int_{\Omega} (\rho\bar{\phi}' + \mathbf{P}_o\cdot\nabla\bar{\phi}')\mathrm{d}\Omega + \int_{\Gamma} (\rho\bar{\phi} + \mathbf{P}_o\cdot\nabla\bar{\phi})V_n\mathrm{d}\Gamma \\
&= \int_{\Omega} \Big[\rho\dot{\bar{\phi}} - \rho(\mathbf{V}\cdot\nabla\bar{\phi}) + \mathbf{P}_o\cdot\nabla\dot{\bar{\phi}} - \mathbf{P}_o\cdot\nabla(\mathbf{V}\cdot\nabla\bar{\phi})\Big]\mathrm{d}\Omega \\
&\quad + \int_{\Gamma} (\rho\bar{\phi} + \mathbf{P}_o\cdot\nabla\bar{\phi})V_n\mathrm{d}\Gamma
\end{aligned} \tag{3.2.7}
$$

$$
\begin{aligned}
\dot{F} &= \int_{\Omega} g' m_p\mathrm{d}\Omega + \int_{\Gamma} g m_p V_n\mathrm{d}\Gamma \\
&= \int_{\Omega} (g_\phi\phi' + \mathbf{g}_\mathbf{E}\cdot\nabla\phi')m_p\mathrm{d}\Omega + \int_{\Gamma} g m_p V_n\mathrm{d}\Gamma \\
&= \int_{\Omega} \Big[g_\phi\dot{\phi} - g_\phi(\mathbf{V}\cdot\nabla\phi) + \mathbf{g}_\mathbf{E}\cdot\nabla\dot{\phi} - \mathbf{g}_\mathbf{E}\cdot\nabla(\mathbf{V}\cdot\nabla\phi)\Big]m_p\mathrm{d}\Omega \\
&\quad + \int_{\Gamma} g m_p V_n\mathrm{d}\Gamma
\end{aligned} \tag{3.2.8}
$$

For g' in (3.2.8), we used the relation (3.1.23):

$$
g' = \frac{\partial g}{\partial\phi}\phi' + \frac{\partial g}{\partial\nabla\phi}\cdot\nabla\phi' = g_\phi\phi' + \mathbf{g}_\mathbf{E}\cdot\nabla\phi' \tag{3.2.9}
$$

where

$$
g_\phi \equiv \frac{\partial g}{\partial\phi} \quad \text{and} \quad \mathbf{g}_\mathbf{E} \equiv \frac{\partial g}{\partial\nabla\phi} = \left[\frac{\partial g}{\partial(\nabla\phi)_x}, \frac{\partial g}{\partial(\nabla\phi)_y}, \frac{\partial g}{\partial(\nabla\phi)_z}\right]^T \tag{3.2.10}
$$

In derivation of (3.2.6) and (3.2.7), $\varepsilon' = 0$ and $\rho' = 0$ were used (3.2.6)–(3.2.8) are inserted into (3.2.5) to provide

$$\begin{aligned}
\dot{G} = &-\int_{\Omega} \varepsilon\nabla\dot{\phi}\cdot\nabla\bar{\phi}\mathrm{d}\Omega - \int_{\Omega} \varepsilon\nabla\phi\cdot\nabla\dot{\bar{\phi}}\mathrm{d}\Omega \\
&+\int_{\Omega} \left(\rho\dot{\bar{\phi}} + \mathbf{P}_o\cdot\nabla\dot{\bar{\phi}}\right)\mathrm{d}\Omega \\
&+\int_{\Omega} \left(g_{\phi}\dot{\phi} + \mathbf{g}_{\mathbf{E}}\cdot\nabla\dot{\phi}\right)m_p\mathrm{d}\Omega \\
&+\int_{\Omega} \varepsilon\nabla(\mathbf{V}\cdot\nabla\phi)\cdot\nabla\bar{\phi}\mathrm{d}\Omega + \int_{\Omega} \varepsilon\nabla\phi\cdot\nabla(\mathbf{V}\cdot\nabla\bar{\phi})\mathrm{d}\Omega \\
&-\int_{\Omega} \left[\rho(\mathbf{V}\cdot\nabla\bar{\phi}) + \mathbf{P}_o\cdot\nabla(\mathbf{V}\cdot\nabla\bar{\phi})\right]\mathrm{d}\Omega \\
&-\int_{\Omega} \left[g_{\phi}(\mathbf{V}\cdot\nabla\phi) + \mathbf{g}_{\mathbf{E}}\cdot\nabla(\mathbf{V}\cdot\nabla\phi)\right]m_p\mathrm{d}\Omega \\
&-\int_{\Gamma} \left(\varepsilon\nabla\phi\cdot\nabla\bar{\phi} - \mathbf{P}_o\cdot\nabla\bar{\phi} - \rho\bar{\phi} - gm_p\right)V_n\mathrm{d}\Gamma \quad \forall\bar{\phi}\in\Phi
\end{aligned} \tag{3.2.11}$$

Since $\dot{\bar{\phi}}$ belongs to Φ, the variational state Eq. (2.1.17) gives the following relation:

$$\int_{\Omega} \varepsilon\nabla\phi\cdot\nabla\dot{\bar{\phi}}\mathrm{d}\Omega = \int_{\Omega} \left(\rho\dot{\bar{\phi}} + \mathbf{P}_0\cdot\nabla\dot{\bar{\phi}}\right)\mathrm{d}\Omega \tag{3.2.12}$$

Thus, all terms containing $\dot{\bar{\phi}}$ in (3.2.11) are canceled out and (3.2.11) becomes

$$\begin{aligned}
\dot{G} = &-\int_{\Omega} \varepsilon\nabla\dot{\phi}\cdot\nabla\bar{\phi}\mathrm{d}\Omega + \int_{\Omega} \left(g_{\phi}\dot{\phi} + \mathbf{g}_{\mathbf{E}}\cdot\nabla\dot{\phi}\right)m_p\mathrm{d}\Omega \\
&+\int_{\Omega} \varepsilon\nabla(\mathbf{V}\cdot\nabla\phi)\cdot\nabla\bar{\phi}d\Omega + \int_{\Omega} \varepsilon\nabla\phi\cdot\nabla(\mathbf{V}\cdot\nabla\bar{\phi})\mathrm{d}\Omega \\
&-\int_{\Omega} \left[\rho(\mathbf{V}\cdot\nabla\bar{\phi}) + \mathbf{P}_o\cdot\nabla(\mathbf{V}\cdot\nabla\bar{\phi})\right]\mathrm{d}\Omega \\
&-\int_{\Omega} \left[g_{\phi}(\mathbf{V}\cdot\nabla\phi) + \mathbf{g}_{\mathbf{E}}\cdot\nabla(\mathbf{V}\cdot\nabla\phi)\right]m_p\mathrm{d}\Omega \\
&-\int_{\Gamma} \left(\varepsilon\nabla\phi\cdot\nabla\bar{\phi} - \mathbf{P}_o\cdot\nabla\bar{\phi} - \rho\bar{\phi} - gm_p\right)V_n\mathrm{d}\Gamma \quad \forall\bar{\phi}\in\Phi
\end{aligned} \tag{3.2.13}$$

Recall that $\dot{\phi}$ and $\nabla\dot{\phi}$ depend on the design variable. The objective here is to obtain an explicit expression of $\dot{G}$ in terms of the design variation, which is represented with the velocity field $\mathbf{V}$. Thus, all the terms with $\dot{\phi}$ and $\nabla\dot{\phi}$ need to be replaced in order that $\dot{G}$ is explicitly expressed in terms of the velocity field $\mathbf{V}$.

3.2.3 Adjoint Variable Method for Sensitivity Analysis

In order to avoid calculation of the term of $\dot{\phi}$ in the sensitivity (3.2.13) and express explicitly it with the velocity field $\mathbf{V}$, we introduce an adjoint equation, which is paralleled with the method used for the discrete method in Sect. 1.5.2. The adjoint equation is obtained by replacing $\dot{\phi}$ in the g-related terms ($g_\phi\dot{\phi}$ and $\mathbf{g}_\mathbf{E}\cdot\nabla\dot{\phi}$) of (3.2.13) with a virtual potential $\bar{\lambda}$ and by equating the terms to the energy bilinear form $a(\lambda,\bar{\lambda})$ [8, 9]. The adjoint equation so obtained is written as

$$\int_\Omega \varepsilon\nabla\lambda\cdot\nabla\bar{\lambda}\mathrm{d}\Omega = \int_\Omega (g_\phi\bar{\lambda} + \mathbf{g}_\mathbf{E}\cdot\nabla\bar{\lambda})m_p\mathrm{d}\Omega \quad \forall\bar{\lambda}\in\Phi \tag{3.2.14}$$

where λ is the adjoint variable and its solution is desired, and Φ is the space of admissible potential: $\Phi = \{\bar{\lambda}\in H^1(\Omega)|\bar{\lambda}=0 \quad \text{on} \quad \mathbf{x}\in\Gamma^0\}$.

The objective is to express (3.2.13) in terms of the adjoint variable λ obtained from (3.2.14). To take advantage of the adjoint equation, we evaluate (3.2.14) at a specific $\bar{\lambda}=\dot{\phi}$ since (3.2.14) holds for all $\bar{\lambda}\in\Phi$:

$$\int_\Omega \varepsilon\nabla\lambda\cdot\nabla\dot{\phi}\mathrm{d}\Omega = \int_\Omega \left(g_\phi\dot{\phi} + \mathbf{g}_\mathbf{E}\cdot\nabla\dot{\phi}\right)m_p\mathrm{d}\Omega \tag{3.2.15}$$

Similarly, the sensitivity Eq. (3.2.13) is evaluated at the specific $\bar{\phi}=\lambda$ since the λ belongs to the admissible space Φ;

$$\begin{aligned}
\dot{G} = &-\int_\Omega \varepsilon\nabla\dot{\phi}\cdot\nabla\lambda\mathrm{d}\Omega + \int_\Omega \left(g_\phi\dot{\phi} + \mathbf{g}_\mathbf{E}\cdot\nabla\dot{\phi}\right)m_p\mathrm{d}\Omega \\
&+ \int_\Omega \varepsilon\nabla(\mathbf{V}\cdot\nabla\phi)\cdot\nabla\lambda\mathrm{d}\Omega + \int_\Omega \varepsilon\nabla\phi\cdot\nabla(\mathbf{V}\cdot\nabla\lambda)\mathrm{d}\Omega \\
&- \int_\Omega [\rho(\mathbf{V}\cdot\nabla\lambda) + \mathbf{P}_o\cdot\nabla(\mathbf{V}\cdot\nabla\lambda)]\mathrm{d}\Omega \\
&- \int_\Omega [g_\phi(\mathbf{V}\cdot\nabla\phi) + \mathbf{g}_\mathbf{E}\cdot\nabla(\mathbf{V}\cdot\nabla\phi)]m_p\mathrm{d}\Omega \\
&- \int_\Gamma (\varepsilon\nabla\phi\cdot\nabla\lambda - \mathbf{P}_o\cdot\nabla\lambda - \rho\lambda - gm_p)V_n\mathrm{d}\Gamma
\end{aligned} \tag{3.2.16}$$

The energy bilinear form $a(\bullet, \bullet)$ is symmetric in its arguments;

$$\int_{\Omega} \varepsilon \nabla \lambda \cdot \nabla \dot{\phi} \mathrm{d}\Omega = \int_{\Omega} \varepsilon \nabla \dot{\phi} \cdot \nabla \lambda \mathrm{d}\Omega \tag{3.2.17}$$

By using (3.2.15) and (3.2.17), (3.2.16) becomes the desired one;

$$\begin{aligned} \dot{G} = & \int_{\Omega} [\varepsilon \nabla \phi \cdot \nabla (\mathbf{V} \cdot \nabla \lambda) - \rho (\mathbf{V} \cdot \nabla \lambda) - \mathbf{P}_o \cdot \nabla (\mathbf{V} \cdot \nabla \lambda)] \mathrm{d}\Omega \\ & + \int_{\Omega} \left[\varepsilon \nabla \lambda \cdot \nabla (\mathbf{V} \cdot \nabla \phi) - g_{\phi} (\mathbf{V} \cdot \nabla \phi) m_p - \mathbf{g}_{\mathbf{E}} \cdot \nabla (\mathbf{V} \cdot \nabla \phi) m_p\right] \mathrm{d}\Omega \\ & - \int_{\Gamma} \left(\varepsilon \nabla \phi \cdot \nabla \lambda - \mathbf{P}_o \cdot \nabla \lambda - \rho \lambda - g m_p\right) V_n \mathrm{d}\Gamma \end{aligned} \tag{3.2.18}$$

where all the terms are expressed with the velocity field $\mathbf{V}$. Once the state variable ϕ and the adjoint variable λ are determined to be the solutions to (2.1.17) and (3.2.14), respectively, the design sensitivity Eq. (3.2.18) can be evaluated.

3.2.4 Boundary Expression of Shape Sensitivity

The domain integrals in the design sensitivity (3.2.18) can be expressed in boundary integrals by using the variational identities. The fact that the sensitivity is expressed as a boundary integral provides not only good physical insight of the design sensitivity but also an advantage in numerical implementation.

To express the domain integral of the sensitivity (3.2.18) as a boundary integral, we need two variational identities for the state and the adjoint equations. First, the variational identity for the state equation was given as (2.1.13) in Sect. 2.1 and it is written again as

$$\int_{\Omega} \left(\varepsilon \nabla \phi \cdot \nabla \bar{\phi} - \rho \bar{\phi} - \mathbf{P}_o \cdot \nabla \bar{\phi}\right) \mathrm{d}\Omega = - \int_{\Gamma} D_n(\phi) \bar{\phi} \mathrm{d}\Gamma \quad \forall \bar{\phi} \in \Phi \tag{3.2.19}$$

Next, the needed variational identity for the adjoint equation can be derived from a differential adjoint equation, which is obtained by comparing the variational adjoint equation (3.2.14) with the variational state Eq. (2.1.17). The two variational equations are written again for convenience;

$$\int_{\Omega} \varepsilon \nabla \lambda \cdot \nabla \bar{\lambda} \mathrm{d}\Omega = \int_{\Omega} \left(g_{\phi} \bar{\lambda} + \mathbf{g}_{\mathbf{E}} \cdot \nabla \bar{\lambda}\right) m_p \mathrm{d}\Omega \quad \forall \bar{\lambda} \in \boldsymbol{\Phi} \tag{3.2.20}$$

$$\int_{\Omega} \varepsilon \nabla \phi \cdot \nabla \bar{\phi} \mathrm{d}\Omega = \int_{\Omega} \left(\rho \bar{\phi} + \mathbf{P}_o \cdot \nabla \bar{\phi}\right) \mathrm{d}\Omega \quad \forall \bar{\phi} \in \boldsymbol{\Phi} \tag{3.2.21}$$

These equations have the same form except the different source terms. The sources $g_{\phi} m_p$ and $\mathbf{g}_{\mathbf{E}} m_p$ in the adjoint equation correspond to the ρ and $\mathbf{P}_o$ in the state equation, respectively. Thus, just as the variational state Eq. (3.2.21) is equivalent to the differential state Eq. (2.1.5), the variational adjoint Eq. (3.2.20) is equivalent to a differential adjoint equation:

$$-\nabla \cdot \varepsilon \nabla \lambda = \left(g_{\phi} - \nabla \cdot \mathbf{g}_{\mathbf{E}}\right) m_p \tag{3.2.22}$$

with boundary conditions

$$\lambda = 0 \quad \text{on } \Gamma^0 \quad \text{(homogeneous Dirichlet boundary condition)} \tag{3.2.23}$$

$$\frac{\partial \lambda}{\partial n} = 0 \quad \text{on } \Gamma^1 \quad \text{(homogeneous Neumann boundary condition)} \tag{3.2.24}$$

Since the adjoint sources $g_{\phi} m_p$ and $\mathbf{g}_{\mathbf{E}} m_p$ exist only in the domain Ω as the original sources ρ and $\mathbf{P}_o$, there is no adjoint surface source equivalent to Dirichlet boundary condition. That is, while the original state equation may have surface sources equivalent to its Dirichlet boundary condition, the adjoint equation has no surface source equivalent to its Dirichlet boundary condition. Hence, the Dirichlet boundary condition of the adjoint equation is given as zero. In addition, since the structure symmetry is maintained in the adjoint system as well, the homogeneous Neumann condition is imposed on Γ^1 of the adjoint system.

In the same way that the variational identity (3.2.19) for the state equation is obtained from the differential state Eq. (2.1.5), a variational identity for the adjoint equation is obtained as

$$\int_{\Omega} \left(\varepsilon \nabla \lambda \cdot \nabla \bar{\lambda} - g_{\phi} \bar{\lambda} m_p - \mathbf{g}_{\mathbf{E}} \cdot \nabla \bar{\lambda} m_p\right) \mathrm{d}\Omega = -\int_{\Gamma} D_n(\lambda) \bar{\lambda} \mathrm{d}\Gamma \quad \forall \bar{\lambda} \in \boldsymbol{\Phi} \tag{3.2.25}$$

where

$$D_n(\lambda) = \left(-\varepsilon \nabla \lambda + \mathbf{g}_{\mathbf{E}} m_p\right) \cdot \mathbf{n} = -\varepsilon \frac{\partial \lambda}{\partial n} + \mathbf{g}_{\mathbf{E}} \cdot \mathbf{n} m_p \tag{3.2.26}$$

Note that imposing the boundary conditions (3.2.23) and (3.2.24) on (3.2.25) provides the variational adjoint Eq. (3.2.20).

The variational identities of (3.2.19) and (3.2.25) are used to express the domain integrals in (3.2.18) as boundary integrals. Choosing $\bar{\phi} = \mathbf{V} \cdot \nabla\lambda$ in (3.2.19) yields

$$\begin{aligned}&\int_\Omega [\varepsilon\nabla\phi \cdot \nabla(\mathbf{V} \cdot \nabla\lambda) - \rho(\mathbf{V} \cdot \nabla\lambda) - \mathbf{P}_o \cdot \nabla(\mathbf{V} \cdot \nabla\lambda)]\mathrm{d}\Omega \\ &\quad = -\int_\Gamma D_n(\phi)(\mathbf{V} \cdot \nabla\lambda)\mathrm{d}\Gamma\end{aligned} \tag{3.2.27}$$

and choosing $\bar{\lambda} = \mathbf{V} \cdot \nabla\phi$ in (3.2.25) yields

$$\begin{aligned}&\int_\Omega \left[\varepsilon\nabla\lambda \cdot \nabla(\mathbf{V} \cdot \nabla\phi) - g_\phi(\mathbf{V} \cdot \nabla\phi)m_p - \mathbf{g}_\mathbf{E} \cdot \nabla(\mathbf{V} \cdot \nabla\phi)m_p\right]\mathrm{d}\Omega \\ &\quad = -\int_\Gamma D_n(\lambda)(\mathbf{V} \cdot \nabla\phi)\mathrm{d}\Gamma\end{aligned} \tag{3.2.28}$$

By substituting (3.2.27) and (3.2.28) into (3.2.18), the domain integrals in (3.2.18) become a boundary integral;

$$\begin{aligned}\dot{G} = &-\int_\Gamma [D_n(\phi)(\mathbf{V} \cdot \nabla\lambda) + D_n(\lambda)(\mathbf{V} \cdot \nabla\phi)]\mathrm{d}\Gamma \\ &-\int_\Gamma \left(\varepsilon\nabla\phi \cdot \nabla\lambda - \mathbf{P}_o \cdot \nabla\lambda - \rho\lambda - gm_p\right)V_n\mathrm{d}\Gamma\end{aligned} \tag{3.2.29}$$

which is the desired expression.

Furthermore, this sensitivity can be expressed as a simpler form using the boundary conditions, the boundary relations of the velocity $\mathbf{V}$, etc. On the boundary, the gradients of the state and adjoint variables can be expressed with the normal and tangential components as

$$\nabla\phi = \frac{\partial\phi}{\partial n}\mathbf{n} + \frac{\partial\phi}{\partial t}\mathbf{t} \quad \text{on } \Gamma \tag{3.2.30}$$

$$\nabla\lambda = \frac{\partial\lambda}{\partial n}\mathbf{n} + \frac{\partial\lambda}{\partial t}\mathbf{t} \quad \text{on } \Gamma \tag{3.2.31}$$

where $\mathbf{n}$ and $\mathbf{t}$ are the normal and tangential unit vectors on the boundary, respectively. Only the tangential component of the velocity vector $\mathbf{V}$, which can have both the normal and tangential components as $\mathbf{V} = V_n\mathbf{n} + V_t\mathbf{t}$, does not contribute to the domain deformation. Thus, we can let $\mathbf{V} = V_n\mathbf{n}$ to provide

$$\mathbf{V}\cdot\nabla\phi = V_n\frac{\partial\phi}{\partial n}\quad \text{on } \Gamma \tag{3.2.32}$$

$$\mathbf{V}\cdot\nabla\lambda = V_n\frac{\partial\lambda}{\partial n}\quad \text{on } \Gamma \tag{3.2.33}$$

The term $\nabla\phi\cdot\nabla\lambda$ in (3.2.29) is written as

$$\nabla\phi\cdot\nabla\lambda = \frac{\partial\phi}{\partial n}\frac{\partial\lambda}{\partial n} + \frac{\partial\phi}{\partial t}\frac{\partial\lambda}{\partial t}\quad \text{on } \Gamma \tag{3.2.34}$$

By inserting (3.2.32)–(3.2.34) into (3.2.29), the sensitivity (3.2.29) becomes

$$\begin{aligned}\dot{G} = \int_{\Gamma}\Bigg(&-D_n(\phi)\frac{\partial\lambda}{\partial n} - D_n(\lambda)\frac{\partial\phi}{\partial n} - \varepsilon\frac{\partial\phi}{\partial n}\frac{\partial\lambda}{\partial n}\\ &-\varepsilon\frac{\partial\phi}{\partial t}\frac{\partial\lambda}{\partial t} + \mathbf{P}_o\cdot\nabla\lambda + \rho\lambda + g m_p\Bigg)V_n\mathrm{d}\Gamma\end{aligned} \tag{3.2.35}$$

where

$$\mathbf{P}_o\cdot\nabla\lambda = \mathbf{P}_o\cdot\mathbf{n}\frac{\partial\lambda}{\partial n} + \mathbf{P}_o\cdot\mathbf{t}\frac{\partial\lambda}{\partial t} \tag{3.2.36}$$

By using the relations of (2.1.14), (3.2.26) and (3.2.36), the sensitivity (3.2.35) is rewritten as

$$\dot{G} = \int_{\Gamma}\left[\varepsilon\frac{\partial\phi}{\partial n}\frac{\partial\lambda}{\partial n} - \varepsilon\frac{\partial\phi}{\partial t}\frac{\partial\lambda}{\partial t} + \mathbf{P}_o\cdot\mathbf{t}\frac{\partial\lambda}{\partial t} + \rho\lambda + \left(g - \mathbf{g}_{\mathbf{E}}\cdot\mathbf{n}\frac{\partial\phi}{\partial n}\right)m_p\right]V_n\mathrm{d}\Gamma \tag{3.2.37}$$

The homogeneous Neumann condition in the electrostatic system appears on the plane of symmetry. Thus, the Neumann boundary Γ^1 is not taken as a design variable; $V_n = 0$ on Γ^1. Consequently, the integral in (3.2.37) is taken only on the Dirichlet boundary Γ^0. Moreover, the following conditions hold on the Dirichlet boundary Γ^0:

$$\frac{\partial\phi}{\partial t} = 0\quad \text{on } \Gamma^0 \tag{3.2.38}$$

$$\frac{\partial\lambda}{\partial t} = 0\quad \text{on } \Gamma^0 \tag{3.2.39}$$

Inserting (3.2.38) and (3.2.39) into the sensitivity (3.2.37) yields

$$\dot{G} = \int_{\Gamma^0} \left[\varepsilon \frac{\partial \phi}{\partial n} \frac{\partial \lambda}{\partial n} + \rho \lambda + \left(g - \mathbf{g}_{\mathbf{E}} \cdot \mathbf{n} \frac{\partial \phi}{\partial n} \right) m_p \right] V_n \mathrm{d}\Gamma \tag{3.2.40}$$

The distribution of the space charge in the electrostatic systems is neither stable nor controllable; the space charge density ρ is not normally taken as a design variable. It was assumed in Sect. 3.2.1 that the space charge is fixed and constant in the space. It results in $\rho = 0$ on Γ^0; the sensitivity (3.2.40) becomes

$$\dot{G} = \int_{\Gamma^0} \left[\varepsilon \frac{\partial \phi}{\partial n} \frac{\partial \lambda}{\partial n} + \left(g - \mathbf{g}_{\mathbf{E}} \cdot \mathbf{n} \frac{\partial \phi}{\partial n} \right) m_p \right] V_n \mathrm{d}\Gamma \tag{3.2.41}$$

When the objective function is defined on the inner area in the domain Ω that does not intersect with the design boundary, $m_p = 0$ on Γ. Hence, the sensitivity (3.2.41) becomes

$$\dot{G} = \int_{\Gamma^0} \varepsilon \frac{\partial \phi}{\partial n} \frac{\partial \lambda}{\partial n} V_n \mathrm{d}\Gamma \tag{3.2.42}$$

When an operator $\mathbf{E}$ is defined as

$$\mathbf{E}(\psi) \equiv -\nabla \psi \tag{3.2.43}$$

, the electric field $\mathbf{E}(\phi)$ and the adjoint field $\mathbf{E}(\lambda)$ are written, respectively, as

$$\mathbf{E}(\phi) = -\nabla \phi \quad \text{and} \quad \mathbf{E}(\lambda) = -\nabla \lambda \tag{3.2.44}$$

The sensitivity formula (3.2.42) is rewritten as

$$\dot{G} = \int_{\Gamma^0} \varepsilon E_n(\phi) E_n(\lambda) V_n \mathrm{d}\Gamma \tag{3.2.45}$$

This sensitivity formula for the variation of Dirichlet boundary can be used for designing the electrode on which a constant voltage is imposed. In particular, when the electrode surface comes into contacts with the air, the sensitivity formula (3.2.45) is expressed as

$$\dot{G} = \int_{\Gamma^0} \varepsilon_0 E_n(\phi) E_n(\lambda) V_n \mathrm{d}\Gamma \tag{3.2.46}$$

3.2.5 Analytical Example

In the previous Sect. 3.2.4, the sensitivity formula for the objective function defined on the inner area was derived as (3.2.45). To show that the sensitivity formula is correct, one-dimensional analytical example, which has the analytic field solution, is taken. The sensitivity result, which is analytically calculated in the example, is compared with the result of the sensitivity formula, ensuring that it gives the same result. The analytical example is a cylindrical coaxial capacitor, which is the one-dimensional capacitor model in the cylindrical coordinates.

3.2.5.1 Cylindrical Coaxial Capacitor

For an example that can be analytically calculated, consider a cylindrical capacitor in Fig. 3.4, where two cylindrical conducting electrodes are separated by a dielectric of permittivity ε and a voltage ϕ_o is applied between the two electrodes. The design objective is to obtain a target electric field $\mathbf{E}_o$ in Ω_p by moving the outer electrode Γ^{0d}. The design variable is the radius a of the outer electrode. The design sensitivity with respect to the design variable a is analytically calculated, and then, its result is compared with the result by the sensitivity formula.

The objective function is defined as

$$F = \int_{\Omega} (\mathbf{E}(\phi) - \mathbf{E}_o)^2 m_p \mathrm{d}\Omega \tag{3.2.47}$$

where

$$\mathbf{E}(\phi) = \frac{\phi_o}{\ln a/b} \frac{1}{r} \mathbf{r} \tag{3.2.48}$$

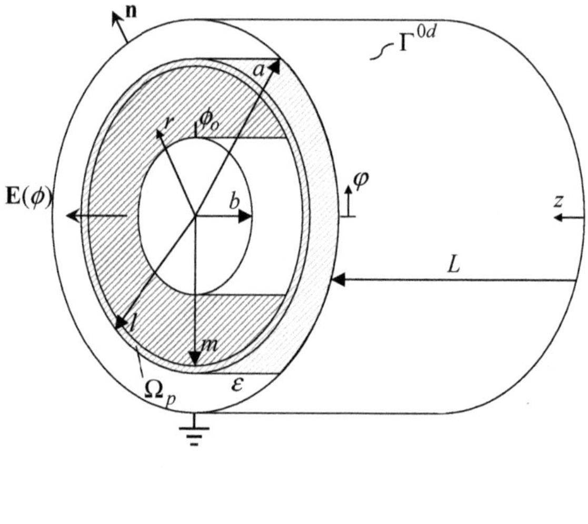

Fig. 3.4 Cylindrical coaxial capacitor-outer boundary design

$$\mathbf{E}_o = \frac{E_o}{r}\mathbf{r} \tag{3.2.49}$$

The objective function is rewritten by using the fields (3.2.47) and (3.2.48):

$$F = 2\pi \ln\frac{l}{m}\left(\frac{\phi_o}{\ln a/b} - E_o\right)^2 L \tag{3.2.50}$$

The analytical sensitivity per unit length is obtained by differentiating the objective function (3.2.50) with respect to the radius a of the outer electrode;

$$\frac{\mathrm{d}F}{\mathrm{d}b} = -4\pi\phi_o \frac{\ln l/m}{a(\ln a/b)^2}\left(\frac{\phi_o}{\ln a/b} - E_o\right) \tag{3.2.51}$$

Alternatively, the sensitivity can be calculated by using the sensitivity formula (3.2.45) in Sect. 3.2.4:

$$\dot{G} = \int_{\Gamma^{0d}} \varepsilon E_n(\phi) E_n(\lambda) V_n \mathrm{d}\Gamma \tag{3.2.52}$$

This sensitivity formula requires an adjoint variable λ, which can be obtained in the adjoint variable system in Fig. 3.5.

The differential adjoint equation is given as

$$-\nabla \cdot \varepsilon \nabla \lambda = -\nabla \cdot \mathbf{g}_{\mathbf{E}} m_p \tag{3.2.53}$$

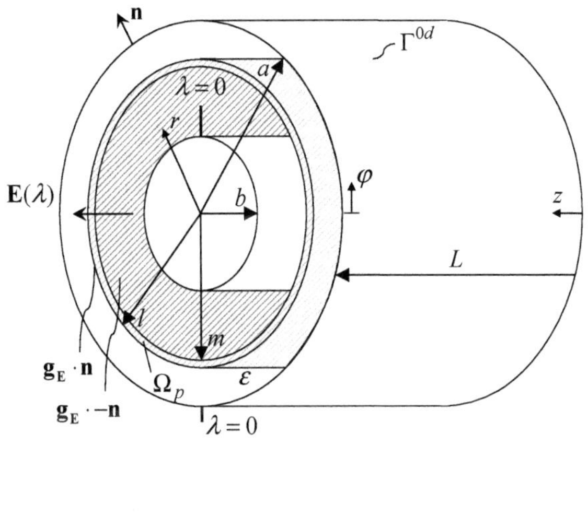

Fig. 3.5 Cylindrical coaxial capacitor: adjoint variable system

Inserting the electric fields (3.2.48), (3.2.49) into (3.2.53) provides

$$\varepsilon \frac{1}{r} \frac{\mathrm{d}}{\mathrm{d}r}\left(r \frac{\mathrm{d}\lambda}{\mathrm{d}r}\right) = -2\left(\frac{\phi_o}{\ln a/b} \frac{1}{r} - \frac{E_o}{r}\right)(\delta(r-m) - \delta(r-l)) \tag{3.2.54}$$

where $\delta(r)$ is the Dirac delta function. Integrating (3.2.54) yields the solution of the adjoint field as

$$E(\lambda) = -\frac{\mathrm{d}\lambda}{\mathrm{d}r} = \frac{2}{\varepsilon}\left(\frac{\phi_o}{\ln a/b} - E_o\right)(H(r-m) - H(r-l))\frac{1}{r} - \frac{k}{r} \tag{3.2.55}$$

where $H(r)$ is the Heaviside function and k, the integral constant, is obtained by integrating (3.2.55) and applying the boundary condition in Fig. 3.5;

$$k = \frac{2 \ln l/m}{\varepsilon \ln a/b}\left(\frac{\phi_o}{\ln /ab} - E_o\right) \tag{3.2.56}$$

Inserting (3.2.56) into (3.2.55) yields the adjoint field;

$$\mathbf{E}(\lambda) = \frac{2}{\varepsilon}\left(\frac{\phi_o}{\ln a/b} - E_o\right)\left(H(r-m) - H(r-l) - \frac{\ln l/m}{\ln a/b}\right)\frac{1}{r}\mathbf{r} \tag{3.2.57}$$

(3.2.48) and (3.2.57) are inserted into the sensitivity formula (3.2.52);

$$\dot{G} = -4\pi\phi_o \frac{\ln l/m}{a(\ln a/b)^2}\left(\frac{\phi_o}{\ln a/b} - E_o\right)V_n L \tag{3.2.58}$$

Using $V_n = \frac{da}{dt}$, the design sensitivity per unit length is obtained as

$$\frac{\mathrm{d}G}{\mathrm{d}a} = -4\pi\phi_o \frac{\ln l/m}{a(\ln a/b)^2}\left(\frac{\phi_o}{\ln a/b} - E_o\right) \tag{3.2.59}$$

which is the same as the analytical result in (3.2.51).

3.2.6 *Numerical Examples*

Here, the sensitivity formula (3.2.45) derived in the Sect. 3.2.4 is applied to shape optimization problems of two-dimensional design model, of which the analytic field solutions are not given. These design models are taken to illustrate how well the sensitivity formula is applied to the shape design problem of the outer boundary in the electrostatic system. The numerical examples are an axi-symmetric capacitor and a pin-plate electrode. The optimal design for the first example is known, but the optimal design for the second one is not known. If the result of the first example is obtained as the expected optimal design, it can be said that the shape optimization using the sensitivity formula is feasible for the design of the outer boundary. The result of the second example shows that this optimization method is useful for the design of the outer boundary and applicable to any shaped electrodes.

The sensitivity formula in these examples requires the state and the adjoint variables, which are numerically calculated by the finite element method. The sensitivity information obtained is used for the optimization algorithm to provide the evolution of the electrode shape. For the optimization algorithm, the level set method is used to represent the shape evolution of the design model. The level set method is described in Chap. 7, where the shape evolution is expressed by the parameter t of unit s for the amount of shape change.

3.2.6.1 Axi-Symmetric Capacitor

For an example with a known optimal design, a capacitor is given as in Fig. 3.6, where the dielectric of permittivity ε is between two electrodes. The inner electrode is cylindrical, but the outer electrode is not cylindrical in the form. When a voltage ϕ_o is applied between the two electrodes, the distribution of the electric field is not uniform along the axial direction. When the shape of the outer electrode is changed to be a cylinder, the capacitor becomes a coaxial capacitor and the electric field between the two electrodes becomes uniform along the axial direction, which is known.

The design objective is to obtain a uniform field $\mathbf{E}_o$ in the region Ω_p, which is analytically given in the coaxial capacitor. The design variable is the shape of the outer electrode Γ^{0d}, of which the optimal shape is the cylinder. The objective function to be minimized is defined in the integral in Ω_p as

$$F = \int_{\Omega} (\mathbf{E}(\phi) - \mathbf{E}_o)^2 m_p \mathrm{d}\Omega \tag{3.2.60}$$

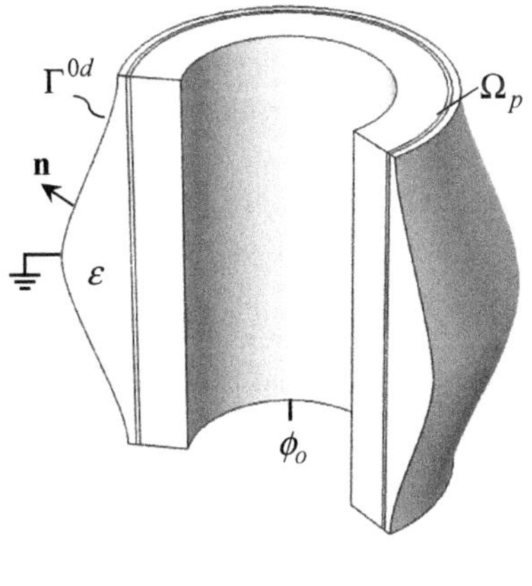

Fig. 3.6 Axi-symmetric capacitor-outer boundary design

where

$$\mathbf{E}_o = \frac{E_o}{r}\mathbf{r} \tag{3.2.61}$$

The variational adjoint equation for (3.2.60) is obtained as

$$a(\lambda, \bar{\lambda}) = \int_{\Omega} 2(\mathbf{E}(\phi) - \mathbf{E}_o) \cdot \mathbf{E}(\bar{\lambda}) m_p \mathrm{d}\Omega \quad \forall \bar{\lambda} \in \Phi \tag{3.2.62}$$

The shape sensitivity for this outer boundary design is the sensitivity formula (3.2.45):

$$\dot{G} = \int_{\Gamma^{0d}} \varepsilon E_n(\phi) E_n(\lambda) V_n \mathrm{d}\Gamma \tag{3.2.63}$$

The design velocity for this minimization problem is taken as

$$V_n = -\varepsilon E_n(\phi) E_n(\lambda) \tag{3.2.64}$$

Figure 3.7 illustrates the design results for the shape evolution of the outer electrode, where the outer electrode becomes gradually a cylinder as expected. The final design is obtained as the cylindrical shape at the 25 s, where the objective function converges to 0 as in Fig. 3.8. The result of this example shows that the shape optimization using the sensitivity formula is feasible for the design of the outer boundary in the electrostatic system.

3.2.6.2 Pin-Plate Electrode

The shape optimization using the sensitivity formula for the design of the outer boundary is applied to a shape design problem, which has neither the analytical field solution nor a known optimal shape. Consider a pin-plate electrode as in Fig. 3.9, where the sharp pin electrode is above the plate electrode. When a voltage ϕ_o is applied between the two electrodes, the electric field is concentrated on the pin tip and is also not uniform near the center of the plate electrode.

The objective of this example is to obtain a uniform electric field in the central region Ω_p on the plane electrode by deforming the shape of the pin tip. The design variable is the shape of the pin tip Γ^{0d}, the optimal shape of which is not known. But it is expected that the sharp pin tip will become rounded for a uniform field on the plate electrode. The objective function to be minimized is defined as the integration of the field difference in Ω_p;

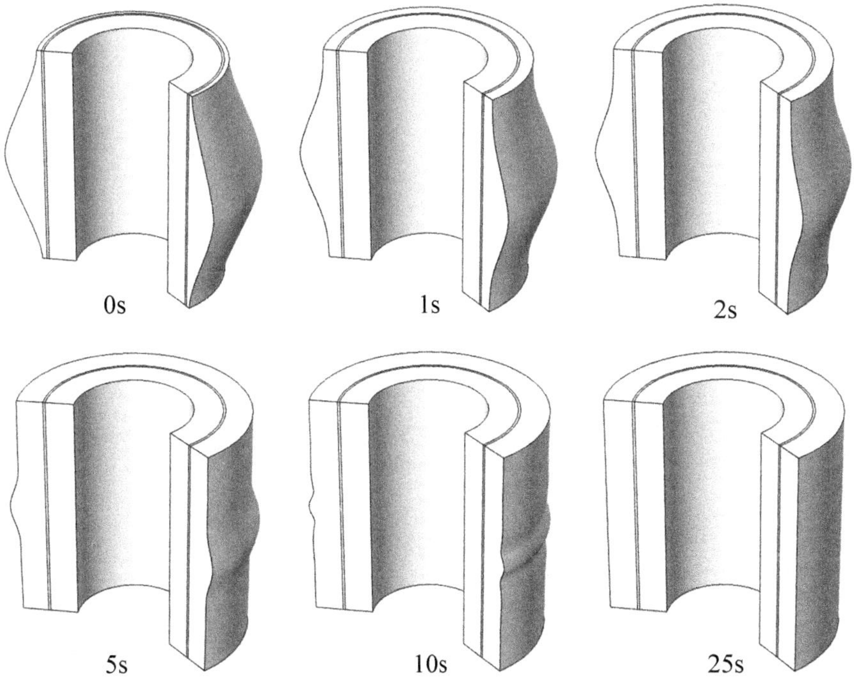

Fig. 3.7 Axi-symmetric capacitor: shape variation

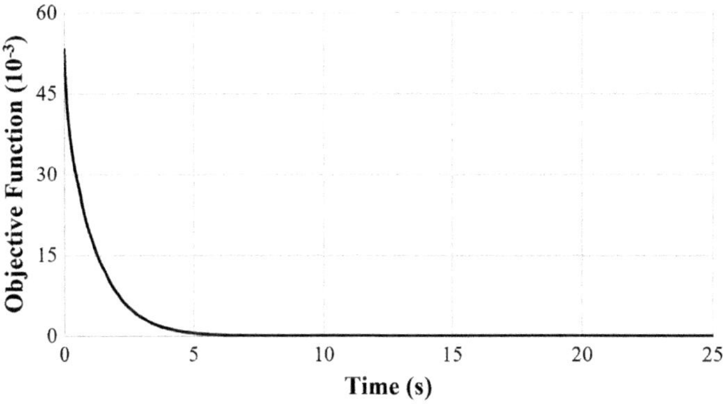

Fig. 3.8 Axi-symmetric capacitor: evolution of objective function

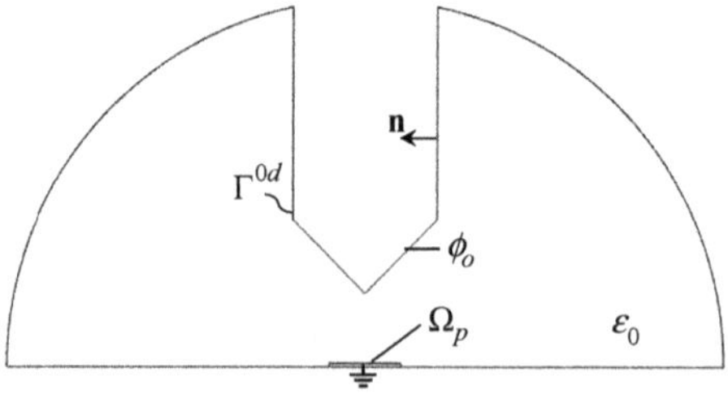

Fig. 3.9 Pin-plate electrode-outer boundary design

$$F = \int_{\Omega} (\mathbf{E}(\phi) - \mathbf{E}_o)^2 m_p \mathrm{d}\Omega \tag{3.2.65}$$

where the target field $\mathbf{E}_o$ is taken to be the median value of the field distribution in Ω_p in the initial design. The variational adjoint equation for (3.2.65) is obtained as

$$a(\lambda, \bar{\lambda}) = \int_{\Omega} 2(\mathbf{E}(\phi) - \mathbf{E}_o) \cdot \mathbf{E}(\bar{\lambda}) m_p \mathrm{d}\Omega \quad \forall \bar{\lambda} \in \mathbf{\Phi} \tag{3.2.66}$$

The shape sensitivity for this outer boundary design is the sensitivity formula (3.2.45):

$$\dot{G} = \int_{\Gamma^{0d}} \varepsilon_0 E_n(\phi) E_n(\lambda) V_n \mathrm{d}\Gamma \tag{3.2.67}$$

The design velocity for this minimization problem is taken as

$$V_n = -\varepsilon_0 E_n(\phi) E_n(\lambda) \tag{3.2.68}$$

The results of the shape evolution in the pin tip during the optimization are shown in Fig. 3.10, where, with the increase in the number of iteration, the pin tip moves up and the pin sides move down. According to this shape evolution, the electric field distribution in Ω_p is changed as in Fig. 3.11. In the earlier stage of the optimization, the electric field intensity in the center is much higher than the one in the both sides of Ω_p, but the electric field in the center becomes lower and the one near the pin sides becomes higher with the iteration. When the objective function value converges at 250 s as in Fig. 3.12, the final design of the pin tip is obtained as a concave shape and the deviation of the field distribution decreases by 75%. The results of this example show that the shape sensitivity method is feasible for the shape design of the outer boundary and useful for the optimal design of the electrode shape.

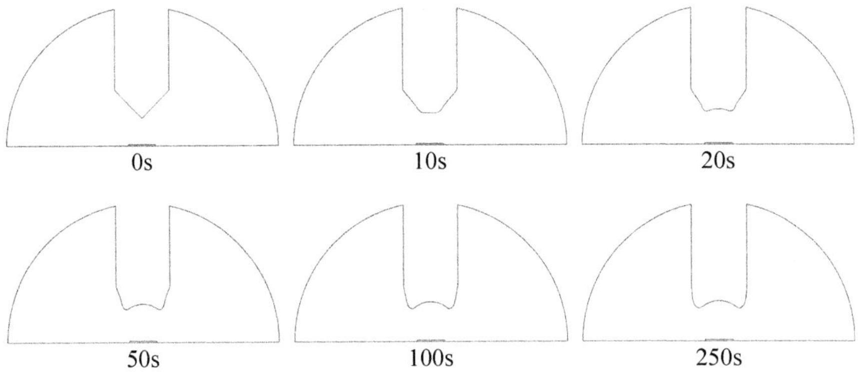

Fig. 3.10 Pin-plate electrode: shape variation

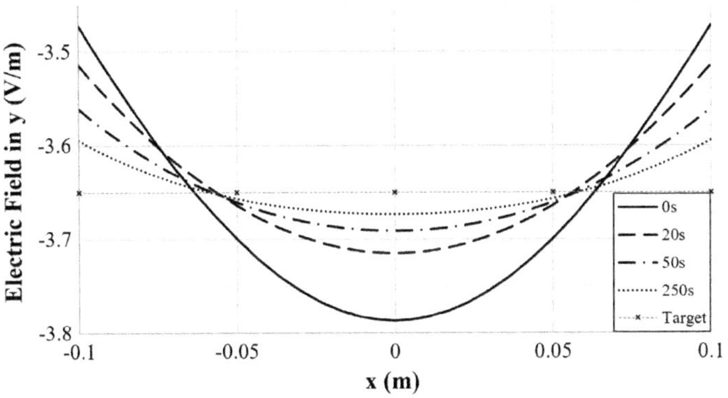

Fig. 3.11 Pin-plate electrode: electric field distribution in Ω_p

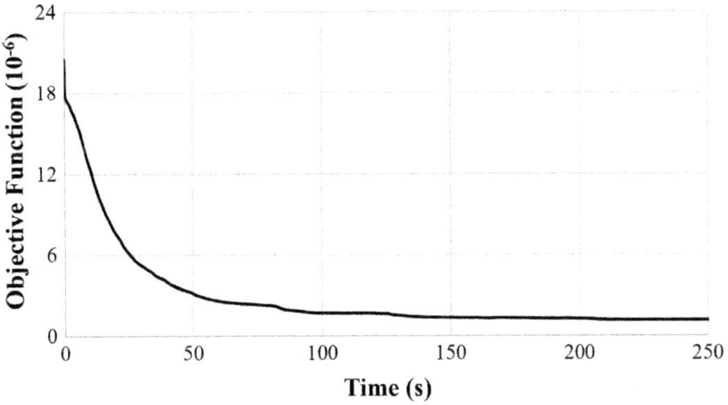

Fig. 3.12 Pin-plate electrode: evolution of objective function

3.3 Shape Sensitivity of Outer Boundary for System Energy

In this section, the three-dimensional shape sensitivity for the outer boundary design is also derived in the electrostatic system as in the Sect. 3.2; but the objective function is the energy of electrostatic system. The system energy of the electrostatic system is related to the equivalent capacitance, and it can be used for alleviation of the electric field intensity on the electrode surface and for calculating the electrostatic force. The derivation procedure is almost the same as the one in Sect. 3.2. The difference is that the adjoint variable for the system energy is obtained as the half of the state variable. Thus, solving the adjoint variable equation is not necessary. The derived sensitivity formula is tested and validated with analytical and numerical examples.

3.3.1 Problem Definition

An electrostatic system is given as in Fig. 3.13, where the domain Ω has arbitrary distribution of ε and ρ. The domain Ω is surrounded by the outer boundary Γ, where the $\mathbf{n}$ is an outward normal vector. The electrostatic system is almost the same as Fig. 3.3 in Sect. 3.2 except that the permanent polarization is excluded. Most of the design problems of the electrostatic system do not include the permanent polarization; it is not taken into account for the system energy.

The objective function is the system energy of the electrostatic system;

$$W_e = \int_{\Omega} \frac{1}{2} \rho \phi \mathrm{d}\Omega \tag{3.3.1}$$

where the charge density ρ is assumed to be fixed and constant in the domain.

The state equations of the electrostatic system are the same as the ones in Sect. 2.1 except that the permanent polarization is excluded. The governing differential equation with the state variable of the electric scalar potential ϕ is written as

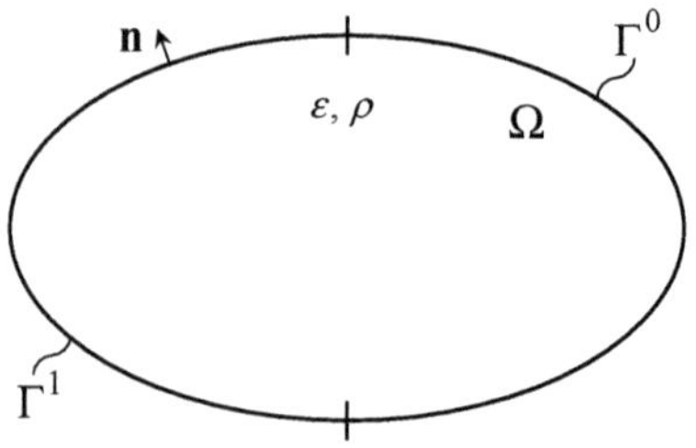

Fig. 3.13 Outer boundary design of electrostatic system for system energy

$$-\nabla \cdot \varepsilon \nabla \phi = \rho \tag{3.3.2}$$

The boundary conditions are the same as in Sect. 2.1;

$$\phi = C(\mathbf{x}) \quad \text{on } \Gamma^0 \quad \text{(Dirichlet boundary condition)} \tag{3.3.3}$$

$$\frac{\partial \phi}{\partial n} = 0 \quad \text{on } \Gamma^1 \quad \text{(homogeneous Neumann boundary condition)} \tag{3.3.4}$$

The variational identity for the state equation is also obtained by multiplying both sides of (3.3.2) by an arbitrary virtual potential $\bar{\phi}$ and by using the vector identity and the divergence theorem;

$$\int_{\Omega} (\varepsilon \nabla \phi \cdot \nabla \bar{\phi} - \rho \bar{\phi}) \mathrm{d}\Omega = \int_{\Gamma} \varepsilon \frac{\partial \phi}{\partial n} \bar{\phi} \mathrm{d}\Gamma \quad \forall \bar{\phi} \in \Phi \tag{3.3.5}$$

where Φ is the space of admissible potential: $\Phi = \{\bar{\phi} \in H^1(\Omega) | \bar{\phi} = 0 \text{ on } \mathbf{x} \in \Gamma^0.\}$. Imposing the boundary conditions (2.1.15) and (2.1.16) yields the variational state equation corresponding to the differential Eq. (3.3.2);

$$\int_{\Omega} \varepsilon \nabla \phi \cdot \nabla \bar{\phi} \mathrm{d}\Omega = \int_{\Omega} \rho \bar{\phi} \mathrm{d}\Omega \quad \forall \bar{\phi} \in \Phi \tag{3.3.6}$$

The energy bilinear form $a(\phi, \bar{\phi})$ and the source linear form $l(\bar{\phi})$ are defined as

$$a(\phi, \bar{\phi}) \equiv \int_{\Omega} \varepsilon \nabla \phi \cdot \nabla \bar{\phi} \mathrm{d}\Omega \tag{3.3.7}$$

$$l(\bar{\phi}) \equiv \int_{\Omega} \rho \bar{\phi} d\Omega \tag{3.3.8}$$

The variational Eq. (3.3.6) is rewritten with the energy bilinear form and the source linear form as

$$a(\phi, \bar{\phi}) = l(\bar{\phi}) \quad \forall \bar{\phi} \in \Phi \tag{3.3.9}$$

3.3.2 *Lagrange Multiplier Method for Energy Sensitivity*

The variational state Eq. (3.3.9), which is always valid regardless of the variation of the system geometry, can be thought to be as an equality constraint. For the sensitivity of the system energy F (3.3.1) subject to the constraint (3.3.9), the

Lagrange multiplier method is employed for the implicit constraint of the variational state equation to provide an augmented objective function G as

$$G = W_e + l(\bar{\phi}) - a(\phi, \bar{\phi}) \quad \forall \bar{\phi} \in \Phi \tag{3.3.10}$$

where the arbitrary virtual potential $\bar{\phi}$ plays the role of Lagrange multipliers and Φ is the space of admissible potential: $\Phi = \{\bar{\phi} \in H^1(\Omega) | \bar{\phi} = 0 \quad \text{on} \quad \mathbf{x} \in \Gamma^0\}$.

The sensitivity, the material derivative of the augmented objective function G, is written as

$$\dot{G} = -\dot{a}(\phi, \bar{\phi}) + \dot{l}(\bar{\phi}) + \dot{W}_e \quad \forall \bar{\phi} \in \Phi \tag{3.3.11}$$

By applying the material derivative formula (3.1.22) to the variational state Eq. (3.3.6) and the objective function (3.3.1) and using the relation (3.1.15), the first term in (3.3.11) is the same as (3.2.6) in Sect. 3.2, and the last two terms are obtained as followings:

$$\begin{aligned} \dot{l}(\bar{\phi}) &= \int_\Omega \rho \bar{\phi}' \mathrm{d}\Omega + \int_\Gamma \rho \bar{\phi} V_n \mathrm{d}\Gamma \\ &= \int_\Omega \left[\rho \dot{\bar{\phi}} - \rho(\mathbf{V} \cdot \nabla \bar{\phi}) \right] \mathrm{d}\Omega + \int_\Gamma \rho \bar{\phi} V_n \mathrm{d}\Gamma \end{aligned} \tag{3.3.12}$$

$$\begin{aligned} \dot{W}_e &= \int_\Omega \frac{1}{2} \rho \phi' \mathrm{d}\Omega + \int_\Gamma \frac{1}{2} \rho \phi V_n \mathrm{d}\Gamma \\ &= \int_\Omega \left[\frac{1}{2} \rho \dot{\phi} - \frac{1}{2} \rho(\mathbf{V} \cdot \nabla \phi) \right] \mathrm{d}\Omega + \int_\Gamma \frac{1}{2} \rho \phi V_n \mathrm{d}\Gamma \end{aligned} \tag{3.3.13}$$

Inserting (3.2.6), (3.3.12) and (3.3.13) into (3.3.11) provides

$$\begin{aligned} \dot{G} = &- \int_\Omega \varepsilon \nabla \dot{\phi} \cdot \nabla \bar{\phi} \mathrm{d}\Omega - \int_\Omega \varepsilon \nabla \phi \cdot \nabla \dot{\bar{\phi}} \mathrm{d}\Omega + \int_\Omega \rho \dot{\bar{\phi}} \mathrm{d}\Omega + \int_\Omega \frac{1}{2} \rho \dot{\phi} d\Omega \\ &+ \int_\Omega \varepsilon \nabla(\mathbf{V} \cdot \nabla \phi) \cdot \nabla \bar{\phi} \mathrm{d}\Omega + \int_\Omega \varepsilon \nabla \phi \cdot \nabla(\mathbf{V} \cdot \nabla \bar{\phi}) \mathrm{d}\Omega - \int_\Omega \rho(\mathbf{V} \cdot \nabla \bar{\phi}) \mathrm{d}\Omega \\ &- \int_\Omega \frac{1}{2} \rho(\mathbf{V} \cdot \nabla \phi) \mathrm{d}\Omega - \int_\Gamma \left(\varepsilon \nabla \phi \cdot \nabla \bar{\phi} - \rho \bar{\phi} - \frac{1}{2} \rho \phi \right) V_n \mathrm{d}\Gamma \quad \forall \bar{\phi} \in \Phi \end{aligned} \tag{3.3.14}$$

Since $\dot{\bar{\phi}} \in \Phi$, the variational state Eq. (3.3.6) provides

$$\int_{\Omega} \varepsilon \nabla \phi \cdot \nabla \dot{\bar{\phi}} \mathrm{d}\Omega = \int_{\Omega} \rho \dot{\bar{\phi}} \mathrm{d}\Omega \tag{3.3.15}$$

The relation (3.3.15) cancels out all terms containing $\dot{\bar{\phi}}$ in (3.3.14);

$$\begin{aligned} \dot{G} = &- \int_{\Omega} \varepsilon \nabla \dot{\phi} \cdot \nabla \bar{\phi} \mathrm{d}\Omega + \int_{\Omega} \frac{1}{2} \rho \dot{\phi} \mathrm{d}\Omega \\ &+ \int_{\Omega} \varepsilon \nabla (\mathbf{V} \cdot \nabla \phi) \cdot \nabla \bar{\phi} \mathrm{d}\Omega + \int_{\Omega} \varepsilon \nabla \phi \cdot \nabla (\mathbf{V} \cdot \nabla \bar{\phi}) \mathrm{d}\Omega \\ &- \int_{\Omega} \rho (\mathbf{V} \cdot \nabla \bar{\phi}) \mathrm{d}\Omega - \int_{\Omega} \frac{1}{2} \rho (\mathbf{V} \cdot \nabla \phi) \mathrm{d}\Omega \\ &- \int_{\Gamma} \left(\varepsilon \nabla \phi \cdot \nabla \bar{\phi} - \rho \bar{\phi} - \frac{1}{2} \rho \phi \right) V_n \mathrm{d}\Gamma \quad \forall \bar{\phi} \in \Phi \end{aligned} \tag{3.3.16}$$

3.3.3 Adjoint Variable Method for Sensitivity Analysis

An adjoint equation is introduced to avoid calculation of the term of $\dot{\phi}$ in (3.3.16) and to obtain an explicit expression of (3.3.16) in terms of the velocity field **V**. The adjoint equation is obtained by replacing $\dot{\phi}$ in the second integral of (3.3.16) with a virtual potential $\bar{\lambda}$ and by equating the integral to the energy bilinear form $a(\lambda, \bar{\lambda})$. The adjoint equation so obtained is written as

$$\int_{\Omega} \varepsilon \nabla \lambda \cdot \nabla \bar{\lambda} \mathrm{d}\Omega = \int_{\Omega} \frac{1}{2} \rho \bar{\lambda} \mathrm{d}\Omega \quad \forall \bar{\lambda} \in \Phi \tag{3.3.17}$$

where λ is the adjoint variable.

This adjoint equation is evaluated at a specific $\bar{\lambda} = \dot{\phi}$ to provide

$$\int_{\Omega} \varepsilon \nabla \lambda \cdot \nabla \dot{\phi} \mathrm{d}\Omega = \int_{\Omega} \frac{1}{2} \rho \dot{\phi} \mathrm{d}\Omega \tag{3.3.18}$$

Similarly, the sensitivity (3.3.16) is evaluated at a specific $\bar{\phi} = \lambda$ to yield

$$\begin{aligned} \dot{G} = & -\int_{\Omega} \varepsilon \nabla \dot{\phi} \cdot \nabla \lambda \mathrm{d}\Omega + \int_{\Omega} \frac{1}{2} \rho \dot{\phi} \mathrm{d}\Omega \\ & + \int_{\Omega} \varepsilon \nabla (\mathbf{V} \cdot \nabla \phi) \cdot \nabla \lambda \mathrm{d}\Omega + \int_{\Omega} \varepsilon \nabla \phi \cdot \nabla (\mathbf{V} \cdot \nabla \lambda) \mathrm{d}\Omega \\ & - \int_{\Omega} \rho (\mathbf{V} \cdot \nabla \lambda) \mathrm{d}\Omega - \int_{\Omega} \frac{1}{2} \rho (\mathbf{V} \cdot \nabla \phi) \mathrm{d}\Omega \\ & - \int_{\Gamma} \left(\varepsilon \nabla \phi \cdot \nabla \lambda - \rho \lambda - \frac{1}{2} \rho \phi \right) V_n \mathrm{d}\Gamma \end{aligned} \tag{3.3.19}$$

The energy bilinear form is symmetric in its arguments;

$$\int_{\Omega} \varepsilon \nabla \lambda \cdot \nabla \dot{\phi} \mathrm{d}\Omega = \int_{\Omega} \varepsilon \nabla \dot{\phi} \cdot \nabla \lambda \mathrm{d}\Omega \tag{3.3.20}$$

By using the relations (3.3.18) and (3.3.20), all terms with $\dot{\phi}$ in (3.3.19) are canceled out;

$$\begin{aligned} \dot{G} = & \int_{\Omega} \left[\varepsilon \nabla (\mathbf{V} \cdot \nabla \phi) \cdot \nabla \lambda + \varepsilon \nabla \phi \cdot \nabla (\mathbf{V} \cdot \nabla \lambda) - \rho (\mathbf{V} \cdot \nabla \lambda) - \frac{1}{2} \rho (\mathbf{V} \cdot \nabla \phi) \right] \mathrm{d}\Omega \\ & - \int_{\Gamma} \left(\varepsilon \nabla \phi \cdot \nabla \lambda - \rho \lambda - \frac{1}{2} \rho \phi \right) V_n \mathrm{d}\Gamma \end{aligned} \tag{3.3.21}$$

Next, the adjoint Eq. (3.3.17) is compared with the original state Eq. (3.3.6). Only the difference between the two equations is that the source of the adjoint equation is the half of the original state equation. If the boundary condition is also given as the half of the original state equation, the adjoint variable is obtained as the half of the state variable in the whole field region. Thus, the boundary conditions for the adjoint equation are given as

$$\lambda = \tfrac{1}{2} C(\mathbf{x}) \quad \text{on } \Gamma^0 \quad \text{(Dirichlet boundary condition)} \tag{3.3.22}$$

$$\tfrac{\partial \lambda}{\partial n} = 0 \quad \text{on } \Gamma^1 \quad \text{(homogeneous Neumann boundary condition)} \tag{3.3.23}$$

That is, the adjoint variable, which is determined from the adjoint Eq. (3.3.17) with the boundary conditions of (3.3.22) and (3.3.23), is simply the half of the state variable;

$$\lambda = \frac{1}{2}\phi \quad \text{in} \quad \Omega \quad \text{and on} \quad \Gamma \tag{3.3.24}$$

Consequently, for the sensitivity (3.3.21) of the system energy, solving the adjoint equation is not necessary. By inserting (3.3.24) into (3.3.21), the sensitivity (3.3.21) becomes

$$\dot{G} = \int_{\Omega} [\varepsilon \nabla\phi \cdot \nabla(\mathbf{V} \cdot \nabla\phi) - \rho(\mathbf{V} \cdot \nabla\phi)] \mathrm{d}\Omega - \int_{\Gamma} \left(\frac{1}{2}\varepsilon \nabla\phi \cdot \nabla\phi - \rho\phi \right) V_n \mathrm{d}\Gamma \tag{3.3.25}$$

3.3.4 Boundary Expression of Shape Sensitivity

The variational identity (3.3.5) is used to express the domain integrals in (3.3.25) as boundary integrals. Moreover, the sensitivity expression of the boundary integral becomes simpler by using the boundary conditions: Dirichlet and Neumann boundary conditions.

The variational identity (3.3.5) is written again;

$$\int_{\Omega} (\varepsilon \nabla\phi \cdot \nabla\bar{\phi} - \rho\bar{\phi}) \mathrm{d}\Omega = \int_{\Gamma} \varepsilon \frac{\partial \phi}{\partial n} \bar{\phi} \mathrm{d}\Gamma \quad \forall \bar{\phi} \in \Phi \tag{3.3.26}$$

Choosing $\bar{\phi} = \mathbf{V} \cdot \nabla\phi$ in this equation yields

$$\int_{\Omega} [\varepsilon \nabla\phi \cdot \nabla(\mathbf{V} \cdot \nabla\phi) - \rho(\mathbf{V} \cdot \nabla\phi)] \mathrm{d}\Omega = \int_{\Gamma} \varepsilon \frac{\partial \phi}{\partial n} (\mathbf{V} \cdot \nabla\phi) \mathrm{d}\Gamma \tag{3.3.27}$$

By substituting (3.3.27) into (3.3.25), the sensitivity of (3.3.25) is expressed as the desired boundary integrals;

$$\dot{G} = \int_{\Gamma} \varepsilon \frac{\partial \phi}{\partial n} (\mathbf{V} \cdot \nabla\phi) \mathrm{d}\Gamma - \int_{\Gamma} \left(\frac{1}{2}\varepsilon \nabla\phi \cdot \nabla\phi - \rho\phi \right) V_n \mathrm{d}\Gamma \tag{3.3.28}$$

Using the relations (3.2.32) and (3.2.34) in Sect. 3.2, this sensitivity (3.3.28) is rewritten as

$$\dot{G} = \int_{\Gamma} \left(\frac{1}{2}\varepsilon \frac{\partial \phi}{\partial n}\frac{\partial \phi}{\partial n} - \frac{1}{2}\varepsilon \frac{\partial \phi}{\partial t}\frac{\partial \phi}{\partial t} + \rho\phi \right) V_n \mathrm{d}\Gamma \tag{3.3.29}$$

The space charge, which is assumed to be fixed in the domain, is not taken as a design variable; so $\rho = 0$ on Γ and the sensitivity (3.3.29) becomes

$$\dot{G} = \int_{\Gamma} \left(\frac{1}{2} \varepsilon \frac{\partial \phi}{\partial n} \frac{\partial \phi}{\partial n} - \frac{1}{2} \varepsilon \frac{\partial \phi}{\partial t} \frac{\partial \phi}{\partial t} \right) V_n \mathrm{d}\Gamma \tag{3.3.30}$$

The Dirichlet boundary condition is usually imposed on the electrode conductor, where the electric field has only the normal component;

$$\frac{\partial \phi}{\partial t} = 0 \quad \text{on } \Gamma^0 \tag{3.3.31}$$

On the homogeneous Neumann boundary,

$$\frac{\partial \phi}{\partial n} = 0 \quad \text{on } \Gamma^1 \tag{3.3.32}$$

Using these boundary conditions, the integral of (3.3.30) is decomposed into two integrals on Dirichlet and Neumann boundaries;

$$\dot{G} = \int_{\Gamma^0} \frac{1}{2} \varepsilon \frac{\partial \phi}{\partial n} \frac{\partial \phi}{\partial n} V_n \mathrm{d}\Gamma - \int_{\Gamma^1} \frac{1}{2} \varepsilon \frac{\partial \phi}{\partial t} \frac{\partial \phi}{\partial t} V_n \mathrm{d}\Gamma \tag{3.3.33}$$

The Neumann boundary Γ^1 in the electrostatic system, which appears on the plane of symmetry, is not taken as a design variable. Hence, $V_n = 0$ on Γ^1;

$$\dot{G} = \int_{\Gamma^0} \frac{1}{2} \varepsilon \frac{\partial \phi}{\partial n} \frac{\partial \phi}{\partial n} V_n \mathrm{d}\Gamma \tag{3.3.34}$$

In most of the electrostatic systems, the design variables are taken as the electrode shape, on which Dirichlet boundary condition of constant voltage is imposed. Using (3.2.44), the sensitivity formula (3.3.34) is expressed as

$$\dot{G} = \int_{\Gamma^0} \frac{1}{2} \varepsilon E_n^2 V_n \mathrm{d}\Gamma \tag{3.3.35}$$

where the integrand is the field energy density on the boundary surface.

When the electrode surface comes in contact with the air, this sensitivity formula becomes

$$\dot{G} = \int_{\Gamma^0} \frac{1}{2} \varepsilon_0 E_n^2 V_n \mathrm{d}\Gamma \tag{3.3.36}$$

3.3.5 Source Condition and Capacitance Sensitivity

In this section, the sign of the energy sensitivity is examined in the electric-circuit point of view and the capacitance sensitivity is derived using the energy sensitivity obtained in Sect. 3.3.4. First, we examine how the sign of the energy sensitivity changes according to the condition of source application. This phenomenon occurs also in the electromechanical systems, where the sign of force changes according to the source condition. The force on the moving part in the electromechanical system is obtained by applying the energy conservation law and using the relation of the input energy and the stored field energy due to an infinitesimal displacement of the moving part. The sign of the force changes according to the external source connected to the circuit terminal [10]. What is common between these two problems is that they are both related to the energy variation with respect to the geometry variation.

An electric circuit model of capacitor is employed to examine how the sign of the energy sensitivity depends on the source condition. A capacitor has two electrodes on the surface of dielectric material as in Fig. 3.14. While the first source is a voltage source, the second source is a current source. With a current source I given, a charge Q on the electrodes is given since the current is the time derivative of the charge. Even when the shape of the electrode is changed, the first and second conditions are maintained.

Under the first condition of voltage source, the stored energy of the capacitor is written with the capacitance C and the given voltage V;

$$W_e = \frac{1}{2} C V^2 \tag{3.3.37}$$

The shape variation of the electrodes causes the variation of the capacitance, which is determined only by its geometry and material property. It results in the variation of the system energy (3.3.37). This energy variation can be expressed by taking the total derivative of (3.3.37) as

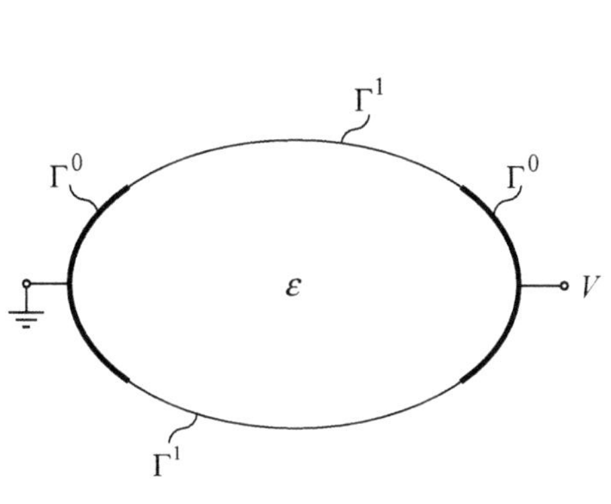

Fig. 3.14 Capacitor model for system energy

$$\dot{W}_e = \frac{1}{2}\dot{C}V^2 \tag{3.3.38}$$

With the voltage V given, the variation of the capacitance C is proportional to the variation of the system energy. Since $Q = CV$, the variation of the capacitance causes the variation of the accumulated charges on the electrodes, which is a current flow.

Under the second condition of current source, the stored energy of the capacitor is written with the capacitance C and the given charge Q;

$$W_e = \frac{1}{2}\frac{Q^2}{C} \tag{3.3.39}$$

The capacitance variation due to the shape variation of the electrode results in the variation of the system energy (3.3.39). This energy variation can be expressed by taking the total derivative of (3.3.39) as

$$\dot{W}_e = -\frac{1}{2}\frac{Q^2}{C^2}\dot{C} \tag{3.3.40}$$

With the charge Q given, the increase of the capacitance C results in the decrease of the system energy, and vice versa. Using $Q = CV$, (3.3.40) can be rewritten as

$$\dot{W}_e = -\frac{1}{2}\dot{C}V^2 \tag{3.3.41}$$

Comparing the two energy sensitivities of (3.3.38) and (3.3.41) shows that they have the opposite sign. That is, the sign of the energy sensitivity changes according to the condition of external source.

In Sect. 3.3.1, the objective functions of system energy were defined with the fixed charge and its sensitivity was also derived from the state equations with the fixed charge distribution. That is, the final sensitivity formulas (3.3.35) were derived under the second condition of the given charges. But the electrodes of the electrostatic system are usually connected to the voltage source. Thus, when the electrostatic system of voltage source is designed, the sign of the sensitivity formulas (3.3.35) should be changed.

The sensitivity formula for the voltage-source electrostatic system is, therefore, written as

$$\dot{G} = -\int_{\Gamma^0} \frac{1}{2}\varepsilon E_n^2 V_n \mathrm{d}\Gamma \tag{3.3.42}$$

This sensitivity formula for the voltage-source electrostatic system can be used to obtain the capacitance sensitivity.

The total derivative of the capacitor-stored energy (3.3.38) is equal to the energy sensitivity with the voltage source condition (3.3.42);

$$\frac{1}{2}\dot{C}V^2 = -\int_{\Gamma^0} \frac{1}{2}\varepsilon E_n^2 V_n \mathrm{d}\Gamma \tag{3.3.43}$$

From this relation, the capacitance sensitivity $\dot{C}$ for the outer boundary problem is obtained;

$$\dot{C} = -\frac{1}{V^2}\int_{\Gamma^0} \varepsilon E_n^2 V_n \mathrm{d}\Gamma \tag{3.3.44}$$

3.3.6 Analytical Example

When the objective function is defined as the system energy, the sensitivity formula was derived as (3.3.42) in Sect. 3.3.5. One-dimensional example with the analytic field solution is taken to show that the sensitivity formula is correct. The objective is to compare the analytical sensitivity result with the result by the sensitivity formula to ensure that the two results are the same. The analytical example is a cylindrical coaxial capacitor, which is the one-dimensional capacitor model in the cylindrical coordinates.

3.3.6.1 Cylindrical Coaxial Capacitor

As an analytical example, a cylindrical capacitor is given in Fig. 3.15, where two cylindrical electrodes are separated by a dielectric of permittivity ε and a voltage ϕ_o is applied between the two electrodes. The design objective is to obtain a desired system energy by moving the outer electrode Γ^{0d}, so the design variable is the radius a of the outer electrode. The design sensitivity with respect to the design variable a is analytically calculated, and then its result is compared with the result by the sensitivity formula.

The objective function is the system energy;

$$F = W_e = \int_{\Omega} \frac{1}{2}\varepsilon \mathbf{E}^2(\phi)\mathrm{d}\Omega \tag{3.3.45}$$

Fig. 3.15 Cylindrical coaxial capacitor-outer boundary design, system energy

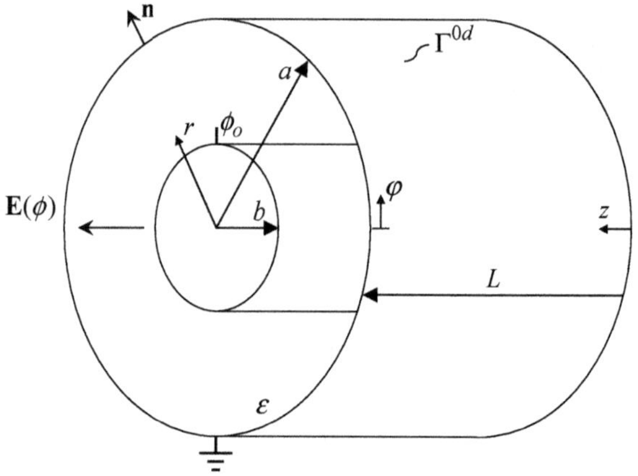

where

$$\mathbf{E}(\phi) = \frac{\phi_o}{\ln a/b}\frac{1}{r}\mathbf{r} \tag{3.3.46}$$

The objective function is rewritten by using the field (3.3.46):

$$F = \pi\varepsilon\frac{\phi_o^2}{\ln a/b}L \tag{3.3.47}$$

The analytical sensitivity per unit length of the cylindrical capacitor is obtained by differentiating the objective function (3.3.47) with respect to the radius a of the outer electrode;

$$\frac{\mathrm{d}F}{\mathrm{d}a} = -\pi\frac{\varepsilon}{a}\left(\frac{\phi_o}{\ln a/b}\right)^2 \tag{3.3.48}$$

This analytical sensitivity is compared with the result obtained from the sensitivity formula (3.3.42) in Sect. 3.3.5:

$$\dot{G} = -\int_{\Gamma^{0d}} \frac{1}{2}\varepsilon E_n^2(\phi)V_n \mathrm{d}\Gamma \tag{3.3.49}$$

(3.3.46) is inserted into the sensitivity formula (3.3.49);

$$\dot{G} = -\pi\frac{\varepsilon}{a}\left(\frac{\phi_o}{\ln a/b}\right)^2 V_n L \tag{3.3.50}$$

Using $V_n = \frac{da}{dt}$, the design sensitivity per unit length is obtained as

$$\frac{\mathrm{d}G}{\mathrm{d}a} = -\pi \frac{\varepsilon}{a} \left(\frac{\phi_o}{\ln a/b} \right)^2 \quad (3.3.51)$$

which is identical to the analytical sensitivity in (3.3.48).

3.3.7 *Numerical Examples*

The sensitivity formula (3.3.42) in Sect. 3.3.5 is applied to two shape optimization problems of two-dimensional design model without the analytic field solutions. These design models show that the sensitivity formula is well applied to the shape design of the outer boundary for the system energy in the electrostatic system. The numerical examples are an axi-symmetric capacitor and a three-phase cable. While the first example has the known optimal design, the second one does not have the known optimal design. If the result of the first example is obtained as the expected optimal design, the shape optimization using the sensitivity formula is feasible for the design of the outer boundary for the system energy in the electrostatic system. The result of the second example shows that this optimization method is useful for the design of the outer boundary.

In these two-dimensional examples the state variable, which is numerically calculated by the finite element method, is required to evaluate the sensitivity formula. The sensitivity evaluated is used for the optimization algorithm to evolve the electrode shape. The level set method is used to represent the shape evolution of the design model. The level set method is described in Chap. 7, where the shape evolution is expressed by the parameter t of unit s for the amount of shape change.

3.3.7.1 Axi-Symmetric Coaxial Capacitor

As an example of which the optimal design is known, consider a capacitor in Fig. 3.16, where the dielectric of permittivity ε is between two electrodes and a voltage ϕ_0 is applied between the two electrodes. The inner electrode is cylindrical but the outer electrode is not cylindrical. If the outer electrode is changed to become a cylinder, the capacitor becomes a coaxial capacitor, of which the system energy has the minimum value under the constraint of constant dielectric volume.

The objective function to be minimized is the system energy;

$$F = W_e = \int_{\Omega} \frac{1}{2} \varepsilon \mathbf{E}^2(\phi) \mathrm{d}\Omega \quad (3.3.52)$$

In this design problem, the shape of the outer electrode Γ^{0d} is the design variable, which has a constraint of constant dielectric volume;

$$\int_{\Omega} \mathrm{d}\Omega = C \tag{3.3.53}$$

where the constant C is a given volume per unit length. By using the material derivative formula (3.1.22) in Sect. 3.1.2, the material derivative of the constraint (3.3.53) is obtained as

$$\dot{C} = \int_{\Omega} 1' \mathrm{d}\Omega + \int_{\Gamma} V_n \mathrm{d}\Gamma = \int_{\Gamma^{0d}} V_n \mathrm{d}\Gamma = 0 \tag{3.3.54}$$

which is a different form of the constraint (3.3.53) expressed with the design velocity field V_n. The shape sensitivity for this outer boundary design is the sensitivity formula (3.3.42):

$$\dot{G} = -\int_{\Gamma^{0d}} \frac{1}{2} \varepsilon E_n^2(\phi) V_n \mathrm{d}\Gamma \tag{3.3.55}$$

The design velocity for this minimization problem is taken as

$$V_n = \frac{1}{2} \varepsilon E_n^2(\phi) \tag{3.3.56}$$

In order that the velocity field satisfies the constraint (3.3.54), the design velocity (3.3.56) is modified by subtracting its average V_{na} to become U_n as

$$U_n = V_n - V_{na} \tag{3.3.57}$$

where

$$V_{na} = \int_{\Gamma^{0d}} \frac{1}{2} \varepsilon E_n^2(\phi) \mathrm{d}\Gamma \Big/ \int_{\Gamma^{0d}} \mathrm{d}\Gamma \tag{3.3.58}$$

The design result is shown in Fig. 3.17, where the shape of the outer electrode becomes, as expected, gradually a cylinder with the iteration. The final design of the cylindrical shape is obtained at the 12 s, when the system energy converges as in Fig. 3.18. The result of this example shows that the shape sensitivity analysis for the outer boundary in the axi-symmetric electrostatic system is feasible.

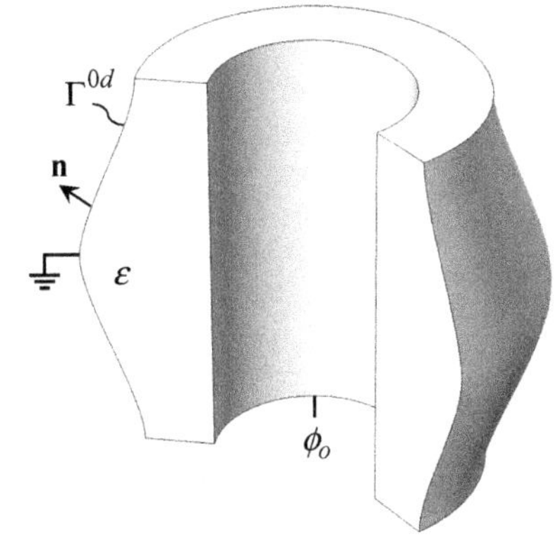

Fig. 3.16 Axi-symmetric capacitor-outer boundary design, system energy

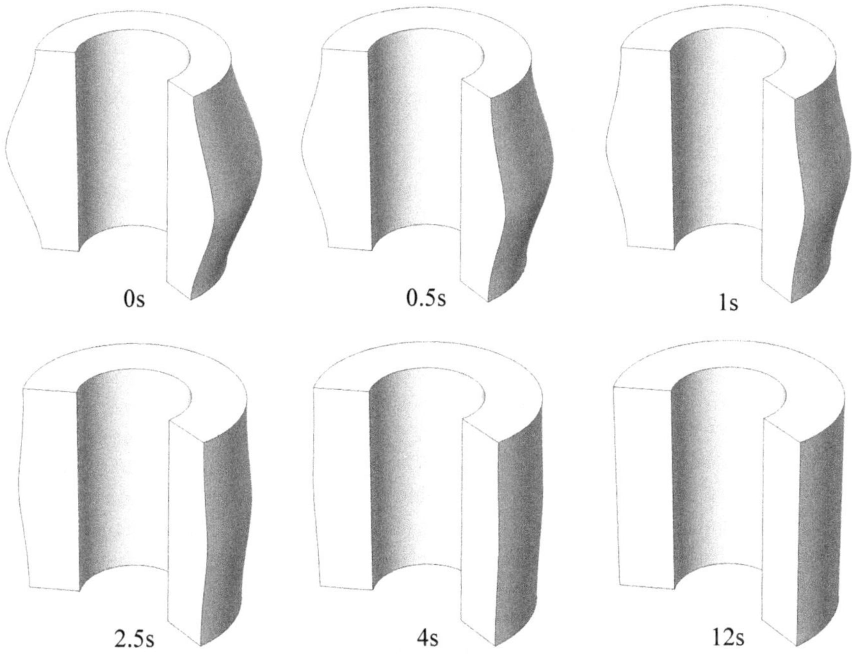

Fig. 3.17 Axi-symmetric capacitor: shape variation of design variation

3.3.7.2 Three-Phase Cable

The shape optimization using the sensitivity formula for the design of the outer boundary is applied to a shape design problem, which has neither the analytical field solution nor a known optimal shape [11]. Consider a three-phase power cable

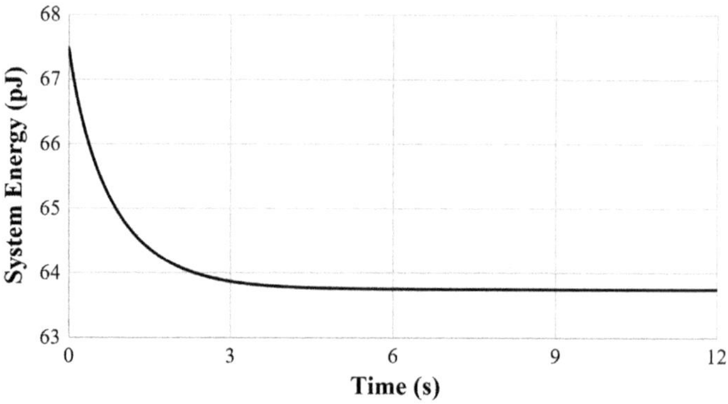

Fig. 3.18 Axi-symmetric capacitor: evolution of objective function

Fig. 3.19 Three-phase cable-outer boundary design, system energy

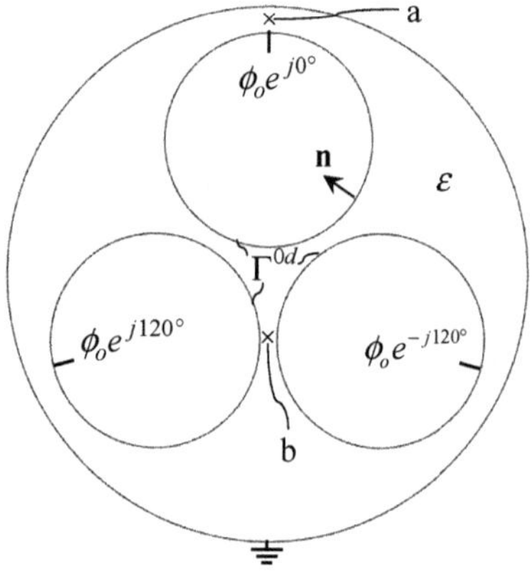

in Fig. 3.19, where the balanced three-phase voltages are imposed on the three conductors and the conductor enclosure is grounded. In the dielectric region between the three cables and the enclosure, the electric field is highest between two conductors or between conductors and the enclosure, such as the points a and b. The field concentration in the high-voltage power cable can cause the partial electric discharge and the deterioration of the dielectric material to the extent that the dielectric breakdown leads to the serious accident. In this example, the shape of the three conductors is optimized to alleviate the field concentration in the cable. In the electrostatic system with the voltage source, as the capacitance C becomes lower, the system energy $W_e = \frac{1}{2}CV^2$ decreases. The decrease of the capacitance is obtained by increasing the distances between the electrodes: the three cables and the enclosure. In other words, the decrease of the capacitance results in the decrease of the accumulated charges $Q = CV$ on the electrodes surface, which determines the electric field concentration near the conductors. For the state variable in the state

Eq. (3.3.9) for the electric field analysis, the complex variable of the three-phase voltage is applied to the cable as shown in Fig. 3.19.

The objective function is defined as the electric energy;

$$F = W_e = \int_{\Omega} \frac{1}{2}\varepsilon \mathbf{E}(\phi) \cdot \mathbf{E}^*(\phi) \mathrm{d}\Omega \tag{3.3.59}$$

where $*$ means the conjugate of complex variable. This system energy is the integral of the time average field energy density. In this design problem, the shape of the surfaces of the three cables Γ^{0d} is the design variable, which has a constraint of constant conductor volume;

$$\int_{\Omega} \mathrm{d}\Omega = C \tag{3.3.60}$$

where C is constant and it is the volume of the three conductors per unit length. The material derivative of the constraint (3.3.60) is obtained as

$$\int_{\Gamma^{0d}} V_n \mathrm{d}\Gamma = 0 \tag{3.3.61}$$

The sensitivity formula (3.3.42) for the energy objective function is rewritten with the complex state variable;

$$\dot{G} = -\int_{\Gamma^{0d}} \frac{1}{2}\varepsilon E_n(\phi) E_n^*(\phi) V_n \mathrm{d}\Gamma \tag{3.3.62}$$

The design velocity with the constraint is taken as

$$U_n = V_n - V_{na} \tag{3.3.63}$$

where

$$V_n = \frac{1}{2}\varepsilon E_n(\phi) E_n^*(\phi) \tag{3.3.64}$$

$$V_{na} = \int_{\Gamma^{0d}} \frac{1}{2}\varepsilon E_n(\phi) E_n^*(\phi) \mathrm{d}\Gamma \Big/ \int_{\Gamma^{0d}} \mathrm{d}\Gamma \tag{3.3.65}$$

Figure 3.20 shows the evolution of the inner conductor surfaces during the optimization process. The parts of the cable conductors, which are close to each other or the grounded enclosure, become rounded or flat, while the rest parts are expanded to the vacant region. At 8s in Fig. 3.20, the optimized cable conductor is

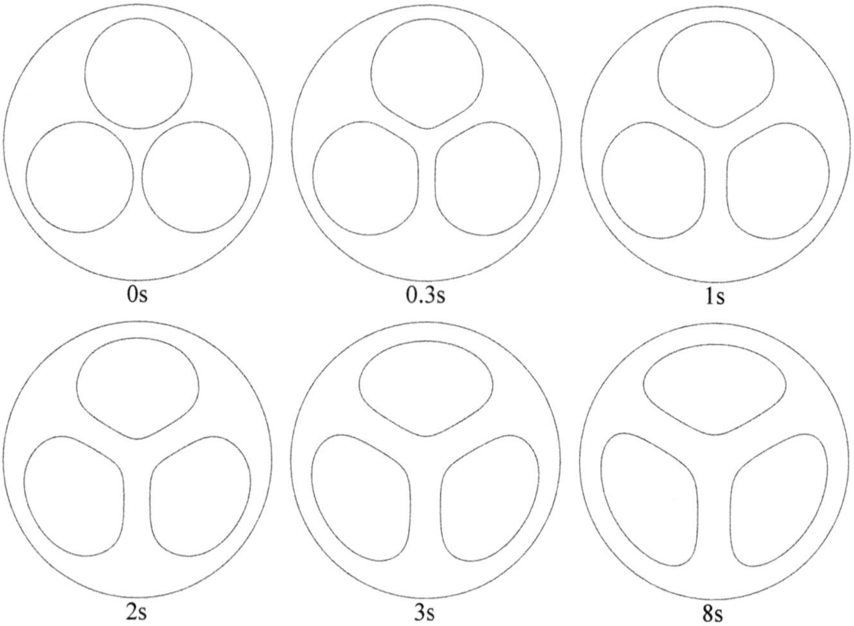

Fig. 3.20 Three-phase cable: shape variation

obtained as a fan-shaped one. The variation of the electric field with the iteration is shown in Fig. 3.21, where the final electric field intensities at the points *a* and *b* decrease by 35% and 75%, respectively, compared with those of the initial design. The objective function rapidly decreases in the earlier stage of the optimization process and then gradually converges to the minimum value as shown in Fig. 3.22.

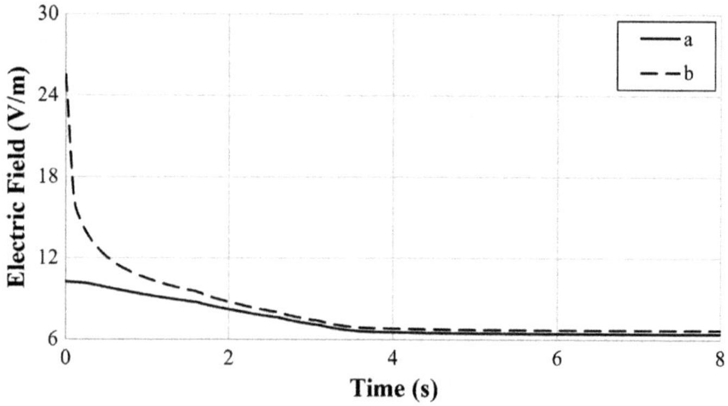

Fig. 3.21 Three-phase cable: electric field at points a, b

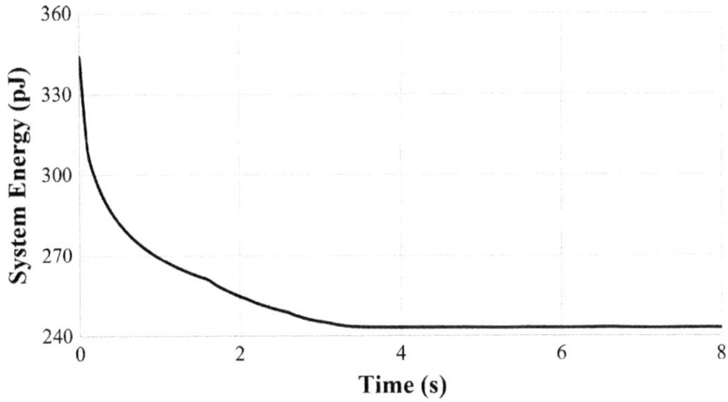

Fig. 3.22 Three-phase cable: evolution of objective function

The results of this example show that the sensitivity method is useful for the shape design of the electrode. This shape sensitivity method may be applied to the design problems for various high-voltage apparatus.

3.4 Shape Sensitivity of Interface

As mentioned in Sect. 3.2, the electrostatic system also has the interface design problem. A typical interface is the dielectric material versus the air. The shape variation of the interface results in the variation in the electric field distribution in the domain. In this section, the general three-dimensional sensitivity for the interface variation is derived. First, the electrostatic system for the interface design is depicted, and a general objective function is defined as a domain integral. Second, the Lagrange multiplier method is introduced to handle the equality constraint of the variational state equation. Third, the adjoint variable method is used to express explicitly the sensitivity in terms of design variation. Fourth, the variational identities are used to transform the domain integral of the sensitivity into the interface integral, which provides the general three-dimensional sensitivity formula for the interface design. Finally, the obtained sensitivity formula is tested and validated with analytical and numerical examples.

3.4.1 Problem Definition and Objective Function

An electrostatic system for the interface design is given as in Fig. 3.23, where the whole domain Ω comprises two domains Ω_1 and Ω_2 that are divided by an interface γ. The domains Ω_1 have a distribution of ε_1, ρ_1, and $\mathbf{P}_{o1}$ and the domains Ω_2 have a

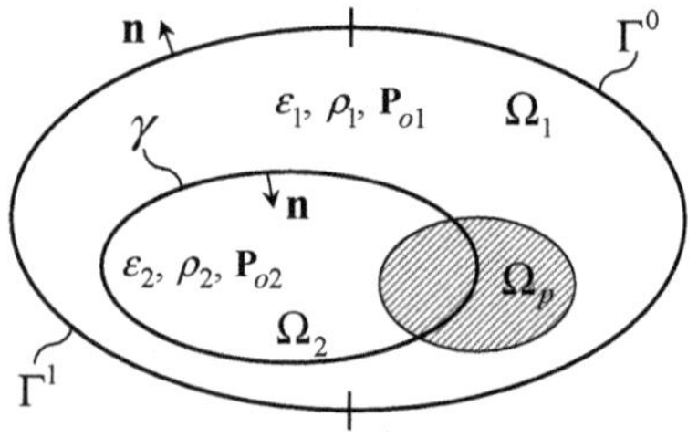

Fig. 3.23 Interface design of electrostatic system

distribution of ε_2, ρ_2, and $\mathbf{P}_{o2}$. The charge density ρ_1 and ρ_2 are assumed to be fixed and constant in the domain. The domain Ω_1 has two boundaries of the outer boundary Γ and the interface γ, where $\mathbf{n}$ is defined as the outward normal vector on the two boundaries. The outer boundary consists of the Dirichlet boundary Γ^0 and the Neumann boundary Γ^1. In this shape sensitivity analysis for the interface design, the interface γ is taken as design variable.

A general objective function is defined in integral form;

$$\begin{aligned} F &= \int_{\Omega} g(\phi, \nabla\phi) m_p \mathrm{d}\Omega \\ &= \int_{\Omega_1} g_1 m_p \mathrm{d}\Omega + \int_{\Omega_2} g_2 m_p \mathrm{d}\Omega \end{aligned} \tag{3.4.1}$$

where the g_1 and g_2 are any functions continuously differentiable to their arguments;

$$g_1 = g(\phi_1, \nabla\phi_1) \text{and}\, g_2 = g(\phi_2, \nabla\phi_2) \tag{3.4.2}$$

and m_p is a characteristic function that is defined as

$$m_p = \begin{cases} 1 & \mathbf{x} \in \Omega_p \\ 0 & \mathbf{x} \notin \Omega_p \end{cases} \tag{3.4.3}$$

The region Ω_p , the integral domain for the objective function, can include the interface as shown in Fig. 3.23. The objective function in the form of (3.4.1) can represent a wide range of design objectives for the electrostatic system because g_1 and g_2 can be any function of electric potential ϕ and electric field $\mathbf{E}$ in the two domains Ω_1 and Ω_2.

The governing differential equations for the state variables of the electric scalar potential ϕ_1 and ϕ_2 are given as;

$$-\nabla \cdot \varepsilon_1 \nabla\phi_1 = \rho_1 - \nabla \cdot \mathbf{P}_{o1} \quad \text{in}\, \Omega_1 \tag{3.4.4}$$

$$-\nabla \cdot \varepsilon_2 \nabla \phi_2 = \rho_2 - \nabla \cdot \mathbf{P}_{o2} \quad \text{in}\,\Omega_2 \tag{3.4.5}$$

where $-\nabla \cdot \mathbf{P}_{o1} = \rho_{P1}$, $-\nabla \cdot \mathbf{P}_{o2} = \rho_{P2}$, and ρ_{P1}, ρ_{P2} are the bound charge densities of permanent polarization of $\mathbf{P}_{o1}$ and $\mathbf{P}_{o2}$, respectively. The governing differential equations of the electrostatic system (3.4.4) and (3.4.5) have a unique solution with boundary conditions. We employ most common boundary conditions:

$$\phi_1 = C(\mathbf{x}) \quad \text{on}\,\Gamma^0 \quad \text{(Dirichlet boundary condition)} \tag{3.4.6}$$

$$\frac{\partial \phi_1}{\partial n} = 0 \quad \text{on}\,\Gamma^1 \quad \text{(homogeneous Neumann boundary condition)} \tag{3.4.7}$$

The variational identities for the state equations are obtained by multiplying both sides of (3.4.4) and (3.4.5) by an arbitrary virtual potential $\bar{\phi}_1$ and $\bar{\phi}_2$, and by using the vector identity and the divergence theorem;

$$\int_{\Omega_1} (\varepsilon_1 \nabla \phi_1 \cdot \nabla \bar{\phi}_1 - \rho_1 \bar{\phi}_1 - \mathbf{P}_{o1} \cdot \nabla \bar{\phi}_1) \mathrm{d}\Omega = - \int_{\gamma + \Gamma} D_n(\phi_1) \bar{\phi}_1 \mathrm{d}\Gamma \quad \forall \bar{\phi}_1 \in \Phi \tag{3.4.8}$$

$$\int_{\Omega_2} (\varepsilon_2 \nabla \phi_2 \cdot \nabla \bar{\phi}_2 - \rho_2 \bar{\phi}_2 - \mathbf{P}_{o2} \cdot \nabla \bar{\phi}_2) \mathrm{d}\Omega = \int_{\gamma} D_n(\phi_2) \bar{\phi}_2 \mathrm{d}\Gamma \quad \forall \bar{\phi}_2 \in \Phi \tag{3.4.9}$$

where

$$D_n(\phi_1) = -\varepsilon_1 \frac{\partial \phi_1}{\partial n} + \mathbf{P}_{o1} \cdot \mathbf{n} \tag{3.4.10}$$

$$D_n(\phi_2) = -\varepsilon_2 \frac{\partial \phi_2}{\partial n} + \mathbf{P}_{o2} \cdot \mathbf{n} \tag{3.4.11}$$

and Φ is the space of admissible potential: $\Phi = \{\bar{\phi} \in H^1(\Omega) | \bar{\phi} = 0 \text{ on } \mathbf{x} \in \Gamma^0.\}$. The boundary conditions of (3.4.6) and (3.4.7) can be rewritten for the variational equation;

$$\bar{\phi}_1 = 0 \quad \text{on}\,\Gamma^0 \quad \text{(Dirichlet boundary condition)} \tag{3.4.12}$$

$$D_n(\phi_1) = 0 \quad \text{on}\,\Gamma^1 \quad \text{(homogeneous Neumann boundary condition)} \tag{3.4.13}$$

and the interface condition is

$$D_n(\phi_1) = D_n(\phi_2) \quad \text{on}\,\gamma \quad \text{(interface condition)} \tag{3.4.14}$$

By summing (3.4.8) and (3.4.9) and imposing the boundary conditions and the interface condition, the variational state equation is obtained as

$$\begin{aligned} &\int_{\Omega_1} \varepsilon_1 \nabla\phi_1 \cdot \nabla\bar{\phi}_1 \mathrm{d}\Omega + \int_{\Omega_2} \varepsilon_2 \nabla\phi_2 \cdot \nabla\bar{\phi}_2 \mathrm{d}\Omega \\ &\quad = \int_{\Omega_1} \left(\rho_1\bar{\phi}_1 + \mathbf{P}_{o1} \cdot \nabla\bar{\phi}_1\right)\mathrm{d}\Omega \\ &\quad + \int_{\Omega_2} \left(\rho_2\bar{\phi}_2 + \mathbf{P}_{o2} \cdot \nabla\bar{\phi}_2\right)\mathrm{d}\Omega \quad \forall\bar{\phi}_1, \bar{\phi}_2 \in \Phi \end{aligned} \tag{3.4.15}$$

The energy bilinear form $a(\phi, \bar{\phi})$ and the source linear form $l(\bar{\phi})$ are defined as

$$a(\phi, \bar{\phi}) \equiv \int_{\Omega_1} \varepsilon_1 \nabla\phi_1 \cdot \nabla\bar{\phi}_1 \mathrm{d}\Omega + \int_{\Omega_2} \varepsilon_2 \nabla\phi_2 \cdot \nabla\bar{\phi}_2 \mathrm{d}\Omega \tag{3.4.16}$$

$$l(\bar{\phi}) \equiv \int_{\Omega_1} \left(\rho_1\bar{\phi}_1 + \mathbf{P}_{o1} \cdot \nabla\bar{\phi}_1\right)\mathrm{d}\Omega + \int_{\Omega_2} \left(\rho_2\bar{\phi}_2 + \mathbf{P}_{o2} \cdot \nabla\bar{\phi}_2\right)\mathrm{d}\Omega \tag{3.4.17}$$

The variational Eq. (3.4.15) is rewritten with the energy bilinear form and the source linear form as

$$a(\phi, \bar{\phi}) = l(\bar{\phi}) \quad \forall\bar{\phi} \in \Phi \tag{3.4.18}$$

where

$$\phi = \phi_1 \cup \phi_2 \quad \text{and} \quad \bar{\phi} = \bar{\phi}_1 \cup \bar{\phi}_2. \tag{3.4.19}$$

3.4.2 Lagrange Multiplier Method for Sensitivity Derivation

The Lagrange multiplier method is applied to this interface problem for the implicit equality constraint of the variational state equation (3.4.18). When the domain Ω is perturbed by the interface shape, the state variable ϕ in the whole domain Ω is also perturbed through the state equation.

The method of Lagrange multiplier provides an augmented objective function G as

$$G = F + l(\bar{\phi}) - a(\phi, \bar{\phi}) \quad \forall\bar{\phi} \in \Phi \tag{3.4.20}$$

where the arbitrary virtual potential $\bar{\phi}$ plays the role of Lagrange multipliers and Φ is the space of admissible potential: $\Phi = \left\{\bar{\phi} \in H^1(\Omega) \middle| \bar{\phi} = 0 \, \text{on} \, \mathbf{x} \in \Gamma^0\right\}$.

The sensitivity, the material derivative of the augmented objective function, is written as

$$\dot{G} = -\dot{a}(\phi, \bar{\phi}) + \dot{l}(\bar{\phi}) + \dot{F} \quad \forall \bar{\phi} \in \Phi \tag{3.4.21}$$

By applying the material derivative formula (3.1.22) to the variational state Eq. (3.4.15) and the objective function (3.4.1) and using the relation (3.1.15), each term in (3.4.21) is obtained as the followings (3.4.22)–(3.4.24). In this interface sensitivity problem, when the material derivative formula is applied, only the integrals on the interface remain since the outer boundary is not taken as design variable ($V_n = 0$ on Γ).

$$\begin{aligned}
\dot{a}(\phi, \bar{\phi}) = & \int_{\Omega_1} (\varepsilon_1 \nabla \phi_1' \cdot \nabla \bar{\phi}_1 + \varepsilon_1 \nabla \phi_1 \cdot \nabla \bar{\phi}_1') \mathrm{d}\Omega + \int_{\gamma} \varepsilon_1 \nabla \phi_1 \cdot \nabla \bar{\phi}_1 V_n \mathrm{d}\Gamma \\
& + \int_{\Omega_2} (\varepsilon_2 \nabla \phi_2' \cdot \nabla \bar{\phi}_2 + \varepsilon_2 \nabla \phi_2 \cdot \nabla \bar{\phi}_2') \mathrm{d}\Omega - \int_{\gamma} \varepsilon_2 \nabla \phi_2 \cdot \nabla \bar{\phi}_2 V_n \mathrm{d}\Gamma \\
= & \int_{\Omega_1} \Big(\varepsilon_1 \nabla \dot{\phi}_1 \cdot \nabla \bar{\phi}_1 - \varepsilon_1 \nabla (\mathbf{V} \cdot \nabla \phi_1) \cdot \nabla \bar{\phi}_1 \\
& + \varepsilon_1 \nabla \phi_1 \cdot \nabla \dot{\bar{\phi}}_1 - \varepsilon_1 \nabla \phi_1 \cdot \nabla (\mathbf{V} \cdot \nabla \bar{\phi}_1) \Big) \mathrm{d}\Omega \\
& + \int_{\Omega_2} \Big(\varepsilon_2 \nabla \dot{\phi}_2 \cdot \nabla \bar{\phi}_2 - \varepsilon_2 \nabla (\mathbf{V} \cdot \nabla \phi_2) \cdot \nabla \bar{\phi}_2 \\
& + \varepsilon_2 \nabla \phi_2 \cdot \nabla \dot{\bar{\phi}}_2 - \varepsilon_2 \nabla \phi_2 \cdot \nabla (\mathbf{V} \cdot \nabla \bar{\phi}_2) \Big) \mathrm{d}\Omega \\
& + \int_{\gamma} (\varepsilon_1 \nabla \phi_1 \cdot \nabla \bar{\phi}_1 - \varepsilon_2 \nabla \phi_2 \cdot \nabla \bar{\phi}_2) V_n \mathrm{d}\Gamma
\end{aligned} \tag{3.4.22}$$

$$\begin{aligned}
\dot{l}(\bar{\phi}) = & \int_{\Omega_1} (\rho_1 \bar{\phi}_1' + \mathbf{P}_{o1} \cdot \nabla \bar{\phi}_1') \mathrm{d}\Omega + \int_{\gamma} (\rho_1 \bar{\phi}_1 + \mathbf{P}_{o1} \cdot \nabla \bar{\phi}_1) V_n \mathrm{d}\Gamma \\
& + \int_{\Omega_2} (\rho_2 \bar{\phi}_2' + \mathbf{P}_{o2} \cdot \nabla \bar{\phi}_2') \mathrm{d}\Omega - \int_{\gamma} (\rho_2 \bar{\phi}_2 + \mathbf{P}_{o2} \cdot \nabla \bar{\phi}_2) V_n \mathrm{d}\Gamma \\
= & \int_{\Omega_1} \Big[\rho_1 \dot{\bar{\phi}}_1 - \rho_1 (\mathbf{V} \cdot \nabla \bar{\phi}_1) + \mathbf{P}_{o1} \cdot \nabla \dot{\bar{\phi}}_1 - \mathbf{P}_{o1} \cdot \nabla (\mathbf{V} \cdot \nabla \bar{\phi}_1) \Big] \mathrm{d}\Omega \\
& + \int_{\Omega_2} \Big[\rho_2 \dot{\bar{\phi}}_2 - \rho_2 (\mathbf{V} \cdot \nabla \bar{\phi}_2) + \mathbf{P}_{o2} \cdot \nabla \dot{\bar{\phi}}_2 - \mathbf{P}_{o2} \cdot \nabla (\mathbf{V} \cdot \nabla \bar{\phi}_2) \Big] \mathrm{d}\Omega \\
& + \int_{\gamma} (\rho_1 \bar{\phi}_1 - \rho_2 \bar{\phi}_2 + \mathbf{P}_{o1} \cdot \nabla \bar{\phi}_1 - \mathbf{P}_{o2} \cdot \nabla \bar{\phi}_2) V_n \mathrm{d}\Gamma
\end{aligned} \tag{3.4.23}$$

$$\begin{aligned}\dot{F} &= \int_{\Omega_1} g_1' m_p \mathrm{d}\Omega + \int_{\gamma} g_1 m_p V_n \mathrm{d}\Gamma + \int_{\Omega_2} g_2' m_p \mathrm{d}\Omega - \int_{\gamma} g_2 m_p V_n \mathrm{d}\Gamma \\ &= \int_{\Omega_1} \left(g_{\phi_1}\phi_1' + \mathbf{g}_{\mathbf{E}_1} \cdot \nabla\phi_1'\right) m_p \mathrm{d}\Omega + \int_{\gamma} g_1 m_p V_n \mathrm{d}\Gamma \\ &\quad + \int_{\Omega_2} \left(g_{\phi_2}\phi_2' + \mathbf{g}_{\mathbf{E}_2} \cdot \nabla\phi_2'\right) m_p \mathrm{d}\Omega - \int_{\gamma} g_2 m_p V_n \mathrm{d}\Gamma \\ &= \int_{\Omega_1} \left[g_{\phi_1}\dot{\phi}_1 - g_{\phi_1}(\mathbf{V}\cdot\nabla\phi_1) + \mathbf{g}_{\mathbf{E}_1}\cdot\nabla\dot{\phi}_1 - \mathbf{g}_{\mathbf{E}_1}\cdot\nabla(\mathbf{V}\cdot\nabla\phi_1)\right] m_p \mathrm{d}\Omega \\ &\quad + \int_{\Omega_2} \left[g_{\phi_2}\dot{\phi}_2 - g_{\phi_2}(\mathbf{V}\cdot\nabla\phi_2) + \mathbf{g}_{\mathbf{E}_2}\cdot\nabla\dot{\phi}_2 - \mathbf{g}_{\mathbf{E}_2}\cdot\nabla(\mathbf{V}\cdot\nabla\phi_2)\right] m_p \mathrm{d}\Omega \\ &\quad + \int_{\gamma} (g_1 - g_2) m_p V_n \mathrm{d}\Gamma\end{aligned} \tag{3.4.24}$$

For g_1' and g_2' in (3.4.24), we used the relation (3.1.23):

$$g' = \frac{\partial g}{\partial \phi}\phi' + \frac{\partial g}{\partial \nabla\phi}\cdot\nabla\phi' = g_\phi\phi' + \mathbf{g}_\mathbf{E}\cdot\nabla\phi' \tag{3.4.25}$$

where

$$g_\phi \equiv \frac{\partial g}{\partial \phi} \quad \text{and} \quad \mathbf{g}_\mathbf{E} \equiv \frac{\partial g}{\partial \nabla\phi} = \left[\frac{\partial g}{\partial(\nabla\phi)_x}, \frac{\partial g}{\partial(\nabla\phi)_y}, \frac{\partial g}{\partial(\nabla\phi)_z}\right]^T \tag{3.4.26}$$

In derivation of (3.4.22) and (3.4.23), $\varepsilon_1', \varepsilon_2' = 0$, $\rho_1', \rho_2' = 0$, and $\mathbf{P}_{o1}', \mathbf{P}_{o2}' = 0$ were used. (3.4.22)–(3.4.24) are inserted into (3.4.21) to provide

$$
\begin{aligned}
\dot{G} = & -\int_{\Omega_1} \varepsilon_1 \nabla \dot{\phi}_1 \cdot \nabla \bar{\phi}_1 \mathrm{d}\Omega - \int_{\Omega_2} \varepsilon_2 \nabla \dot{\phi}_2 \cdot \nabla \bar{\phi}_2 \mathrm{d}\Omega - \int_{\Omega_1} \varepsilon_1 \nabla \phi_1 \cdot \nabla \dot{\bar{\phi}}_1 \mathrm{d}\Omega - \int_{\Omega_2} \varepsilon_2 \nabla \phi_2 \cdot \nabla \dot{\bar{\phi}}_2 \mathrm{d}\Omega \\
& + \int_{\Omega_1} \left(\rho_1 \dot{\bar{\phi}}_1 + \mathbf{P}_{o1} \cdot \nabla \dot{\bar{\phi}}_1 \right) \mathrm{d}\Omega + \int_{\Omega_2} \left(\rho_2 \dot{\bar{\phi}}_2 + \mathbf{P}_{o2} \cdot \nabla \dot{\bar{\phi}}_2 \right) \mathrm{d}\Omega \\
& + \int_{\Omega_1} \left(g_{\phi_1} \dot{\phi}_1 + \mathbf{g}_{\mathbf{E}_1} \cdot \nabla \dot{\phi}_1 \right) m_p \mathrm{d}\Omega + \int_{\Omega_2} \left(g_{\phi_2} \dot{\phi}_2 + \mathbf{g}_{\mathbf{E}_2} \cdot \nabla \dot{\phi}_2 \right) m_p \mathrm{d}\Omega \\
& + \int_{\Omega_1} \varepsilon_1 \nabla (\mathbf{V} \cdot \nabla \phi_1) \cdot \nabla \bar{\phi}_1 \mathrm{d}\Omega + \int_{\Omega_2} \varepsilon_2 \nabla (\mathbf{V} \cdot \nabla \phi_2) \cdot \nabla \bar{\phi}_2 \mathrm{d}\Omega \\
& + \int_{\Omega_1} \varepsilon_1 \nabla \phi_1 \cdot \nabla (\mathbf{V} \cdot \nabla \bar{\phi}_1) \mathrm{d}\Omega + \int_{\Omega_2} \varepsilon_2 \nabla \phi_2 \cdot \nabla (\mathbf{V} \cdot \nabla \bar{\phi}_2) \mathrm{d}\Omega \\
& - \int_{\Omega_1} \left[\rho_1 (\mathbf{V} \cdot \nabla \bar{\phi}_1) + \mathbf{P}_{o1} \cdot \nabla (\mathbf{V} \cdot \nabla \bar{\phi}_1) \right] \mathrm{d}\Omega \\
& - \int_{\Omega_2} \left[\rho_2 (\mathbf{V} \cdot \nabla \bar{\phi}_2) + \mathbf{P}_{o2} \cdot \nabla (\mathbf{V} \cdot \nabla \bar{\phi}_2) \right] \mathrm{d}\Omega \\
& - \int_{\Omega_1} \left[g_{\phi_1} (\mathbf{V} \cdot \nabla \phi_1) + \mathbf{g}_{\mathbf{E}_1} \cdot \nabla (\mathbf{V} \cdot \nabla \phi_1) \right] m_p \mathrm{d}\Omega \\
& - \int_{\Omega_2} \left[g_{\phi_2} (\mathbf{V} \cdot \nabla \phi_2) + \mathbf{g}_{\mathbf{E}_2} \cdot \nabla (\mathbf{V} \cdot \nabla \phi_2) \right] m_p \mathrm{d}\Omega \\
& - \int_{\gamma} \left(\varepsilon_1 \nabla \phi_1 \cdot \nabla \bar{\phi}_1 - \varepsilon_2 \nabla \phi_2 \cdot \nabla \bar{\phi}_2 \right) V_n \mathrm{d}\Gamma + \int_{\gamma} \left(\mathbf{P}_{o1} \cdot \nabla \bar{\phi}_1 - \mathbf{P}_{o2} \cdot \nabla \bar{\phi}_2 \right) V_n \mathrm{d}\Gamma \\
& + \int_{\gamma} \left[\rho_1 \bar{\phi}_1 - \rho_2 \bar{\phi}_2 + (g_1 - g_2) m_p \right] V_n \mathrm{d}\Gamma \quad \forall \bar{\phi}_1, \bar{\phi}_2 \in \Phi
\end{aligned}
\tag{3.4.27}
$$

Since $\dot{\bar{\phi}}_1$ and $\dot{\bar{\phi}}_2$ belong to Φ, the variational state equation of (3.4.15) gives the following relation:

$$
\begin{aligned}
& \int_{\Omega_1} \varepsilon_1 \nabla \phi_1 \cdot \nabla \dot{\bar{\phi}}_1 \mathrm{d}\Omega + \int_{\Omega_2} \varepsilon_2 \nabla \phi_2 \cdot \nabla \dot{\bar{\phi}}_2 \mathrm{d}\Omega \\
& \quad = \int_{\Omega_1} \left(\rho_1 \dot{\bar{\phi}}_1 + \mathbf{P}_{o1} \cdot \nabla \dot{\bar{\phi}}_1 \right) \mathrm{d}\Omega + \int_{\Omega_2} \left(\rho_2 \dot{\bar{\phi}}_2 + \mathbf{P}_{o2} \cdot \nabla \dot{\bar{\phi}}_2 \right) \mathrm{d}\Omega
\end{aligned}
\tag{3.4.28}
$$

Hence, all terms containing $\dot{\bar{\phi}}_1$ and $\dot{\bar{\phi}}_2$ in (3.4.27) are canceled out;

$$
\begin{aligned}
\dot{G} = & -\int_{\Omega_1} \varepsilon_1 \nabla\dot{\phi}_1 \cdot \nabla\bar{\phi}_1 \mathrm{d}\Omega - \int_{\Omega_2} \varepsilon_2 \nabla\dot{\phi}_2 \cdot \nabla\bar{\phi}_2 \mathrm{d}\Omega \\
& + \int_{\Omega_1} \left(g_{\phi_1}\dot{\phi}_1 + \mathbf{g}_{\mathbf{E}_1} \cdot \nabla\dot{\phi}_1\right) m_p \mathrm{d}\Omega + \int_{\Omega_2} \left(g_{\phi_2}\dot{\phi}_2 + \mathbf{g}_{\mathbf{E}_2} \cdot \nabla\dot{\phi}_2\right) m_p \mathrm{d}\Omega \\
& + \int_{\Omega_1} \varepsilon_1 \nabla(\mathbf{V}\cdot\nabla\phi_1) \cdot \nabla\bar{\phi}_1 \mathrm{d}\Omega + \int_{\Omega_2} \varepsilon_2 \nabla(\mathbf{V}\cdot\nabla\phi_2) \cdot \nabla\bar{\phi}_2 \mathrm{d}\Omega \\
& + \int_{\Omega_1} \varepsilon_1 \nabla\phi_1 \cdot \nabla(\mathbf{V}\cdot\nabla\bar{\phi}_1) \mathrm{d}\Omega + \int_{\Omega_2} \varepsilon_2 \nabla\phi_2 \cdot \nabla(\mathbf{V}\cdot\nabla\bar{\phi}_2) \mathrm{d}\Omega \\
& - \int_{\Omega_1} \left[\rho_1(\mathbf{V}\cdot\nabla\bar{\phi}_1) + \mathbf{P}_{o1}\cdot\nabla(\mathbf{V}\cdot\nabla\bar{\phi}_1)\right] d\Omega \\
& - \int_{\Omega_2} \left[\rho_2(\mathbf{V}\cdot\nabla\bar{\phi}_2) + \mathbf{P}_{o2}\cdot\nabla(\mathbf{V}\cdot\nabla\bar{\phi}_2)\right] \mathrm{d}\Omega \\
& - \int_{\Omega_1} \left[g_{\phi_1}(\mathbf{V}\cdot\nabla\phi_1) + \mathbf{g}_{\mathbf{E}_1}\cdot\nabla(\mathbf{V}\cdot\nabla\phi_1)\right] m_p \mathrm{d}\Omega \\
& - \int_{\Omega_2} \left[g_{\phi_2}(\mathbf{V}\cdot\nabla\phi_2) + \mathbf{g}_{\mathbf{E}_2}\cdot\nabla(\mathbf{V}\cdot\nabla\phi_2)\right] m_p \mathrm{d}\Omega \\
& - \int_{\gamma} \left(\varepsilon_1 \nabla\phi_1 \cdot \nabla\bar{\phi}_1 - \varepsilon_2 \nabla\phi_2 \cdot \nabla\bar{\phi}_2\right) V_n \mathrm{d}\Gamma + \int_{\gamma} \left(\mathbf{P}_{o1}\cdot\nabla\bar{\phi}_1 - \mathbf{P}_{o2}\cdot\nabla\bar{\phi}_2\right) V_n d\Gamma \\
& + \int_{\gamma} \left[\rho_1\bar{\phi}_1 - \rho_2\bar{\phi}_2 + (g_1 - g_2)m_p\right] V_n \mathrm{d}\Gamma \quad \forall\bar{\phi}_1, \bar{\phi}_2 \in \Phi
\end{aligned}
\tag{3.4.29}
$$

3.4.3 Adjoint Variable Method for Sensitivity Analysis

In order to avoid the term of $\dot{\phi}$ in the sensitivity (3.4.29) and express explicitly the sensitivity with the velocity field $\mathbf{V}$, an adjoint equation is introduced. The adjoint equation is obtained by replacing $\dot{\phi}_1$ and $\dot{\phi}_2$ in the g-related terms of (3.4.29) with a virtual potential $\bar{\lambda}_1$ and $\bar{\lambda}_2$, respectively, and by equating the integrals to the energy bilinear form $a(\lambda, \bar{\lambda})$. The adjoint equation so obtained is written as

$$\begin{aligned}
&\int\limits_{\Omega_1} \varepsilon_1 \nabla\lambda_1 \cdot \nabla\bar{\lambda}_1 \mathrm{d}\Omega + \int\limits_{\Omega_2} \varepsilon_2 \nabla\lambda_2 \cdot \nabla\bar{\lambda}_2 \mathrm{d}\Omega \\
&\quad = \int\limits_{\Omega_1} \left(g_{\phi_1}\bar{\lambda}_1 + \mathbf{g}_{\mathbf{E}_1} \cdot \nabla\bar{\lambda}_1\right) m_p \mathrm{d}\Omega \\
&\qquad + \int\limits_{\Omega_2} \left(g_{\phi_2}\bar{\lambda}_2 + \mathbf{g}_{\mathbf{E}_2} \cdot \nabla\bar{\lambda}_2\right) m_p \mathrm{d}\Omega \quad \forall \bar{\lambda}_1, \bar{\lambda}_2 \in \Phi
\end{aligned} \tag{3.4.30}$$

where λ_1 and λ_2 are the adjoint variables and their solution is desired, and Φ is the space of admissible potential: $\Phi = \left\{\bar{\lambda} \in H^1(\Omega)\middle|\bar{\lambda} = 0 \quad \text{on} \quad \mathbf{x} \in \Gamma^0\right\}$.

This adjoint equation is evaluated at specific $\bar{\lambda}_1 = \dot{\phi}_1$ and $\bar{\lambda}_2 = \dot{\phi}_2$ since it holds for all $\bar{\lambda}_1, \bar{\lambda}_2 \in \Phi$, to yield the relation:

$$\begin{aligned}
&\int\limits_{\Omega_1} \varepsilon_1 \nabla\lambda_1 \cdot \nabla\dot{\phi}_1 \mathrm{d}\Omega + \int\limits_{\Omega_2} \varepsilon_2 \nabla\lambda_2 \cdot \nabla\dot{\phi}_2 \mathrm{d}\Omega \\
&\quad = \int\limits_{\Omega_1} \left(g_{\phi_1}\dot{\phi}_1 + \mathbf{g}_{\mathbf{E}_1} \cdot \nabla\dot{\phi}_1\right) m_p \mathrm{d}\Omega \\
&\qquad + \int\limits_{\Omega_2} \left(g_{\phi_2}\dot{\phi}_2 + \mathbf{g}_{\mathbf{E}_2} \cdot \nabla\dot{\phi}_2\right) m_p \mathrm{d}\Omega
\end{aligned} \tag{3.4.31}$$

Similarly, the sensitivity Eq. (3.4.29) is evaluated at specific $\bar{\phi}_1 = \lambda_1$, $\bar{\phi}_2 = \lambda_2$ since the λ_1 and λ_2 belong to the admissible space Φ, to yield

$$
\begin{aligned}
\dot{G} = & -\int_{\Omega_1} \varepsilon_1 \nabla \dot{\phi}_1 \cdot \nabla \lambda_1 \mathrm{d}\Omega - \int_{\Omega_2} \varepsilon_2 \nabla \dot{\phi}_2 \cdot \nabla \lambda_2 \mathrm{d}\Omega \\
& + \int_{\Omega_1} \left(g_{\phi_1} \dot{\phi}_1 + \mathbf{g}_{\mathbf{E}_1} \cdot \nabla \dot{\phi}_1 \right) m_p \mathrm{d}\Omega + \int_{\Omega_2} \left(g_{\phi_2} \dot{\phi}_2 + \mathbf{g}_{\mathbf{E}_2} \cdot \nabla \dot{\phi}_2 \right) m_p \mathrm{d}\Omega \\
& + \int_{\Omega_1} \varepsilon_1 \nabla (\mathbf{V} \cdot \nabla \phi_1) \cdot \nabla \lambda_1 \mathrm{d}\Omega + \int_{\Omega_2} \varepsilon_2 \nabla (\mathbf{V} \cdot \nabla \phi_2) \cdot \nabla \lambda_2 \mathrm{d}\Omega \\
& + \int_{\Omega_1} \varepsilon_1 \nabla \phi_1 \cdot \nabla (\mathbf{V} \cdot \nabla \lambda_1) \mathrm{d}\Omega + \int_{\Omega_2} \varepsilon_2 \nabla \phi_2 \cdot \nabla (\mathbf{V} \cdot \nabla \lambda_2) \mathrm{d}\Omega \\
& - \int_{\Omega_1} [\rho_1 (\mathbf{V} \cdot \nabla \lambda_1) + \mathbf{P}_{o1} \cdot \nabla (\mathbf{V} \cdot \nabla \lambda_1)] d\Omega \\
& - \int_{\Omega_2} [\rho_2 (\mathbf{V} \cdot \nabla \lambda_2) + \mathbf{P}_{o2} \cdot \nabla (\mathbf{V} \cdot \nabla \lambda_2)] \mathrm{d}\Omega \\
& - \int_{\Omega_1} \left[g_{\phi_1} (\mathbf{V} \cdot \nabla \phi_1) + \mathbf{g}_{\mathbf{E}_1} \cdot \nabla (\mathbf{V} \cdot \nabla \phi_1) \right] m_p \mathrm{d}\Omega \\
& - \int_{\Omega_2} \left[g_{\phi_2} (\mathbf{V} \cdot \nabla \phi_2) + \mathbf{g}_{\mathbf{E}_2} \cdot \nabla (\mathbf{V} \cdot \nabla \phi_2) \right] m_p \mathrm{d}\Omega \\
& - \int_{\gamma} (\varepsilon_1 \nabla \phi_1 \cdot \nabla \lambda_1 - \varepsilon_2 \nabla \phi_2 \cdot \nabla \lambda_2) V_n \mathrm{d}\Gamma + \int_{\gamma} (\mathbf{P}_{o1} \cdot \nabla \lambda_1 - \mathbf{P}_{o2} \cdot \nabla \lambda_2) V_n \mathrm{d}\Gamma \\
& + \int_{\gamma} \left[\rho_1 \lambda_1 - \rho_2 \lambda_2 + (g_1 - g_2) m_p \right] V_n \mathrm{d}\Gamma
\end{aligned} \tag{3.4.32}
$$

The energy bilinear form is symmetric in its arguments;

$$\int_{\Omega_1} \varepsilon_1 \nabla \lambda_1 \cdot \nabla \dot{\phi}_1 \mathrm{d}\Omega = \int_{\Omega_1} \varepsilon_1 \nabla \dot{\phi}_1 \cdot \nabla \lambda_1 \mathrm{d}\Omega \tag{3.4.33}$$

$$\int_{\Omega_2} \varepsilon_2 \nabla \lambda_2 \cdot \nabla \dot{\phi}_2 \mathrm{d}\Omega = \int_{\Omega_2} \varepsilon_2 \nabla \dot{\phi}_2 \cdot \nabla \lambda_2 \mathrm{d}\Omega \tag{3.4.34}$$

By using the relations (3.4.31), (3.4.33), and (3.4.34), all terms with $\dot{\phi}_1$, $\dot{\phi}_2$ in (3.4.32) are canceled out and all terms are expressed with the velocity field **V**;

$$
\begin{aligned}
\dot{G} = & \int_{\Omega_1} [\varepsilon_1 \nabla\phi_1 \cdot \nabla(\mathbf{V}\cdot\nabla\lambda_1) - \rho_1(\mathbf{V}\cdot\nabla\lambda_1) - \mathbf{P}_{o1}\cdot\nabla(\mathbf{V}\cdot\nabla\lambda_1)]\mathrm{d}\Omega \\
& + \int_{\Omega_2} [\varepsilon_2 \nabla\phi_2 \cdot \nabla(\mathbf{V}\cdot\nabla\lambda_2) - \rho_2(\mathbf{V}\cdot\nabla\lambda_2) - \mathbf{P}_{o2}\cdot\nabla(\mathbf{V}\cdot\nabla\lambda_2)]\mathrm{d}\Omega \\
& - \int_{\Omega_1} \left[\varepsilon_1 \nabla\lambda_1 \cdot \nabla(\mathbf{V}\cdot\nabla\phi_1) - g_{\phi_1}(\mathbf{V}\cdot\nabla\phi_1) + \mathbf{g}_{\mathbf{E}_1}\cdot\nabla(\mathbf{V}\cdot\nabla\phi_1)\right] m_p \mathrm{d}\Omega \\
& - \int_{\Omega_2} \left[\varepsilon_2 \nabla\lambda_2 \cdot \nabla(\mathbf{V}\cdot\nabla\phi_2) - g_{\phi_2}(\mathbf{V}\cdot\nabla\phi_2) + \mathbf{g}_{\mathbf{E}_2}\cdot\nabla(\mathbf{V}\cdot\nabla\phi_2)\right] m_p \mathrm{d}\Omega \\
& - \int_{\gamma} (\varepsilon_1 \nabla\phi_1\cdot\nabla\lambda_1 - \varepsilon_2\nabla\phi_2\cdot\nabla\lambda_2) V_n \mathrm{d}\Gamma + \int_{\gamma} (\mathbf{P}_{o1}\cdot\nabla\lambda_1 - \mathbf{P}_{o2}\cdot\nabla\lambda_2) V_n \mathrm{d}\Gamma \\
& + \int_{\gamma} \left[\rho_1\lambda_1 - \rho_2\lambda_2 + (g_1 - g_2)m_p\right] V_n \mathrm{d}\Gamma
\end{aligned}
\tag{3.4.35}
$$

3.4.4 Boundary Expression of Shape Sensitivity

The domain integrals in the design sensitivity (3.4.35) can be expressed in boundary integrals by using the variational identities. For this purpose, two variational identities for the state and the adjoint equations are needed.

First, the variational identities for the state equation were given as (3.4.8) and (3.4.9), which are written again;

$$
\int_{\Omega_1} (\varepsilon_1 \nabla\phi_1 \cdot \nabla\bar{\phi}_1 - \rho\bar{\phi}_1 - \mathbf{P}_{o1}\cdot\nabla\bar{\phi}_1)\mathrm{d}\Omega = -\int_{\gamma+\Gamma} D_n(\phi_1)\bar{\phi}_1 \mathrm{d}\Gamma \quad \forall\bar{\phi}_1 \in \Phi
\tag{3.4.36}
$$

$$
\int_{\Omega_2} (\varepsilon_2 \nabla\phi_2 \cdot \nabla\bar{\phi}_2 - \rho\bar{\phi}_2 - \mathbf{P}_{o2}\cdot\nabla\bar{\phi}_2)\mathrm{d}\Omega = \int_{\gamma} D_n(\phi_2)\bar{\phi}_2 \mathrm{d}\Gamma \quad \forall\bar{\phi}_2 \in \Phi
\tag{3.4.37}
$$

Next, the variational identities for the adjoint equation can be derived from a differential adjoint equation, which is obtained by comparing the variational adjoint equation (3.4.30) with the variational state Eq. (3.4.15). The two variational equations are written again for convenience;

$$\int_{\Omega_1} \varepsilon_1 \nabla \lambda_1 \cdot \nabla \bar{\lambda}_1 \mathrm{d}\Omega + \int_{\Omega_2} \varepsilon_2 \nabla \lambda_2 \cdot \nabla \bar{\lambda}_2 \mathrm{d}\Omega$$
$$= \int_{\Omega_1} \left(g_{\phi_1} \bar{\lambda}_1 + \mathbf{g}_{\mathbf{E}_1} \cdot \nabla \bar{\lambda}_1\right) m_p \mathrm{d}\Omega$$
$$+ \int_{\Omega_2} \left(g_{\phi_2} \bar{\lambda}_2 + \mathbf{g}_{\mathbf{E}_2} \cdot \nabla \bar{\lambda}_2\right) m_p \mathrm{d}\Omega \quad \forall \bar{\lambda}_1, \bar{\lambda}_2 \in \Phi \tag{3.4.38}$$

$$\int_{\Omega_1} \varepsilon_1 \nabla \phi_1 \cdot \nabla \bar{\phi}_1 \mathrm{d}\Omega + \int_{\Omega_2} \varepsilon_2 \nabla \phi_2 \cdot \nabla \bar{\phi}_2 \mathrm{d}\Omega$$
$$= \int_{\Omega_1} \left(\rho_1 \bar{\phi}_1 + \mathbf{P}_{o1} \cdot \nabla \bar{\phi}_1\right) \mathrm{d}\Omega + \int_{\Omega_2} \left(\rho_2 \bar{\phi}_2 + \mathbf{P}_{o2} \cdot \nabla \bar{\phi}_2\right) \mathrm{d}\Omega \quad \forall \bar{\phi}_1, \bar{\phi}_2 \in \Phi \tag{3.4.39}$$

These two equations have the same form except the different source terms. The sources $g_{\phi_1} m_p$, $g_{\phi_2} m_p$ and $\mathbf{g}_{\mathbf{E}_1} m_p$, $\mathbf{g}_{\mathbf{E}_2} m_p$ of the adjoint equation correspond to the ones ρ_1, ρ_2 and $\mathbf{P}_{o1}$, $\mathbf{P}_{o2}$ of the state equation, respectively. Just as the variational state Eq. (3.4.39) is equivalent to the differential state equations of (3.4.4) and (3.4.5), the variational adjoint Eq. (3.4.38) is equivalent to differential adjoint equations:

$$-\nabla \cdot \varepsilon_1 \nabla \lambda_1 = \left(g_{\phi_1} - \nabla \cdot \mathbf{g}_{\mathbf{E}_1}\right) m_p \quad \text{in } \Omega_1 \tag{3.4.40}$$

$$-\nabla \cdot \varepsilon_2 \nabla \lambda_2 = \left(g_{\phi_2} - \nabla \cdot \mathbf{g}_{\mathbf{E}_2}\right) m_p \quad \text{in } \Omega_2 \tag{3.4.41}$$

with the boundary condition:

$$\lambda_1 = 0 \quad \text{on } \Gamma^0 \quad \text{(homogeneous Dirichlet boundary condition)} \tag{3.4.42}$$

$$\frac{\partial \lambda_1}{\partial n} = 0 \quad \text{on } \Gamma^1 \quad \text{(homogeneous Neumann boundary condition)} \tag{3.4.43}$$

The adjoint sources $g_{\phi_1} m_p$, $g_{\phi_2} m_p$ and $\mathbf{g}_{\mathbf{E}_1} m_p$, $\mathbf{g}_{\mathbf{E}_2} m_p$ exist only in the domain Ω as the original sources ρ_1, ρ_2 and $\mathbf{P}_{o1}$, $\mathbf{P}_{o2}$, respectively. Thus, there is no adjoint surface source equivalent to Dirichlet boundary condition. That is, while the original state equation may have surface sources equivalent to its Dirichlet boundary condition, the adjoint equation has no surface source equivalent to its Dirichlet boundary condition. Hence, the Dirichlet boundary condition of the adjoint equation is given as zero. Since the structure symmetry is maintained in the adjoint system as well, the homogeneous Neumann condition is imposed on Γ^1 of the adjoint system.

Just as the variational identities of (3.4.36) and (3.4.37) for the state equation are obtained from the differential state Eqs. (3.4.4) and (3.4.5), the variational identities for the adjoint equation are obtained as

$$\int_{\Omega_1} \left(\varepsilon_1 \nabla \lambda_1 \cdot \nabla \bar{\lambda}_1 - g_{\phi_1} \bar{\lambda}_1 m_p - \mathbf{g}_{\mathbf{E}_1} \cdot \nabla \bar{\lambda}_1 m_p\right) \mathrm{d}\Omega = - \int_{\gamma + \Gamma} D_n(\lambda_1) \bar{\lambda}_1 \mathrm{d}\Gamma \quad \forall \bar{\lambda}_1 \in \Phi \tag{3.4.44}$$

$$\int_{\Omega_2} \left(\varepsilon_2 \nabla \lambda_2 \cdot \nabla \bar{\lambda}_2 - g_{\phi_2} \bar{\lambda}_2 m_p - \mathbf{g}_{\mathbf{E}_2} \cdot \nabla \bar{\lambda}_2 m_p\right) \mathrm{d}\Omega = \int_{\gamma} D_n(\lambda_2) \bar{\lambda}_2 \mathrm{d}\Gamma \quad \forall \bar{\lambda}_2 \in \Phi \tag{3.4.45}$$

where

$$D_n(\lambda_1) = -\varepsilon_1 \frac{\partial \lambda_1}{\partial n} + \mathbf{g}_{\mathbf{E}_1} \cdot \mathbf{n} m_p \tag{3.4.46}$$

$$D_n(\lambda_2) = -\varepsilon_2 \frac{\partial \lambda_2}{\partial n} + \mathbf{g}_{\mathbf{E}_2} \cdot \mathbf{n} m_p \tag{3.4.47}$$

The variational identities of (3.4.36), (3.4.37) and (3.4.44), (3.4.45) are used to express the domain integrals in (3.4.35) as boundary integrals. First, (3.4.36) and (3.4.37) are evaluated at $\bar{\phi}_1 = \mathbf{V} \cdot \nabla \lambda_1$ and $\bar{\phi}_2 = \mathbf{V} \cdot \nabla \lambda_2$, respectively;

$$\int_{\Omega_1} [\varepsilon_1 \nabla \phi_1 \cdot \nabla(\mathbf{V} \cdot \nabla \lambda_1) - \rho_1(\mathbf{V} \cdot \nabla \lambda_1) - \mathbf{P}_{o1} \cdot \nabla(\mathbf{V} \cdot \nabla \lambda_1)] \mathrm{d}\Omega = - \int_{\gamma} D_n(\phi_1)(\mathbf{V} \cdot \nabla \lambda_1) \mathrm{d}\Gamma \tag{3.4.48}$$

$$\int_{\Omega_2} [\varepsilon_2 \nabla \phi_2 \cdot \nabla(\mathbf{V} \cdot \nabla \lambda_2) - \rho_2(\mathbf{V} \cdot \nabla \lambda_2) - \mathbf{P}_{o2} \cdot \nabla(\mathbf{V} \cdot \nabla \lambda_2)] \mathrm{d}\Omega = \int_{\gamma} D_n(\phi_2)(\mathbf{V} \cdot \nabla \lambda_2) \mathrm{d}\Gamma \tag{3.4.49}$$

Second, (3.4.44) and (3.4.45) are evaluated at $\bar{\lambda}_1 = \mathbf{V}\cdot\nabla\phi_1$ and $\bar{\lambda}_2 = \mathbf{V}\cdot\nabla\phi_2$, respectively;

$$\int_{\Omega_1} \left[\varepsilon_1\nabla\lambda_1\cdot\nabla(\mathbf{V}\cdot\nabla\phi_1) - g_{\phi_1}(\mathbf{V}\cdot\nabla\phi_1)m_p - \mathbf{g}_{\mathbf{E}_1}\cdot\nabla(\mathbf{V}\cdot\nabla\phi_1)m_p\right]\mathrm{d}\Omega = -\int_{\Gamma} D_n(\lambda_1)(\mathbf{V}\cdot\nabla\phi_1)\mathrm{d}\Gamma \tag{3.4.50}$$

$$\int_{\Omega_2} \left[\varepsilon_2\nabla\lambda_2\cdot\nabla(\mathbf{V}\cdot\nabla\phi_2) - g_{\phi_2}(\mathbf{V}\cdot\nabla\phi_2)m_p - \mathbf{g}_{\mathbf{E}_2}\cdot\nabla(\mathbf{V}\cdot\nabla\phi_2)m_p\right]\mathrm{d}\Omega = \int_{\Gamma} D_n(\lambda_2)(\mathbf{V}\cdot\nabla\phi_2)\mathrm{d}\Gamma \tag{3.4.51}$$

By substituting (3.4.48)–(3.4.51) into (3.4.35), the domain integrals in (3.4.35) become the boundary integrals;

$$\begin{aligned}\dot{G} = &\int_{\gamma} [-D_n(\phi_1)(\mathbf{V}\cdot\nabla\lambda_1) + D_n(\phi_2)(\mathbf{V}\cdot\nabla\lambda_1)\\ &-D_n(\lambda_1)(\mathbf{V}\cdot\nabla\phi_1) + D_n(\lambda_2)(\mathbf{V}\cdot\nabla\phi_2)]\mathrm{d}\Gamma\\ &-\int_{\gamma}(\varepsilon_1\nabla\phi_1\cdot\nabla\lambda_1 - \varepsilon_2\nabla\phi_2\cdot\nabla\lambda_2)V_n\mathrm{d}\Gamma\\ &+\int_{\gamma}(\mathbf{P}_{o1}\cdot\nabla\lambda_1 - \mathbf{P}_{o2}\cdot\nabla\lambda_2)V_n\mathrm{d}\Gamma\\ &+\int_{\gamma}\left[\rho_1\lambda_1 - \rho_2\lambda_2 + (g_1 - g_2)m_p\right]V_n\mathrm{d}\Gamma\end{aligned} \tag{3.4.52}$$

There is no surface charge on the interface γ; we have the continuity condition for the normal component of the $\mathbf{D}(\phi)$ and $\mathbf{D}(\lambda)$:

$$D_n(\phi_1) = D_n(\phi_2) \quad \text{on}\,\gamma \quad \text{(interface condition)} \tag{3.4.53}$$

$$D_n(\lambda_1) = D_n(\lambda_2) \quad \text{on}\,\gamma \quad \text{(interface condition)} \tag{3.4.54}$$

Using these interface conditions, (3.4.52) is rewritten as

$$\begin{aligned}\dot{G} = & \int_{\gamma} [D_n(\phi_1)(\mathbf{V}\cdot\nabla\lambda_2 - \mathbf{V}\cdot\nabla\lambda_1) + D_n(\lambda_2)(\mathbf{V}\cdot\nabla\phi_2 - \mathbf{V}\cdot\nabla\phi_1)]\mathrm{d}\Gamma \\ & - \int_{\gamma} (\varepsilon_1\nabla\phi_1\cdot\nabla\lambda_1 - \varepsilon_2\nabla\phi_2\cdot\nabla\lambda_2)V_n\mathrm{d}\Gamma \\ & + \int_{\gamma} (\mathbf{P}_{o1}\cdot\nabla\lambda_1 - \mathbf{P}_{o2}\cdot\nabla\lambda_2)V_n\mathrm{d}\Gamma \\ & + \int_{\gamma} \left[\rho_1\lambda_1 - \rho_2\lambda_2 + (g_1 - g_2)m_p\right]V_n\mathrm{d}\Gamma\end{aligned} \tag{3.4.55}$$

This sensitivity can be expressed as a simpler form using the interface conditions, the boundary relations of the velocity **V**, etc. By using the relations (3.2.32) and (3.2.33) in Sect. 3.2, the integrand of the first integral in (3.4.55) is written as

$$\begin{aligned}& D_n(\phi_1)(\mathbf{V}\cdot\nabla\lambda_2 - \mathbf{V}\cdot\nabla\lambda_1) + D_n(\lambda_2)(\mathbf{V}\cdot\nabla\phi_2 - \mathbf{V}\cdot\nabla\phi_1) \\ & = \left[D_n(\phi_1)\left(\frac{\partial\lambda_2}{\partial n} - \frac{\partial\lambda_1}{\partial n}\right) + D_n(\lambda_2)\left(\frac{\partial\phi_2}{\partial n} - \frac{\partial\phi_1}{\partial n}\right)\right]V_n\end{aligned} \tag{3.4.56}$$

By using (3.4.10) and (3.4.47) this is rewritten without V_n as

$$\begin{aligned}& \left(-\varepsilon_1\frac{\partial\phi_1}{\partial n} + P_{o1n}\right)\left(\frac{\partial\lambda_2}{\partial n} - \frac{\partial\lambda_1}{\partial n}\right) + \left(-\varepsilon_2\frac{\partial\lambda_2}{\partial n} + g_{\mathrm{E}_2 n}m_p\right)\left(\frac{\partial\phi_2}{\partial n} - \frac{\partial\phi_1}{\partial n}\right) \\ & = -\varepsilon_1\frac{\partial\phi_1}{\partial n}\left(\frac{\partial\lambda_2}{\partial n} - \frac{\partial\lambda_1}{\partial n}\right) - \varepsilon_2\frac{\partial\lambda_2}{\partial n}\left(\frac{\partial\phi_2}{\partial n} - \frac{\partial\phi_1}{\partial n}\right) + P_{o1n}\left(\frac{\partial\lambda_2}{\partial n} - \frac{\partial\lambda_1}{\partial n}\right) \\ & + g_{\mathrm{E}_2 n}\left(\frac{\partial\phi_2}{\partial n} - \frac{\partial\phi_1}{\partial n}\right)m_p\end{aligned} \tag{3.4.57}$$

where

$$P_{o1n} = \mathbf{P}_{o1}\cdot\mathbf{n} \quad \text{and} \quad g_{\mathrm{E}_2 n} = \mathbf{g}_{\mathbf{E}_2}\cdot\mathbf{n} \tag{3.4.58}$$

By the use of the relations (3.2.34) in Sect. 3.2, the integrand of the second integral in (3.4.55) is written without V_n as

$$\begin{aligned}
&-\varepsilon_1\nabla\phi_1\cdot\nabla\lambda_1+\varepsilon_2\nabla\phi_2\cdot\nabla\lambda_2=-\varepsilon_1\frac{\partial\phi_1}{\partial n}\frac{\partial\lambda_1}{\partial n}\\
&-\varepsilon_1\frac{\partial\phi_1}{\partial t}\frac{\partial\lambda_1}{\partial t}+\varepsilon_2\frac{\partial\phi_2}{\partial n}\frac{\partial\lambda_2}{\partial n}+\varepsilon_2\frac{\partial\phi_2}{\partial t}\frac{\partial\lambda_2}{\partial t}
\end{aligned}\tag{3.4.59}$$

The integrand of the third integral in (3.4.55) is written without V_n as

$$\mathbf{P}_{o1}\cdot\nabla\lambda_1-\mathbf{P}_{o2}\cdot\nabla\lambda_2=P_{o1n}\frac{\partial\lambda_1}{\partial n}+P_{o1t}\frac{\partial\lambda_1}{\partial t}-P_{o2n}\frac{\partial\lambda_2}{\partial n}-P_{o2t}\frac{\partial\lambda_2}{\partial t}\tag{3.4.60}$$

where

$$P_{o2n}=\mathbf{P}_{o2}\cdot\mathbf{n},P_{o1t}=\mathbf{P}_{o1}\cdot\mathbf{t}\quad\text{and}\quad P_{o2t}=\mathbf{P}_{o2}\cdot\mathbf{t}\tag{3.4.61}$$

The expression (3.4.59) and the first two terms of (3.4.57) are summed and it is arranged with the interface condition;

$$\begin{aligned}
&-\varepsilon_1\frac{\partial\phi_1}{\partial n}\left(\frac{\partial\lambda_2}{\partial n}-\frac{\partial\lambda_1}{\partial n}\right)-\varepsilon_2\frac{\partial\lambda_2}{\partial n}\left(\frac{\partial\phi_2}{\partial n}-\frac{\partial\phi_1}{\partial n}\right)-\varepsilon_1\frac{\partial\phi_1}{\partial n}\frac{\partial\lambda_1}{\partial n}\\
&-\varepsilon_1\frac{\partial\phi_1}{\partial t}\frac{\partial\lambda_1}{\partial t}+\varepsilon_2\frac{\partial\phi_2}{\partial n}\frac{\partial\lambda_2}{\partial n}+\varepsilon_2\frac{\partial\phi_2}{\partial t}\frac{\partial\lambda_2}{\partial t}\\
&=(\varepsilon_2-\varepsilon_1)\frac{\partial\phi_1}{\partial n}\frac{\partial\lambda_2}{\partial n}+(\varepsilon_2-\varepsilon_1)\frac{\partial\phi_1}{\partial t}\frac{\partial\lambda_2}{\partial t}\\
&=(\varepsilon_2-\varepsilon_1)\nabla\phi_1\cdot\nabla\lambda_2
\end{aligned}\tag{3.4.62}$$

The sum of (3.4.60) and the third term of (3.4.57) is arranged with the interface condition;

$$\begin{aligned}
&P_{o1n}\left(\frac{\partial\lambda_2}{\partial n}-\frac{\partial\lambda_1}{\partial n}\right)+P_{o1n}\frac{\partial\lambda_1}{\partial n}+P_{o1t}\frac{\partial\lambda_1}{\partial t}-P_{o2n}\frac{\partial\lambda_2}{\partial n}-P_{o2t}\frac{\partial\lambda_2}{\partial t}\\
&=-(P_{o2n}-P_{o1n})\frac{\partial\lambda_2}{\partial n}-(P_{o2t}-P_{o1t})\frac{\partial\lambda_2}{\partial t}\\
&=-(\mathbf{P}_{o2}-\mathbf{P}_{o1})\cdot\nabla\lambda_2
\end{aligned}\tag{3.4.63}$$

Consequently, the sum of (3.4.57), (3.4.59), and (3.4.60) results in

$$(\varepsilon_2-\varepsilon_1)\nabla\phi_1\cdot\nabla\lambda_2-(\mathbf{P}_{o2}-\mathbf{P}_{o1})\cdot\nabla\lambda_2+g_{\mathbf{E}_2n}\left(\frac{\partial\phi_2}{\partial n}-\frac{\partial\phi_1}{\partial n}\right)m_p\tag{3.4.64}$$

Using (3.4.64), the sensitivity (3.4.55) for the interface variation is obtained:

$$\dot{G} = \int_{\gamma} \left[(\varepsilon_2 - \varepsilon_1)\nabla\phi_1 \cdot \nabla\lambda_2 - (\mathbf{P}_{o2} - \mathbf{P}_{o1}) \cdot \nabla\lambda_2 \right. \\ \left. -(\rho_2 - \rho_1)\lambda_2 - (g_2 - g_1)m_p + g_{\mathbf{E}_2 n}\left(\frac{\partial\phi_2}{\partial n} - \frac{\partial\phi_1}{\partial n}\right)m_p \right] V_n \mathrm{d}\Gamma \tag{3.4.65}$$

This general sensitivity for the interface variation is arranged by considering characteristics of the electrostatic system and design problems.

It was assumed in Sect. 3.2 that the fixed space charge in the domain is not taken as a design variable. Thus, with $\rho_1 = 0$ and $\rho_2 = 0$ on γ, (3.4.65) becomes

$$\dot{G} = \int_{\gamma} \left[(\varepsilon_2 - \varepsilon_1)\nabla\phi_1 \cdot \nabla\lambda_2 - (\mathbf{P}_{o2} - \mathbf{P}_{o1}) \cdot \nabla\lambda_2 \right. \\ \left. -(g_2 - g_1)m_p + g_{\mathbf{E}_2 n}\left(\frac{\partial\phi_2}{\partial n} - \frac{\partial\phi_1}{\partial n}\right)m_p \right] V_n \mathrm{d}\Gamma \tag{3.4.66}$$

Each integrand in this sensitivity formula represents exchange of both sides by the interface variation. Each term means the exchanges of dielectric constant, permanent polarization, the objective function due to variation of electric field, and the objective function by variation of integral region, respectively.

When the objective function is defined on the inner area in the domain Ω that does not intersect with the design boundary, $m_p = 0$ on γ. Hence, the sensitivity (3.4.66) becomes

$$\dot{G} = \int_{\Gamma} [(\varepsilon_2 - \varepsilon_1)\nabla\phi_1 \cdot \nabla\lambda_2 - (\mathbf{P}_{o2} - \mathbf{P}_{o1}) \cdot \nabla\lambda_2] V_n \mathrm{d}\Gamma \tag{3.4.67}$$

When the permanent polarization is not taken as design variable, $\mathbf{P}_{o1} = 0$ and $\mathbf{P}_{o2} = 0$ on γ. Thus, (3.4.67) becomes

$$\dot{G} = \int_{\gamma} (\varepsilon_2 - \varepsilon_1)\nabla\phi_1 \cdot \nabla\lambda_2 V_n \mathrm{d}\Gamma \tag{3.4.68}$$

By using (3.2.44), this sensitivity formula is rewritten as

$$\dot{G} = \int_{\gamma} (\varepsilon_2 - \varepsilon_1)\mathbf{E}(\phi_1) \cdot \mathbf{E}(\lambda_2) V_n \mathrm{d}\Gamma \tag{3.4.69}$$

This sensitivity formula represents the exchange of dielectric constant by the variation of the interface γ. It can be applied to the design problem of insulator. In particular, when the insulator surface come into contacts with the air, this sensitivity formula becomes

$$\dot{G} = \int_{\gamma} \varepsilon_0(\varepsilon_r - 1)\mathbf{E}(\phi_1) \cdot \mathbf{E}(\lambda_2) V_n \mathrm{d}\Gamma \tag{3.4.70}$$

3.4.5 Analytical Example

For the interface design problem in the Sect. 3.4.4, the sensitivity formula, which is for the objective function defined in the inner area, was derived as (3.4.69). One-dimensional example, which has the analytic field solution, is taken to show that the sensitivity formula is correct. The sensitivity result analytically calculated in the example is compared with the result of the sensitivity formula. The analytical example is a cylindrical coaxial capacitor, which is the one-dimensional capacitor model in the cylindrical coordinate.

3.4.5.1 Cylindrical Coaxial Capacitor

As an analytical example, a cylindrical capacitor is given as in Fig. 3.24, where two cylindrical conducting electrodes are separated by two dielectrics of permittivity ε_1 and ε_2, and a voltage ϕ_o is applied between the two electrodes. The design objective is to obtain a target electric field $\mathbf{E}_o$ in Ω_p by moving the interface γ, where the two dielectrics meet; the design variable is the radius b of the interface γ. The design sensitivity with respect to the design variable b is analytically calculated, and then its result is compared with the result by the sensitivity formula.

Fig. 3.24 Cylindrical coaxial capacitor-interface design

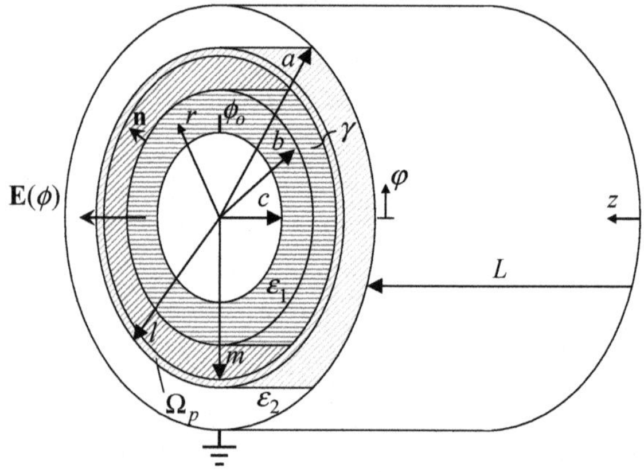

The objective function is defined as

$$F = \int_{\Omega} (\mathbf{E}(\phi) - \mathbf{E}_o)^2 m_p \mathrm{d}\Omega \tag{3.4.71}$$

where

$$\mathbf{E}(\phi_1) = \frac{\phi_o}{\frac{\varepsilon_1}{\varepsilon_2}\ln a/b + \ln b/c}\frac{1}{r}\mathbf{r} \quad \text{for } c \le r \le b \tag{3.4.72}$$

$$\mathbf{E}(\phi_2) = \frac{\phi_o}{\ln a/b + \frac{\varepsilon_2}{\varepsilon_1}\ln b/c}\frac{1}{r}\mathbf{r} \quad \text{for } b \le r \le a \tag{3.4.73}$$

$$\mathbf{E}_o = \frac{E_o}{r}\mathbf{r} \tag{3.4.74}$$

The objective function is rewritten by using the fields (3.4.73) and (3.4.74):

$$F = 2\pi \ln\frac{l}{m}\left(\frac{\phi_o}{\ln a/b + \frac{\varepsilon_2}{\varepsilon_1}\ln b/c} - E_o\right)^2 L \tag{3.4.75}$$

The analytical sensitivity per unit length is obtained by differentiating the objective function (3.4.75) with respect to the radius b of the interface;

$$\frac{\mathrm{d}F}{\mathrm{d}b} = -4\pi\phi_o \frac{\left(\frac{\varepsilon_2}{\varepsilon_1} - 1\right)\ln l/m}{b\left(\ln a/b + \frac{\varepsilon_2}{\varepsilon_1}\ln b/c\right)^2}\left(\frac{\phi_o}{\ln a/b + \frac{\varepsilon_2}{\varepsilon_1}\ln b/c} - E_o\right) \tag{3.4.76}$$

Alternatively, the sensitivity can be calculated by using the sensitivity formula (3.4.69):

$$\dot{G} = \int_{\gamma} (\varepsilon_2 - \varepsilon_1)\mathbf{E}(\phi_1) \cdot \mathbf{E}(\lambda_2) V_n \mathrm{d}\Gamma \tag{3.4.77}$$

This sensitivity formula requires an adjoint variable λ, which can be obtained in the adjoint variable system in Fig. 3.25.

The differential adjoint equations are given as

$$-\nabla \cdot \varepsilon_1 \nabla \lambda_1 = \mathbf{0} \quad \text{in}\, \Omega_1 \tag{3.4.78}$$

$$-\nabla \cdot \varepsilon_2 \nabla \lambda_2 = -\nabla \cdot \mathbf{g}_{\mathbf{E}_2} m_p \quad \text{in}\, \Omega_2 \tag{3.4.79}$$

The differential adjoint equations are obtained by using the electric fields (3.4.73) and (3.4.74);

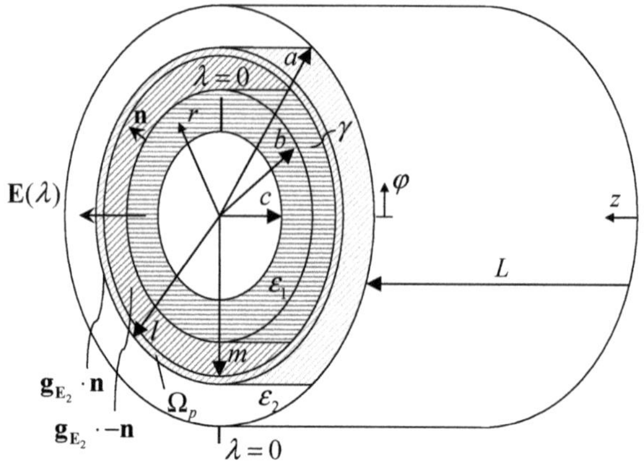

Fig. 3.25 Cylindrical coaxial capacitor-interface design: adjoint variable system

$$\varepsilon_1 \frac{1}{r}\frac{d}{dr}\left(r\frac{d\lambda_1}{\mathrm{d}r}\right) = 0 \quad \text{for } c \le r \le b \tag{3.4.80}$$

$$\varepsilon_2 \frac{1}{r}\frac{\mathrm{d}}{\mathrm{d}r}\left(r\frac{\mathrm{d}\lambda_2}{\mathrm{d}r}\right) = -2\left(\frac{\phi_o}{\ln {}^{a}\!/_{b} + \frac{\varepsilon_2}{\varepsilon_1}\ln {}^{b}\!/_{c}}\frac{1}{r} - \frac{E_o}{r}\right) (\delta(r-m) - \delta(r-l)) \quad \text{for } b \le r \le a \tag{3.4.81}$$

Integrating (3.4.80) and (3.4.81) yields the solution of the adjoint fields as

$$E(\lambda_1) = -\frac{\mathrm{d}\lambda_1}{\mathrm{d}r} = -\frac{k_1}{r} \quad \text{for } c \le r \le b \tag{3.4.82}$$

$$E(\lambda_2) = -\frac{\mathrm{d}\lambda_2}{\mathrm{d}r} = \frac{2}{\varepsilon_2}\left(\frac{\phi_o}{\ln {}^{a}\!/_{b} + \frac{\varepsilon_2}{\varepsilon_1}\ln {}^{b}\!/_{c}} - E_o\right) (H(r-m) - H(r-l))\frac{1}{r} - \frac{k_2}{r} \quad \text{for } b \le r \le a \tag{3.4.83}$$

where k_1, k_2, the integral constants, are obtained by integrating (3.4.82), (3.4.83) and applying the boundary and interface conditions in Fig. 3.25;

$$k_1 = \frac{2}{\varepsilon_1}\frac{\ln l/m}{\ln a/b + \frac{\varepsilon_2}{\varepsilon_1}\ln b/c}\left(\frac{\phi_o}{\ln a/b + \frac{\varepsilon_2}{\varepsilon_1}\ln b/c} - E_o\right) \tag{3.4.84}$$

$$k_2 = \frac{2}{\varepsilon_2}\frac{\ln l/m}{\ln a/b + \frac{\varepsilon_2}{\varepsilon_1}\ln b/c}\left(\frac{\phi_o}{\ln a/b + \frac{\varepsilon_2}{\varepsilon_1}\ln b/c} - E_o\right) \tag{3.4.85}$$

Inserting (3.4.84) and (3.4.85) into (3.4.82) and (3.4.83), respectively, yields the adjoint fields;

$$\mathbf{E}(\lambda_1) = -\frac{2}{\varepsilon_1}\frac{\ln l/m}{\ln a/b+\frac{\varepsilon_2}{\varepsilon_1}\ln b/c}\left(\frac{\phi_o}{\ln a/b+\frac{\varepsilon_2}{\varepsilon_1}\ln b/c} - E_o\right)\frac{1}{r}\mathbf{r} \quad \text{for}\; c \le r \le b \tag{3.4.86}$$

$$\begin{aligned}\mathbf{E}(\lambda_2) = \frac{2}{\varepsilon_2}&\left(\frac{\phi_o}{\ln {}^{a}\!/_{b}+\frac{\varepsilon_2}{\varepsilon_1}\ln {}^{b}\!/_{c}} - E_o\right)\\ &\left(H(r-m) - H(r-l) - \frac{\ln {}^{l}\!/_{m}}{\ln {}^{a}\!/_{b}+\frac{\varepsilon_2}{\varepsilon_1}\ln {}^{b}\!/_{c}}\right)\frac{1}{r}\mathbf{r} \quad \text{for}\; b \le r \le a\end{aligned} \tag{3.4.87}$$

(3.4.72) and (3.4.87) are inserted into the sensitivity formula (3.4.77);

$$\dot{G} = -4\pi\phi_o \frac{\left(\frac{\varepsilon_2}{\varepsilon_1} - 1\right)\ln l/m}{b\left(\ln a/b+\frac{\varepsilon_2}{\varepsilon_1}\ln b/c\right)^2}\left(\frac{\phi_o}{\ln a/b+\frac{\varepsilon_2}{\varepsilon_1}\ln b/c} - E_o\right)V_n L \tag{3.4.88}$$

Using $V_n = \frac{db}{dt}$, the design sensitivity per unit length is obtained as

$$\frac{\mathrm{d}G}{\mathrm{d}b} = -4\pi\phi_o \frac{\left(\frac{\varepsilon_2}{\varepsilon_1} - 1\right)\ln l/m}{b\left(\ln a/b+\frac{\varepsilon_2}{\varepsilon_1}\ln b/c\right)^2}\left(\frac{\phi_o}{\ln a/b+\frac{\varepsilon_2}{\varepsilon_1}\ln b/c} - E_o\right) \tag{3.4.89}$$

which is the same as the analytical result in (3.4.76).

3.4.6 Numerical Example

The sensitivity formula (3.4.69) derived in Sect. 3.4.4 is applied to a shape optimization problem of two-dimensional model, of which the analytic field solution is not given. This design model shows that the sensitivity formula is well applied to the shape design problem of the interface in the electrostatic system. The numerical example is an axi-symmetric capacitor, whose optimal design is known. If the result of the example is obtained as the expected optimal design, the shape optimization using the sensitivity formula can be said to be feasible for the shape design of the interface.

In this example the sensitivity formula requires the state and the adjoint variables,which are numerically calculated by the finite element method. The sensitivity information obtained is used for the optimization algorithm to provide the evolution of the electrode shape. For the optimization algorithm, the level set method is used to represent the shape evolution of the design model.

3.4.6.1 Axi-symmetric Capacitor

For an example with a known optimal design, consider an axi-symmetric capacitor in Fig. 3.26, where the inner electrode and the outer electrode are both cylindrical, but the interface between the two dielectrics is not cylindrical. When a voltage ϕ_o is applied between the two electrodes, the distribution of the electric field is not uniform along the axial direction. When the shape of the interface is changed to be cylindrical, the capacitor becomes a coaxial capacitor and the electric field between the two electrodes becomes uniform along the axial direction.

The design objective is to obtain a uniform field $\mathbf{E}_o$ in the region Ω_p, which is analytically given by the coaxial capacitor. The design variable is the shape of the interface γ, of which the optimal shape is a cylinder for the uniform field. The objective function to be minimized is defined as the integral of the field difference in Ω_p.

$$F = \int_{\Omega} (\mathbf{E}(\phi) - \mathbf{E}_o)^2 m_p \mathrm{d}\Omega \tag{3.4.90}$$

where

$$\mathbf{E}_o = \frac{E_o}{r}\mathbf{r} \tag{3.4.91}$$

The variational adjoint equation for (3.4.90) is obtained as

$$a(\lambda, \bar{\lambda}) = \int_{\Omega} 2(\mathbf{E}(\phi) - \mathbf{E}_o) \cdot \mathbf{E}(\bar{\lambda}) m_p \mathrm{d}\Omega \quad \forall \bar{\lambda} \in \Phi \tag{3.4.92}$$

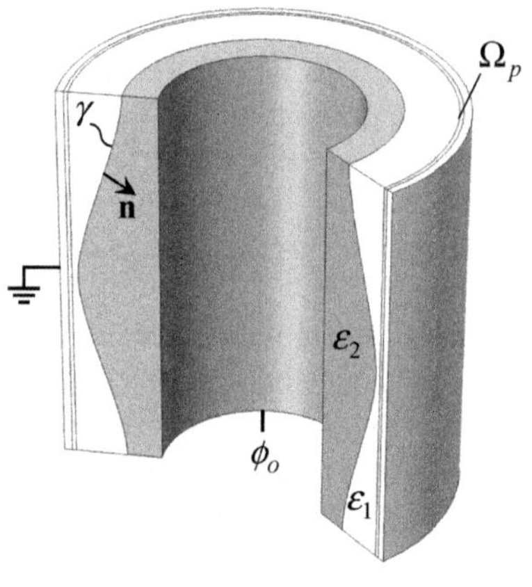

Fig. 3.26 Axi-symmetric capacitor-interface design

The shape sensitivity for this interface design is the formula (3.4.69):

$$\dot{G} = \int_{\gamma} (\varepsilon_2 - \varepsilon_1)\mathbf{E}(\phi_1) \cdot \mathbf{E}(\lambda_2) V_n \mathrm{d}\Gamma \qquad (3.4.93)$$

The design velocity for this minimization problem is taken as

$$V_n = -(\varepsilon_2 - \varepsilon_1)\mathbf{E}(\phi_1) \cdot \mathbf{E}(\lambda_2) \qquad (3.4.94)$$

The design result is shown in Fig. 3.27, where the shape of the interface becomes gradually a cylinder with the increase of the iteration number as expected. The final design of the cylindrical shape is obtained at the 15 s, when the objective function value converges to 0 as in Fig. 3.28. The optimal design by the shape sensitivity is well applied to the interface design problem in the two-dimensional axi-symmetric electrostatic system.

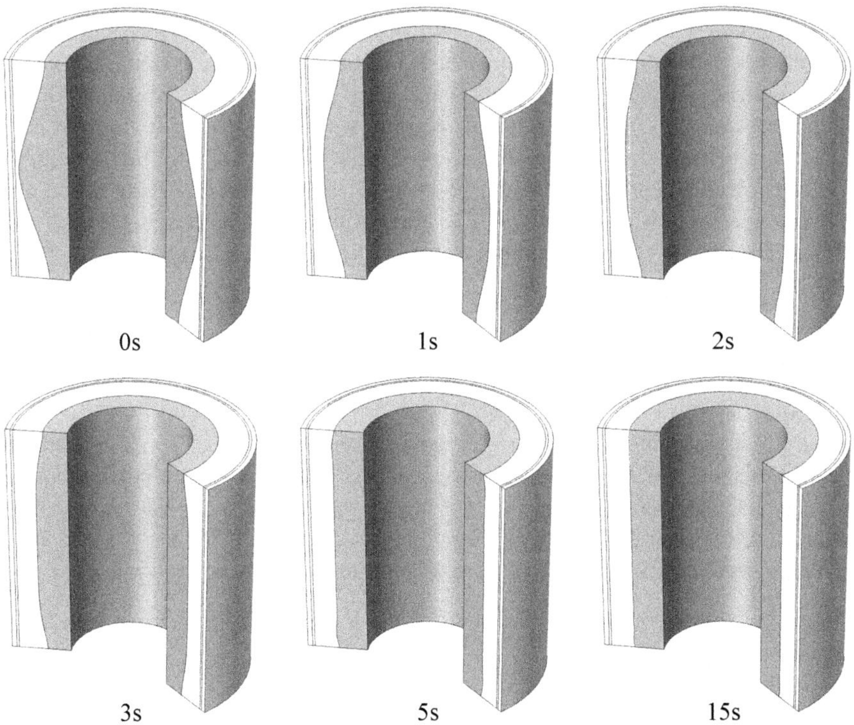

Fig. 3.27 Axi-symmetric capacitor: shape variation

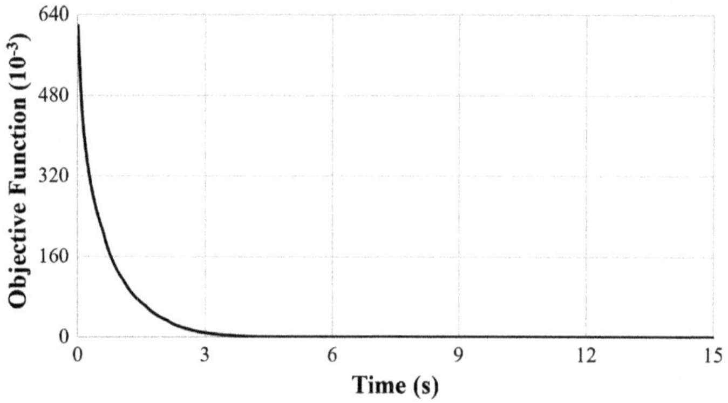

Fig. 3.28 Axi-symmetric capacitor: evolution of objective function

3.5 Shape Sensitivity of Interface for System Energy

In this section, the shape sensitivity for the interface design is developed in the electrostatic system as in Sect. 3.4; but the objective function is the system energy. As mentioned in Sect. 3.3, the system energy of the electrostatic system is related to the capacitance of the electric circuit and it may be utilized for designing various electrostatic systems. The derivation procedure for the sensitivity is almost the same as the one in Sect. 3.4. The difference is that the adjoint variable for the system energy is obtained simply as the half of the state variable. Thus, solving the adjoint variable equation is not necessary. The derived three-dimensional sensitivity formula is tested and validated with analytical and numerical examples.

3.5.1 Problem Definition

An electrostatic system for interface design is given as in Fig. 3.29, which is almost the same as Fig. 3.23 in Sect. 3.4 except that the permanent polarization is excluded. Since the design problem with the permanent polarization is hardly found, it is not included for the system energy.

The objective function is the system energy of the electrostatic system;

$$\begin{aligned} W_e &= \int_{\Omega} \frac{1}{2} \rho \phi \mathrm{d}\Omega \\ &= \int_{\Omega_1} \frac{1}{2} \rho_1 \phi_1 \mathrm{d}\Omega + \int_{\Omega_2} \frac{1}{2} \rho_2 \phi_2 \mathrm{d}\Omega \end{aligned} \tag{3.5.1}$$

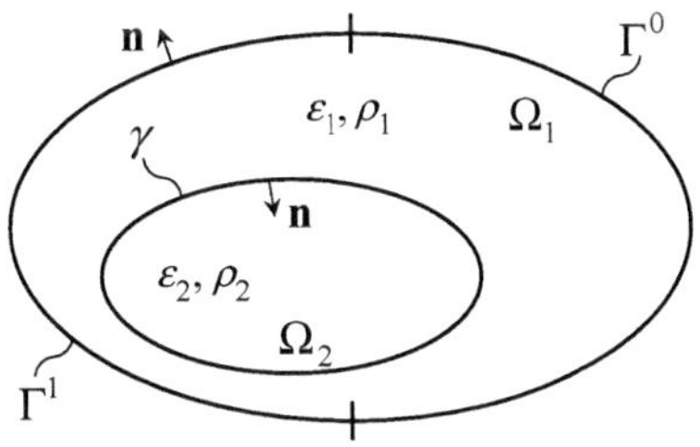

Fig. 3.29 Interface design of electrostatic system for system energy

where the charge densities ρ_1 and ρ_2 are assumed to be fixed and constant in the domain.

The governing differential equations for the state variables of the electric scalar potential ϕ_1 and ϕ_2 are the same as the ones in Sect. 3.4 except that the permanent polarization is excluded;

$$-\nabla \cdot \varepsilon_1 \nabla \phi_1 = \rho_1 \quad \text{in } \Omega_1 \tag{3.5.2}$$

$$-\nabla \cdot \varepsilon_2 \nabla \phi_2 = \rho_2 \quad \text{in } \Omega_2 \tag{3.5.3}$$

The boundary conditions are also the same as in Sect. 3.4;

$$\phi_1 = C(\mathbf{x}) \quad \text{on } \Gamma^0 \quad \text{(Dirichlet boundary condition)} \tag{3.5.4}$$

$$\frac{\partial \phi_1}{\partial n} = 0 \quad \text{on } \Gamma^1 \quad \text{(homogeneous Neumann boundary condition)} \tag{3.5.5}$$

The variational identities for the state equations are obtained by multiplying both sides of (3.5.2) and (3.5.3) by an arbitrary virtual potential $\bar{\phi}_1$ and $\bar{\phi}_2$ and by using the vector identity and the divergence theorem;

$$\int_{\Omega_1} (\varepsilon_1 \nabla \phi_1 \cdot \nabla \bar{\phi}_1 - \rho_1 \bar{\phi}_1) d\Omega = \int_{\gamma + \Gamma} \varepsilon_1 \frac{\partial \phi_1}{\partial n} \bar{\phi}_1 d\Gamma \quad \forall \bar{\phi}_1 \in \Phi \tag{3.5.6}$$

$$\int_{\Omega_2} (\varepsilon_2 \nabla \phi_2 \cdot \nabla \bar{\phi}_2 - \rho_2 \bar{\phi}_2) d\Omega = -\int_{\gamma} \varepsilon_2 \frac{\partial \phi_2}{\partial n} \bar{\phi}_2 d\Gamma \quad \forall \bar{\phi}_2 \in \Phi \tag{3.5.7}$$

where Φ is the space of admissible potential: $\Phi = \{ \bar{\phi} \in H^1(\Omega) | \bar{\phi} = 0 \text{ on } \mathbf{x} \in \Gamma^0 \}$. The boundary conditions of (3.5.4) and (3.5.5) are written for the variational equation as

$$\bar{\phi}_1 = 0 \quad \text{on } \Gamma^0 \quad \text{(Dirichlet boundary condition)} \tag{3.5.8}$$

$$\frac{\partial \phi_1}{\partial n} = 0 \quad \text{on } \Gamma^1 \quad \text{(homogeneous Neumann boundary condition)} \tag{3.5.9}$$

and the interface condition is

$$\varepsilon_1 \frac{\partial \phi_1}{\partial n} = \varepsilon_2 \frac{\partial \phi_2}{\partial n} \quad \text{on}\, \gamma \quad \text{(interface condition)} \tag{3.5.10}$$

By summing (3.5.6) and (3.5.7) and by imposing the boundary conditions and the interface condition, the variational state equation reduced from the differential Eqs. (3.5.4) and (3.5.5) is obtained as

$$\begin{aligned} &\int_{\Omega_1} \varepsilon_1 \nabla\phi_1 \cdot \nabla\bar{\phi}_1 \mathrm{d}\Omega + \int_{\Omega_2} \varepsilon_2 \nabla\phi_2 \cdot \nabla\bar{\phi}_2 \mathrm{d}\Omega \\ &= \int_{\Omega_1} \rho_1 \bar{\phi}_1 \mathrm{d}\Omega + \int_{\Omega_2} \rho_2 \bar{\phi}_2 \mathrm{d}\Omega \quad \forall \bar{\phi}_1, \bar{\phi}_2 \in \Phi \end{aligned} \tag{3.5.11}$$

We define the energy bilinear form $a(\phi, \bar{\phi})$ and the source linear form $l(\bar{\phi})$ as

$$a(\phi, \bar{\phi}) \equiv \int_{\Omega_1} \varepsilon_1 \nabla\phi_1 \cdot \nabla\bar{\phi}_1 \mathrm{d}\Omega + \int_{\Omega_2} \varepsilon_2 \nabla\phi_2 \cdot \nabla\bar{\phi}_2 \mathrm{d}\Omega \tag{3.5.12}$$

$$l(\bar{\phi}) \equiv \int_{\Omega_1} \rho_1 \bar{\phi}_1 \mathrm{d}\Omega + \int_{\Omega_2} \rho_2 \bar{\phi}_2 \mathrm{d}\Omega \tag{3.5.13}$$

The variational Eq. (3.5.11) is rewritten with the energy bilinear form and the source linear form as

$$a(\phi, \bar{\phi}) = l(\bar{\phi}) \quad \forall \bar{\phi} \in \Phi \tag{3.5.14}$$

where

$$\phi = \phi_1 \cup \phi_2 \quad \text{and} \quad \bar{\phi} = \bar{\phi}_1 \cup \bar{\phi}_2. \tag{3.5.15}$$

3.5.2 *Lagrange Multiplier Method for Energy Sensitivity*

To take the variational state equation (3.5.14) as an equality constraint for the shape sensitivity analysis, the method of Lagrange multipliers is used to provide an augmented objective function G as

$$G = W_e + l(\bar{\phi}) - a(\phi, \bar{\phi}) \quad \forall \bar{\phi} \in \Phi \tag{3.5.16}$$

where the arbitrary virtual potential $\bar{\phi}$ plays the role of Lagrange multipliers.

The sensitivity, the material derivative of the augmented objective function, is written as

$$\dot{G} = -\dot{a}(\phi, \bar{\phi}) + \dot{l}(\bar{\phi}) + \dot{W}_e \quad \forall \bar{\phi} \in \Phi \tag{3.5.17}$$

By applying the material derivative formula (3.1.22) to the variational state equation (3.5.11) and the objective function (3.5.1) and by using the relation (3.1.15), the first term in (3.5.17) is obtained as the same as (3.4.22), and the second and third terms are obtained below as (3.5.18) and (3.5.19), respectively. In this interface problem, when the material derivative formula is applied, only the integrals on the interface remain since the outer boundary is not taken as design variable ($V_n = 0$ on Γ).

$$\begin{aligned}
\dot{l}(\bar{\phi}) &= \int_{\Omega_1} \rho_1 \bar{\phi}_1' \mathrm{d}\Omega + \int_{\gamma} \rho_1 \bar{\phi}_1 V_n \mathrm{d}\Gamma + \int_{\Omega_2} \rho_2 \bar{\phi}_2' d\Omega - \int_{\gamma} \rho_2 \bar{\phi}_2 V_n d\Gamma \\
&= \int_{\Omega_1} \left[\rho_1 \dot{\bar{\phi}}_1 - \rho_1 (\mathbf{V} \cdot \nabla \bar{\phi}_1)\right] \mathrm{d}\Omega + \int_{\Omega_2} \left[\rho_2 \dot{\bar{\phi}}_2 - \rho_2 (\mathbf{V} \cdot \nabla \bar{\phi}_2)\right] \mathrm{d}\Omega \\
&\quad + \int_{\gamma} \left(\rho_1 \bar{\phi}_1 - \rho_2 \bar{\phi}_2\right) V_n \mathrm{d}\Gamma
\end{aligned} \tag{3.5.18}$$

$$\begin{aligned}
\dot{W}_e &= \int_{\Omega_1} \frac{1}{2} \rho_1 \phi_1' \mathrm{d}\Omega + \int_{\gamma} \frac{1}{2} \rho_1 \phi_1 V_n \mathrm{d}\Gamma + \int_{\Omega_2} \frac{1}{2} \rho_2 \phi_2' \mathrm{d}\Omega - \int_{\gamma} \frac{1}{2} \rho_2 \phi_2 V_n \mathrm{d}\Gamma \\
&= \int_{\Omega_1} \left[\frac{1}{2} \rho_1 \dot{\phi}_1 - \frac{1}{2} \rho_1 (\mathbf{V} \cdot \nabla \phi_1)\right] \mathrm{d}\Omega \\
&\quad + \int_{\Omega_2} \left[\frac{1}{2} \rho_2 \dot{\phi}_2 - \frac{1}{2} \rho_2 (\mathbf{V} \cdot \nabla \phi_2)\right] \mathrm{d}\Omega + \int_{\gamma} \left(\frac{1}{2} \rho_1 \phi_1 - \frac{1}{2} \rho_2 \phi_2\right) V_n \mathrm{d}\Gamma
\end{aligned} \tag{3.5.19}$$

(3.4.22), (3.5.18), and (3.5.19) are substituted into (3.5.17) to provide

$$
\begin{aligned}
\dot{G} = & -\int_{\Omega_1} \varepsilon_1 \nabla\dot{\phi}_1 \cdot \nabla\bar{\phi}_1 \mathrm{d}\Omega - \int_{\Omega_2} \varepsilon_2 \nabla\dot{\phi}_2 \cdot \nabla\bar{\phi}_2 \mathrm{d}\Omega \\
& - \int_{\Omega_1} \varepsilon_1 \nabla\phi_1 \cdot \nabla\dot{\bar{\phi}}_1 \mathrm{d}\Omega - \int_{\Omega_2} \varepsilon_2 \nabla\phi_2 \cdot \nabla\dot{\bar{\phi}}_2 \mathrm{d}\Omega \\
& + \int_{\Omega_1} \rho_1 \dot{\bar{\phi}}_1 \mathrm{d}\Omega + \int_{\Omega_2} \rho_2 \dot{\bar{\phi}}_2 \mathrm{d}\Omega + \int_{\Omega_1} \frac{1}{2}\rho_1 \dot{\phi}_1 \mathrm{d}\Omega + \int_{\Omega_2} \frac{1}{2}\rho_2 \dot{\phi}_2 \mathrm{d}\Omega \\
& + \int_{\Omega_1} \varepsilon_1 \nabla(\mathbf{V} \cdot \nabla\phi_1) \cdot \nabla\bar{\phi}_1 \mathrm{d}\Omega + \int_{\Omega_2} \varepsilon_2 \nabla(\mathbf{V} \cdot \nabla\phi_2) \cdot \nabla\bar{\phi}_2 \mathrm{d}\Omega \\
& + \int_{\Omega_1} \varepsilon_1 \nabla\phi_1 \cdot \nabla(\mathbf{V} \cdot \nabla\bar{\phi}_1) \mathrm{d}\Omega + \int_{\Omega_2} \varepsilon_2 \nabla\phi_2 \cdot \nabla(\mathbf{V} \cdot \nabla\bar{\phi}_2) \mathrm{d}\Omega \\
& - \int_{\Omega_1} \rho_1 (\mathbf{V} \cdot \nabla\bar{\phi}_1) \mathrm{d}\Omega - \int_{\Omega_2} \rho_2 (\mathbf{V} \cdot \nabla\bar{\phi}_2) d\Omega \\
& - \int_{\Omega_1} \frac{1}{2}\rho_1 (\mathbf{V} \cdot \nabla\phi_1) \mathrm{d}\Omega - \int_{\Omega_2} \frac{1}{2}\rho_2 (\mathbf{V} \cdot \nabla\phi_2) \mathrm{d}\Omega \\
& - \int_{\gamma} \left(\varepsilon_1 \nabla\phi_1 \cdot \nabla\bar{\phi}_1 - \varepsilon_2 \nabla\phi_2 \cdot \nabla\bar{\phi}_2 - \rho_1\bar{\phi}_1 + \rho_2\bar{\phi}_2 - \frac{1}{2}\rho_1\phi_1 + \frac{1}{2}\rho_2\phi_2 \right) \\
& V_n \mathrm{d}\Gamma \quad \forall \bar{\phi}_1, \bar{\phi}_2 \in \Phi
\end{aligned}
\tag{3.5.20}
$$

Since $\dot{\bar{\phi}}_1$ and $\dot{\bar{\phi}}_2$ belong to Φ, the variational state equation of (3.5.11) provides the following relation:

$$
\int_{\Omega_1} \varepsilon_1 \nabla\phi_1 \cdot \nabla\dot{\bar{\phi}}_1 \mathrm{d}\Omega + \int_{\Omega_2} \varepsilon_2 \nabla\phi_2 \cdot \nabla\dot{\bar{\phi}}_2 \mathrm{d}\Omega = \int_{\Omega_1} \rho_1 \dot{\bar{\phi}}_1 \mathrm{d}\Omega + \int_{\Omega_2} \rho_2 \dot{\bar{\phi}}_2 \mathrm{d}\Omega \tag{3.5.21}
$$

Hence, all terms with $\dot{\bar{\phi}}$ in (3.5.20) are canceled out;

$$
\begin{aligned}
\dot{G} = & -\int_{\Omega_1} \varepsilon_1 \nabla \not{\phi}_{\infty} \cdot \nabla \not{\Phi}_{\infty} \lceil \Omega - \int_{\Omega_2} \varepsilon_2 \nabla \not{\phi}_{\in} \cdot \nabla \not{\Phi}_{\in} \lceil \Omega \\
& + \int_{\Omega_1} \frac{1}{2} \rho_1 \dot{\phi}_1 \mathrm{d}\Omega + \int_{\Omega_2} \frac{1}{2} \rho_2 \dot{\phi}_2 \mathrm{d}\Omega \\
& + \int_{\Omega_1} \varepsilon_1 \nabla (\mathbf{V} \cdot \nabla \phi_{\infty}) \cdot \nabla \not{\Phi}_{\infty} \lceil \Omega + \int_{\Omega_2} \varepsilon_2 \nabla (\mathbf{V} \cdot \nabla \phi_{\in}) \cdot \nabla \not{\Phi}_{\in} \lceil \Omega \\
& + \int_{\Omega_1} \varepsilon_1 \nabla \phi_{\infty} \cdot \nabla (\mathbf{V} \cdot \nabla \not{\Phi}_{\infty}) \lceil \Omega + \int_{\Omega_2} \varepsilon_2 \nabla \phi_{\in} \cdot \nabla (\mathbf{V} \cdot \nabla \not{\Phi}_{\in}) \lceil \Omega \\
& - \int_{\Omega_1} \rho_1 (\mathbf{V} \cdot \nabla \not{\Phi}_{\infty}) \lceil \Omega - \int_{\Omega_2} \rho_2 (\mathbf{V} \cdot \nabla \not{\Phi}_{\in}) \lceil \Omega \\
& - \int_{\Omega_1} \frac{1}{2} \rho_1 (\mathbf{V} \cdot \nabla \phi_{\infty}) \lceil \Omega - \int_{\Omega_2} \frac{1}{2} \rho_2 (\mathbf{V} \cdot \nabla \phi_{\in}) \lceil \Omega \\
& - \int_{\gamma} \left(\varepsilon_1 \nabla \phi_{\infty} \cdot \nabla \not{\Phi}_{\infty} - \varepsilon_{\in} \nabla \phi_{\in} \cdot \nabla \not{\Phi}_{\in} - \rho_{\infty} \not{\Phi}_{\infty} + \rho_{\in} \not{\Phi}_{\in} - \frac{\infty}{\in} \rho_{\infty} \phi_{\infty} + \frac{\infty}{\in} \rho_{\in} \phi_{\in} \right) \\
& V_n d\Gamma \quad \forall \bar{\phi}_1, \bar{\phi}_2 \in \Phi
\end{aligned}
\tag{3.5.22}
$$

3.5.3 Adjoint Variable Method for Sensitivity Analysis

To obtain an explicit expression of (3.5.22) in terms of the velocity field $\mathbf{V}$, an adjoint equation is introduced. The adjoint equation is obtained by replacing $\dot{\phi}_1$, $\dot{\phi}_2$ in the ninth and tenth integrals of (3.5.22) with a virtual potential $\bar{\lambda}_1$, $\bar{\lambda}_2$, respectively, and by equating the integrals to the energy bilinear form $a(\lambda, \bar{\lambda})$. The adjoint equation so obtained is written as

$$
\int_{\Omega_1} \varepsilon_1 \nabla \lambda_1 \cdot \nabla \bar{\lambda}_1 \mathrm{d}\Omega + \int_{\Omega_2} \varepsilon_2 \nabla \lambda_2 \cdot \nabla \bar{\lambda}_2 \mathrm{d}\Omega = \int_{\Omega_1} \tfrac{1}{2} \rho_1 \bar{\lambda}_1 \mathrm{d}\Omega + \int_{\Omega_2} \tfrac{1}{2} \rho_2 \bar{\lambda}_2 \mathrm{d}\Omega \quad \forall \bar{\lambda}_1, \bar{\lambda}_2 \in \Phi
\tag{3.5.23}
$$

where λ_1 and λ_2 are the adjoint variables and their solution is desired, and Φ is the space of admissible potential defined: $\Phi = \left\{ \bar{\lambda} \in H^1(\Omega) \middle| \bar{\lambda} = 0 \quad \text{on} \quad \mathbf{x} \in \Gamma^0 \right\}$.

This adjoint Eq. (3.5.23) is evaluated at the specific $\bar{\lambda}_1 = \dot{\phi}_1$, $\bar{\lambda}_2 = \dot{\phi}_2$ since (3.5.23) holds for all $\bar{\lambda}_1, \bar{\lambda}_2 \in \Phi$, to yield

$$\int_{\Omega_1} \varepsilon_1 \nabla\lambda_1 \cdot \nabla\dot{\phi}_1 \mathrm{d}\Omega + \int_{\Omega_2} \varepsilon_2 \nabla\lambda_2 \cdot \nabla\dot{\phi}_2 \mathrm{d}\Omega = \int_{\Omega_1} \frac{1}{2}\rho_1\dot{\phi}_1 \mathrm{d}\Omega + \int_{\Omega_2} \frac{1}{2}\rho_2\dot{\phi}_2 \mathrm{d}\Omega \tag{3.5.24}$$

Similarly, the sensitivity Eq. (3.5.22) is evaluated at the specific $\bar{\phi}_1 = \lambda_1, \bar{\phi}_2 = \lambda_2$ since the λ_1 and λ_2 belong to the admissible space Φ, to yield

$$\begin{aligned}
\dot{G} = &- \int_{\Omega_1} \varepsilon_1 \nabla\dot{\phi}_1 \cdot \nabla\lambda_1 \mathrm{d}\Omega - \int_{\Omega_2} \varepsilon_2 \nabla\dot{\phi}_2 \cdot \nabla\lambda_2 \mathrm{d}\Omega + \int_{\Omega_1} \frac{1}{2}\rho_1\dot{\phi}_1 \mathrm{d}\Omega + \int_{\Omega_2} \frac{1}{2}\rho_2\dot{\phi}_2 \mathrm{d}\Omega \\
&+ \int_{\Omega_1} \varepsilon_1 \nabla(\mathbf{V}\cdot\nabla\phi_1) \cdot \nabla\lambda_1 \mathrm{d}\Omega + \int_{\Omega_2} \varepsilon_2 \nabla(\mathbf{V}\cdot\nabla\phi_2) \cdot \nabla\lambda_2 \mathrm{d}\Omega \\
&+ \int_{\Omega_1} \varepsilon_1 \nabla\phi_1 \cdot \nabla(\mathbf{V}\cdot\nabla\lambda_1) \mathrm{d}\Omega + \int_{\Omega_2} \varepsilon_2 \nabla\phi_2 \cdot \nabla(\mathbf{V}\cdot\nabla\lambda_2) \mathrm{d}\Omega \\
&- \int_{\Omega_1} \rho_1(\mathbf{V}\cdot\nabla\lambda_1) \mathrm{d}\Omega - \int_{\Omega_2} \rho_2(\mathbf{V}\cdot\nabla\lambda_2) \mathrm{d}\Omega \\
&- \int_{\Omega_1} \frac{1}{2}\rho_1(\mathbf{V}\cdot\nabla\phi_1) \mathrm{d}\Omega - \int_{\Omega_2} \frac{1}{2}\rho_2(\mathbf{V}\cdot\nabla\phi_2) \mathrm{d}\Omega \\
&- \int_{\gamma} \left(\varepsilon_1 \nabla\phi_1 \cdot \nabla\lambda_1 - \varepsilon_2 \nabla\phi_2 \cdot \nabla\lambda_2 - \rho_1\lambda_1 + \rho_2\lambda_2 - \frac{1}{2}\rho_1\phi_1 + \frac{1}{2}\rho_2\phi_2\right) V_n \mathrm{d}\Gamma
\end{aligned} \tag{3.5.25}$$

The energy bilinear form is symmetric in its arguments;

$$\int_{\Omega_1} \varepsilon_1 \nabla\lambda_1 \cdot \nabla\dot{\phi}_1 \mathrm{d}\Omega = \int_{\Omega_1} \varepsilon_1 \nabla\dot{\phi}_1 \cdot \nabla\lambda_1 \mathrm{d}\Omega \tag{3.5.26}$$

$$\int_{\Omega_2} \varepsilon_2 \nabla\lambda_2 \cdot \nabla\dot{\phi}_2 \mathrm{d}\Omega = \int_{\Omega_2} \varepsilon_2 \nabla\dot{\phi}_2 \cdot \nabla\lambda_2 \mathrm{d}\Omega \tag{3.5.27}$$

By using the relations (3.5.24), (3.5.26), and (3.5.27), all terms with $\dot{\phi}_1$, $\dot{\phi}_2$ in (3.5.25) are canceled out and all terms are expressed with the velocity field $\mathbf{V}$;

$$\dot{G} = \int_{\Omega_1} \left[\varepsilon_1 \nabla(\mathbf{V} \cdot \nabla\phi_1) \cdot \nabla\lambda_1 + \varepsilon_1 \nabla\phi_1 \cdot \nabla(\mathbf{V} \cdot \nabla\lambda_1) - \rho_1(\mathbf{V} \cdot \nabla\lambda_1) - \frac{1}{2}\rho_1(\mathbf{V} \cdot \nabla\phi_1) \right] \mathrm{d}\Omega$$
$$+ \int_{\Omega_2} \left[\varepsilon_2 \nabla(\mathbf{V} \cdot \nabla\phi_2) \cdot \nabla\lambda_2 + \varepsilon_2 \nabla\phi_2 \cdot \nabla(\mathbf{V} \cdot \nabla\lambda_2) - \rho_2(\mathbf{V} \cdot \nabla\lambda_2) - \frac{1}{2}\rho_2(\mathbf{V} \cdot \nabla\phi_2) \right] \mathrm{d}\Omega$$
$$- \int_{\gamma} \left(\varepsilon_1 \nabla\phi_1 \cdot \nabla\lambda_1 - \varepsilon_2 \nabla\phi_2 \cdot \nabla\lambda_2 - \rho_1\lambda_1 + \rho_2\lambda_2 - \frac{1}{2}\rho_1\phi_1 + \frac{1}{2}\rho_2\phi_2 \right) V_n \mathrm{d}\Gamma \tag{3.5.28}$$

Next, the adjoint Eq. (3.5.23) and the original state Eq. (3.5.11) are written again to be are compared;

$$\int_{\Omega_1} \varepsilon_1 \nabla\lambda_1 \cdot \nabla\bar{\lambda}_1 \mathrm{d}\Omega + \int_{\Omega_2} \varepsilon_2 \nabla\lambda_2 \cdot \nabla\bar{\lambda}_2 \mathrm{d}\Omega = \int_{\Omega_1} \tfrac{1}{2}\rho_1\bar{\lambda}_1 \mathrm{d}\Omega + \int_{\Omega_2} \tfrac{1}{2}\rho_2\bar{\lambda}_2 \mathrm{d}\Omega \quad \forall\bar{\lambda}_1, \bar{\lambda}_2 \in \Phi \tag{3.5.29}$$

$$\int_{\Omega_1} \varepsilon_1 \nabla\phi_1 \cdot \nabla\bar{\phi}_1 \mathrm{d}\Omega + \int_{\Omega_2} \varepsilon_2 \nabla\phi_2 \cdot \nabla\bar{\phi}_2 \mathrm{d}\Omega = \int_{\Omega_1} \rho_1\bar{\phi}_1 \mathrm{d}\Omega + \int_{\Omega_2} \rho_2\bar{\phi}_2 \mathrm{d}\Omega \quad \forall\bar{\phi}_1, \bar{\phi}_2 \in \Phi \tag{3.5.30}$$

Only the difference between these two equations is that the source of the adjoint equation is the half of the original state equation. If the boundary condition is also given as the half of the original state equation, the adjoint variable is obtained as the half of the state variable. Thus, the boundary conditions for the adjoint equation are given as

$$\lambda_1 = \tfrac{1}{2}C(\mathbf{x}) \quad \text{on } \Gamma^0 \quad \text{(Dirichlet boundary condition)} \tag{3.5.31}$$

$$\frac{\partial\lambda_1}{\partial n} = 0 \quad \text{on } \Gamma^1 \quad \text{(homogeneous Neumann boundary condition)} \tag{3.5.32}$$

Hence, the adjoint variable in the adjoint Eq. (3.5.29) with the boundary conditions of (3.5.31) and (3.5.32) is the half of the state variable;

$$\lambda = \tfrac{1}{2}\phi \quad \text{in} \quad \Omega \quad \text{and on} \quad \Gamma \tag{3.5.33}$$

Consequently, solving the adjoint equation for the objective function of the system energy is not necessary. By inserting (3.5.33) into (3.5.28), the sensitivity (3.5.28) becomes

$$\begin{aligned}\dot{G} = & \int_{\Omega_1} [\varepsilon_1 \nabla\phi_1 \cdot \nabla(\mathbf{V} \cdot \nabla\phi_1) - \rho_1(\mathbf{V} \cdot \nabla\phi_1)]\mathrm{d}\Omega \\ & + \int_{\Omega_2} [\varepsilon_2 \nabla\phi_2 \cdot \nabla(\mathbf{V} \cdot \nabla\phi_2) - \rho_2(\mathbf{V} \cdot \nabla\phi_2)]\mathrm{d}\Omega \\ & - \int_{\gamma} \left(\frac{1}{2}\varepsilon_1 \nabla\phi_1 \cdot \nabla\phi_1 - \frac{1}{2}\varepsilon_2 \nabla\phi_2 \cdot \nabla\phi_2 - \rho_1\phi_1 + \rho_2\phi_2\right) V_n \mathrm{d}\Gamma\end{aligned} \tag{3.5.34}$$

3.5.4 Boundary Expression of Shape Sensitivity

By using the variational identities of (3.5.6) and (3.5.7), the domain integrals in the sensitivity (3.5.34) can be expressed in boundary integrals. The variational identities (3.5.6) and (3.5.7) are written again;

$$\int_{\Omega_1} (\varepsilon_1 \nabla\phi_1 \cdot \nabla\bar{\phi}_1 - \rho_1\bar{\phi}_1)\mathrm{d}\Omega = \int_{\gamma+\Gamma} \varepsilon_1 \frac{\partial\phi_1}{\partial n}\bar{\phi}_1 \mathrm{d}\Gamma \quad \forall\bar{\phi}_1 \in \Phi \tag{3.5.35}$$

$$\int_{\Omega_2} (\varepsilon_2 \nabla\phi_2 \cdot \nabla\bar{\phi}_2 - \rho_2\bar{\phi}_2)\mathrm{d}\Omega = -\int_{\gamma} \varepsilon_2 \frac{\partial\phi_2}{\partial n}\bar{\phi}_2 \mathrm{d}\Gamma \quad \forall\bar{\phi}_2 \in \Phi \tag{3.5.36}$$

Choosing $\bar{\phi}_1 = \mathbf{V} \cdot \nabla\phi_1$ and $\bar{\phi}_2 = \mathbf{V} \cdot \nabla\phi_2$ in these equations yields

$$\int_{\Omega_1} [\varepsilon_1 \nabla\phi_1 \cdot \nabla(\mathbf{V} \cdot \nabla\phi_1) - \rho_1(\mathbf{V} \cdot \nabla\phi_1)]\mathrm{d}\Omega = \int_{\gamma} \varepsilon_1 \frac{\partial\phi_1}{\partial n}(\mathbf{V} \cdot \nabla\phi_1)\mathrm{d}\Gamma \tag{3.5.37}$$

$$\int_{\Omega_2} [\varepsilon_2 \nabla\phi_2 \cdot \nabla(\mathbf{V} \cdot \nabla\phi_2) - \rho_2(\mathbf{V} \cdot \nabla\phi_2)]\mathrm{d}\Omega = -\int_{\gamma} \varepsilon_2 \frac{\partial\phi_2}{\partial n}(\mathbf{V} \cdot \nabla\phi_2)\mathrm{d}\Gamma \tag{3.5.38}$$

By inserting (3.5.37) and (3.5.38) into (3.5.34), the sensitivity (3.5.34) is expressed as the desired boundary integrals;

$$\begin{aligned}\dot{G} = & \int_{\gamma} \left[\varepsilon_1 \frac{\partial\phi_1}{\partial n}(\mathbf{V} \cdot \nabla\phi_1) - \varepsilon_2 \frac{\partial\phi_2}{\partial n}(\mathbf{V} \cdot \nabla\phi_2)\right]\mathrm{d}\Gamma \\ & - \int_{\gamma} \left(\frac{1}{2}\varepsilon_1 \nabla\phi_1 \cdot \nabla\phi_1 - \frac{1}{2}\varepsilon_2 \nabla\phi_2 \cdot \nabla\phi_2 - \rho_1\phi_1 + \rho_2\phi_2\right) V_n \mathrm{d}\Gamma\end{aligned} \tag{3.5.39}$$

Using the relations (3.2.32) and (3.2.34) in Sect. 3.2, this sensitivity is rewritten as

$$\dot{G} = \int_{\gamma} \left[\frac{1}{2} \left(\varepsilon_1 \frac{\partial \phi_1}{\partial n} \frac{\partial \phi_2}{\partial n} - \varepsilon_1 \frac{\partial \phi_1}{\partial t} \frac{\partial \phi_2}{\partial t} - \varepsilon_2 \frac{\partial \phi_1}{\partial n} \frac{\partial \phi_2}{\partial n} + \varepsilon_2 \frac{\partial \phi_1}{\partial t} \frac{\partial \phi_2}{\partial t} \right) + \rho_1 \phi_1 - \rho_2 \phi_2 \right] V_n \mathrm{d}\Gamma \tag{3.5.40}$$

The interface conditions for the continuity of the normal component of **D** and the tangential component of **E** on the interface γ are written as

$$\varepsilon_1 \frac{\partial \phi_1}{\partial n} = \varepsilon_2 \frac{\partial \phi_2}{\partial n} \quad \text{on } \gamma \tag{3.5.41}$$

$$\frac{\partial \phi_1}{\partial t} = \frac{\partial \phi_2}{\partial t} \quad \text{on } \gamma \tag{3.5.42}$$

These interface conditions are used to rewrite (3.5.40);

$$\dot{G} = \int_{\gamma} \left[\frac{1}{2} \left((\varepsilon_2 - \varepsilon_1) \frac{\partial \phi_1}{\partial n} \frac{\partial \phi_2}{\partial n} + (\varepsilon_2 - \varepsilon_1) \frac{\partial \phi_1}{\partial t} \frac{\partial \phi_2}{\partial t} \right) - (\rho_2 - \rho_1) \phi_2 \right] V_n \mathrm{d}\Gamma \tag{3.5.43}$$

By the relation (3.2.34), this sensitivity formula is expressed as

$$\dot{G} = \int_{\gamma} \left[\frac{1}{2} (\varepsilon_2 - \varepsilon_1) \nabla \phi_1 \cdot \nabla \phi_2 - (\rho_2 - \rho_1) \phi_2 \right] V_n \mathrm{d}\Gamma \tag{3.5.44}$$

With (3.2.44), this sensitivity formula is written as

$$\dot{G} = \int_{\gamma} \left[\frac{1}{2} (\varepsilon_2 - \varepsilon_1) \mathbf{E}(\phi_1) \cdot \mathbf{E}(\phi_2) - (\rho_2 - \rho_1) \phi_2 \right] V_n \mathrm{d}\Gamma \tag{3.5.45}$$

This general three-dimensional sensitivity formula for the interface variation becomes simpler according to specific conditions of a given problem.

With the space charge assumed fixed in the domain, $\rho_1 = 0$ and $\rho_2 = 0$ on γ. Thus, the sensitivity formula (3.5.45) becomes

$$\dot{G} = \int_{\gamma} \frac{1}{2} (\varepsilon_2 - \varepsilon_1) \mathbf{E}(\phi_1) \cdot \mathbf{E}(\phi_2) V_n \mathrm{d}\Gamma \tag{3.5.46}$$

This sensitivity formula represents exchange of dielectric materials by the variation of the interface γ.

When the dielectric material comes into contact with the air, this sensitivity formula becomes

$$\dot{G} = \int_{\gamma} \frac{1}{2}\varepsilon_0(\varepsilon_r - 1)\mathbf{E}(\phi_1) \cdot \mathbf{E}(\phi_2)V_n \mathrm{d}\Gamma \tag{3.5.47}$$

3.5.5 Source Condition and Capacitance Sensitivity

In Sect. 3.3.5 for the outer boundary design problem, it was examined how the sign of the energy sensitivity depends on the source condition in the electric-circuit point of view, and the capacitance sensitivity was derived using the obtained energy sensitivity. This principle is equally applied to the interface problem.

Under the condition of voltage source, the energy variation is:

$$\dot{W}_e = \frac{1}{2}\dot{C}V^2 \tag{3.5.48}$$

Under the condition of current source, the energy variation is:

$$\dot{W}_e = -\frac{1}{2}\dot{C}V^2 \tag{3.5.49}$$

In Sect. 3.5.1, the objective function of system energy was defined with a fixed charge and its sensitivities were also derived from the state equations with fixed charge distribution. That is, the final sensitivity formulas (3.5.46) were derived with the fixed charges. The electrostatic systems are usually connected to the voltage source through the electrodes; for the electrostatic systems with voltage source, the sign of the sensitivity formulas (3.5.46) should be changed.

The sensitivity formula for the voltage-source system is, therefore, written as

$$\dot{G} = -\int_{\gamma} \frac{1}{2}(\varepsilon_2 - \varepsilon_1)\mathbf{E}(\phi_1) \cdot \mathbf{E}(\phi_2)V_n \mathrm{d}\Gamma \tag{3.5.50}$$

This sensitivity formula for the voltage-source system is used to obtain the capacitance sensitivity.

The total derivative of the capacitor-stored energy (3.5.48) is equal to the energy sensitivity with the voltage source condition (3.5.50);

$$\frac{1}{2}\dot{C}V^2 = -\int_{\gamma} \frac{1}{2}(\varepsilon_2 - \varepsilon_1)\mathbf{E}(\phi_1) \cdot \mathbf{E}(\phi_2)V_n \mathrm{d}\Gamma \tag{3.5.51}$$

From this relation, the capacitance sensitivity $\dot{C}$ for the interface problem is obtained as

$$\dot{C} = -\frac{1}{V^2}\int_{\gamma} (\varepsilon_2 - \varepsilon_1)\mathbf{E}(\phi_1)\cdot\mathbf{E}(\phi_2)V_n \mathrm{d}\Gamma \tag{3.5.52}$$

3.5.6 *Analytical Example*

For the objective function of the system energy, the sensitivity formula for the interface variation was derived as (3.5.50) in Sect. 3.5.5. A one-dimensional example with the analytic field solution is taken to show that the sensitivity formula is correct. Comparison of the result by the analytical sensitivity result and the result and the one by the sensitivity formula shows that the two results are the same. The analytical example is a cylindrical coaxial capacitor, which is the one-dimensional capacitor model in the cylindrical coordinates.

3.5.6.1 Cylindrical Coaxial Capacitor

For the analytical example a cylindrical capacitor is considered in Fig. 3.30, where two cylindrical conducting electrodes are separated by two dielectrics of permittivity ε_1 and ε_2, and a voltage ϕ_o is applied between the two electrodes. The design objective is to obtain a desired system energy by moving the interface γ, where the two dielectrics meet. The design sensitivity with respect to the design variable b is analytically calculated, and then its result is compared with the one by the sensitivity formula.

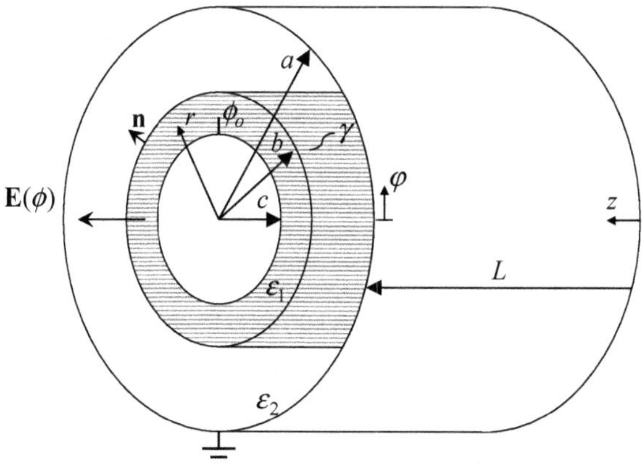

Fig. 3.30 Cylindrical coaxial capacitor-interface design, system energy

The objective function is the system energy;

$$F = W_e = \int_{\Omega} \frac{1}{2} \varepsilon \mathbf{E}^2(\phi) \mathrm{d}\Omega \tag{3.5.53}$$

where

$$\mathbf{E}(\phi_1) = \frac{\phi_o}{\frac{\varepsilon_1}{\varepsilon_2} \ln a/b + \ln b/c} \frac{1}{r} \mathbf{r} \quad \text{for}\, c \le r \le b \tag{3.5.54}$$

$$\mathbf{E}(\phi_2) = \frac{\phi_o}{\ln a/b + \frac{\varepsilon_2}{\varepsilon_1} \ln b/c} \frac{1}{r} \mathbf{r} \quad \text{for}\, b \le r \le a \tag{3.5.55}$$

The objective function is rewritten by using the fields (3.5.54) and (3.5.55):

$$F = \pi \varepsilon_2 \frac{\phi_o^2}{\ln a/b + \frac{\varepsilon_2}{\varepsilon_1} \ln b/c} L \tag{3.5.56}$$

The analytical sensitivity per unit length is obtained by differentiating the objective function (3.5.56) with respect to the radius b of the interface;

$$\frac{\mathrm{d}F}{\mathrm{d}b} = -\pi \frac{\varepsilon_2}{b} \left(\frac{\varepsilon_2}{\varepsilon_1} - 1 \right) \left(\frac{\phi_o}{\ln a/b + \frac{\varepsilon_2}{\varepsilon_1} \ln b/c} \right)^2 \tag{3.5.57}$$

The objective of this example is to compare this analytical sensitivity with the result by the sensitivity formula;

$$\dot{G} = -\int_{\gamma} \frac{1}{2} (\varepsilon_2 - \varepsilon_1) \mathbf{E}(\phi_1) \cdot \mathbf{E}(\phi_2) V_n \mathrm{d}\Gamma \tag{3.5.58}$$

(3.5.54) and (3.5.55) are inserted into the sensitivity formula (3.5.58);

$$\dot{G} = -\pi \frac{\varepsilon_2}{b} \left(\frac{\varepsilon_2}{\varepsilon_1} - 1 \right) \left(\frac{\phi_o}{\ln a/b + \frac{\varepsilon_2}{\varepsilon_1} \ln b/c} \right)^2 V_n L \tag{3.5.59}$$

Using $V_n = \frac{\mathrm{d}_b}{\mathrm{d}_t}$, the design sensitivity per unit length is obtained as

$$\frac{\mathrm{d}G}{\mathrm{d}b} = -\pi \frac{\varepsilon_2}{b} \left(\frac{\varepsilon_2}{\varepsilon_1} - 1 \right) \left(\frac{\phi_o}{\ln a/b + \frac{\varepsilon_2}{\varepsilon_1} \ln b/c} \right)^2 \tag{3.5.60}$$

which is the same as the analytical sensitivity result in (3.5.57).

3.5.7 *Numerical Examples*

The sensitivity formula (3.5.50) derived in Sect. 3.5.5 is applied to two shape optimization problems of two-dimensional model, of which the analytic field solutions are not given. These design models show that the sensitivity formula is well applied to the shape design problem of the interface for the system energy in the electrostatic system. The numerical examples are an axi-symmetric capacitor and a microelectromechanical system (MEMS) motor. While the first example has the known optimal design, the second one does not have the known optimal design. If the result of the first example is obtained as the expected optimal design, it can be said that the shape optimization using the sensitivity formula is feasible for the shape design of the interface. The result of the second example shows that this optimization method is useful for the design of the outer boundary.

In these examples, the evaluation of the sensitivity formula needs the state variable, which is numerically calculated by the finite element method. The sensitivity evaluated is used for the optimization algorithm, which provides the evolution of the electrode shape. The level set method is used as an optimization algorithm to provide the shape evolution of the design model.

3.5.7.1 Axi-Symmetric Coaxial Capacitor

As an example with a known optimal design, an axi-symmetric capacitor is given in Fig. 3.31, where the inner electrode and the outer electrode are both cylindrical, but the interface between the two dielectrics is not cylindrical. When a voltage ϕ_0 is applied between the two electrodes, the distribution of the electric field is not uniform along the axial direction. If the interface is changed to be a cylinder, the

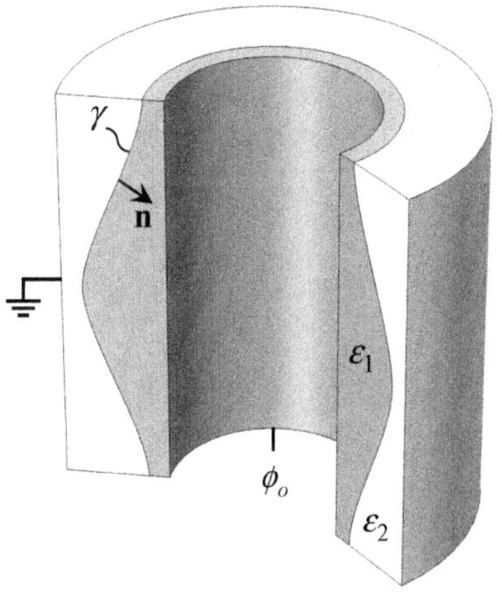

Fig. 3.31 Axi-symmetric capacitor-interface design, system energy minimization

axi-symmetric capacitor becomes a coaxial capacitor, of which the system energy has the minimum value under the constraint of constant dielectric volume.

The design objective is to minimize the system energy; the objective function is defined as

$$F = W_e = \int_{\Omega} \frac{1}{2} \varepsilon \mathbf{E}^2(\phi) \mathrm{d}\Omega \tag{3.5.61}$$

and the design variable is the shape of the interface γ. The shape sensitivity for this interface design is the sensitivity formula (3.5.50):

$$\dot{G} = -\int_{\gamma} \frac{1}{2} (\varepsilon_2 - \varepsilon_1) \mathbf{E}(\phi_1) \cdot \mathbf{E}(\phi_2) V_n \mathrm{d}\Gamma \tag{3.5.62}$$

The design velocity for this minimization problem is taken as

$$V_n = \frac{1}{2} (\varepsilon_2 - \varepsilon_1) \mathbf{E}(\phi_1) \cdot \mathbf{E}(\phi_2) \tag{3.5.63}$$

In this problem, the constraint of constant dielectric volume is required;

$$\int_{\Omega_1} \mathrm{d}\Omega = C \tag{3.5.64}$$

where the constant C is a given dielectric volume per unit length. The constant volume (3.5.64) is equivalent to the zero sum of the design velocity over the interface, which is obtained by differentiating (3.5.64);

$$\dot{C} = \int_{\Omega_1} 1' \mathrm{d}\Omega + \int_{\gamma} V_n \mathrm{d}\Gamma = \int_{\gamma} V_n \mathrm{d}\Gamma = 0 \tag{3.5.65}$$

For the constraint of the constant volume, the modified design velocity U_n is obtained by subtracting its average V_{na};

$$U_n = V_n - V_{na} \tag{3.5.66}$$

where

$$V_{na} = \int_{\gamma} \frac{1}{2} (\varepsilon_2 - \varepsilon_1) \mathbf{E}(\phi_1) \cdot \mathbf{E}(\phi_2) \mathrm{d}\Gamma \Big/ \int_{\gamma} \mathrm{d}\Gamma \tag{3.5.67}$$

Figure 3.32 shows design result that the shape of the interface becomes gradually a cylinder with the increase of the iteration number as expected. The final design of the cylindrical shape is obtained at the 15, when the system energy converges as in Fig. 3.33. The result shows that the shape sensitivity analysis is feasible for the interface design problem with the system energy objective function in the axi-symmetric electrostatic system.

3.5.7.2 MEMS Motor

The shape optimization using the sensitivity formula for the interface design is applied to a shape design problem which does not have the analytical field solution nor the known optimal shape. As an example, consider a MEMS motor in Fig. 3.34, where the initial dielectric rotor is at the center and the eight electrodes on the outer surface are connected with the exciting source voltage for driving the dielectric rotor [12]. As the voltage on the electrodes is switched in the clockwise direction as in Fig. 3.34, a rotating electric field is generated in the air gap. If the shape of the rotor is properly designed, a reluctance torque on the rotor will be generated. The

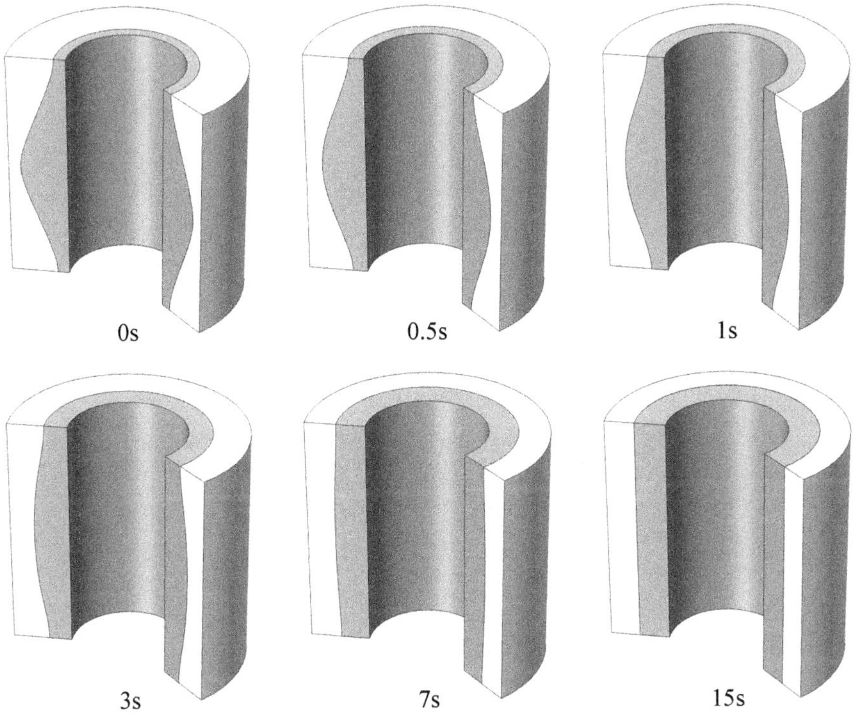

Fig. 3.32 Axi-symmetric capacitor: shape variation

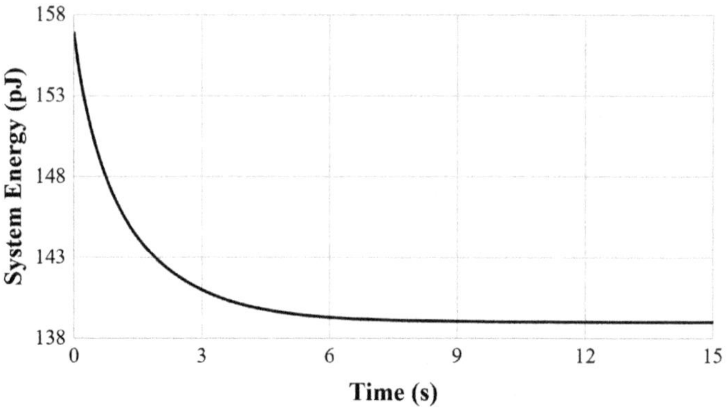

Fig. 3.33 Axi-symmetric capacitor: evolution of objective function

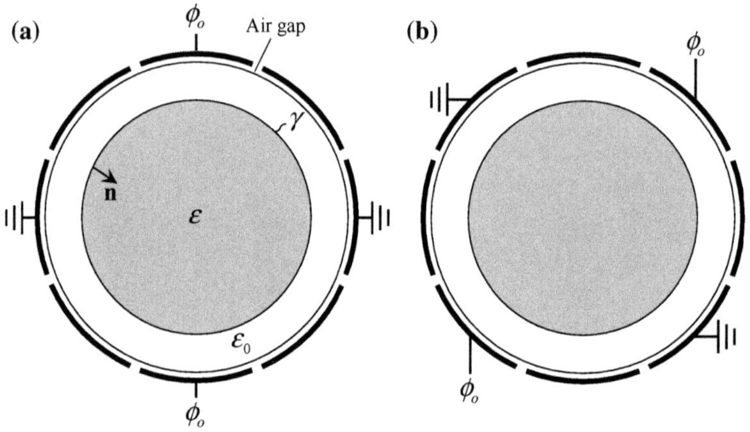

Fig. 3.34 MEMS motor-interface design, torque

design objective is to find a rotor shape which maximizes the reluctance torque. The initial design of the dielectric rotor is taken as a circular shape as in Fig. 3.34.

The torque on the dielectric rotor can be expressed as a derivative of the system energy W_e with respect to the angular position θ;

$$T_e = \frac{\mathrm{d}W_e}{\mathrm{d}\theta} \tag{3.5.68}$$

As the energy difference between the two positions by 90 electrical degrees increases, the generated torque also increases. To maximize the torque on the dielectric rotor, the objective function to be maximized is defined as the energy difference between two positions in Fig. 3.34.

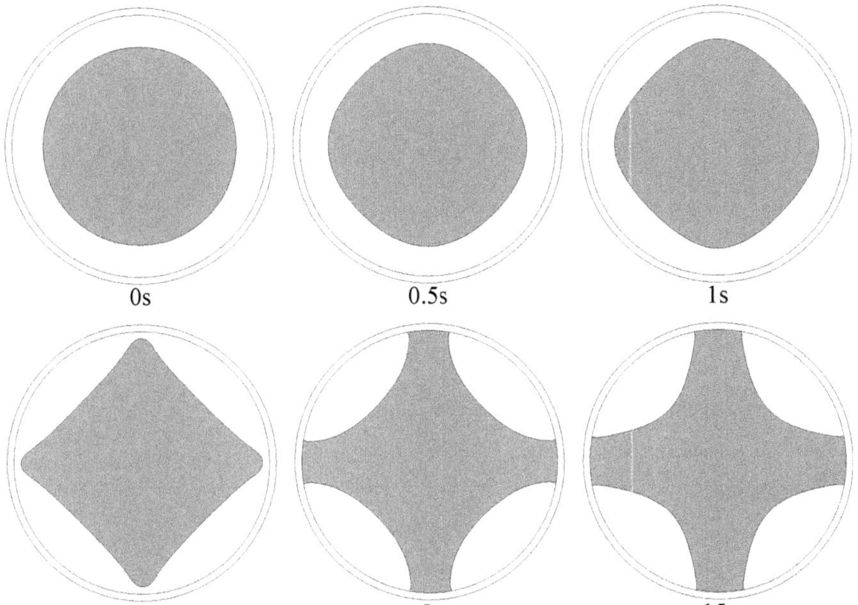

Fig. 3.35 MEMS motor: shape variation of rotor

$$F = W_{ea} - W_{eb} \tag{3.5.69}$$

where the subscripts a and b denote the switching positions a and b, respectively. In this design example, the design variable is the shape of the dielectric rotor and the constraint on the dielectric volume is not necessary. The shape sensitivity formula for the objective function (3.5.69) is obtained as

$$\dot{G} = -\int_{\gamma} \frac{1}{2}\varepsilon_0(\varepsilon_r - 1)(\mathbf{E}_a(\phi_1)\cdot\mathbf{E}_a(\phi_2) - \mathbf{E}_b(\phi_1)\cdot\mathbf{E}_b(\phi_2))V_n \mathrm{d}\Gamma \tag{3.5.70}$$

For this maximization problem, the design velocity on the interface is taken as

$$V_n = -\varepsilon_0(\varepsilon_r - 1)(\mathbf{E}_a(\phi_1)\cdot\mathbf{E}_a(\phi_2) - \mathbf{E}_b(\phi_1)\cdot\mathbf{E}_b(\phi_2)) \tag{3.5.71}$$

Figure 3.35 shows the rotor shape variation during the optimization. In the earlier stage of the optimization process, the dielectric rotor becomes gradually squared. As the shape of the dielectric rotor approaches the boundary of the air gap, the objective function sharply increases. Finally, it becomes cross-shaped near 15 s, when the objective function of the system energy difference converges as in Fig. 3.36.

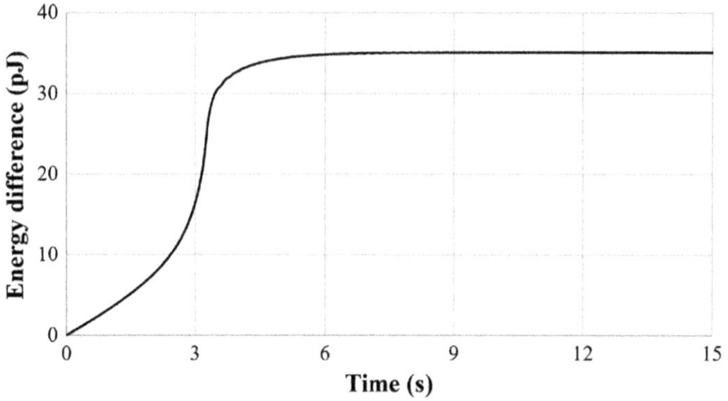

Fig. 3.36 MEMS motor: evolution of objective function

References

1. Choi, K.K., Kim, N.H.: Structural Sensitivity Analysis and Optimization 1: Linear Systems. Springer, New York (2005)
2. Haug, E.J., Choi, K.K., Komkov, V.: Design Sensitivity Analysis of Structural Systems. Academic Press, Orlando (1988)
3. Lai, W.M., Rubin, D., Krempl, E.: Introduction to Continuum Mechanics. Pergamon Press, Oxford (1978)
4. Neittaanmäki, P., Rudnicki, M., Savini, A.: Inverse Problems and Optimal Design in Electricity and Magnetism. Clarendon Press, Oxford (1996)
5. Korovkin, N.V., Chechurin, V.L., Hayakawa, M.: Inverse Problems in Electric Circuits and Electromagnetics. Springer, Boston (2007)
6. Hinze, M., Pinnau, R., Ulbrich, M., Ulbrich, S.: Optimization with PDE Constraints. Springer, Heidelberg (2009)
7. Choi, K.K., Seong, H.G.: Design component method for sensitivity analysis of built-up structures. J. Struct. Mech. **14**, 379–399 (1986)
8. Dems, K., Mróz, Z.: Variational approach by means of adjoint systems to structural optimization and sensitivity analysis—I: variation of material parameters within fixed domain. Int. J. Solids Struct. **19**, 677–692 (1983)
9. Dems, K., Mróz, Z.: Variational approach by means of adjoint systems to structural optimization and sensitivity analysis—II: structure shape variation. Int. J. Solids Struct. **20**, 527–552 (1984)
10. Woodson, H.H., Melcher, J.R.: Electromechanical Dynamics. Wiley, New York (1968)
11. Lee, K.H., Choi, C.Y., Park, I.H.: Continuum sensitivity analysis and shape optimization of Dirichlet conductor boundary in electrostatic system. IEEE Trans. Magn. **54**, 9400804 (2018)
12. Lee, K.H., Seo, K.S., Choi, H.S., Park, I.H.: Multiple level set method for multi-material shape optimization in electromagnetic system. Int. J. Appl. Electromagn. Mech. **56**, 183–193 (2018)

Chapter 4
Continuum Shape Design Sensitivity of Magnetostatic System

The magnetostatic system is composed of ferromagnetic material, permanent magnet, and source current. There are examples of the design objective such as improving device performance, obtaining a desired distribution of magnetic field, obtaining a desired inductance. For these objectives, the shape of the composing materials is optimally designed. The shape design sensitivity for the magnetostatic system can be utilized by designers to improve such performances.

We have various magnetostatic systems such as power apparatus, electric machines, magnetic devices, magnet, examples of which are transformer, generator, motor (brushless DC motor, switched reluctance motor, permanent magnet motor), actuator, magnetic bearing, magnetic levitator, inductor, speaker, MRI magnet, accelerator magnet, magnetic fluid, magnetic shielding, etc.

In this chapter, the continuum shape sensitivity for the linear magnetostatic system is derived by using the material derivative concept. In the same way as in the previous Chapter 3, the Lagrange multiplier method, an adjoint variable technique, and the variational identities are used to obtain a general three-dimensional shape sensitivity of boundary integral.

The design variable of the magnetostatic system, unlike in the case of the electrostatic system, is only the interface. Thus, the shape sensitivity for the magnetostatic system is classified into two categories: the domain integral objective function and the system energy objective function. Since the system energy objective function does not require solving for the adjoint variable and is used for the equivalent inductance, it is taken as another objective function.

In Sect. 4.1, for the domain integral objective function, a general three-dimensional sensitivity of the interface design is derived. Since the magnetostatic system is expressed with three material properties: permeability, permanent magnetization and current density, the derived sensitivity formula has the interface integral of three terms. If the air is treated as another material, it has four material properties. The magnetostatic system, therefore, has nine kinds of interfaces, each sensitivity of which are arranged in Sect. 4.1.5.

I. H. Park, *Design Sensitivity Analysis and Optimization of Electromagnetic Systems*, Mathematical and Analytical Techniques with Applications to Engineering,
https://doi.org/10.1007/978-981-13-0230-5_4

In Sect. 4.2, for the system energy objective functions, a general three-dimensional sensitivity of the interface design is derived. It also has the nine interfaces, each sensitivity of which are arranged in Sect. 4.2.5. The energy sensitivity is examined for the sign dependence on the source condition in the electric-circuit point of view, and then the inductance sensitivity is derived using the energy sensitivity.

At the ends of Sects. 4.1 and 4.2, the general sensitivity formulas are applied to design examples to be validated. The design examples are divided into two kinds. One is one-dimensional simple examples with the analytic solution. The other is two-dimensional numerical examples that do not have the analytic solution. In the former examples, the results of the analytical sensitivity are compared with the results by the sensitivity formulas in order to show the correctness of the derived sensitivity formula. In the latter examples, the sensitivity evaluated using the finite element method is applied to optimal shape design problems in order to show that the shape optimization method using the sensitivity formulas works well for the magnetostatic systems.

4.1 Interface Shape Sensitivity

The magnetostatic system is one of the open boundary problems. Since the boundary conditions (homogeneous Dirichlet and homogeneous Neumann condition) in the magnetostatic system are applied to the symmetry surface or the infinite boundary, its outer boundaries are not taken as the design variable. The magnetostatic system, therefore, has only the interface design problem [1, 2]. The shape variation of the interface results in the variation in the magnetic field distribution in the domain. In this section, the general three-dimensional sensitivity for the interface variation in the linear magnetostatic system is derived. First, the magnetostatic system for the interface design is depicted and a general objective function is defined as a domain integral. Second, the Lagrange multiplier method is introduced to handle the equality constraint of the variational state equation. Third, the adjoint variable method is used to explicitly express the sensitivity in terms of design variation. Fourth, the variational identities are used to transform the domain integral of the sensitivity into the interface integral, which provides the general three-dimensional sensitivity formula for the interface design. Finally, the obtained sensitivity formula is tested and validated with analytical and numerical examples [2].

4.1.1 Problem Definition and Objective Function

A linear magnetostatic system for the interface design is given as in Fig. 4.1, where the whole domain Ω comprises two domains Ω_1 and Ω_2 divided by an interface γ. The domain Ω_1 has a distribution of ν_1, $\mathbf{J}_1$ and $\mathbf{M}_{o1}$ and the domain Ω_2 has a

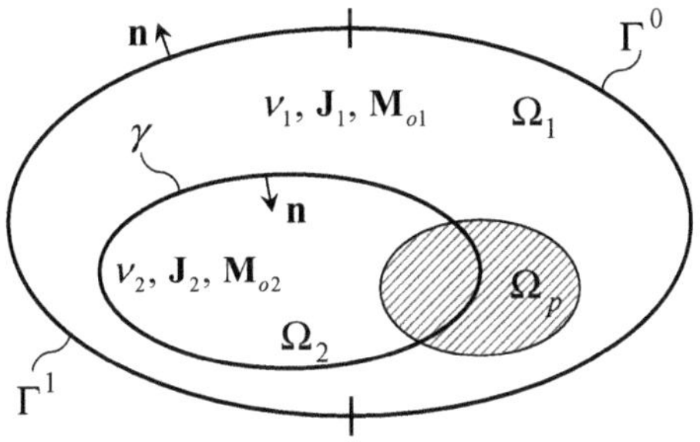

Fig. 4.1 Interface design of linear magnetostatic system

distribution of ν_2, $\mathbf{J}_2$ and $\mathbf{M}_{o2}$. The domain Ω_1 has the outer boundary Γ where **n** is defined as the outward normal vector on the outer boundary. The outer boundary consists of the Dirichlet boundary Γ^0 and the Neumann boundary Γ^1. In this shape sensitivity analysis, the interface is taken as design variable.

A general objective function is defined in integral form as

$$F = \int_{\Omega} g(\mathbf{A}, \mathbf{B}(\mathbf{A}))m_p \mathrm{d}\Omega$$
$$= \int_{\Omega_1} g_1 m_p \mathrm{d}\Omega + \int_{\Omega_2} g_2 m_p \mathrm{d}\Omega \tag{4.1.1}$$

where g_1 and g_2 are any functions that are continuously differentiable with respect to their arguments;

$$g_1 \equiv g(\mathbf{A}_1, \mathbf{B}(\mathbf{A}_1)) \quad \text{and} \quad g_2 \equiv g(\mathbf{A}_2, \mathbf{B}(\mathbf{A}_2)) \tag{4.1.2}$$

and m_p is a characteristic function that is defined as

$$m_p = \begin{cases} 1 & \mathbf{x} \in \Omega_p \\ 0 & \mathbf{x} \notin \Omega_p \end{cases} \tag{4.1.3}$$

The region Ω_p, the integral domain for the objective function, can include the interface as shown in Fig. 4.1. Since g_1 and g_2 in (4.1.1) are any function of magnetic vector potential **A** and magnetic flux density **B**, the objective function can represent a wide range of design objectives.

The governing differential equations for the state variables of the magnetic vector potential $\mathbf{A}_1$ and $\mathbf{A}_2$ are given as

$$\nabla \times \nu_1 \nabla \times \mathbf{A}_1 = \mathbf{J}_1 + \nabla \times \mathbf{M}_{o1} \quad \text{in } \Omega_1 \tag{4.1.4}$$

$$\nabla \times \nu_2 \nabla \times \mathbf{A}_2 = \mathbf{J}_2 + \nabla \times \mathbf{M}_{o2} \quad \text{in } \Omega_2 \tag{4.1.5}$$

where $\nabla \times \mathbf{M}_{o1} = \mathbf{J}_{m1}$, $\nabla \times \mathbf{M}_{o2} = \mathbf{J}_{m2}$, and $\mathbf{J}_{m1}, \mathbf{J}_{m2}$ are the magnetization current densities of permanent magnetization of $\mathbf{M}_{o1}$ and $\mathbf{M}_{o2}$, respectively. These equations have a unique solution with the boundary conditions:

$$\mathbf{A}_1 = \mathbf{0} \quad \text{on } \Gamma^0 \quad \text{(homogeneous Dirichlet boundary condition)} \tag{4.1.6}$$

$$\frac{\partial \mathbf{A}_1}{\partial n} = \mathbf{0} \quad \text{on } \Gamma^1 \quad \text{(homogeneous Neumann boundary condition)} \tag{4.1.7}$$

The variational identities for the state equations are obtained by multiplying both sides of (4.1.4) and (4.1.5) by arbitrary virtual vector potentials $\bar{\mathbf{A}}_1$ and $\bar{\mathbf{A}}_2$, and by using the vector identity and the divergence theorem;

$$\begin{aligned}&\int_{\Omega_1} \left(v_1 \mathbf{B}(\mathbf{A}_1) \cdot \mathbf{B}(\bar{\mathbf{A}}_1) - \mathbf{J}_1 \cdot \bar{\mathbf{A}}_1 - \mathbf{M}_{o1} \cdot \mathbf{B}(\bar{\mathbf{A}}_1)\right) d\Omega \\ &= -\int_{\gamma+\Gamma} (\mathbf{n} \times \mathbf{H}(\mathbf{A}_1)) \cdot \bar{\mathbf{A}}_1 d\Gamma \quad \forall \bar{\mathbf{A}}_1 \in \Phi\end{aligned} \tag{4.1.8}$$

$$\begin{aligned}&\int_{\Omega_2} \left(v_2 \mathbf{B}(\mathbf{A}_2) \cdot \mathbf{B}(\bar{\mathbf{A}}_2) - \mathbf{J}_2 \cdot \bar{\mathbf{A}}_2 - \mathbf{M}_{o2} \cdot \mathbf{B}(\bar{\mathbf{A}}_2)\right) d\Omega \\ &= \int_{\gamma} (\mathbf{n} \times \mathbf{H}(\mathbf{A}_2)) \cdot \bar{\mathbf{A}}_2 d\Gamma \quad \forall \bar{\mathbf{A}}_2 \in \Phi\end{aligned} \tag{4.1.9}$$

where

$$\mathbf{H}(\mathbf{A}_1) = v_1 \mathbf{B}(\mathbf{A}_1) - \mathbf{M}_{o1} \tag{4.1.10}$$

$$\mathbf{H}(\mathbf{A}_2) = v_2 \mathbf{B}(\mathbf{A}_2) - \mathbf{M}_{o2} \tag{4.1.11}$$

and Φ is the space of admissible vector potential defined in Sect. 2.2.2;

$$\Phi = \left\{ \bar{\mathbf{A}} \in \left[H^1(\Omega)\right]^3 | \bar{\mathbf{A}} = \mathbf{0} \quad \text{on} \quad \mathbf{x} \in \Gamma^0 \right\} \tag{4.1.12}$$

The boundary conditions of (4.1.6) and (4.1.7) can be rewritten for the variational equation;

$$\bar{\mathbf{A}}_1 = \mathbf{0} \quad \text{on} \quad \Gamma^0 \quad \text{(homogeneous Dirichlet boundary condition)} \tag{4.1.13}$$

$$\mathbf{n} \times \mathbf{H}(\mathbf{A}_1) = \mathbf{0} \quad \text{on} \quad \Gamma^1 \quad \text{(homogeneous Neumann boundary condition)} \tag{4.1.14}$$

and the interface condition is

$$\mathbf{n} \times \mathbf{H}(\mathbf{A}_1) = \mathbf{n} \times \mathbf{H}(\mathbf{A}_2) \quad \text{on} \quad \gamma \quad \text{(interface condition)} \tag{4.1.15}$$

By summing (4.1.8) and (4.1.9) and imposing the boundary conditions and the interface condition, the variational state equation reduced from the differential Eqs. (4.1.4) and (4.1.5) is obtained as

$$\begin{aligned}
&\int_{\Omega_1} \nu_1 \mathbf{B}(\mathbf{A}_1) \cdot \mathbf{B}(\bar{\mathbf{A}}_1) d\Omega + \int_{\Omega_2} \nu_2 \mathbf{B}(\mathbf{A}_2) \cdot \mathbf{B}(\bar{\mathbf{A}}_2) d\Omega \\
&= \int_{\Omega_1} (\mathbf{J}_1 \cdot \bar{\mathbf{A}}_1 + \mathbf{M}_{o1} \cdot \mathbf{B}(\bar{\mathbf{A}}_1)) d\Omega \\
&+ \int_{\Omega_2} (\mathbf{J}_2 \cdot \bar{\mathbf{A}}_2 + \mathbf{M}_{o2} \cdot \mathbf{B}(\bar{\mathbf{A}}_2)) d\Omega \quad \forall \bar{\mathbf{A}}_1, \bar{\mathbf{A}}_2 \in \Phi
\end{aligned} \tag{4.1.16}$$

The energy bilinear form $a(\mathbf{A}, \bar{\mathbf{A}})$ and the source linear form $l(\bar{\mathbf{A}})$ are defined as

$$a(\mathbf{A}, \bar{\mathbf{A}}) \equiv \int_{\Omega_1} \nu_1 \mathbf{B}(\mathbf{A}_1) \cdot \mathbf{B}(\bar{\mathbf{A}}_1) d\Omega + \int_{\Omega_2} \nu_2 \mathbf{B}(\mathbf{A}_2) \cdot \mathbf{B}(\bar{\mathbf{A}}_2) d\Omega \tag{4.1.17}$$

$$l(\bar{\mathbf{A}}) \equiv \int_{\Omega_1} (\mathbf{J}_1 \cdot \bar{\mathbf{A}}_1 + \mathbf{M}_{o1} \cdot \mathbf{B}(\bar{\mathbf{A}}_1)) d\Omega + \int_{\Omega_2} (\mathbf{J}_2 \cdot \bar{\mathbf{A}}_2 + \mathbf{M}_{o2} \cdot \mathbf{B}(\bar{\mathbf{A}}_2)) d\Omega \tag{4.1.18}$$

The variational Eq. (4.1.16) is rewritten with the energy bilinear form and the source linear form as

$$a(\mathbf{A}, \bar{\mathbf{A}}) = l(\bar{\mathbf{A}}) \quad \forall \bar{\mathbf{A}} \in \Phi \tag{4.1.19}$$

where

$$\mathbf{A} = \mathbf{A}_1 \cup \mathbf{A}_2 \quad \text{and} \quad \bar{\mathbf{A}} = \bar{\mathbf{A}}_1 \cup \bar{\mathbf{A}}_2 \tag{4.1.20}$$

4.1.2 Lagrange Multiplier Method for Sensitivity Derivation

The Lagrange multiplier method is introduced for the implicit equality constraint of the variational state equation. The objective function F depends on both the integral domain Ω_p and the state variable $\mathbf{A}$. While its dependency on the integral domain is explicit, the dependency on the state variable $\mathbf{A}$ is implicit since it is expressed in the state equation (4.1.19). That is, when the domain Ω is perturbed by the boundary shape, the state variable $\mathbf{A}$ in the whole domain Ω is also perturbed through the state equation.

The method of Lagrange multipliers, which is employed for the implicit constraint of the variational state equation, provides an augmented objective function G;

$$G = F + l(\bar{\mathbf{A}}) - a(\mathbf{A}, \bar{\mathbf{A}}) \quad \forall \bar{\mathbf{A}} \in \Phi \tag{4.1.21}$$

where the arbitrary virtual potential $\bar{\mathbf{A}}$ plays the role of Lagrange multipliers and Φ is the space of admissible vector potential: $\Phi = \left\{ \bar{\mathbf{A}} \in [H^1(\Omega)]^3 \middle| \bar{\mathbf{A}} = \mathbf{0} \quad \text{on} \quad \mathbf{x} \in \Gamma^0 \right\}$.

The sensitivity, the material derivative of the augmented objective function , is written as

$$\dot{G} = -\dot{a}(\mathbf{A}, \bar{\mathbf{A}}) + \dot{l}(\bar{\mathbf{A}}) + \dot{F} \quad \forall \bar{\mathbf{A}} \in \Phi \tag{4.1.22}$$

The differentiability of the bilinear form $a(\mathbf{A}, \bar{\mathbf{A}})$ and the source linear form $l(\bar{\mathbf{A}})$, which was proved in [3], is used here to derive the sensitivity. By applying the material derivative formula (3.1.22) to the variational state equation Eq. (4.1.16) and the objective function (4.4.1) and using the relation (3.1.17), each term in (4.1.22) is obtained below as (4.1.23), (4.1.24), and (4.1.25). In this interface sensitivity problem, when the material derivative formula is applied, only the integrals on the interface remain since the outer boundary is not taken as design variable ($V_n = 0$ on Γ).

$$\begin{aligned}
\dot{a}(\mathbf{A}, \bar{\mathbf{A}}) &= \int_{\Omega_1} \left(v_1 \mathbf{B}(\mathbf{A}_1') \cdot \mathbf{B}(\bar{\mathbf{A}}_1) + v_1 \mathbf{B}(\mathbf{A}_1) \cdot \mathbf{B}(\bar{\mathbf{A}}_1') \right) \mathrm{d}\Omega + \int_{\gamma} v_1 \mathbf{B}(\mathbf{A}_1) \cdot \mathbf{B}(\bar{\mathbf{A}}_1) V_n \mathrm{d}\Gamma \\
&\quad + \int_{\Omega_2} \left(v_2 \mathbf{B}(\mathbf{A}_2') \cdot \mathbf{B}(\bar{\mathbf{A}}_2) + v_2 \mathbf{B}(\mathbf{A}_2) \cdot \mathbf{B}(\bar{\mathbf{A}}_2') \right) \mathrm{d}\Omega - \int_{\gamma} v_2 \mathbf{B}(\mathbf{A}_2) \cdot \mathbf{B}(\bar{\mathbf{A}}_2) V_n \mathrm{d}\Gamma \\
&= \int_{\Omega_1} \left(v_1 \mathbf{B}(\dot{\mathbf{A}}_1) \cdot \mathbf{B}(\bar{\mathbf{A}}_1) - v_1 \mathbf{B}(\mathbf{V} \cdot \nabla \mathbf{A}_1) \cdot \mathbf{B}(\bar{\mathbf{A}}_1) + v_1 \mathbf{B}(\mathbf{A}_1) \cdot \mathbf{B}(\dot{\bar{\mathbf{A}}}_1) - v_1 \mathbf{B}(\mathbf{A}_1) \cdot \mathbf{B}(\mathbf{V} \cdot \nabla \mathbf{A}_1) \right) \mathrm{d}\Omega \\
&\quad + \int_{\Omega_2} \left(v_2 \mathbf{B}(\dot{\mathbf{A}}_2) \cdot \mathbf{B}(\bar{\mathbf{A}}_2) - v_2 \mathbf{B}(\mathbf{V} \cdot \nabla \mathbf{A}_2) \cdot \mathbf{B}(\bar{\mathbf{A}}_2) + v_2 \mathbf{B}(\mathbf{A}_2) \cdot \mathbf{B}(\dot{\bar{\mathbf{A}}}_2) - v_2 \mathbf{B}(\mathbf{A}_2) \cdot \mathbf{B}(\mathbf{V} \cdot \nabla \mathbf{A}_2) \right) \mathrm{d}\Omega \\
&\quad + \int_{\gamma} \left(v_1 \mathbf{B}(\mathbf{A}_1) \cdot \mathbf{B}(\bar{\mathbf{A}}_1) - v_2 \mathbf{B}(\mathbf{A}_2) \cdot \mathbf{B}(\bar{\mathbf{A}}_2) \right) V_n \mathrm{d}\Gamma
\end{aligned} \tag{4.1.23}$$

$$\begin{aligned}
\dot{l}(\bar{\mathbf{A}}) &= \int_{\Omega_1} \left(\mathbf{J}_1 \cdot \bar{\mathbf{A}}_1' + \mathbf{M}_{o1} \cdot \mathbf{B}(\bar{\mathbf{A}}_1')\right)\mathrm{d}\Omega + \int_{\gamma} (\mathbf{J}_1 \cdot \bar{\mathbf{A}}_1 + \mathbf{M}_{o1} \cdot \mathbf{B}(\bar{\mathbf{A}}_1))V_n \mathrm{d}\Gamma \\
&+ \int_{\Omega_2} \left(\mathbf{J}_2 \cdot \bar{\mathbf{A}}_2' + \mathbf{M}_{o2} \cdot \mathbf{B}(\bar{\mathbf{A}}_2')\right)\mathrm{d}\Omega - \int_{\gamma} (\mathbf{J}_2 \cdot \bar{\mathbf{A}}_2 + \mathbf{M}_{o2} \cdot \mathbf{B}(\bar{\mathbf{A}}_2))V_n \mathrm{d}\Gamma \\
&= \int_{\Omega_1} \left[\mathbf{J}_1 \cdot \dot{\bar{\mathbf{A}}}_1 - \mathbf{J}_1 \cdot (\mathbf{V} \cdot \nabla\bar{\mathbf{A}}_1) + \mathbf{M}_{o1} \cdot \mathbf{B}(\dot{\bar{\mathbf{A}}}_1) - \mathbf{M}_{o1} \cdot \mathbf{B}(\mathbf{V} \cdot \nabla\bar{\mathbf{A}}_1)\right]\mathrm{d}\Omega \\
&+ \int_{\Omega_2} \left[\mathbf{J}_2 \cdot \dot{\bar{\mathbf{A}}}_2 - \mathbf{J}_2 \cdot (\mathbf{V} \cdot \nabla\bar{\mathbf{A}}_2) + \mathbf{M}_{o2} \cdot \mathbf{B}(\dot{\bar{\mathbf{A}}}_2) - \mathbf{M}_{o2} \cdot \mathbf{B}(\mathbf{V} \cdot \nabla\bar{\mathbf{A}}_2)\right]\mathrm{d}\Omega \\
&+ \int_{\gamma} (\mathbf{J}_1 \cdot \bar{\mathbf{A}}_1 - \mathbf{J}_2 \cdot \bar{\mathbf{A}}_2 + \mathbf{M}_{o1} \cdot \mathbf{B}(\bar{\mathbf{A}}_1) - \mathbf{M}_{o2} \cdot \mathbf{B}(\bar{\mathbf{A}}_2))V_n \mathrm{d}\Gamma
\end{aligned} \tag{4.1.24}$$

$$\begin{aligned}
\dot{F} &= \int_{\Omega_1} g_1' m_p \mathrm{d}\Omega + \int_{\gamma} g_1 m_p V_n \mathrm{d}\Gamma + \int_{\Omega_2} g_2' m_p \mathrm{d}\Omega - \int_{\gamma} g_2 m_p V_n \mathrm{d}\Gamma \\
&= \int_{\Omega_1} (\mathbf{g}_{\mathbf{A}_1} \cdot \mathbf{A}'_1 + \mathbf{g}_{\mathbf{B}_1} \cdot \mathbf{B}(\mathbf{A}'_1))m_p \mathrm{d}\Omega + \int_{\gamma} g_1 m_p V_n \mathrm{d}\Gamma \\
&+ \int_{\Omega_2} (\mathbf{g}_{\mathbf{A}_2} \cdot \mathbf{A}'_2 + \mathbf{g}_{\mathbf{B}_2} \cdot \mathbf{B}(\mathbf{A}'_2))m_p \mathrm{d}\Omega - \int_{\gamma} g_2 m_p V_n \mathrm{d}\Gamma \\
&= \int_{\Omega_1} [\mathbf{g}_{\mathbf{A}_1} \cdot \dot{\mathbf{A}}_1 - \mathbf{g}_{\mathbf{A}_1} \cdot (\mathbf{V} \cdot \nabla\mathbf{A}_1) + \mathbf{g}_{\mathbf{B}_1} \cdot \mathbf{B}(\dot{\mathbf{A}}_1) - \mathbf{g}_{\mathbf{B}_1} \cdot \mathbf{B}(\mathbf{V} \cdot \nabla\mathbf{A}_1)]m_p \mathrm{d}\Omega \\
&+ \int_{\Omega_2} [\mathbf{g}_{\mathbf{A}_2} \cdot \dot{\mathbf{A}}_2 - \mathbf{g}_{\mathbf{A}_2} \cdot (\mathbf{V} \cdot \nabla\mathbf{A}_2) + \mathbf{g}_{\mathbf{B}_2} \cdot \mathbf{B}(\dot{\mathbf{A}}_2) - \mathbf{g}_{\mathbf{B}_2} \cdot \mathbf{B}(\mathbf{V} \cdot \nabla\mathbf{A}_2)]m_p \mathrm{d}\Omega \\
&+ \int_{\gamma} (g_1 - g_2)m_p V_n \mathrm{d}\Gamma
\end{aligned} \tag{4.1.25}$$

For g_1' and g_2' in (4.1.25), we used the relation (3.1.27):

$$g' = \frac{\partial g}{\partial \mathbf{A}} \cdot \mathbf{A}' + \frac{\partial g}{\partial \mathrm{B}} \cdot \mathbf{B}(\mathbf{A}') = \mathbf{g}_{\mathbf{A}} \cdot \mathbf{A}' + \mathbf{g}_{\mathbf{B}} \cdot \mathbf{B}(\mathbf{A}') \tag{4.1.26}$$

where

$$\mathbf{g}_{\mathbf{A}} \equiv \frac{\partial g}{\partial \mathbf{A}} = \left[\frac{\partial g}{\partial A_x}, \frac{\partial g}{\partial A_y}, \frac{\partial g}{\partial A_z}\right]^T \text{ and } \mathbf{g}_{\mathbf{B}} \equiv \frac{\partial g}{\partial \mathbf{B}} = \left[\frac{\partial g}{\partial B_x}, \frac{\partial g}{\partial B_y}, \frac{\partial g}{\partial B_z}\right]^T \tag{4.1.27}$$

In derivation of (4.1.23) and (4.1.24), $v_1', v_2' = 0$, $\mathbf{J}_1', \mathbf{J}_2' = \mathbf{0}$ and $\mathbf{M}_{o1}', \mathbf{M}_{o2}' = \mathbf{0}$ were used. (4.1.23), (4.1.24) and (4.1.25) are inserted into (4.1.22) to provide

$$
\begin{aligned}
\dot{G} = & -\int_{\Omega_1} \nu_1 \mathbf{B}(\dot{\mathbf{A}}_1) \cdot \mathbf{B}(\bar{\mathbf{A}}_1) \mathrm{d}\Omega - \int_{\Omega_2} \nu_2 \mathbf{B}(\dot{\mathbf{A}}_2) \cdot \mathbf{B}(\bar{\mathbf{A}}_2) \mathrm{d}\Omega \\
& - \int_{\Omega_1} \nu_1 \mathbf{B}(\mathbf{A}_1) \cdot \mathbf{B}(\dot{\bar{\mathbf{A}}}_1) \mathrm{d}\Omega - \int_{\Omega_2} \nu_2 \mathbf{B}(\mathbf{A}_2) \cdot \mathbf{B}(\dot{\bar{\mathbf{A}}}_2) \mathrm{d}\Omega \\
& + \int_{\Omega_1} \left(\mathbf{J}_1 \cdot \dot{\bar{\mathbf{A}}}_1 + \mathbf{M}_{o1} \cdot \mathbf{B}(\dot{\bar{\mathbf{A}}}_1) \right) \mathrm{d}\Omega + \int_{\Omega_2} \left(\mathbf{J}_2 \cdot \dot{\bar{\mathbf{A}}}_2 + \mathbf{M}_{o2} \cdot \mathbf{B}(\dot{\bar{\mathbf{A}}}_2) \right) \mathrm{d}\Omega \\
& + \int_{\Omega_1} (\mathbf{g}_{\mathbf{A}_1} \cdot \dot{\mathbf{A}}_1 + \mathbf{g}_{\mathbf{B}_1} \cdot \mathbf{B}(\dot{\mathbf{A}}_1)) m_p \mathrm{d}\Omega + \int_{\Omega_2} (\mathbf{g}_{\mathbf{A}_2} \cdot \dot{\mathbf{A}}_2 + \mathbf{g}_{\mathbf{B}_2} \cdot \mathbf{B}(\dot{\mathbf{A}}_2)) m_p \mathrm{d}\Omega \\
& + \int_{\Omega_1} \nu_1 \mathbf{B}(\mathbf{V} \cdot \nabla \mathbf{A}_1) \cdot \mathbf{B}(\bar{\mathbf{A}}_1) \mathrm{d}\Omega + \int_{\Omega_2} \nu_2 \mathbf{B}(\mathbf{V} \cdot \nabla \mathbf{A}_2) \cdot \mathbf{B}(\bar{\mathbf{A}}_2) \mathrm{d}\Omega \\
& + \int_{\Omega_1} \nu_1 \mathbf{B}(\mathbf{A}_1) \cdot \mathbf{B}(\mathbf{V} \cdot \nabla \bar{\mathbf{A}}_1) \mathrm{d}\Omega + \int_{\Omega_2} \nu_2 \mathbf{B}(\mathbf{A}_2) \cdot \mathbf{B}(\mathbf{V} \cdot \nabla \bar{\mathbf{A}}_2) \mathrm{d}\Omega \\
& - \int_{\Omega_1} \left[\mathbf{J}_1 \cdot (\mathbf{V} \cdot \nabla \bar{\mathbf{A}}_1) + \mathbf{M}_{o1} \cdot \mathbf{B}(\mathbf{V} \cdot \nabla \bar{\mathbf{A}}_1) \right] \mathrm{d}\Omega \\
& - \int_{\Omega_2} \left[\mathbf{J}_2 \cdot (\mathbf{V} \cdot \nabla \bar{\mathbf{A}}_2) + \mathbf{M}_{o2} \cdot \mathbf{B}(\mathbf{V} \cdot \nabla \bar{\mathbf{A}}_2) \right] \mathrm{d}\Omega \\
& - \int_{\Omega_1} \left[\mathbf{g}_{\mathbf{A}_1} \cdot (\mathbf{V} \cdot \nabla \mathbf{A}_1) + \mathbf{g}_{\mathbf{B}_1} \cdot \mathbf{B}(\mathbf{V} \cdot \nabla \mathbf{A}_1) \right] m_p \mathrm{d}\Omega \\
& - \int_{\Omega_2} \left[\mathbf{g}_{\mathbf{A}_2} \cdot (\mathbf{V} \cdot \nabla \mathbf{A}_2) + \mathbf{g}_{\mathbf{B}_2} \cdot \mathbf{B}(\mathbf{V} \cdot \nabla \mathbf{A}_2) \right] m_p \mathrm{d}\Omega \\
& - \int_{\gamma} \left(\nu_1 \mathbf{B}(\mathbf{A}_1) \cdot \mathbf{B}(\bar{\mathbf{A}}_1) - \nu_2 \mathbf{B}(\mathbf{A}_2) \cdot \mathbf{B}(\bar{\mathbf{A}}_2) \right) V_n \mathrm{d}\Gamma \\
& + \int_{\gamma} \left(\mathbf{M}_{o1} \cdot \mathbf{B}(\bar{\mathbf{A}}_1) - \mathbf{M}_{o2} \cdot \mathbf{B}(\bar{\mathbf{A}}_2) \right) V_n \mathrm{d}\Gamma \\
& + \int_{\gamma} \left[\mathbf{J}_1 \cdot \bar{\mathbf{A}}_1 - \mathbf{J}_2 \cdot \bar{\mathbf{A}}_2 + (g_1 - g_2) m_p \right] V_n \mathrm{d}\Gamma \quad \forall \bar{\mathbf{A}}_1, \bar{\mathbf{A}}_2 \in \Phi
\end{aligned}
\tag{4.1.28}
$$

Since $\dot{\bar{\mathbf{A}}}_1$ and $\dot{\bar{\mathbf{A}}}_2$ belong to Φ, the variational state equation of (4.1.16) provides

$$\begin{aligned}
&\int_{\Omega_1} \nu_1 \mathbf{B}(\mathbf{A}_1)\cdot\mathbf{B}(\dot{\bar{\mathbf{A}}}_1)\mathrm{d}\Omega + \int_{\Omega_2} \nu_2 \mathbf{B}(\mathbf{A}_2)\cdot\mathbf{B}(\dot{\bar{\mathbf{A}}}_2)\mathrm{d}\Omega \\
&= \int_{\Omega_1} \left(\mathbf{J}_1\cdot\dot{\bar{\mathbf{A}}}_1 + \mathbf{M}_{o1}\cdot\mathbf{B}(\dot{\bar{\mathbf{A}}}_1)\right)\mathrm{d}\Omega + \int_{\Omega_2} \left(\mathbf{J}_2\cdot\dot{\bar{\mathbf{A}}}_2 + \mathbf{M}_{o2}\cdot\mathbf{B}(\dot{\bar{\mathbf{A}}}_2)\right)\mathrm{d}\Omega
\end{aligned} \tag{4.1.29}$$

Inserting (4.1.29) into (4.1.28) canceled out all terms with $\dot{\bar{\mathbf{A}}}_1$ and $\dot{\bar{\mathbf{A}}}_2$;

$$\begin{aligned}
\dot{G} = &-\int_{\Omega_1} \nu_1 \mathbf{B}(\dot{\mathbf{A}}_1)\cdot\mathbf{B}(\bar{\mathbf{A}}_1)\mathrm{d}\Omega - \int_{\Omega_2} \nu_2 \mathbf{B}(\dot{\mathbf{A}}_2)\cdot\mathbf{B}(\bar{\mathbf{A}}_2)\mathrm{d}\Omega \\
&+ \int_{\Omega_1} (\mathbf{g}_{\mathbf{A}_1}\cdot\dot{\mathbf{A}}_1 + \mathbf{g}_{\mathbf{B}_1}\cdot\mathbf{B}(\dot{\mathbf{A}}_1))m_p\mathrm{d}\Omega + \int_{\Omega_2} (\mathbf{g}_{\mathbf{A}_2}\cdot\dot{\mathbf{A}}_2 + \mathbf{g}_{\mathbf{B}_2}\cdot\mathbf{B}(\dot{\mathbf{A}}_2))m_p\mathrm{d}\Omega \\
&+ \int_{\Omega_1} \nu_1 \mathbf{B}(\mathbf{V}\cdot\nabla\mathbf{A}_1)\cdot\mathbf{B}(\bar{\mathbf{A}}_1)\mathrm{d}\Omega + \int_{\Omega_2} \nu_2 \mathbf{B}(\mathbf{V}\cdot\nabla\mathbf{A}_2)\cdot\mathbf{B}(\bar{\mathbf{A}}_2)\mathrm{d}\Omega \\
&+ \int_{\Omega_1} \nu_1 \mathbf{B}(\mathbf{A}_1)\cdot\mathbf{B}(\mathbf{V}\cdot\nabla\bar{\mathbf{A}}_1)\mathrm{d}\Omega + \int_{\Omega_2} \nu_2 \mathbf{B}(\mathbf{A}_2)\cdot\mathbf{B}(\mathbf{V}\cdot\nabla\bar{\mathbf{A}}_2)\mathrm{d}\Omega \\
&- \int_{\Omega_1} \left[\mathbf{J}_1\cdot(\mathbf{V}\cdot\nabla\bar{\mathbf{A}}_1) + \mathbf{M}_{o1}\cdot\mathbf{B}(\mathbf{V}\cdot\nabla\bar{\mathbf{A}}_1)\right]\mathrm{d}\Omega \\
&- \int_{\Omega_2} \left[\mathbf{J}_2\cdot(\mathbf{V}\cdot\nabla\bar{\mathbf{A}}_2) + \mathbf{M}_{o2}\cdot\mathbf{B}(\mathbf{V}\cdot\nabla\bar{\mathbf{A}}_2)\right]\mathrm{d}\Omega \\
&- \int_{\Omega_1} \left[\mathbf{g}_{\mathbf{A}_1}\cdot(\mathbf{V}\cdot\nabla\mathbf{A}_1) + \mathbf{g}_{\mathbf{B}_1}\cdot\mathbf{B}(\mathbf{V}\cdot\nabla\mathbf{A}_1)\right]m_p\mathrm{d}\Omega \\
&- \int_{\Omega_2} \left[\mathbf{g}_{\mathbf{A}_2}\cdot(\mathbf{V}\cdot\nabla\mathbf{A}_2) + \mathbf{g}_{\mathbf{B}_2}\cdot\mathbf{B}(\mathbf{V}\cdot\nabla\mathbf{A}_2)\right]m_p\mathrm{d}\Omega \\
&- \int_{\gamma} \left(\nu_1 \mathbf{B}(\mathbf{A}_1)\cdot\mathbf{B}(\bar{\mathbf{A}}_1) - \nu_2 \mathbf{B}(\mathbf{A}_2)\cdot\mathbf{B}(\bar{\mathbf{A}}_2)\right)V_n\mathrm{d}\Gamma \\
&+ \int_{\gamma} \left(\mathbf{M}_{o1}\cdot\mathbf{B}(\bar{\mathbf{A}}_1) - \mathbf{M}_{o2}\cdot\mathbf{B}(\bar{\mathbf{A}}_2)\right)V_n\mathrm{d}\Gamma \\
&+ \int_{\gamma} \left[\mathbf{J}_1\cdot\bar{\mathbf{A}}_1 - \mathbf{J}_2\cdot\bar{\mathbf{A}}_2 + (g_1 - g_2)m_p\right]V_n\mathrm{d}\Gamma \quad \forall\bar{\mathbf{A}}_1,\bar{\mathbf{A}}_2\in\Phi
\end{aligned} \tag{4.1.30}$$

4.1.3 Adjoint Variable Method for Sensitivity Analysis

An adjoint equation is introduced to avoid calculation of the term of $\dot{\mathbf{A}}$ in the sensitivity (4.1.30) and to obtain an explicit expression of (4.1.30) in terms of the velocity field $\mathbf{V}$. The adjoint equation is obtained by replacing $\dot{\mathbf{A}}_1$ and $\dot{\mathbf{A}}_2$ in the g-related terms of (4.1.30) with a virtual potential $\bar{\boldsymbol{\lambda}}_1$ and $\bar{\boldsymbol{\lambda}}_2$, respectively, and by equating the integrals to the energy bilinear form $a(\boldsymbol{\lambda}, \bar{\boldsymbol{\lambda}})$. The adjoint equation so obtained is written as

$$\int_{\Omega_1} \nu_1 \mathbf{B}(\boldsymbol{\lambda}_1) \cdot \mathbf{B}(\bar{\lambda}_1) d\Omega + \int_{\Omega_2} \nu_2 \mathbf{B}(\boldsymbol{\lambda}_2) \cdot \mathbf{B}(\bar{\lambda}_2) d\Omega$$
$$= \int_{\Omega_1} \left(\mathbf{g}_{\mathbf{A}_1} \cdot \bar{\lambda}_1 + \mathbf{g}_{\mathbf{B}_1} \cdot \mathbf{B}(\bar{\lambda}_1)\right) m_p d\Omega$$
$$+ \int_{\Omega_2} \left(\mathbf{g}_{\mathbf{A}_2} \cdot \bar{\lambda}_2 + \mathbf{g}_{\mathbf{B}_2} \cdot \mathbf{B}(\bar{\lambda}_2)\right) m_p d\Omega \quad \forall \bar{\lambda}_1, \bar{\lambda}_2 \in \Phi. \tag{4.1.31}$$

where $\boldsymbol{\lambda}_1$ and $\boldsymbol{\lambda}_2$ are the adjoint variables and its solution is desired, and Φ is the space of admissible vector potential: $\Phi = \left\{\bar{\boldsymbol{\lambda}} \in [H^1(\Omega)]^3 \middle| \bar{\boldsymbol{\lambda}} = \mathbf{0} \quad \text{on} \quad \mathbf{x} \in \Gamma^0\right\}$.

To take advantage of the adjoint equation, (4.1.31) is evaluated at the specific $\bar{\boldsymbol{\lambda}}_1 = \dot{\mathbf{A}}_1$ and $\bar{\boldsymbol{\lambda}}_2 = \dot{\mathbf{A}}_2$ since (4.1.31) holds for all $\bar{\boldsymbol{\lambda}}_1, \bar{\boldsymbol{\lambda}}_2 \in \Phi$, to yield

$$\int_{\Omega_1} \nu_1 \mathbf{B}(\boldsymbol{\lambda}_1) \cdot \mathbf{B}(\dot{\mathbf{A}}_1) d\Omega + \int_{\Omega_2} \nu_2 \mathbf{B}(\boldsymbol{\lambda}_2) \cdot \mathbf{B}(\dot{\mathbf{A}}_2) d\Omega$$
$$= \int_{\Omega_1} \left(\mathbf{g}_{\mathbf{A}_1} \cdot \dot{\mathbf{A}}_1 + \mathbf{g}_{\mathbf{B}_1} \cdot \mathbf{B}(\dot{\mathbf{A}}_1)\right) m_p d\Omega + \int_{\Omega_2} \left(\mathbf{g}_{\mathbf{A}_2} \cdot \dot{\mathbf{A}}_2 + \mathbf{g}_{\mathbf{B}_2} \cdot \mathbf{B}(\dot{\mathbf{A}}_2)\right) m_p d\Omega \tag{4.1.32}$$

Similarly, the sensitivity Eq. (4.1.30) is evaluated at the specific $\bar{\mathbf{A}}_1 = \boldsymbol{\lambda}_1$, $\bar{\mathbf{A}}_2 = \boldsymbol{\lambda}_2$ since the $\boldsymbol{\lambda}_1$ and $\boldsymbol{\lambda}_2$ belong to the admissible space Φ, to yield

$$
\begin{aligned}
\dot{G} = & -\int_{\Omega_1} \nu_1 \mathbf{B}(\dot{\mathbf{A}}_1) \cdot \mathbf{B}(\boldsymbol{\lambda}_1) \mathrm{d}\Omega - \int_{\Omega_2} \nu_2 \mathbf{B}(\dot{\mathbf{A}}_2) \cdot \mathbf{B}(\boldsymbol{\lambda}_2) \mathrm{d}\Omega \\
& + \int_{\Omega_1} \left(\mathbf{g}_{\mathbf{A}_1} \cdot \dot{\mathbf{A}}_1 + \mathbf{g}_{\mathbf{B}_1} \cdot \mathbf{B}(\dot{\mathbf{A}}_1)\right) m_p \mathrm{d}\Omega + \int_{\Omega_2} \left(\mathbf{g}_{\mathbf{A}_2} \cdot \dot{\mathbf{A}}_2 + \mathbf{g}_{\mathbf{B}_2} \cdot \mathbf{B}(\dot{\mathbf{A}}_2)\right) m_p \mathrm{d}\Omega \\
& + \int_{\Omega_1} \nu_1 \mathbf{B}(\mathbf{V} \cdot \nabla \mathbf{A}_1) \cdot \mathbf{B}(\boldsymbol{\lambda}_1) \mathrm{d}\Omega + \int_{\Omega_2} \nu_2 \mathbf{B}(\mathbf{V} \cdot \nabla \mathbf{A}_2) \cdot \mathbf{B}(\boldsymbol{\lambda}_2) \mathrm{d}\Omega \\
& + \int_{\Omega_1} \nu_1 \mathbf{B}(\mathbf{A}_1) \cdot \mathbf{B}(\mathbf{V} \cdot \nabla \boldsymbol{\lambda}_1) \mathrm{d}\Omega + \int_{\Omega_2} \nu_2 \mathbf{B}(\mathbf{A}_2) \cdot \mathbf{B}(\mathbf{V} \cdot \nabla \boldsymbol{\lambda}_2) \mathrm{d}\Omega \\
& - \int_{\Omega_1} [\mathbf{J}_1 \cdot (\mathbf{V} \cdot \nabla \boldsymbol{\lambda}_1) + \mathbf{M}_{o1} \cdot \mathbf{B}(\mathbf{V} \cdot \nabla \boldsymbol{\lambda}_1)] \mathrm{d}\Omega \\
& - \int_{\Omega_2} [\mathbf{J}_2 \cdot (\mathbf{V} \cdot \nabla \boldsymbol{\lambda}_2) + \mathbf{M}_{o2} \cdot \mathbf{B}(\mathbf{V} \cdot \nabla \boldsymbol{\lambda}_2)] \mathrm{d}\Omega \\
& - \int_{\Omega_1} \left[\mathbf{g}_{\mathbf{A}_1} \cdot (\mathbf{V} \cdot \nabla \mathbf{A}_1) + \mathbf{g}_{\mathbf{B}_1} \cdot \mathbf{B}(\mathbf{V} \cdot \nabla \mathbf{A}_1)\right] m_p \mathrm{d}\Omega \\
& - \int_{\Omega_2} \left[\mathbf{g}_{\mathbf{A}_2} \cdot (\mathbf{V} \cdot \nabla \mathbf{A}_2) + \mathbf{g}_{\mathbf{B}_2} \cdot \mathbf{B}(\mathbf{V} \cdot \nabla \mathbf{A}_2)\right] m_p \mathrm{d}\Omega \\
& - \int_{\gamma} (\nu_1 \mathbf{B}(\mathbf{A}_1) \cdot \mathbf{B}(\boldsymbol{\lambda}_1) - \nu_2 \mathbf{B}(\mathbf{A}_2) \cdot \mathbf{B}(\boldsymbol{\lambda}_2)) V_n \mathrm{d}\Gamma \\
& + \int_{\gamma} (\mathbf{M}_{o1} \cdot \mathbf{B}(\boldsymbol{\lambda}_1) - \mathbf{M}_{o2} \cdot \mathbf{B}(\boldsymbol{\lambda}_2)) V_n \mathrm{d}\Gamma \\
& + \int_{\gamma} \left[\mathbf{J}_1 \cdot \boldsymbol{\lambda}_1 - \mathbf{J}_2 \cdot \boldsymbol{\lambda}_2 + (g_1 - g_2) m_p\right] V_n \mathrm{d}\Gamma
\end{aligned} \tag{4.1.33}
$$

The energy bilinear form is symmetric in its arguments;

$$\int_{\Omega_1} \nu_1 \mathbf{B}(\boldsymbol{\lambda}_1) \cdot \mathbf{B}(\dot{\mathbf{A}}_1) \mathrm{d}\Omega = \int_{\Omega_1} \nu_1 \mathbf{B}(\dot{\mathbf{A}}_1) \cdot \mathbf{B}(\boldsymbol{\lambda}_1) \mathrm{d}\Omega \tag{4.1.34}$$

$$\int_{\Omega_2} \nu_2 \mathbf{B}(\boldsymbol{\lambda}_2) \cdot \mathbf{B}(\dot{\mathbf{A}}_2) \mathrm{d}\Omega = \int_{\Omega_2} \nu_2 \mathbf{B}(\dot{\mathbf{A}}_2) \cdot \mathbf{B}(\boldsymbol{\lambda}_2) \mathrm{d}\Omega \tag{4.1.35}$$

By using the relations (4.1.32), (4.1.34), and (4.1.35), all terms with $\dot{\mathbf{A}}_1$, $\dot{\mathbf{A}}_2$ in (4.1.33) are canceled out and all terms are expressed with the velocity field $\mathbf{V}$;

$$
\begin{aligned}
\dot{G} = & \int_{\Omega_1} [\nu_1 \mathbf{B}(\mathbf{A}_1)\cdot\mathbf{B}(\mathbf{V}\cdot\nabla\boldsymbol{\lambda}_1) - \mathbf{J}_1\cdot(\mathbf{V}\cdot\nabla\boldsymbol{\lambda}_1) - \mathbf{M}_{o1}\cdot\mathbf{B}(\mathbf{V}\cdot\nabla\boldsymbol{\lambda}_1)]\mathrm{d}\Omega \\
& + \int_{\Omega_2} [\nu_2 \mathbf{B}(\mathbf{A}_2)\cdot\mathbf{B}(\mathbf{V}\cdot\nabla\boldsymbol{\lambda}_2) - \mathbf{J}_2\cdot(\mathbf{V}\cdot\nabla\boldsymbol{\lambda}_2) - \mathbf{M}_{o2}\cdot\mathbf{B}(\mathbf{V}\cdot\nabla\boldsymbol{\lambda}_2)]\mathrm{d}\Omega \\
& + \int_{\Omega_1} \left[\nu_1 \mathbf{B}(\boldsymbol{\lambda}_1)\cdot\mathbf{B}(\mathbf{V}\cdot\nabla\mathbf{A}_1) - \mathbf{g}_{\mathbf{A}_1}\cdot(\mathbf{V}\cdot\nabla\mathrm{A}_1)m_p - \mathbf{g}_{\mathbf{B}_1}\cdot\mathbf{B}(\mathbf{V}\cdot\nabla\mathrm{A}_1)m_p\right]\mathrm{d}\Omega \\
& + \int_{\Omega_2} \left[\nu_2 \mathbf{B}(\boldsymbol{\lambda}_2)\cdot\mathbf{B}(\mathbf{V}\cdot\nabla\mathbf{A}_2) - \mathbf{g}_{\mathbf{A}_2}\cdot(\mathbf{V}\cdot\nabla\mathrm{A}_2)m_p - \mathbf{g}_{\mathbf{B}_2}\cdot\mathbf{B}(\mathbf{V}\cdot\nabla\mathrm{A}_2)m_p\right]\mathrm{d}\Omega \\
& - \int_{\gamma} (\nu_1 \mathbf{B}(\mathbf{A}_1)\cdot\mathbf{B}(\boldsymbol{\lambda}_1) - \nu_2 \mathbf{B}(\mathbf{A}_2)\cdot\mathbf{B}(\boldsymbol{\lambda}_2))V_n\mathrm{d}\Gamma \\
& + \int_{\gamma} (\mathbf{M}_{o1}\cdot\mathbf{B}(\boldsymbol{\lambda}_1) - \mathbf{M}_{o2}\cdot\mathbf{B}(\boldsymbol{\lambda}_2))V_n\mathrm{d}\Gamma \\
& + \int_{\gamma} \left[\mathbf{J}_1\cdot\boldsymbol{\lambda}_1 - \mathbf{J}_2\cdot\boldsymbol{\lambda}_2 + (g_1 - g_2)m_p\right]V_n\mathrm{d}\Gamma
\end{aligned}
\tag{4.1.36}
$$

4.1.4 Boundary Expression of Shape Sensitivity

The domain integrals in the design sensitivity (4.1.36) can be transformed into boundary integrals by using the variational identities. The boundary integral expression of the sensitivity leads to an advantage in its numerical implementation. For this purpose, two variational identities for the state and the adjoint equations are needed.

First, the variational identity for the state equation was given as (4.1.8) and (4.1.9);

$$
\begin{aligned}
& \int_{\Omega_1} \left(\nu_1 \mathbf{B}(\mathbf{A}_1)\cdot\mathbf{B}(\bar{\mathbf{A}}_1) - \mathbf{J}_1\cdot\bar{\mathbf{A}}_1 - \mathbf{M}_{o1}\cdot\mathbf{B}(\bar{\mathbf{A}}_1)\right)\mathrm{d}\Omega \\
& = -\int_{\gamma+\Gamma} (\mathbf{n}\times\mathbf{H}(\mathbf{A}_1))\cdot\bar{\mathbf{A}}_1\mathrm{d}\Gamma \quad \forall\bar{\mathbf{A}}_1\in\Phi
\end{aligned}
\tag{4.1.37}
$$

$$\int_{\Omega_2} \left(\nu_2 \mathbf{B}(\mathbf{A}_2) \cdot \mathbf{B}(\bar{\mathbf{A}}_2) - \mathbf{J}_2 \cdot \bar{\mathbf{A}}_2 - \mathbf{M}_{o2} \cdot \mathbf{B}(\bar{\mathbf{A}}_2)\right) \mathrm{d}\Omega = \int_{\gamma} (\mathbf{n} \times \mathbf{H}(\mathbf{A}_2)) \cdot \bar{\mathbf{A}}_2 \mathrm{d}\Gamma \quad \forall \bar{\mathbf{A}}_2 \in \Phi \tag{4.1.38}$$

Next, the variational identity for the adjoint equation can be derived from the differential adjoint equation, which is obtained by comparing the variational adjoint equation (4.1.31) with the variational state equation (4.1.16). The two variational equations are written again for convenience;

$$\int_{\Omega_1} \nu_1 \mathbf{B}(\boldsymbol{\lambda}_1) \cdot \mathbf{B}(\bar{\boldsymbol{\lambda}}_1) \mathrm{d}\Omega + \int_{\Omega_2} \nu_2 \mathbf{B}(\boldsymbol{\lambda}_2) \cdot \mathbf{B}(\bar{\boldsymbol{\lambda}}_2) \mathrm{d}\Omega = \int_{\Omega_1} \left(\mathbf{g}_{\mathbf{A}_1} \cdot \bar{\boldsymbol{\lambda}}_1 + \mathbf{g}_{\mathbf{B}_1} \cdot \mathbf{B}(\bar{\boldsymbol{\lambda}}_1)\right) m_p \mathrm{d}\Omega + \int_{\Omega_2} \left(\mathbf{g}_{\mathbf{A}_2} \cdot \bar{\boldsymbol{\lambda}}_2 + \mathbf{g}_{\mathbf{B}_2} \cdot \mathbf{B}(\bar{\boldsymbol{\lambda}}_2)\right) m_p \mathrm{d}\Omega \quad \forall \bar{\boldsymbol{\lambda}}_1, \bar{\boldsymbol{\lambda}}_2 \in \Phi \tag{4.1.39}$$

$$\int_{\Omega_1} \nu_1 \mathbf{B}(\mathbf{A}_1) \cdot \mathbf{B}(\overline{\mathbf{A}}_1) \mathrm{d}\Omega + \int_{\Omega_2} \nu_2 \mathbf{B}(\mathbf{A}_2) \cdot \mathbf{B}(\overline{\mathbf{A}}_2) \mathrm{d}\Omega = \int_{\Omega_1} \left(\mathbf{J}_1 \cdot \overline{\mathbf{A}}_1 + \mathbf{M}_{o1} \cdot \mathbf{B}(\overline{\mathbf{A}}_1)\right) \mathrm{d}\Omega + \int_{\Omega_2} \left(\mathbf{J}_2 \cdot \overline{\mathbf{A}}_2 + \mathbf{M}_{o2} \cdot \mathbf{B}(\overline{\mathbf{A}}_2)\right) \mathrm{d}\Omega \quad \forall \overline{\mathbf{A}}_1, \overline{\mathbf{A}}_2 \in \Phi \tag{4.1.40}$$

These two equations have the same form except different sources. The sources $\mathbf{g}_{\mathbf{A}_1} m_p$, $\mathbf{g}_{\mathbf{A}_2} m_p$ and $\mathbf{g}_{\mathbf{B}_1} m_p$, $\mathbf{g}_{\mathbf{B}_2} m_p$ of the adjoint equation correspond to the ones $\mathbf{J}_1$, $\mathbf{J}_2$ and $\mathbf{M}_{o1}$, $\mathbf{M}_{o2}$ of the state equation, respectively. Just as the variational state equation (4.1.40) is equivalent to the differential state equations of (4.1.4) and (4.1.5), the variational adjoint Eq. (4.1.39) is equivalent to differential adjoint equations:

$$\nabla \times \nu_1 \nabla \times \boldsymbol{\lambda}_1 = \mathbf{g}_{\mathbf{A}_1} m_p + \nabla \times \mathbf{g}_{\mathbf{B}_1} m_p \quad \text{in } \Omega_1 \tag{4.1.41}$$

$$\nabla \times \nu_2 \nabla \times \boldsymbol{\lambda}_2 = \mathbf{g}_{\mathbf{A}_2} m_p + \nabla \times \mathbf{g}_{\mathbf{B}_2} m_p \quad \text{in } \Omega_2 \tag{4.1.42}$$

with the boundary condition:

$$\boldsymbol{\lambda}_1 = \mathbf{0} \quad \text{on } \Gamma^0 \quad \text{(homogeneous Dirichlet boundary condition)} \tag{4.1.43}$$

$$\frac{\partial \boldsymbol{\lambda}_1}{\partial n} = \mathbf{0} \quad \text{on } \Gamma^1 \quad \text{(homogeneous Neumann boundary condition)} \tag{4.1.44}$$

The adjoint sources $\mathbf{g}_{\mathbf{A}_1}m_p$, $\mathbf{g}_{\mathbf{A}_2}m_p$ and $\mathbf{g}_{\mathbf{B}_1}m_p$, $\mathbf{g}_{\mathbf{B}_2}m_p$ exist only in the domain Ω as the original sources $\mathbf{J}_1$, $\mathbf{J}_2$ and $\mathbf{M}_{o1}$, $\mathbf{M}_{o2}$, Thus, there is no adjoint surface source equivalent to Dirichlet boundary condition. That is, while the original state equation may have surface sources equivalent to its Dirichlet boundary condition, the adjoint equation has no surface source equivalent to its Dirichlet boundary condition. Hence, the Dirichlet boundary condition of the adjoint equation is given as zero. Since the structure symmetry is maintained in the adjoint system as well, the homogeneous Neumann condition is imposed on Γ^1 of the adjoint system.

In the same way that the variational identity of (4.1.37) and (4.1.38) for the state equation is obtained from the differential state Eqs. (4.1.4) and (4.1.5), a variational identity for the adjoint equation is obtained as

$$\int_{\Omega_1} \left(\nu_1 \mathbf{B}(\boldsymbol{\lambda}_1) \cdot \mathbf{B}(\bar{\boldsymbol{\lambda}}_1) - \mathbf{g}_{\mathbf{A}_1} \cdot \bar{\boldsymbol{\lambda}}_1 m_p - \mathbf{g}_{\mathbf{B}_1} \cdot \mathbf{B}(\bar{\boldsymbol{\lambda}}_1) m_p\right) \mathrm{d}\Omega = -\int_{\gamma+\Gamma} (\mathbf{n} \times \mathbf{H}(\boldsymbol{\lambda}_1)) \cdot \bar{\boldsymbol{\lambda}}_1 \mathrm{d}\Gamma \quad \forall \bar{\boldsymbol{\lambda}}_1 \in \Phi \tag{4.1.45}$$

$$\int_{\Omega_2} \left(\nu_2 \mathbf{B}(\boldsymbol{\lambda}_2) \cdot \mathbf{B}(\bar{\boldsymbol{\lambda}}_2) - \mathbf{g}_{\mathbf{A}_2} \cdot \bar{\boldsymbol{\lambda}}_2 m_p - \mathbf{g}_{\mathbf{B}_2} \cdot \mathbf{B}(\bar{\boldsymbol{\lambda}}_2) m_p\right) \mathrm{d}\Omega = \int_{\gamma} (\mathbf{n} \times \mathbf{H}(\boldsymbol{\lambda}_2)) \cdot \bar{\boldsymbol{\lambda}}_2 d\Gamma \quad \forall \bar{\boldsymbol{\lambda}}_2 \in \Phi \tag{4.1.46}$$

where

$$\mathbf{H}(\boldsymbol{\lambda}_1) = \nu_1 \mathbf{B}(\boldsymbol{\lambda}_1) - \mathbf{g}_{\mathbf{B}_1} m_p \tag{4.1.47}$$

$$\mathbf{H}(\boldsymbol{\lambda}_2) = \nu_2 \mathbf{B}(\boldsymbol{\lambda}_2) - \mathbf{g}_{\mathbf{B}_2} m_p \tag{4.1.48}$$

and Φ is the space of admissible vector potential: $\Phi = \left\{\bar{\lambda} \in [H^1(\Omega)]^3 \middle| \bar{\lambda} = \mathbf{0} \text{ on } \mathbf{x} \in \Gamma^0.\right\}$. Note that imposing the boundary conditions (4.1.43) and (4.1.44) provides the variational adjoint equation (4.1.39).

The variational identities of (4.1.37), (4.1.38), and (4.1.45), (4.1.46) are used to express the domain integrals in (4.1.36) as boundary integrals. First, (4.1.37) and (4.1.38) are evaluated at $\bar{\mathbf{A}}_1 = \mathbf{V} \cdot \nabla \boldsymbol{\lambda}_1$ and $\bar{\mathbf{A}}_2 = \mathbf{V} \cdot \nabla \boldsymbol{\lambda}_2$, respectively;

$$\int_{\Omega_1} [\nu_1 \mathbf{B}(\mathbf{A}_1) \cdot \mathbf{B}(\mathbf{V} \cdot \nabla \boldsymbol{\lambda}_1) - \mathbf{J}_1 \cdot (\mathbf{V} \cdot \nabla \boldsymbol{\lambda}_1) - \mathbf{M}_{o1} \cdot \mathbf{B}(\mathbf{V} \cdot \nabla \boldsymbol{\lambda}_1)] \mathrm{d}\Omega = -\int_{\gamma} (\mathbf{n} \times \mathbf{H}(\mathbf{A}_1)) \cdot (\mathbf{V} \cdot \nabla \boldsymbol{\lambda}_1) \mathrm{d}\Gamma \tag{4.1.49}$$

$$\int_{\Omega_2} [\nu_2 \mathbf{B}(\mathbf{A}_2) \cdot \mathbf{B}(\mathbf{V} \cdot \nabla \boldsymbol{\lambda}_2) - \mathbf{J}_2 \cdot (\mathbf{V} \cdot \nabla \boldsymbol{\lambda}_2) - \mathbf{M}_{o2} \cdot \mathbf{B}(\mathbf{V} \cdot \nabla \boldsymbol{\lambda}_2)] d\Omega$$
$$= \int_{\gamma} (\mathbf{n} \times \mathbf{H}(\mathbf{A}_2)) \cdot (\mathbf{V} \cdot \nabla \boldsymbol{\lambda}_2) d\Gamma \qquad (4.1.50)$$

Second, (4.1.45) and (4.1.46) are evaluated at $\bar{\boldsymbol{\lambda}}_1 = \mathbf{V} \cdot \nabla \mathrm{A}_1$ and $\bar{\boldsymbol{\lambda}}_2 = \mathbf{V} \cdot \nabla \mathrm{A}_2$, respectively;

$$\int_{\Omega_1} \left[\nu_1 \mathbf{B}(\boldsymbol{\lambda}_1) \cdot \mathbf{B}(\mathbf{V} \cdot \nabla \mathrm{A}_1) - \mathbf{g}_{\mathbf{A}_1} \cdot (\mathbf{V} \cdot \nabla \mathrm{A}_1) m_p - \mathbf{g}_{\mathbf{B}_1} \cdot \mathbf{B}(\mathbf{V} \cdot \nabla \mathrm{A}_1) m_p\right] d\Omega$$
$$= -\int_{\gamma} (\mathbf{n} \times \mathbf{H}(\boldsymbol{\lambda}_1)) \cdot (\mathbf{V} \cdot \nabla \mathrm{A}_1) d\Gamma \qquad (4.1.51)$$

$$\int_{\Omega_2} \left[\nu_2 \mathbf{B}(\boldsymbol{\lambda}_2) \cdot \mathbf{B}(\mathbf{V} \cdot \nabla \mathrm{A}_2) - \mathbf{g}_{\mathbf{A}_2} \cdot (\mathbf{V} \cdot \nabla \mathrm{A}_2) m_p - \mathbf{g}_{\mathbf{B}_2} \cdot \mathbf{B}(\mathbf{V} \cdot \nabla \mathrm{A}_2) m_p\right] d\Omega$$
$$= \int_{\gamma} (\mathbf{n} \times \mathbf{H}(\boldsymbol{\lambda}_2)) \cdot (\mathbf{V} \cdot \nabla \mathrm{A}_2) d\Gamma \qquad (4.1.52)$$

By substituting (4.1.49), (4.1.50), and (4.1.51), (4.1.52) into (4.1.36), the domain integrals in (4.1.36) become boundary integrals;

$$\begin{aligned}\dot{G} = & \int_{\gamma} [-(\mathbf{n} \times \mathbf{H}(\mathbf{A}_1)) \cdot (\mathbf{V} \cdot \nabla \boldsymbol{\lambda}_1) + (\mathbf{n} \times \mathbf{H}(\mathbf{A}_2)) \cdot (\mathbf{V} \cdot \nabla \boldsymbol{\lambda}_2) \\ & -(\mathbf{n} \times \mathbf{H}(\boldsymbol{\lambda}_1)) \cdot (\mathbf{V} \cdot \nabla \mathbf{A}_1) + (\mathbf{n} \times \mathbf{H}(\boldsymbol{\lambda}_2)) \cdot (\mathbf{V} \cdot \nabla \mathbf{A}_2)] d\Gamma \\ & - \int_{\gamma} (\nu_1 \mathbf{B}(\mathbf{A}_1) \cdot \mathbf{B}(\boldsymbol{\lambda}_1) - \nu_2 \mathbf{B}(\mathbf{A}_2) \cdot \mathbf{B}(\boldsymbol{\lambda}_2)) V_n d\Gamma \\ & + \int_{\gamma} (\mathbf{M}_{o1} \cdot \mathbf{B}(\boldsymbol{\lambda}_1) - \mathbf{M}_{o2} \cdot \mathbf{B}(\boldsymbol{\lambda}_2)) V_n d\Gamma \\ & + \int_{\gamma} \left[\mathbf{J}_1 \cdot \boldsymbol{\lambda}_1 - \mathbf{J}_2 \cdot \boldsymbol{\lambda}_2 + (g_1 - g_2) m_p\right] V_n d\Gamma \end{aligned} \qquad (4.1.53)$$

The integrand of the first integral in (4.1.53) is arranged using the interface conditions and a vector identity. The interface conditions for continuity of the tangential component of **H** imply

$$\mathbf{n} \times \mathbf{H}(\mathbf{A}_1) = \mathbf{n} \times \mathbf{H}(\mathbf{A}_2) \quad \text{on } \gamma \quad \text{(interface condition)} \qquad (4.1.54)$$

$$\mathbf{n} \times \mathbf{H}(\boldsymbol{\lambda}_1) = \mathbf{n} \times \mathbf{H}(\boldsymbol{\lambda}_2) \quad \text{on } \gamma \quad \text{(interface condition)} \qquad (4.1.55)$$

With $\mathbf{V} = V_n\mathbf{n}$, the following relations are obtained:

$$\mathbf{V} \cdot \nabla \mathbf{A} = V_n \frac{\partial \mathbf{A}}{\partial n} \quad \text{on}\ \gamma \tag{4.1.56}$$

$$\mathbf{V} \cdot \nabla \boldsymbol{\lambda} = V_n \frac{\partial \boldsymbol{\lambda}}{\partial n} \quad \text{on}\ \gamma \tag{4.1.57}$$

$\frac{\partial \mathbf{A}}{\partial n}$ and $\frac{\partial \boldsymbol{\lambda}}{\partial n}$ in (4.1.56) and (4.1.57) are expressed with $\mathbf{B}(\mathbf{A})$ and $\mathbf{B}(\boldsymbol{\lambda})$ in the following. For the vector identity $\nabla(\mathbf{a} \cdot \mathbf{b}) = \mathbf{a} \times (\nabla \times \mathbf{b}) + \mathbf{b} \times (\nabla \times \mathbf{a}) + (\mathbf{a} \cdot \nabla)\mathbf{b} + (\mathbf{b} \cdot \nabla)\mathbf{a}$, when $\mathbf{a}$ is the magnetic vector potential $\mathbf{A}$ and $\mathbf{b}$ is the normal unit vector $\mathbf{n}$ on the interface, the following relation is obtained:

$$\nabla(\mathbf{A} \cdot \mathbf{n}) = \mathbf{n} \times (\nabla \times \mathbf{A}) + (\mathbf{n} \cdot \nabla)\mathbf{A} \tag{4.1.58}$$

By using the operator $\nabla = \frac{\partial}{\partial n}\mathbf{n}$ on γ, the relation $\mathbf{B} = \nabla \times A$ and the definition $A_n = \mathbf{A}{\cdot}\mathbf{n}$, (4.1.58) is rewritten as

$$\frac{\partial A_n}{\partial n}\mathbf{n} = -\mathbf{B} \times \mathbf{n} + \frac{\partial \mathbf{A}}{\partial n} \quad \text{on} \quad \gamma \tag{4.1.59}$$

The Coulomb gauge $\nabla{\cdot}\mathbf{A} = 0$ is rewritten by the operator $\nabla = \frac{\partial}{\partial n}\mathbf{n}$ as

$$\frac{\partial A_n}{\partial n} = \boldsymbol{0} \quad \text{on}\ \gamma \tag{4.1.60}$$

Hence, the relation between the magnetic vector potential A and the magnetic flux density $\mathbf{B}$ on the interface is obtained as

$$\frac{\partial \mathbf{A}}{\partial n} = \mathbf{B}(\mathbf{A}) \times \mathbf{n} \quad \text{on} \quad \gamma \tag{4.1.61}$$

Likewise, this relation holds for the adjoint variable $\boldsymbol{\lambda}$;

$$\frac{\partial \boldsymbol{\lambda}}{\partial n} = \mathbf{B}(\boldsymbol{\lambda}) \times \mathbf{n} \quad \text{on} \quad \gamma \tag{4.1.62}$$

The interface conditions (4.1.54), (4.1.55) and the relations of (4.1.56), (4.1.57) are applied to the integrand of the first integral in (4.1.53) to provide

$$
\begin{aligned}
&- (\mathbf{n} \times \mathbf{H}(\mathbf{A}_1)) \cdot (\mathbf{V} \cdot \nabla \boldsymbol{\lambda}_1) + (\mathbf{n} \times \mathbf{H}(\mathbf{A}_2)) \cdot (\mathbf{V} \cdot \nabla \boldsymbol{\lambda}_2) \\
&- (\mathbf{n} \times \mathbf{H}(\boldsymbol{\lambda}_1)) \cdot (\mathbf{V} \cdot \nabla \mathbf{A}_1) + (\mathbf{n} \times \mathbf{H}(\boldsymbol{\lambda}_2)) \cdot (\mathbf{V} \cdot \nabla \mathbf{A}_2) \\
&= (\mathbf{n} \times \mathbf{H}(\mathbf{A}_1)) \cdot \left(\frac{\partial \boldsymbol{\lambda}_2}{\partial n} - \frac{\partial \boldsymbol{\lambda}_1}{\partial n} \right) V_n + (\mathbf{n} \times \mathbf{H}(\boldsymbol{\lambda}_2)) \cdot \left(\frac{\partial \mathbf{A}_2}{\partial n} - \frac{\partial \mathbf{A}_1}{\partial n} \right) V_n
\end{aligned} \tag{4.1.63}
$$

The constitutive relations of (4.1.10), (4.1.11) and (4.1.47), (4.1.48), and the relation (4.1.61), (4.1.62) are applied to (4.1.63), which is written without V_n as

$$
\begin{aligned}
&[\mathbf{n} \times (\nu_1 \mathbf{B}(\mathbf{A}_1) - \mathbf{M}_{o1})] \cdot (\mathbf{B}(\boldsymbol{\lambda}_2) \times \mathbf{n} - \mathbf{B}(\boldsymbol{\lambda}_1) \times \mathbf{n}) \\
&+ \left[\mathbf{n} \times \left(\nu_2 \mathbf{B}(\boldsymbol{\lambda}_2) - \mathbf{g}_{\mathbf{B}_2} m_p\right)\right] \cdot (\mathbf{B}(\mathbf{A}_2) \times \mathbf{n} - \mathbf{B}(\mathbf{A}_1) \times \mathbf{n}) \\
&= (\mathbf{n} \times \nu_1 \mathbf{B}(\mathbf{A}_1)) \cdot (\mathbf{B}(\boldsymbol{\lambda}_2) \times \mathbf{n} - \mathbf{B}(\boldsymbol{\lambda}_1) \times \mathbf{n}) + (\mathbf{n} \times \nu_2 \mathbf{B}(\boldsymbol{\lambda}_2)) \cdot (\mathbf{B}(\mathbf{A}_2) \times \mathbf{n} - \mathbf{B}(\mathbf{A}_1) \times \mathbf{n}) \\
&- (\mathbf{n} \times \mathbf{M}_{o1}) \cdot (\mathbf{B}(\boldsymbol{\lambda}_2) \times \mathbf{n} - \mathbf{B}(\boldsymbol{\lambda}_1) \times \mathbf{n}) - \left(\mathbf{n} \times \mathbf{g}_{\mathbf{B}_2} m_p\right) \cdot (\mathbf{B}(\mathbf{A}_2) \times \mathbf{n} - \mathbf{B}(\mathbf{A}_1) \times \mathbf{n})
\end{aligned} \tag{4.1.64}
$$

Consider a vector $\mathbf{S}$ on the interface, which has a normal component $\mathbf{n}$ and two tangential components $\mathbf{t}$ and $\mathbf{t}'$; $\mathbf{S} = S_n \mathbf{n} + S_t \mathbf{t} + S_{t'} \mathbf{t}'$ with $\mathbf{t}' = \mathbf{n} \times \mathbf{t}$. This vector has the relation:

$$
\mathbf{S} \times \mathbf{n} = S_t \mathbf{t}' \tag{4.1.65}
$$

With this relation, (4.1.64) becomes the following (4.1.66), which is the integrand of the first integral without V_n in (4.1.53);

$$
\begin{aligned}
&- \nu_1 B_t(\mathbf{A}_1)\mathbf{t}' \cdot (B_t(\boldsymbol{\lambda}_2)\mathbf{t}' - B_t(\boldsymbol{\lambda}_1)\mathbf{t}') - \nu_2 B_t(\boldsymbol{\lambda}_2)\mathbf{t}' \cdot (B_t(\mathbf{A}_2)\mathbf{t}' - B_t(\mathbf{A}_1)\mathbf{t}') \\
&+ M_{o1t}\mathbf{t}' \cdot (B_t(\boldsymbol{\lambda}_2)\mathbf{t}' - B_t(\boldsymbol{\lambda}_1)\mathbf{t}') + g_{\mathbf{B}_2 t}\mathbf{t}' \cdot (B_t(\mathbf{A}_2)\mathbf{t}' - B_t(\mathbf{A}_1)\mathbf{t}') m_p \\
&= -\nu_1 B_t(\mathbf{A}_1)(B_t(\boldsymbol{\lambda}_2) - B_t(\boldsymbol{\lambda}_1)) - \nu_2 B_t(\boldsymbol{\lambda}_2)(B_t(\mathbf{A}_2) - B_t(\mathbf{A}_1)) \\
&+ M_{o1t}(B_t(\boldsymbol{\lambda}_2) - B_t(\boldsymbol{\lambda}_1)) + g_{\mathbf{B}_2 t}(B_t(\mathbf{A}_2) - B_t(\mathbf{A}_1)) m_p
\end{aligned} \tag{4.1.66}
$$

The integrand of the second integral in (4.1.53) is written without V_n as

$$
\begin{aligned}
&- \nu_1 \mathbf{B}(\mathbf{A}_1) \cdot \mathbf{B}(\boldsymbol{\lambda}_1) + \nu_2 \mathbf{B}(\mathbf{A}_2) \cdot \mathbf{B}(\boldsymbol{\lambda}_2) \\
&= -\nu_1 B_n(\mathbf{A}_1) B_n(\boldsymbol{\lambda}_1) - \nu_1 B_t(\mathbf{A}_1) B_t(\boldsymbol{\lambda}_1) + \nu_2 B_n(\mathbf{A}_2) B_n(\boldsymbol{\lambda}_2) + \nu_2 B_t(\mathbf{A}_2) B_t(\boldsymbol{\lambda}_2)
\end{aligned} \tag{4.1.67}
$$

The integrand of the third integral in (4.1.53) is written without V_n as

$$
\begin{aligned}
\mathbf{M}_{o1} \cdot \mathbf{B}(\boldsymbol{\lambda}_1) - \mathbf{M}_{o2} \cdot \mathbf{B}(\boldsymbol{\lambda}_2) = {} & M_{o1n} B_n(\boldsymbol{\lambda}_1) + M_{o1t} B_t(\boldsymbol{\lambda}_1) - M_{o2n} B_n(\boldsymbol{\lambda}_2) \\
& - M_{o2t} B_t(\boldsymbol{\lambda}_2)
\end{aligned} \tag{4.1.68}
$$

The above (4.1.66), (4.1.67), and (4.1.68), which are each integrand without V_n of the first three integrals, are summed. First, the terms with ν_1 and ν_2 in (4.1.66) and (4.1.67) are summed;

$$\begin{aligned}
&-\nu_1 B_t(\mathbf{A}_1)(B_t(\boldsymbol{\lambda}_2)-B_t(\boldsymbol{\lambda}_1))-\nu_2 B_t(\boldsymbol{\lambda}_2)(B_t(\mathbf{A}_2)-B_t(\mathbf{A}_1))\\
&-\nu_1 B_n(\mathbf{A}_1)B_n(\boldsymbol{\lambda}_1)-\nu_1 B_t(\mathbf{A}_1)B_t(\boldsymbol{\lambda}_1)+\nu_2 B_n(\mathbf{A}_2)B_n(\boldsymbol{\lambda}_2)+\nu_2 B_t(\mathbf{A}_2)B_t(\boldsymbol{\lambda}_2)\\
&=-\nu_1 B_t(\mathbf{A}_1)B_t(\boldsymbol{\lambda}_2)+\nu_2 B_t(\boldsymbol{\lambda}_2)B_t(\mathbf{A}_1)-\nu_1 B_n(\mathbf{A}_1)B_n(\boldsymbol{\lambda}_1)+\nu_2 B_n(\mathbf{A}_2)B_n(\boldsymbol{\lambda}_2)
\end{aligned} \tag{4.1.69}$$

By the interface conditions: $B_n(\boldsymbol{\lambda}_1)=B_n(\boldsymbol{\lambda}_2)$ and $B_n(\mathbf{A}_1)=B_n(\mathbf{A}_2)$, (4.1.69) is rewritten;

$$(\nu_2-\nu_1)(B_t(\mathbf{A}_1)B_t(\boldsymbol{\lambda}_2)+B_n(\mathbf{A}_1)B_n(\boldsymbol{\lambda}_2))=(\nu_2-\nu_1)\mathbf{B}(\mathbf{A}_1)\cdot\mathbf{B}(\boldsymbol{\lambda}_2) \tag{4.1.70}$$

Second, the terms with the permanent magnetization in (4.1.66) and (4.1.68) are summed;

$$\begin{aligned}
&M_{o1t}(B_t(\boldsymbol{\lambda}_2)-B_t(\boldsymbol{\lambda}_1))+M_{o1n}B_n(\boldsymbol{\lambda}_1)+M_{o1t}B_t(\boldsymbol{\lambda}_1)-M_{o2n}B_n(\boldsymbol{\lambda}_2)-M_{o2t}B_t(\boldsymbol{\lambda}_2)\\
&=M_{o1t}B_t(\boldsymbol{\lambda}_2)+M_{o1n}B_n(\boldsymbol{\lambda}_1)-M_{o2n}B_n(\boldsymbol{\lambda}_2)-M_{o2t}B_t(\boldsymbol{\lambda}_2)
\end{aligned} \tag{4.1.71}$$

By the interface condition: $B_n(\boldsymbol{\lambda}_1)=B_n(\boldsymbol{\lambda}_2)$, (4.1.71) is rewritten;

$$-(M_{o2n}-M_{o1n})B_n(\boldsymbol{\lambda}_2)-(M_{o2t}-M_{o1t})B_t(\boldsymbol{\lambda}_2)=-(\mathbf{M}_{o2}-\mathbf{M}_{o1})\cdot\mathbf{B}(\boldsymbol{\lambda}_2) \tag{4.1.72}$$

By using the relations of (4.1.70) and (4.1.72), the sum of the (4.1.66), (4.1.67), and (4.1.68) is obtained as

$$(\nu_2-\nu_1)\mathbf{B}(\mathbf{A}_1)\cdot\mathbf{B}(\boldsymbol{\lambda}_2)-(\mathbf{M}_{o2}-\mathbf{M}_{o1})\cdot\mathbf{B}(\boldsymbol{\lambda}_2)+g_{\mathbf{B}_2 t}(B_t(\mathbf{A}_2)-B_t(\mathbf{A}_1))m_p \tag{4.1.73}$$

which is the sum of three integrand terms without V_n of the first three integrals in (4.1.53). Finally, by substituting (4.1.73) into (4.1.53), the general three-dimensional sensitivity formula by the interface variation in the linear magnetostatic system is obtained as

$$\begin{aligned}
\dot{G}=\int_\gamma &\big[(\nu_2-\nu_1)\mathbf{B}(\mathbf{A}_1)\cdot\mathbf{B}(\boldsymbol{\lambda}_2)-(\mathbf{M}_{o2}-\mathbf{M}_{o1})\cdot\mathbf{B}(\boldsymbol{\lambda}_2)-(\mathbf{J}_2-\mathbf{J}_1)\cdot\boldsymbol{\lambda}_2\\
&-(g_2-g_1)m_p+g_{\mathbf{B}_2 t}(B_t(\mathbf{A}_2)-B_t(\mathbf{A}_1))m_p\big]V_n\mathrm{d}\Gamma
\end{aligned} \tag{4.1.74}$$

In this sensitivity formula, the exchange of both sides by the interface variation is represented by each integrand, which means the exchange of the magnetic permeability, the permanent magnetization and the current density, the objective function due to variation of magnetic flux density and the objective function, respectively.

When the objective function is defined on the inner area in the domain Ω that does not intersect with the design boundary, $m_p = 0$ on γ. Hence, (4.1.74) becomes

$$\dot{G} = \int_{\gamma} [(\nu_2 - \nu_1)\mathbf{B}(\mathbf{A}_1) \cdot \mathbf{B}(\boldsymbol{\lambda}_2) - (\mathbf{M}_{o2} - \mathbf{M}_{o1}) \cdot \mathbf{B}(\boldsymbol{\lambda}_2) - (\mathbf{J}_2 - \mathbf{J}_1) \cdot \boldsymbol{\lambda}_2] V_n \mathrm{d}\Gamma \tag{4.1.75}$$

4.1.5 Interface Problems

The general three-dimensional sensitivity formula (4.1.75) for the interface variation can be expressed in various ways according to the characteristics of the design problems of the magnetostatic system. In fact, the design problem that all terms in (4.1.75) are used is hardly found.

For the first term of (4.1.75), there are some cases of the interfaces: ferromagnetic material versus permanent magnet, ferromagnetic material versus current, ferromagnetic material versus the air, two different ferromagnetic materials. When the other side of the ferromagnetic material is the permanent magnet, the current or the air, their magnetic permeability should be applied.

For the second term, the interface of two different permanent magnets is found in the example of Halbach array magnet, but most cases belong to such interfaces as permanent magnet versus ferromagnetic material, permanent magnet versus the current, or permanent magnet versus the air.

For the third term, the interface of different currents is found in the case of polyphase electrical apparatus such as transformer, motor, and generator, but most cases belong to the interfaces: current versus ferromagnetic material, current versus the air.

The above various cases can simultaneously occur in a design problem. In that case, all the related terms should be used. The general sensitivity formula (4.1.75) for the linear magnetostatic systems is rewritten for the following specific cases.

(1) interface of air and ferromagnetic material

When the air is Ω_1 and the linear ferromagnetic material is Ω_2, (4.1.75) is written as

$$\dot{G} = \int_{\gamma} \nu_0(\nu_r - 1)\mathbf{B}(\mathbf{A}_1) \cdot \mathbf{B}(\boldsymbol{\lambda}_2) V_n \mathrm{d}\Gamma \tag{4.1.76}$$

where $\nu_1 = \nu_0 = 1/\mu_0$, $\nu_2 = \nu_r \nu_0 = 1/\mu_r \mu_0$, $\nu_r = 1/\mu_r$.

(2) interface of magnetic material and permanent magnet

When the linear ferromagnetic material is Ω_1 and the permanent magnet is Ω_2,

$$\dot{G} = \int_{\gamma} [(v_2 - v_1)\mathbf{B}(\mathbf{A}_1) \cdot \mathbf{B}(\boldsymbol{\lambda}_2) - \mathbf{M}_{o2} \cdot \mathbf{B}(\boldsymbol{\lambda}_2)]V_n \mathrm{d}\Gamma \quad (4.1.77)$$

where v_1 is the reluctivity of the linear ferromagnetic material, and v_2 is the reluctivity of the permanent magnet. If the relative permeability of the permanent magnet is assumed to be 1,

$$\dot{G} = \int_{\gamma} [v_0(1 - v_r)\mathbf{B}(\mathbf{A}_1) \cdot \mathbf{B}(\boldsymbol{\lambda}_2) - \mathbf{M}_{o2} \cdot \mathbf{B}(\boldsymbol{\lambda}_2)]V_n \mathrm{d}\Gamma \quad (4.1.78)$$

where $v_1 = v_r v_0 = 1/\mu_r\mu_0$, $v_2 = v_0 = 1/\mu_0$.

(3) interface of magnetic material and current region

When the linear ferromagnetic material is Ω_1 and the current region is Ω_2,

$$\dot{G} = \int_{\gamma} [(v_2 - v_1)\mathbf{B}(\mathbf{A}_1) \cdot \mathbf{B}(\boldsymbol{\lambda}_2) - \mathbf{J}_2 \cdot \boldsymbol{\lambda}_2]V_n \mathrm{d}\Gamma \quad (4.1.79)$$

where v_1 is the reluctivity of the linear ferromagnetic material, and v_2 is the reluctivity of the current region. If the relative permeability of the current region is taken to be 1,

$$\dot{G} = \int_{\gamma} [v_0(1 - v_r)\mathbf{B}(\mathbf{A}_1) \cdot \mathbf{B}(\boldsymbol{\lambda}_2) - \mathbf{J}_2 \cdot \boldsymbol{\lambda}_2]V_n \mathrm{d}\Gamma \quad (4.1.80)$$

where $v_1 = v_r v_0 = 1/\mu_r\mu_0$, $v_2 = v_0 = 1/\mu_0$.

(4) interface of permanent magnet and current region

When the permanent magnet is Ω_1 and the current region is Ω_2,

$$\dot{G} = \int_{\gamma} [(v_2 - v_1)\mathbf{B}(\mathbf{A}_1) \cdot \mathbf{B}(\boldsymbol{\lambda}_2) + \mathbf{M}_{o1} \cdot \mathbf{B}(\boldsymbol{\lambda}_2) - \mathbf{J}_2 \cdot \boldsymbol{\lambda}_2]V_n \mathrm{d}\Gamma \quad (4.1.81)$$

where v_1 is the reluctivity of the permanent magnet, and v_2 is the reluctivity of the current region. If the relative permeabilities of the permanent magnet and the current region are taken to be 1,

$$\dot{G} = \int_{\gamma} (\mathbf{M}_{o1} \cdot \mathbf{B}(\boldsymbol{\lambda}_2) - \mathbf{J}_2 \cdot \boldsymbol{\lambda}_2) V_n \mathrm{d}\Gamma \tag{4.1.82}$$

(5) interface of two permanent magnets with different magnetization

When the permanent magnet 1 is Ω_1 and the permanent magnet 2 is Ω_2,

$$\dot{G} = -\int_{\gamma} (\mathbf{M}_{o2} - \mathbf{M}_{o1}) \cdot \mathbf{B}(\boldsymbol{\lambda}_2) V_n \mathrm{d}\Gamma \tag{4.1.83}$$

(6) interface of permanent magnet and air

When the permanent magnet is Ω_1 and the air is Ω_2,

$$\dot{G} = \int_{\gamma} [(v_2 - v_1)\mathbf{B}(\mathbf{A}_1) \cdot \mathbf{B}(\boldsymbol{\lambda}_2) + \mathbf{M}_{o1} \cdot \mathbf{B}(\boldsymbol{\lambda}_2)] V_n \mathrm{d}\Gamma \tag{4.1.84}$$

where v_1 is the reluctivity of the permanent magnet, and v_2 is the reluctivity of the air. If the relative permeability of the permanent magnet is taken to be 1,

$$\dot{G} = \int_{\gamma} \mathbf{M}_{o1} \cdot \mathbf{B}(\boldsymbol{\lambda}_2) V_n \mathrm{d}\Gamma \tag{4.1.85}$$

(7) interface of two current regions with different current density

When the current region 1 is Ω_1 and the current region 2 is Ω_2,

$$\dot{G} = -\int_{\gamma} (\mathbf{J}_2 - \mathbf{J}_1) \cdot \boldsymbol{\lambda}_2 V_n \mathrm{d}\Gamma \tag{4.1.86}$$

(8) interface of current region and air

When the current region is Ω_1 and the air is Ω_2,

$$\dot{G} = \int_{\gamma} \mathbf{J}_1 \cdot \boldsymbol{\lambda}_2 V_n \mathrm{d}\Gamma \tag{4.1.87}$$

(9) interface of two ferromagnetic materials with different reluctivity

When the ferromagnetic material 1 is Ω_1 and the other ferromagnetic material 2 is Ω_2,

$$\dot{G} = \int_{\gamma} (\nu_2 - \nu_1)\mathbf{B}(\mathbf{A}_1) \cdot \mathbf{B}(\boldsymbol{\lambda}_2)V_n \mathrm{d}\Gamma \quad (4.1.88)$$

4.1.6 Analytical Example

For the interface design problem in Sect. 4.1.4, the sensitivity formula was derived as (4.1.75). A one-dimensional example, which has the analytic field solution, is taken to show the correctness of the derived sensitivity formula. The sensitivity result analytically calculated in the example is compared with the result of the sensitivity formula. Here, the analytical example is an infinite solenoid.

4.1.6.1 Infinite Solenoid

As an example that can be analytically calculated, consider an infinite solenoid shown in Fig. 4.2, where the current density J_ϕ flows in the solenoid coil. The solenoid coil has thickness $b - a$, and the length L in the axial direction is infinite. The objective is to obtain a target magnetic flux density $\mathbf{B}_o$ in region Ω_p by moving interface γ of the outer surface of the solenoid coil. The design variable is the radius b of the outer suface of the coil. The design sensitivity is calculated with respect to the design variable b.

The objective function is defined as

$$F = \int_{\Omega} (\mathbf{B}(\mathbf{A}) - \mathbf{B}_o)^2 m_p \mathrm{d}\Omega \quad (4.1.89)$$

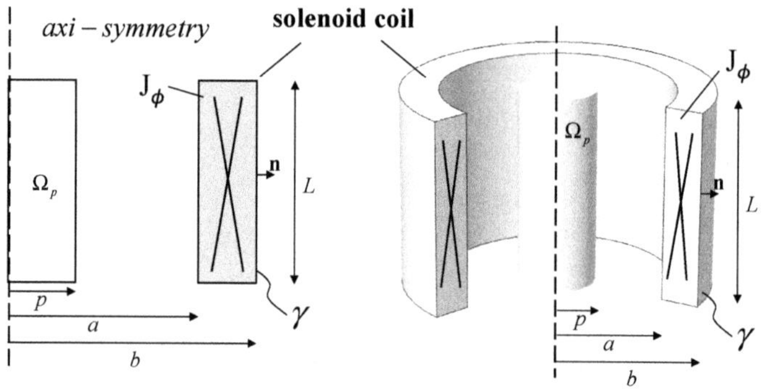

Fig. 4.2 Solenoid model

where

$$\mathbf{B}(\mathbf{A}) = \mu_0 J_\phi (b - a)\mathbf{z} \quad \text{for} \quad 0 \leq s < a \tag{4.1.90}$$

$$\mathbf{B}_o = B_o \mathbf{z} \tag{4.1.91}$$

The objective function is rewritten by using the field (4.1.90) and (4.1.91);

$$F = \left[\mu_0 J_\phi (b - a) - B_o\right]^2 \pi p^2 L \tag{4.1.92}$$

The analytical sensitivity per unit length is obtained by differentiating the objective function (4.1.92) with respect to b;

$$\frac{dF}{db} = 2\pi p^2 \mu_0 J_\phi \left[\mu_0 J_\phi (b - a) - B_o\right] \tag{4.1.93}$$

Alternatively, the sensitivity can be calculated by using the sensitivity formula (4.1.75) in Sect. 4.1.4:

$$\dot{G} = \int_\gamma J_\phi \lambda_\phi V_n \mathrm{d}\Gamma \tag{4.1.94}$$

This formula requires the adjoint variable solution, which can be obtained in the adjoint variable system in Fig. 4.3.

The differential adjoint equation is given as

$$\nabla \times \nu(\nabla \times \boldsymbol{\lambda}) = g_{\mathbf{A}}(\mathbf{A})m_p + (\nabla \times g_{\mathbf{B}}(\mathbf{A}))m_p \tag{4.1.95}$$

The adjoint sources $g_{\mathbf{A}}(\mathbf{A})$ and $\mathbf{g}_{\mathbf{B}}(\mathbf{A})$ are expressed from the objective function (4.1.92).

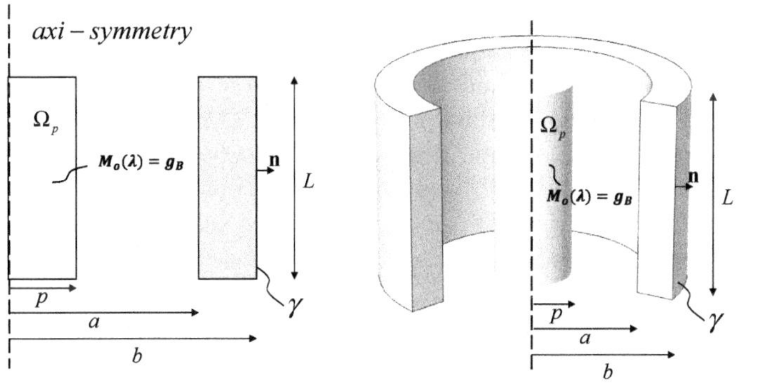

Fig. 4.3 Solenoid model-adjoint variable system

$$\mathbf{g_A}(\mathbf{A}) = \mathbf{0} \tag{4.1.96}$$

$$\mathbf{g_B}(\mathbf{A}) = 2(\mathbf{B}(\mathbf{A}) - \mathbf{B}_o)m_p = 2\left[\mu_0 J_\phi (b-a) - B_o\right] m_p \mathbf{z} \tag{4.1.97}$$

In this adjoint system, the $\mathbf{B}(\lambda_\phi)$ is calculated as

$$\mathbf{B}(\lambda_\phi) = \mu_0 g_B(A_\phi) = 2\mu_0\left[\mu_0 J_\phi (b-a) - B_o\right] m_p \mathbf{z} \tag{4.1.98}$$

By using the notation $\mathbf{B}(\lambda_\phi) = \nabla \times \lambda_\phi$, λ_ϕ on the γ is obtained as

$$\int_0^p \mathbf{B}(\lambda_\phi) r \mathrm{d}r \int_0^{2\pi} \mathrm{d}\phi = \int_0^{2\pi} \lambda_\phi b \mathrm{d}\phi \tag{4.1.99}$$

$$\lambda_\phi = \frac{p^2}{b} \mu_0\left[J_\phi (b-a) - B_o\right] \tag{4.1.100}$$

(4.1.98) is inserted into the sensitivity formula (4.1.94);

$$\dot{G} = 2\pi p^2 \mu_0 J_\phi\left[\mu_0 J_\phi (b-a) - B_o\right] V_n L \tag{4.1.101}$$

Using $V_n = \frac{\mathrm{d}b}{\mathrm{d}t}$, the design sensitivity per unit length is obtained as

$$\frac{\mathrm{d}G}{\mathrm{d}b} = 2\pi p^2 \mu_0 J_\phi\left[\mu_0 J_\phi (b-a) - B_o\right] \tag{4.1.102}$$

which is the correct result when compared with the analytical result in (4.1.93).

4.1.7 Numerical Examples

Here, the sensitivity formula (4.1.75) derived in Sect. 4.1.4 is applied to six shape optimization problems of two-dimensional design model, of which the analytic field solutions are not given. These design models are taken to show that the sensitivity formula is well applied to the shape design problem of the material interface in the magnetostatic system. The optimal designs for the first two examples are known, but the four others are not known. If the results of the first two examples are obtained as the expected optimal designs, it can be said that the shape optimization using the sensitivity formula is feasible for the design of the material interface. The results of the four other examples show that this optimization method is useful for the design of the material interface and applicable to any shaped of magnetostatic models. In these examples, the sensitivity formula requires the state and the adjoint variables, which are numerically calculated by the finite element method. The sensitivity information obtained is used for the optimization algorithm to provide the evolution of the material interface shape. For the optimization algorithm, the level set method is used to represent the shape evolution of the design model. The level set method is described in Chap. 7, where the shape evolution is expressed by the parameter t of unit s for the amount of shape change.

4.1.7.1 Infinite Length Solenoid-Ferromagnetic Interface Design

For an example with a known optimal design, consider an axi-symmetric solenoid in Fig. 4.4, where length in the direction z is infinite and the ferromagnetic material of permeability μ is located inside the solenoid coil. Here, the initial interface between the ferromagnetic material and the air is not cylindrical. When electric current flows in the solenoid coil, the distribution of the magnetic flux density is not uniform along the direction z. If the shape of the interface is changed to be cylindrical, the magnetic flux density in the solenoid coil becomes uniform along the direction z.

The design objective is to obtain a uniform field $\mathbf{B}_o$ in the region Ω_p, which is analytically given by the 1D solenoid. The design variable is the shape of the interface γ, of which the optimal shape is a cylinder for the uniform field. The objective function to be minimized is defined as the integral of the field difference in Ω_p;

$$F = \int_{\Omega} (\mathbf{B}(\mathbf{A}) - \mathbf{B}_o)^2 m_p \mathrm{d}\Omega \tag{4.1.103}$$

where

$$\mathbf{B}_o = B_o\,\mathbf{z} \tag{4.1.104}$$

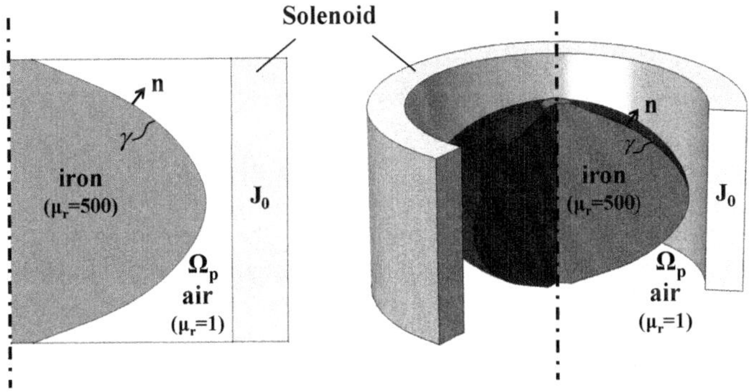

Fig. 4.4 Axi-symmetric solenoid-ferromagnetic material design

The variational adjoint equation for (4.1.103) is obtained as

$$a(\mathbf{A}, \bar{\mathbf{A}}) = \int_{\Omega} 2(\mathbf{B}(\mathbf{A}) - \mathbf{B}_o) \cdot \mathbf{B}(\bar{\mathbf{A}}) m_p \mathrm{d}\Omega \quad \forall \bar{\mathbf{A}} \in \Phi \tag{4.1.105}$$

The shape sensitivity for the ferromagnetic material is the sensitivity formula (4.1.75):

$$\dot{G} = \int_{\gamma} [\nu_0(1 - \nu_r)\mathbf{B}(\mathbf{A}_1) \cdot \mathbf{B}(\boldsymbol{\lambda}_2)] V_n \mathrm{d}\Gamma \tag{4.1.106}$$

For this minimization problem, the design velocity is taken as

$$V_n = -\nu_0(1 - \nu_r)\mathbf{B}(\mathbf{A}_1) \cdot \mathbf{B}(\boldsymbol{\lambda}_2) \tag{4.1.107}$$

This problem has a constraint of constant volume of the ferromagnetic material;

$$\int_{\Omega_1} \mathrm{d}\Omega = C \tag{4.1.108}$$

where C is a given volume of Ω_1. The constant volume (4.1.108) is equivalent to the zero sum of the design velocity over the ferromagnetic interface, which is obtained by differentiating (4.1.108);

$$\int_{\gamma} V_n \mathrm{d}\Gamma = 0 \tag{4.1.109}$$

The design velocity of (4.1.105) is modified to be U_n for the constraint:

$$U_n = V_n - V_{na} \tag{4.1.110}$$

where

$$V_{na} = -\int_{\gamma} \nu_0(1 - \nu_r)\mathbf{B}(\mathbf{A}_1) \cdot \mathbf{B}(\boldsymbol{\lambda}_2) \mathrm{d}\Gamma / \int_{\gamma} \mathrm{d}\Gamma \tag{4.1.111}$$

is the average of the velocity (4.1.107) over the interface.

The design result is shown in Fig. 4.5, where the interface between the ferromagnetic material and the air becomes gradually a cylinder with the increase of the iteration number as expected. The final design of the cylindrical shape is obtained at the 45 s, when the objective function converges to zero as in Fig. 4.6. The result of this example shows the feasibility of the shape sensitivity analysis for the interface of ferromagnetic materials in the two-dimensional axi-symmetric magnetostatic system.

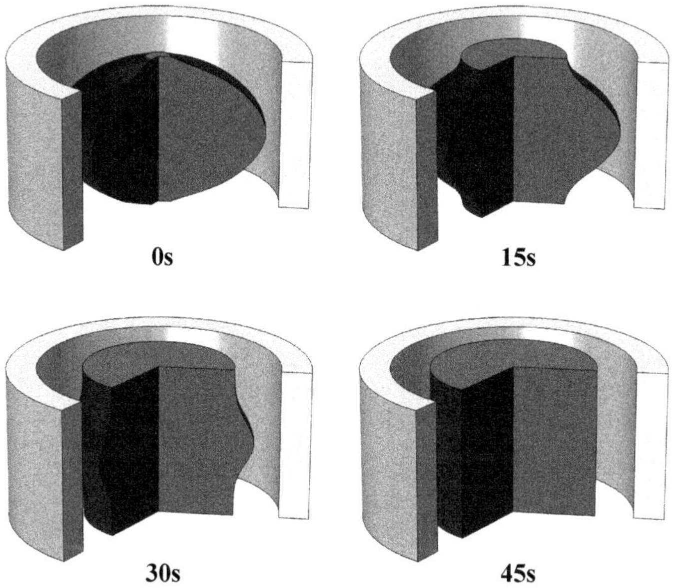

Fig. 4.5 Axi-symmetric solenoid: shape variation

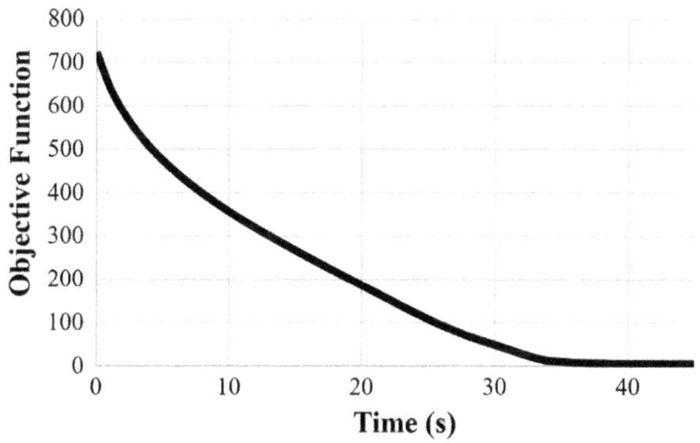

Fig. 4.6 Axi-symmetric solenoid: evolution of objective function

4.1.7.2 Infinite Air Core Solenoid-Current Region Design

To take another example with a known optimal design, consider an axi-symmetric solenoid in Fig. 4.7, where length in the direction z is infinite and the interface between the inner surface of solenoid coil and the air is curved. When a current of density J_1 flows in the solenoid coil, the distribution of the magnetic flux density is not uniform along the direction z. If the shape of the interface is changed to be cylindrical, the magnetic flux density in the solenoid coil becomes uniform along the direction z.

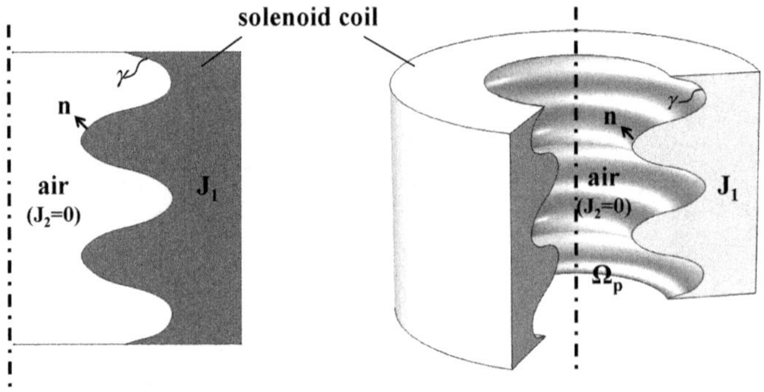

Fig. 4.7 Axi-symmetric solenoid-current region design

The design objective is to obtain a uniform field $\mathbf{B}_o$ in the region Ω_p, which is analytically given by the 1D solenoid. The design variable is the shape of the interface γ, of which the optimal shape is a cylinder. The objective function to be minimized is defined as the integral of the field difference;

$$F = \int_{\Omega} (\mathbf{B}(\mathbf{A}) - \mathbf{B}_o)^2 m_p \mathrm{d}\Omega \tag{4.1.112}$$

where

$$\mathbf{B}_o = B_o\,\mathbf{z} \tag{4.1.113}$$

The variational adjoint equation for (4.1.112) is obtained as

$$a(\mathbf{A}, \bar{\mathbf{A}}) = \int_{\Omega} 2(\mathbf{B}(\mathbf{A}) - \mathbf{B}_o) \cdot \mathbf{B}(\bar{\mathbf{A}}) m_p \mathrm{d}\Omega \quad \forall \bar{\mathbf{A}} \in \Phi \tag{4.1.114}$$

The shape sensitivity for the current region is the sensitivity formula (4.1.75):

$$\dot{G} = \int_{\gamma} \mathbf{J}_1 \cdot \boldsymbol{\lambda}_2 V_n \mathrm{d}\Gamma \tag{4.1.115}$$

This problem has a constraint of constant volume. For this minimization problem with the volume constraint, the design velocity is expressed as

$$U_n = V_n - V_{na} \tag{4.1.116}$$

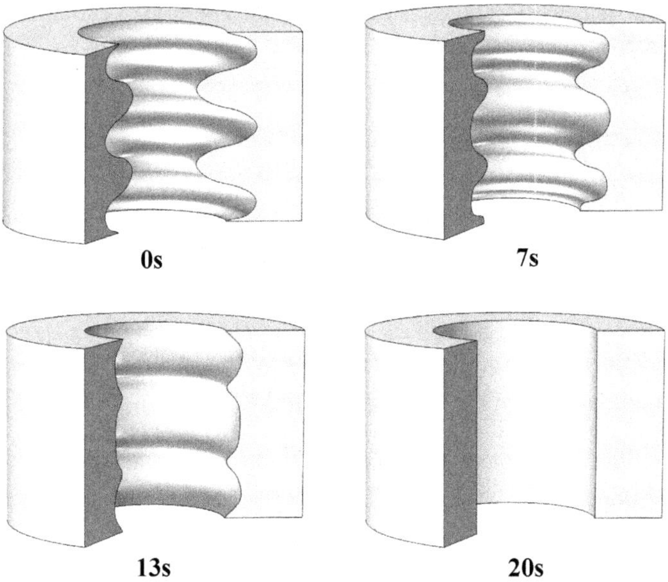

Fig. 4.8 Axi-symmetric solenoid: shape variation

where

$$V_n = -\mathbf{J}_1 \cdot \boldsymbol{\lambda}_2 \tag{4.1.117}$$

$$V_{na} = -\int_{\gamma} \mathbf{J}_1 \cdot \boldsymbol{\lambda}_2 d\Gamma / \int_{\gamma} \mathrm{d}\Gamma \tag{4.1.118}$$

The design result is shown in Fig. 4.8, where the interface γ becomes gradually a cylinder as expected. The final design of the cylindrical shape is obtained at the 20 s, when the objective function converges to zero as in Fig. 4.9. The result shows

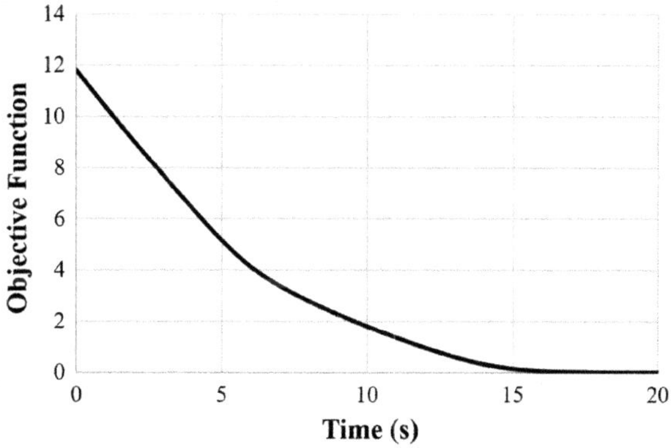

Fig. 4.9 Axi-symmetric solenoid: evolution of objective function

that the shape sensitivity analysis is feasible for the interface of current region in the axi-symmetric magnetostatic system.

4.1.7.3 Monopole Magnet-Ferromagnetic Interface Design

The shape optimization using the sensitivity formula for the ferromagnetic interface is applied to a shape design problem, which has neither the analytical field solution nor a known optimal shape. A monopole magnet is considered in Fig. 4.10, where the current-carrying winding wraps the ferromagnetic core with a squared air gap. The design objective is to obtain a uniform field $\mathbf{B}_o$ along the y-direction in the region Ω_p of the air gap, and $\mathbf{B}_o$ is taken as the average value of the initial magnetic flux density $\mathbf{B}(\mathbf{A})$ in the region Ω_p. The design variable is the shape of the interface between the ferromagnetic material and the air in the air gap.

The objective function to be minimized is defined as the integral of the field difference in

$$F = \int_{\Omega} (\mathbf{B}(\mathbf{A}) - \mathbf{B}_o)^2 m_p \mathrm{d}\Omega \tag{4.1.119}$$

where

$$\mathbf{B}_o = B_o\, \mathbf{y} \tag{4.1.120}$$

$$B_o = \int_{\Omega} |\mathbf{B}(\mathbf{A})| m_p \mathrm{d}\Omega \Big/ \int_{\Omega} m_p \mathrm{d}\Omega \tag{4.1.121}$$

The variational adjoint equation for (4.1.119) is obtained as

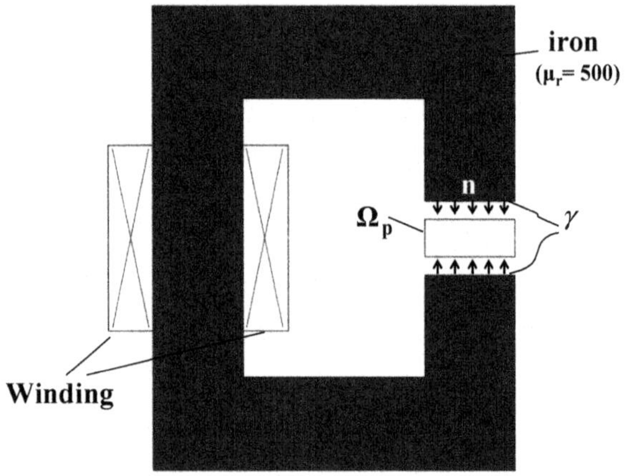

Fig. 4.10 Monopole magnet– ferromagnetic material design

$$a(\mathbf{A}, \bar{\mathbf{A}}) = \int_{\Omega} 2(\mathbf{B}(\mathbf{A}) - \mathbf{B}_o) \cdot \mathbf{B}(\bar{\mathbf{A}}) m_p \mathrm{d}\Omega, \quad \forall \bar{\mathbf{A}} \in \Phi \tag{4.1.122}$$

The sensitivity formula (4.1.75) is used for the shape sensitivity of the ferromagnetic material;

$$\dot{G} = \int_{\gamma} [\nu_0 (1 - \nu_r) \mathbf{B}(\mathbf{A}_1) \cdot \mathbf{B}(\boldsymbol{\lambda}_2)] V_n \mathrm{d}\Gamma \tag{4.1.123}$$

The design velocity for this minimization problem is taken as

$$V_n = -\nu_0 (1 - \nu_r) \mathbf{B}(\mathbf{A}_1) \cdot \mathbf{B}(\boldsymbol{\lambda}_2) \tag{4.1.124}$$

Figure 4.11 shows the initial and the final designs of the monopole electromagnet. The final design is obtained at the 30 s, when the objective function value converges to a minimum value as in Fig. 4.12. At the initial and the final shapes, the flux density distributions along the center line of the air gap are compared in

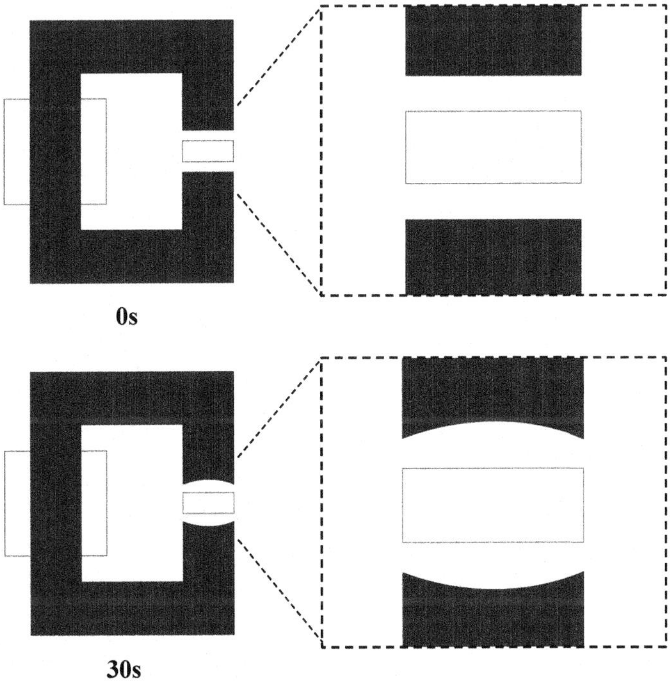

Fig. 4.11 Monopole magnet: initial and final shapes

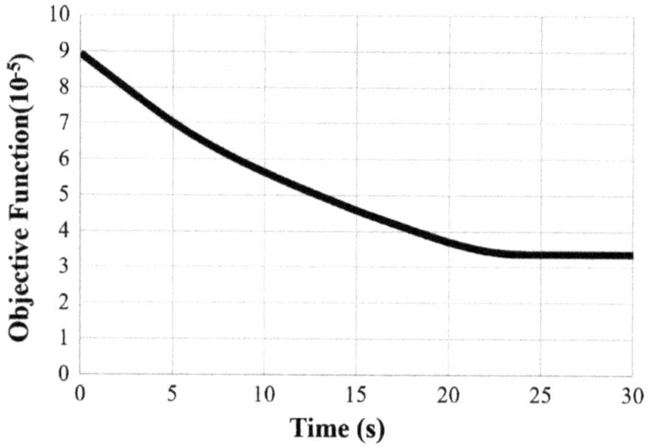

Fig. 4.12 Monopole magnet: evolution of objective function

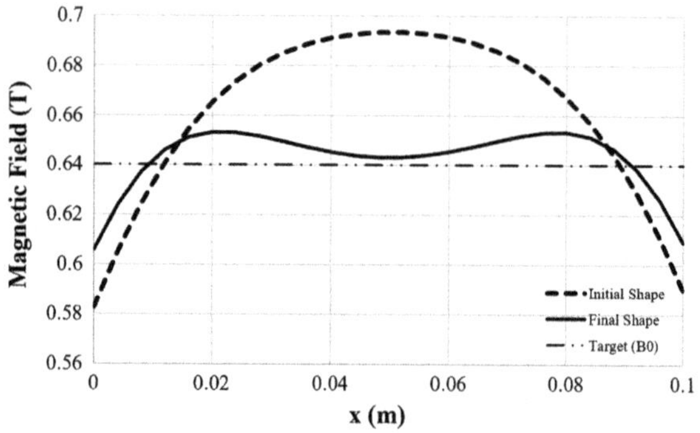

Fig. 4.13 Monopole magnet: magnetic field distribution in Ω_p

Fig. 4.13 where the distribution of the final shape is closer to the target field $\mathbf{B}_o$ than the one of the initial shape.

4.1.7.4 Air Core Solenoid-Current Region Interface Design

The shape optimization using the sensitivity formula for the interface design is applied to a shape design problem without the analytic field solution nor the known optimal shape. As an example, consider an air core solenoid in Fig. 4.14, where the initial shape of the solenoid coil is cylindrical and a current of density J_1 flows in the solenoid coil. The design objective is to obtain a uniform field $\mathbf{B}_o$ along the direction z in the region Ω_p and $\mathbf{B}_o$ is taken as the average value of the initial magnetic flux density $\mathbf{B}(\mathbf{A})$ in the region Ω_p. The design variable is the shape of the interface between the outer surface of the solenoid coil and the air.

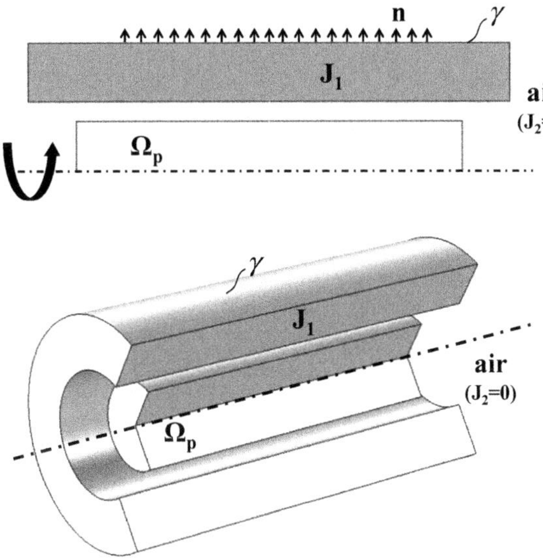

Fig. 4.14 Axi-symmetric solenoid-current region design

The objective function to be minimized is defined as the integral of the field difference;

$$F = \int_{\Omega} (\mathbf{B}(\mathbf{A}) - \mathbf{B}_o)^2 m_p \mathrm{d}\Omega \tag{4.1.125}$$

where

$$\mathbf{B}_o = B_o\,\mathbf{z} \tag{4.1.126}$$

$$B_o = \int_{\Omega} |\mathbf{B}(\mathbf{A})| m_p \mathrm{d}\Omega \Big/ \int_{\Omega} m_p \mathrm{d}\Omega \tag{4.1.127}$$

The variational adjoint equation for (4.1.125) is obtained as

$$a(\mathbf{A}, \overline{\mathbf{A}}) = \int_{\Omega} 2(\mathbf{B}(\mathbf{A}) - \mathbf{B}_o) \cdot \mathbf{B}(\overline{\mathbf{A}}) m_p \mathrm{d}\Omega \quad \forall \overline{\mathbf{A}} \in \Phi \tag{4.1.128}$$

The shape sensitivity by the variation of the current region is the sensitivity formula (4.1.75):

$$\dot{G} = \int_{\gamma} \mathbf{J}_1 \cdot \boldsymbol{\lambda}_2 V_n \mathrm{d}\Gamma \tag{4.1.129}$$

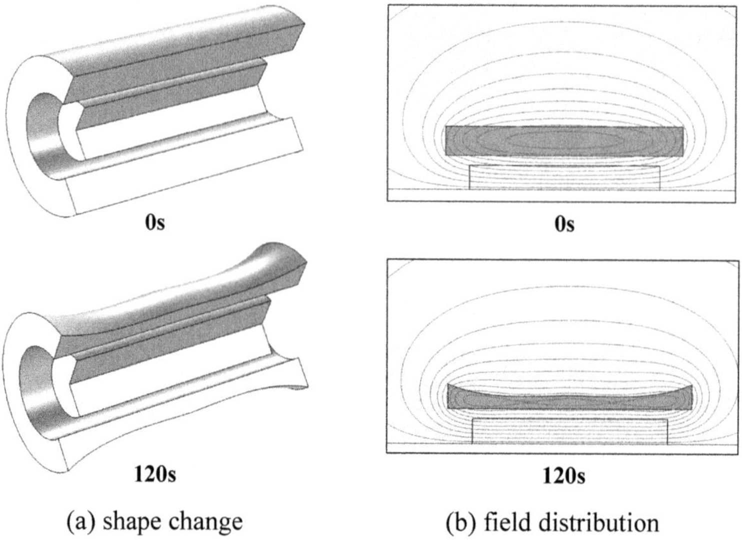

Fig. 4.15 Axi-symmetric solenoid: initial and final designs

Fig. 4.16 Axi-symmetric solenoid: evolution of objective function

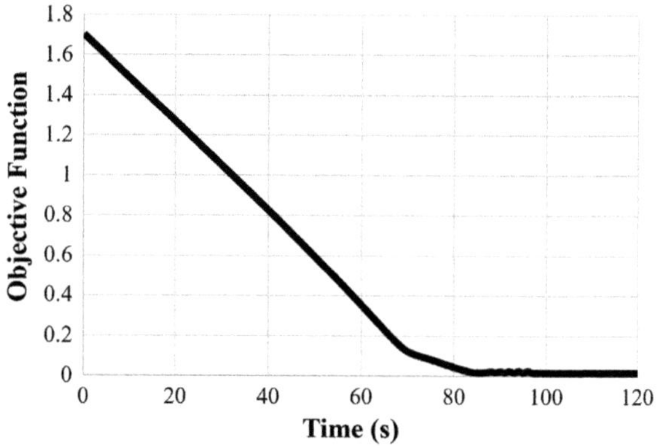

The design velocity for this minimization problem is taken as

$$V_n = -\mathbf{J}_1 \cdot \boldsymbol{\lambda}_2 \tag{4.1.130}$$

Figure 4.15a shows the initial and the final shapes of the air core solenoid, where the middle of the design interface moves down and the edge sides move up in the optimization process. According to this shape evolution, the magnetic field distribution becomes uniform in the final design as in Fig. 4.15b. The final design is obtained at the 120 s, when the objective function converges to almost zero as in Fig. 4.16.

4.1.7.5 Magnetic Shielding-Ferromagnetic Material Design

The shape optimization using the sensitivity formula for the interface design of ferromagnetic material is applied to a shape design problem, which has neither the analytical field solution nor a known optimal shape. Consider a magnetic shielding model in Fig. 4.17, where a ferromagnetic shell encloses a current-carrying coil to reduce the leakage of the magnetic flux. The design objective is to obtain a minimum magnetic flux in the region Ω_p outside of the ferromagnetic shell. The design variable is the shape of the interface between the outer surface of the ferromagnetic material and the surrounding air.

The objective function to be minimized is defined as the integral of the magnetic flux intensity in Ω_p.

$$F = \int_{\Omega} \mathbf{B}(\mathbf{A})^2 m_p \mathrm{d}\Omega \tag{4.1.131}$$

The variational adjoint equation for (4.1.131) is obtained as

$$a(\mathbf{A}, \bar{\mathbf{A}}) = \int_{\Omega} 2(\mathbf{B}(\mathbf{A}) - \mathbf{B}_o) \cdot \mathbf{B}(\bar{\mathbf{A}}) m_p \mathrm{d}\Omega \quad \forall \bar{\mathbf{A}} \in \Phi \tag{4.1.132}$$

The shape sensitivity for the interface of the ferromagnetic material is the sensitivity formula (4.1.75):

$$\dot{G} = \int_{\gamma} [v_0(1 - v_r)\mathbf{B}(\mathbf{A}_1) \cdot \mathbf{B}(\boldsymbol{\lambda}_2)] V_n \mathrm{d}\Gamma \tag{4.1.133}$$

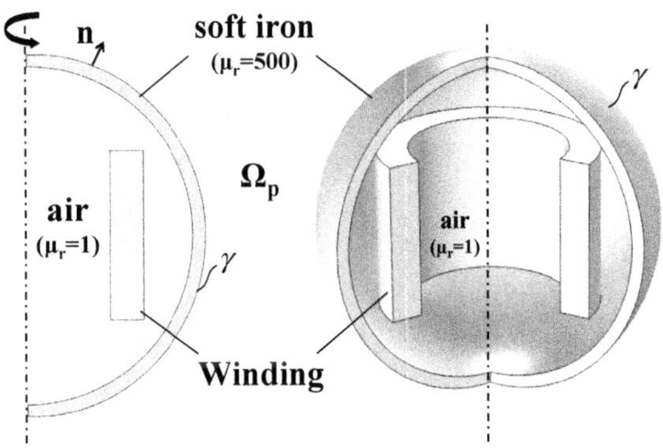

Fig. 4.17 Axi-symmetric solenoid-ferromagnetic material design

This problem has a constraint of constant volume of the ferromagnetic material. The design velocity for this minimization problem with the volume constraint is taken as

$$U_n = V_n - V_{na} \tag{4.1.134}$$

where

$$V_n = -\nu_0(1-\nu_r)\mathbf{B}(\mathbf{A}_1)\cdot\mathbf{B}(\boldsymbol{\lambda}_2) \tag{4.1.135}$$

$$V_{na} = -\int_\gamma \nu_0(1-\nu_r)\mathbf{B}(\mathbf{A}_1)\cdot\mathbf{B}(\boldsymbol{\lambda}_2)\mathrm{d}\Gamma \Big/ \int_\gamma \mathrm{d}\Gamma \tag{4.1.136}$$

Figure 4.18 shows the initial and the final design of the magnetic shielding model. To minimize the leakage flux under the volume constraint, the shell near the axis becomes thinner and finally a hole is generated. According to this shape change, the variation of the objective function is shown in Fig. 4.19, where the final value converges near at the 90s.

Fig. 4.18 Axi-symmetric solenoid: initial and final shapes

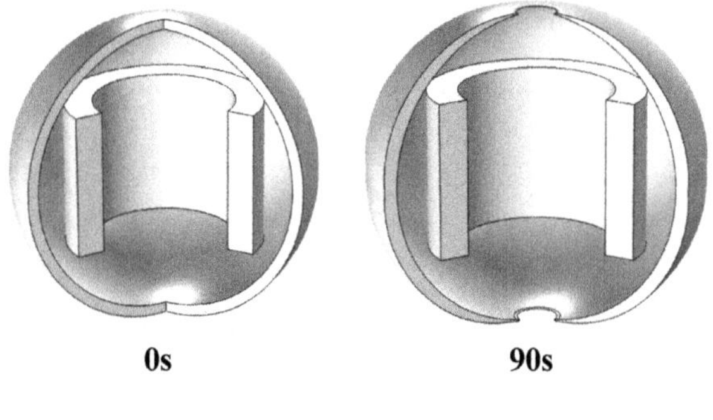

Fig. 4.19 Axi-symmetric solenoid: evolution of objective function

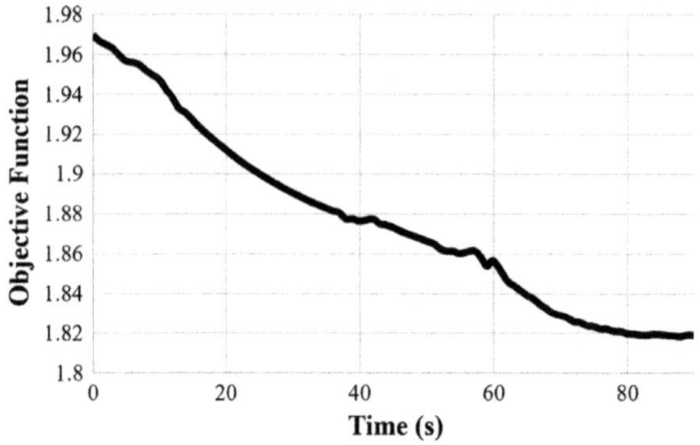

4.1.7.6 C-Shape Permanent Magnet-Magnet Interface Design

The shape optimization using the sensitivity formula for the design of the permanent magnet interface is applied to a shape design problem, which has neither the analytical field solution nor the known optimal shape. Consider a numerical model in Fig. 4.20, where the c-shape permanent magnet with an air gap is magnetized constantly in the azimuthal direction. The design objective is to obtain a uniform field $\mathbf{B}_o$ along the y-direction in the region Ω_p of the air gap. $\mathbf{B}_o$ is taken as the average value of the initial magnetic flux density $\mathbf{B}(\mathbf{A})$ in the region Ω_p. The design variable is the shape of the interface between the permanent magnet and the air in the air gap.

The objective function to be minimized is defined as the integral of the field difference;

$$F = \int_{\Omega} (\mathbf{B}(\mathbf{A}) - \mathbf{B}_o)^2 m_p \mathrm{d}\Omega \tag{4.1.137}$$

where

$$\mathbf{B}_o = B_o \, \mathbf{y} \tag{4.1.138}$$

$$B_o = \int_{\Omega} |\mathbf{B}(\mathbf{A})| m_p d\Omega \; / \int_{\Omega} m_p \mathrm{d}\Omega \tag{4.1.139}$$

The variational adjoint equation for (4.1.137) is obtained as

$$a(\mathbf{A}, \bar{\mathbf{A}}) = \int_{\Omega} 2(\mathbf{B}(\mathbf{A}) - \mathbf{B}_o) \cdot \mathbf{B}(\bar{\mathbf{A}}) m_p \mathrm{d}\Omega \quad \forall \bar{\mathbf{A}} \in \Phi \tag{4.1.140}$$

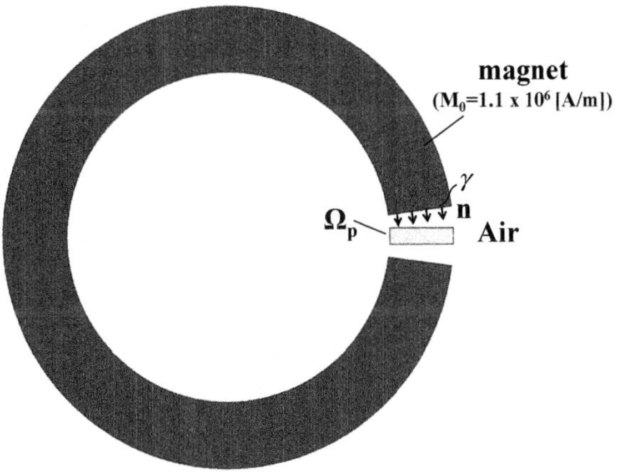

Fig. 4.20 C-shape permanent magnet-magnet design

The shape sensitivity by the variation of the permanent magnetization region is the sensitivity formula (4.1.75):

$$\dot{G} = \int_{\gamma} \mathbf{M}_{o1} \cdot \mathbf{B}(\boldsymbol{\lambda}_2) V_n \mathrm{d}\Gamma \tag{4.1.141}$$

The design velocity for this minimization problem is taken as

$$V_n = -\mathbf{M}_{o1} \cdot \mathbf{B}(\boldsymbol{\lambda}_2) \tag{4.1.142}$$

Figure 4.21 compares the initial and the final designs of the c-shape permanent magnet. The final design is obtained at the 50s, when the objective function converges to minimum value as in Fig. 4.22. For the initial and the final shapes, the flux density distributions along the centerline of the air gap are compared in Fig. 4.23, where the field of the final shape is much closer to the target field $\mathbf{B}_0$ than the one of the initial shape.

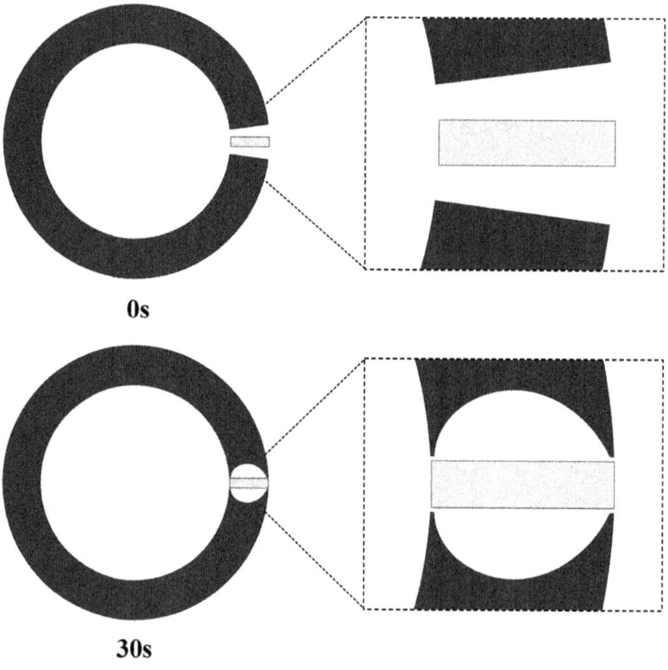

Fig. 4.21 C-shape permanent magnet: initial and final shapes

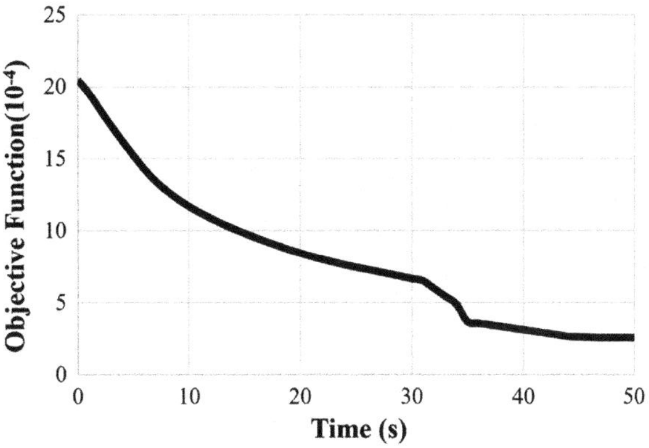

Fig. 4.22 C-shape permanent magnet: evolution of objective function

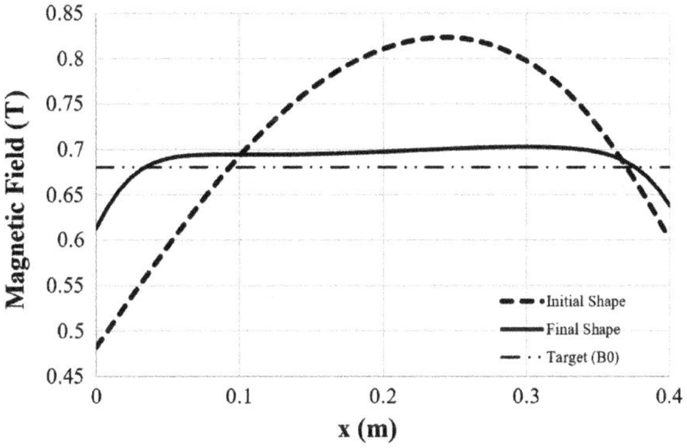

Fig. 4.23 C-shape permanent magnet: magnetic field distribution in Ω_p

4.2 Interface Shape Sensitivity for System Energy

In this section, the three-dimensional shape sensitivity for the interface design is developed in the linear magnetostatic system as in Sect. 4.1, but the objective function is the system energy. The system energy of the magnetostatic system is related to the inductance of electric circuit and it can be also used for designing various magnetostatic systems. The derivation procedure is almost the same as in Sect. 4.1. The difference is that the adjoint variable for the system energy is the same as the state variable. Thus, solving the adjoint variable equation is not necessary. The derived sensitivity formula is tested and validated with analytical and numerical examples.

4.2.1 Problem Definition

A magnetostatic system for interface design is given as in Fig. 4.24, which is the same as Fig. 4.1 in Sect. 4.1. Since there are many magnetostatic problems with

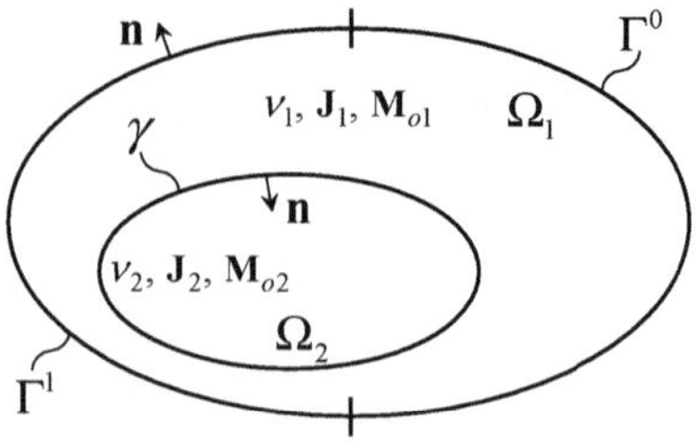

Fig. 4.24 Interface design of linear magnetostatic system for system energy

permanent magnet, the permanent magnet is included for the system energy. There is controversy about the energy density of the permanent magnet, but the expressions for the energy density only differ by constant values. Thus, when the differential of the system energy is used, the use of any energy density does not matter. This appears in the force calculation. It is also assumed that the operating point of the permanent magnet remains in the elastic region on the demagnetization curve. Since the permanent magnet is treated as an equivalent magnetization current for the magnetic vector potential **A**, the system energy is written as the integral of the magnetic field energy density expressed with the magnetic flux density **B**. When the inductance is calculated using the system energy, since the equivalent inductance is not related to the permanent magnet, the permanent magnet is excluded for the system energy.

The objective function is the system energy of the magnetostatic system;

$$
\begin{aligned}
W_m &= \int_{\Omega} \frac{1}{2} \nu \mathbf{B}(\mathbf{A}) \cdot \mathbf{B}(\mathbf{A}) d\Omega \\
&= \int_{\Omega_1} \frac{1}{2} \nu_1 \mathbf{B}(\mathbf{A}_1) \cdot \mathbf{B}(\mathbf{A}_1) d\Omega + \int_{\Omega_2} \frac{1}{2} \nu_2 \mathbf{B}(\mathbf{A}_2) \cdot \mathbf{B}(\mathbf{A}_2) d\Omega
\end{aligned} \tag{4.2.1}
$$

In this problem, the governing differential equations and all the boundary and interface conditions are the same as the ones in Sect. 4.1, and the variational identities and variational state equation for the state equations are also the same as the ones in Sect. 4.1. Thus, they are not written again but referred when needed.

4.2.2 Lagrange Multiplier Method for Energy Sensitivity

The variational state equation (4.1.19) is also treated as an equality constraint in the shape sensitivity analysis. The method of Lagrange multipliers is employed for the implicit constraint of the variational state equation to provide an augmented objective function G as

$$G = W_m + l(\bar{\mathbf{A}}) - a(\mathbf{A}, \bar{\mathbf{A}}) \quad \forall \bar{\mathbf{A}} \in \Phi \tag{4.2.2}$$

where the arbitrary virtual potential $\bar{\mathbf{A}}$ plays the role of Lagrange multipliers and Φ is the space of admissible vector potential: $\Phi = \left\{ \bar{\mathbf{A}} \in [H^1(\Omega)]^3 \middle| \bar{\mathbf{A}} = \mathbf{0} \right.$ on $\mathbf{x} \in \Gamma^0.\}$.

The sensitivity, the material derivative of the augmented objective function G (4.2.2), is written as

$$\dot{G} = -\dot{a}(\mathbf{A}, \bar{\mathbf{A}}) + \dot{l}(\bar{\mathbf{A}}) + \dot{W}_m \quad \forall \bar{\mathbf{A}} \in \Phi \tag{4.2.3}$$

As in Sect. 4.1, each term in (4.2.3) is obtained by applying the material derivative formula. The first and second terms are the same as (4.1.23) and (4.1.24) in Sect. 4.1, respectively. The third term is obtained below as (4.2.4).

$$\begin{aligned}
\dot{W}_m = & \int_{\Omega_1} \nu_1 \mathbf{B}(\mathbf{A}_1) \cdot \mathbf{B}(\mathbf{A}_1') \mathrm{d}\Omega + \int_{\gamma} \frac{1}{2} \nu_1 \mathbf{B}(\mathbf{A}_1) \cdot \mathbf{B}(\mathbf{A}_1) V_n \mathrm{d}\Gamma \\
& + \int_{\Omega_2} \nu_2 \mathbf{B}(\mathbf{A}_2) \cdot \mathbf{B}(\mathbf{A}_2') \mathrm{d}\Omega - \int_{\gamma} \frac{1}{2} \nu_2 \mathbf{B}(\mathbf{A}_2) \cdot \mathbf{B}(\mathbf{A}_2) V_n \mathrm{d}\Gamma \\
= & \int_{\Omega_1} \left(\nu_1 \mathbf{B}(\mathbf{A}_1) \cdot \mathbf{B}(\dot{\mathbf{A}}_1) - \nu_1 \mathbf{B}(\mathbf{A}_1) \cdot \mathbf{B}(\mathbf{V} \cdot \nabla \mathbf{A}_1) \right) \mathrm{d}\Omega \\
& + \int_{\Omega_2} \left(\nu_2 \mathbf{B}(\mathbf{A}_2) \cdot \mathbf{B}(\dot{\mathbf{A}}_2) - \nu_2 \mathbf{B}(\mathbf{A}_2) \cdot \mathbf{B}(\mathbf{V} \cdot \nabla \mathbf{A}_2) \right) \mathrm{d}\Omega \\
& + \int_{\gamma} \left(\frac{1}{2} \nu_1 \mathbf{B}(\mathbf{A}_1) \cdot \mathbf{B}(\mathbf{A}_1) - \frac{1}{2} \nu_2 \mathbf{B}(\mathbf{A}_2) \cdot \mathbf{B}(\mathbf{A}_2) \right) V_n \mathrm{d}\Gamma
\end{aligned} \tag{4.2.4}$$

(4.1.23), (4.1.24), and (4.2.4) are inserted into (4.2.3) to provide

$$
\begin{aligned}
\dot{G} = & -\int_{\Omega_1} \nu_1 \mathbf{B}(\dot{\mathbf{A}}_1) \cdot \mathbf{B}(\bar{\mathbf{A}}_1) \mathrm{d}\Omega - \int_{\Omega_2} \nu_2 \mathbf{B}(\dot{\mathbf{A}}_2) \cdot \mathbf{B}(\bar{\mathbf{A}}_2) \mathrm{d}\Omega \\
& - \int_{\Omega_1} \nu_1 \mathbf{B}(\mathbf{A}_1) \cdot \mathbf{B}(\dot{\bar{\mathbf{A}}}_1) \mathrm{d}\Omega - \int_{\Omega_2} \nu_2 \mathbf{B}(\mathbf{A}_2) \cdot \mathbf{B}(\dot{\bar{\mathbf{A}}}_2) \mathrm{d}\Omega \\
& + \int_{\Omega_1} \left(\mathbf{J}_1 \cdot \dot{\bar{\mathbf{A}}}_1 + \mathbf{M}_{o1} \cdot \mathbf{B}(\dot{\bar{\mathbf{A}}}_1) \right) \mathrm{d}\Omega + \int_{\Omega_2} \left(\mathbf{J}_2 \cdot \dot{\bar{\mathbf{A}}}_2 + \mathbf{M}_{o2} \cdot \mathbf{B}(\dot{\bar{\mathbf{A}}}_2) \right) \mathrm{d}\Omega \\
& + \int_{\Omega_1} \nu_1 \mathbf{B}(\mathbf{A}_1) \cdot \mathbf{B}(\dot{\mathbf{A}}_1) \mathrm{d}\Omega + \int_{\Omega_2} \nu_2 \mathbf{B}(\mathbf{A}_2) \cdot \mathbf{B}(\dot{\mathbf{A}}_2) \mathrm{d}\Omega \\
& + \int_{\Omega_1} \nu_1 \mathbf{B}(\mathbf{V} \cdot \nabla \mathbf{A}_1) \cdot \mathbf{B}(\bar{\mathbf{A}}_1) \mathrm{d}\Omega + \int_{\Omega_2} \nu_2 \mathbf{B}(\mathbf{V} \cdot \nabla \mathbf{A}_2) \cdot \mathbf{B}(\bar{\mathbf{A}}_2) \mathrm{d}\Omega \\
& + \int_{\Omega_1} \nu_1 \mathbf{B}(\mathbf{A}_1) \cdot \mathbf{B}(\mathbf{V} \cdot \nabla \bar{\mathbf{A}}_1) \mathrm{d}\Omega + \int_{\Omega_2} \nu_2 \mathbf{B}(\mathbf{A}_2) \cdot \mathbf{B}(\mathbf{V} \cdot \nabla \bar{\mathbf{A}}_2) \mathrm{d}\Omega \\
& - \int_{\Omega_1} \left[\mathbf{J}_1 \cdot (\mathbf{V} \cdot \nabla \bar{\mathbf{A}}_1) + \mathbf{M}_{o1} \cdot \mathbf{B}(\mathbf{V} \cdot \nabla \bar{\mathbf{A}}_1) \right] \mathrm{d}\Omega \\
& + \int_{\Omega_2} \left[\mathbf{J}_2 \cdot (\mathbf{V} \cdot \nabla \bar{\mathbf{A}}_2) + \mathbf{M}_{o2} \cdot \mathbf{B}(\mathbf{V} \cdot \nabla \bar{\mathbf{A}}_2) \right] \mathrm{d}\Omega \\
& - \int_{\Omega_1} \nu_1 \mathbf{B}(\mathbf{A}_1) \cdot \mathbf{B}(\mathbf{V} \cdot \nabla \mathbf{A}_1) \mathrm{d}\Omega + \int_{\Omega_2} \nu_2 \mathbf{B}(\mathbf{A}_2) \cdot \mathbf{B}(\mathbf{V} \cdot \nabla \mathbf{A}_2) \mathrm{d}\Omega \\
& - \int_{\gamma} \Big(\nu_1 \mathbf{B}(\mathbf{A}_1) \cdot \mathbf{B}(\bar{\mathbf{A}}_1) - \nu_2 \mathbf{B}(\mathbf{A}_2) \cdot \mathbf{B}(\bar{\mathbf{A}}_2) \\
& - \frac{1}{2} \nu_1 \mathbf{B}(\mathbf{A}_1) \cdot \mathbf{B}(\mathbf{A}_1) + \frac{1}{2} \nu_2 \mathbf{B}(\mathbf{A}_2) \cdot \mathbf{B}(\mathbf{A}_2) \Big) V_n \mathrm{d}\Gamma \\
& + \int_{\gamma} \left(\mathbf{M}_{o1} \cdot \mathbf{B}(\bar{\mathbf{A}}_1) - \mathbf{M}_{o2} \cdot \mathbf{B}(\bar{\mathbf{A}}_2) \right) V_n \mathrm{d}\Gamma + \int_{\gamma} \left(\mathbf{J}_1 \cdot \bar{\mathbf{A}}_1 - \mathbf{J}_2 \cdot \bar{\mathbf{A}}_2 \right) V_n \mathrm{d}\Gamma \\
& \forall \bar{\mathbf{A}}_1, \bar{\mathbf{A}}_2 \in \Phi
\end{aligned} \tag{4.2.5}
$$

Since $\dot{\bar{\mathbf{A}}}_1$ and $\dot{\bar{\mathbf{A}}}_2$ belong to Φ, the variational state equation of (4.1.16) provides

$$
\begin{aligned}
& \int_{\Omega_1} \nu_1 \mathbf{B}(\mathbf{A}_1) \cdot \mathbf{B}(\dot{\bar{\mathbf{A}}}_1) \mathrm{d}\Omega + \int_{\Omega_2} \nu_2 \mathbf{B}(\mathbf{A}_2) \cdot \mathbf{B}(\dot{\bar{\mathbf{A}}}_2) \mathrm{d}\Omega \\
& = \int_{\Omega_1} \left(\mathbf{J}_1 \cdot \dot{\bar{\mathbf{A}}}_1 + \mathbf{M}_{o1} \cdot \mathbf{B}(\dot{\bar{\mathbf{A}}}_1) \right) \mathrm{d}\Omega + \int_{\Omega_2} \left(\mathbf{J}_2 \cdot \dot{\bar{\mathbf{A}}}_2 + \mathbf{M}_{o2} \cdot \mathbf{B}(\dot{\bar{\mathbf{A}}}_2) \right) \mathrm{d}\Omega
\end{aligned} \tag{4.2.6}
$$

Hence, all terms with $\dot{\mathbf{A}}_1$ and $\dot{\mathbf{A}}_2$ in (4.2.5) are canceled out;

$$
\begin{aligned}
\dot{G} = & -\int_{\Omega_1} \nu_1 \mathbf{B}(\dot{\mathbf{A}}_1)\cdot\mathbf{B}(\bar{\mathbf{A}}_1)\mathrm{d}\Omega - \int_{\Omega_2} \nu_2 \mathbf{B}(\dot{\mathbf{A}}_2)\cdot\mathbf{B}(\bar{\mathbf{A}}_2)\mathrm{d}\Omega \\
& + \int_{\Omega_1} \nu_1 \mathbf{B}(\mathbf{A}_1)\cdot\mathbf{B}(\dot{\mathbf{A}}_1)\mathrm{d}\Omega + \int_{\Omega_2} \nu_2 \mathbf{B}(\mathbf{A}_2)\cdot\mathbf{B}(\dot{\mathbf{A}}_2)\mathrm{d}\Omega \\
& + \int_{\Omega_1} \nu_1 \mathbf{B}(\mathbf{V}\cdot\nabla\mathbf{A}_1)\cdot\mathbf{B}(\bar{\mathbf{A}}_1)\mathrm{d}\Omega + \int_{\Omega_2} \nu_2 \mathbf{B}(\mathbf{V}\cdot\nabla\mathbf{A}_2)\cdot\mathbf{B}(\bar{\mathbf{A}}_2)\mathrm{d}\Omega \\
& + \int_{\Omega_1} \nu_1 \mathbf{B}(\mathbf{A}_1)\cdot\mathbf{B}(\mathbf{V}\cdot\nabla\bar{\mathbf{A}}_1)\mathrm{d}\Omega + \int_{\Omega_2} \nu_2 \mathbf{B}(\mathbf{A}_2)\cdot\mathbf{B}(\mathbf{V}\cdot\nabla\bar{\mathbf{A}}_2)\mathrm{d}\Omega \\
& - \int_{\Omega_1} \left[\mathbf{J}_1\cdot(\mathbf{V}\cdot\nabla\bar{\mathbf{A}}_1) + \mathbf{M}_{o1}\cdot\mathbf{B}(\mathbf{V}\cdot\nabla\bar{\mathbf{A}}_1)\right]\mathrm{d}\Omega \\
& - \int_{\Omega_2} \left[\mathbf{J}_2\cdot(\mathbf{V}\cdot\nabla\bar{\mathbf{A}}_2) + \mathbf{M}_{o2}\cdot\mathbf{B}(\mathbf{V}\cdot\nabla\bar{\mathbf{A}}_2)\right]\mathrm{d}\Omega \\
& - \int_{\Omega_1} \nu_1 \mathbf{B}(\mathbf{A}_1)\cdot\mathbf{B}(\mathbf{V}\cdot\nabla\mathbf{A}_1)\mathrm{d}\Omega - \int_{\Omega_2} \nu_2 \mathbf{B}(\mathbf{A}_2)\cdot\mathbf{B}(\mathbf{V}\cdot\nabla\mathbf{A}_2)\mathrm{d}\Omega \\
& - \int_{\gamma} \left(\begin{array}{c} \nu_1\mathbf{B}(\mathbf{A}_1)\cdot\mathbf{B}(\bar{\mathbf{A}}_1) - \nu_2\mathbf{B}(\mathbf{A}_2)\cdot\mathbf{B}(\bar{\mathbf{A}}_2) \\ -\frac{1}{2}\nu_1\mathbf{B}(\mathbf{A}_1)\cdot\mathbf{B}(\mathbf{A}_1) + \frac{1}{2}\nu_2\mathbf{B}(\mathbf{A}_2)\cdot\mathbf{B}(\mathbf{A}_2) \end{array} \right) V_n \mathrm{d}\Gamma \\
& + \int_{\gamma} \left(\mathbf{M}_{o1}\cdot\mathbf{B}(\bar{\mathbf{A}}_1) - \mathbf{M}_{o2}\cdot\mathbf{B}(\bar{\mathbf{A}}_2)\right) V_n \mathrm{d}\Gamma \\
& + \int_{\gamma} \left(\mathbf{J}_1\cdot\bar{\mathbf{A}}_1 - \mathbf{J}_2\cdot\bar{\mathbf{A}}_2\right) V_n \mathrm{d}\Gamma \quad \forall \bar{\mathbf{A}}_1, \bar{\mathbf{A}}_2 \in \Phi
\end{aligned}
\tag{4.2.7}
$$

4.2.3 *Adjoint Variable Method for Sensitivity Analysis*

To obtain an explicit expression of (4.2.7) in terms of the velocity field $\mathbf{V}$, an adjoint equation is introduced. The adjoint equation is obtained by replacing $\dot{\mathbf{A}}_1$ and $\dot{\mathbf{A}}_2$ in the third and fourth integrals of (4.2.7) with a virtual potential $\bar{\boldsymbol{\lambda}}_1$ and $\bar{\boldsymbol{\lambda}}_2$, respectively, and by equating the integrals to the energy bilinear form $a(\boldsymbol{\lambda}, \bar{\boldsymbol{\lambda}})$. The adjoint equation so obtained is written as

$$\int_{\Omega_1} \nu_1 \mathbf{B}(\boldsymbol{\lambda}_1) \cdot \mathbf{B}(\bar{\boldsymbol{\lambda}}_1) \mathrm{d}\Omega + \int_{\Omega_2} \nu_2 \mathbf{B}(\boldsymbol{\lambda}_2) \cdot \mathbf{B}(\bar{\boldsymbol{\lambda}}_2) \mathrm{d}\Omega$$
$$= \int_{\Omega_1} \nu_1 \mathbf{B}(\mathbf{A}_1) \cdot \mathbf{B}(\bar{\boldsymbol{\lambda}}_1) \mathrm{d}\Omega + \int_{\Omega_2} \nu_2 \mathbf{B}(\mathbf{A}_2) \cdot \mathbf{B}(\bar{\boldsymbol{\lambda}}_2) \mathrm{d}\Omega \quad \forall \bar{\boldsymbol{\lambda}}_1,\ \bar{\boldsymbol{\lambda}}_2 \in \Phi \tag{4.2.8}$$

where $\boldsymbol{\lambda}_1$ and $\boldsymbol{\lambda}_2$ are the adjoint variables and their solutions are desired, and Φ is the space of admissible vector potential: $\Phi = \left\{ \bar{\boldsymbol{\lambda}} \in [H^1(\Omega)]^3 | \bar{\boldsymbol{\lambda}} = \mathbf{0} \quad \text{on} \quad \mathbf{x} \in \Gamma^0 \right\}$.

The adjoint Eq. (4.2.8) is evaluated at the specific $\bar{\boldsymbol{\lambda}}_1 = \dot{\mathbf{A}}_1$ and $\bar{\boldsymbol{\lambda}}_2 = \dot{\mathbf{A}}_2$, since (4.2.8) holds for all $\bar{\boldsymbol{\lambda}}_1,\ \bar{\boldsymbol{\lambda}}_2 \in \Phi$, to yield

$$\int_{\Omega_1} \nu_1 \mathbf{B}(\boldsymbol{\lambda}_1) \cdot \mathbf{B}(\dot{\mathbf{A}}_1) \mathrm{d}\Omega + \int_{\Omega_2} \nu_2 \mathbf{B}(\boldsymbol{\lambda}_2) \cdot \mathbf{B}(\dot{\mathbf{A}}_2) \mathrm{d}\Omega$$
$$= \int_{\Omega_1} \nu_1 \mathbf{B}(\mathbf{A}_1) \cdot \mathbf{B}(\dot{\mathbf{A}}_1) \mathrm{d}\Omega + \int_{\Omega_2} \nu_2 \mathbf{B}(\mathbf{A}_2) \cdot \mathbf{B}(\dot{\mathbf{A}}_2) \mathrm{d}\Omega \tag{4.2.9}$$

Similarly, the sensitivity Eq. (4.2.7) is evaluated at the specific $\bar{\mathbf{A}}_1 = \boldsymbol{\lambda}_1$, $\bar{\mathbf{A}}_2 = \boldsymbol{\lambda}_2$ since the $\boldsymbol{\lambda}_1$ and $\boldsymbol{\lambda}_2$ belong to the admissible space Φ, to yield

$$\begin{aligned}
\dot{G} = & -\int_{\Omega_1} \nu_1 \mathbf{B}(\dot{\mathbf{A}}_1) \cdot \mathbf{B}(\boldsymbol{\lambda}_1) \mathrm{d}\Omega - \int_{\Omega_2} \nu_2 \mathbf{B}(\dot{\mathbf{A}}_2) \cdot \mathbf{B}(\boldsymbol{\lambda}_2) \mathrm{d}\Omega \\
& + \int_{\Omega_1} \nu_1 \mathbf{B}(\mathbf{A}_1) \cdot \mathbf{B}(\dot{\mathbf{A}}_1) \mathrm{d}\Omega + \int_{\Omega_2} \nu_2 \mathbf{B}(\mathbf{A}_2) \cdot \mathbf{B}(\dot{\mathbf{A}}_2) \mathrm{d}\Omega \\
& + \int_{\Omega_1} \nu_1 \mathbf{B}(\mathbf{V} \cdot \nabla \mathbf{A}_1) \cdot \mathbf{B}(\boldsymbol{\lambda}_1) \mathrm{d}\Omega + \int_{\Omega_2} \nu_2 \mathbf{B}(\mathbf{V} \cdot \nabla \mathbf{A}_2) \cdot \mathbf{B}(\boldsymbol{\lambda}_2) \mathrm{d}\Omega \\
& + \int_{\Omega_1} \nu_1 \mathbf{B}(\mathbf{A}_1) \cdot \mathbf{B}(\mathbf{V} \cdot \nabla \boldsymbol{\lambda}_1) \mathrm{d}\Omega \\
& + \int_{\Omega_2} \nu_2 \mathbf{B}(\mathbf{A}_2) \cdot \mathbf{B}(\mathbf{V} \cdot \nabla \boldsymbol{\lambda}_2) \mathrm{d}\Omega \\
& - \int_{\Omega_1} [\mathbf{J}_1 \cdot (\mathbf{V} \cdot \nabla \boldsymbol{\lambda}_1) + \mathbf{M}_{o1} \cdot \mathbf{B}(\mathbf{V} \cdot \nabla \boldsymbol{\lambda}_1)] \mathrm{d}\Omega - \int_{\Omega_2} [\mathbf{J}_2 \cdot (\mathbf{V} \cdot \nabla \boldsymbol{\lambda}_2) + \mathbf{M}_{o2} \cdot \mathbf{B}(\mathbf{V} \cdot \nabla \boldsymbol{\lambda}_2)] \mathrm{d}\Omega \\
& - \int_{\Omega_1} \nu_1 \mathbf{B}(\mathbf{A}_1) \cdot \mathbf{B}(\mathbf{V} \cdot \nabla \mathbf{A}_1) \mathrm{d}\Omega - \int_{\Omega_2} \nu_2 \mathbf{B}(\mathbf{A}_2) \cdot \mathbf{B}(\mathbf{V} \cdot \nabla \mathbf{A}_2) \mathrm{d}\Omega \\
& - \int_{\gamma} \left(\nu_1 \mathbf{B}(\mathbf{A}_1) \cdot \mathbf{B}(\boldsymbol{\lambda}_1) - \nu_2 \mathbf{B}(\mathbf{A}_2) \cdot \mathbf{B}(\boldsymbol{\lambda}_2) - \frac{1}{2} \nu_1 \mathbf{B}(\mathbf{A}_1) \cdot \mathbf{B}(\mathbf{A}_1) + \frac{1}{2} \nu_2 \mathbf{B}(\mathbf{A}_2) \cdot \mathbf{B}(\mathbf{A}_2) \right) V_n \mathrm{d}\Gamma \\
& + \int_{\gamma} (\mathbf{M}_{o1} \cdot \mathbf{B}(\boldsymbol{\lambda}_1) - \mathbf{M}_{o2} \cdot \mathbf{B}(\boldsymbol{\lambda}_2)) V_n \mathrm{d}\Gamma + \int_{\gamma} (\mathbf{J}_1 \cdot \boldsymbol{\lambda}_1 - \mathbf{J}_2 \cdot \boldsymbol{\lambda}_2) V_n \mathrm{d}\Gamma
\end{aligned} \tag{4.2.10}$$

The energy bilinear form is symmetric in its arguments;

$$\int_{\Omega_1} \nu_1 \mathbf{B}(\boldsymbol{\lambda}_1) \cdot \mathbf{B}(\dot{\mathbf{A}}_1) d\Omega = \int_{\Omega_1} \nu_1 \mathbf{B}(\dot{\mathbf{A}}_1) \cdot \mathbf{B}(\boldsymbol{\lambda}_1) d\Omega \tag{4.2.11}$$

$$\int_{\Omega_2} \nu_2 \mathbf{B}(\boldsymbol{\lambda}_2) \cdot \mathbf{B}(\dot{\mathbf{A}}_2) d\Omega = \int_{\Omega_2} \nu_2 \mathbf{B}(\dot{\mathbf{A}}_2) \cdot \mathbf{B}(\boldsymbol{\lambda}_2) d\Omega \tag{4.2.12}$$

By using the relations (4.2.9), (4.2.11), and (4.2.12), all terms with $\dot{\mathbf{A}}_1$, $\dot{\mathbf{A}}_2$ in (4.2.10) are canceled out and all terms are expressed with the velocity field $\mathbf{V}$;

$$\begin{aligned}
\dot{G} = & \int_{\Omega_1} [\nu_1 \mathbf{B}(\mathbf{V} \cdot \nabla \mathbf{A}_1) \cdot \mathbf{B}(\boldsymbol{\lambda}_1) + \nu_1 \mathbf{B}(\mathbf{A}_1) \cdot \mathbf{B}(\mathbf{V} \cdot \nabla \boldsymbol{\lambda}_1) \\
& - \nu_1 \mathbf{B}(\mathbf{A}_1) \cdot \mathbf{B}(\mathbf{V} \cdot \nabla \mathbf{A}_1) - \mathbf{J}_1 \cdot (\mathbf{V} \cdot \nabla \boldsymbol{\lambda}_1) - \mathbf{M}_{o1} \cdot \mathbf{B}(\mathbf{V} \cdot \nabla \boldsymbol{\lambda}_1)] d\Omega \\
& + \int_{\Omega_2} [\nu_2 \mathbf{B}(\boldsymbol{\lambda}_2) \cdot \mathbf{B}(\mathbf{V} \cdot \nabla \mathbf{A}_2) + \nu_2 \mathbf{B}(\mathbf{A}_2) \cdot \mathbf{B}(\mathbf{V} \cdot \nabla \boldsymbol{\lambda}_2) \\
& - \nu_2 \mathbf{B}(\mathbf{A}_2) \cdot \mathbf{B}(\mathbf{V} \cdot \nabla \mathbf{A}_2) - \mathbf{J}_2 \cdot (\mathbf{V} \cdot \nabla \boldsymbol{\lambda}_2) - \mathbf{M}_{o2} \cdot \mathbf{B}(\mathbf{V} \cdot \nabla \boldsymbol{\lambda}_2)] d\Omega \\
& - \int_{\gamma} \Big(\nu_1 \mathbf{B}(\mathbf{A}_1) \cdot \mathbf{B}(\boldsymbol{\lambda}_1) - \nu_2 \mathbf{B}(\mathbf{A}_2) \cdot \mathbf{B}(\boldsymbol{\lambda}_2) \\
& - \frac{1}{2} \nu_1 \mathbf{B}(\mathbf{A}_1) \cdot \mathbf{B}(\mathbf{A}_1) + \frac{1}{2} \nu_2 \mathbf{B}(\mathbf{A}_2) \cdot \mathbf{B}(\mathbf{A}_2) \Big) V_n d\Gamma \\
& + \int_{\gamma} (\mathbf{M}_{o1} \cdot \mathbf{B}(\boldsymbol{\lambda}_1) - \mathbf{M}_{o2} \cdot \mathbf{B}(\boldsymbol{\lambda}_2)) V_n d\Gamma + \int_{\gamma} (\mathbf{J}_1 \cdot \boldsymbol{\lambda}_1 - \mathbf{J}_2 \cdot \boldsymbol{\lambda}_2) V_n d\Gamma
\end{aligned} \tag{4.2.13}$$

Next, the variational adjoint and state equation (4.1.8) and (4.1.16) are compared;

$$\begin{aligned}
& \int_{\Omega_1} \nu_1 \mathbf{B}(\boldsymbol{\lambda}_1) \cdot \mathbf{B}(\bar{\boldsymbol{\lambda}}_1) d\Omega + \int_{\Omega_2} \nu_2 \mathbf{B}(\boldsymbol{\lambda}_2) \cdot \mathbf{B}(\bar{\boldsymbol{\lambda}}_2) d\Omega \\
& = \int_{\Omega_1} \nu_1 \mathbf{B}(\mathbf{A}_1) \cdot \mathbf{B}(\bar{\boldsymbol{\lambda}}_1) d\Omega + \int_{\Omega_2} \nu_2 \mathbf{B}(\mathbf{A}_2) \cdot \mathbf{B}(\bar{\boldsymbol{\lambda}}_2) d\Omega \quad \forall \bar{\boldsymbol{\lambda}}_1, \bar{\boldsymbol{\lambda}}_2 \in \boldsymbol{\Phi}
\end{aligned} \tag{4.2.14}$$

$$\begin{aligned}
& \int_{\Omega_1} \nu_1 \mathbf{B}(\mathbf{A}_1) \cdot \mathbf{B}(\bar{\mathbf{A}}_1) d\Omega + \int_{\Omega_2} \nu_2 \mathbf{B}(\mathbf{A}_2) \cdot \mathbf{B}(\bar{\mathbf{A}}_2) d\Omega \\
& = \int_{\Omega_1} (\mathbf{J}_1 \cdot \bar{\mathbf{A}}_1 + \mathbf{M}_{o1} \cdot \mathbf{B}(\bar{\mathbf{A}}_1)) d\Omega + \int_{\Omega_2} (\mathbf{J}_2 \cdot \bar{\mathbf{A}}_2 + \mathbf{M}_{o2} \cdot \mathbf{B}(\bar{\mathbf{A}}_2)) d\Omega \quad \forall \bar{\mathbf{A}}_1, \bar{\mathbf{A}}_2 \in \boldsymbol{\Phi}
\end{aligned} \tag{4.2.15}$$

In the variational state equation (4.2.15), when $\bar{\mathbf{A}}_1 = \bar{\boldsymbol{\lambda}}_1$, $\bar{\mathbf{A}}_2 = \bar{\boldsymbol{\lambda}}_2$, (4.2.15) is expressed as

$$\int_{\Omega_1} \nu_1 \mathbf{B}(\mathbf{A}_1) \cdot \mathbf{B}(\bar{\boldsymbol{\lambda}}_1) d\Omega + \int_{\Omega_2} \nu_2 \mathbf{B}(\mathbf{A}_2) \cdot \mathbf{B}(\bar{\boldsymbol{\lambda}}_2) d\Omega$$
$$= \int_{\Omega_1} (\mathbf{J}_1 \cdot \bar{\boldsymbol{\lambda}}_1 + \mathbf{M}_{o1} \cdot \mathbf{B}(\bar{\boldsymbol{\lambda}}_1)) d\Omega + \int_{\Omega_2} (\mathbf{J}_2 \cdot \bar{\boldsymbol{\lambda}}_2 + \mathbf{M}_{o2} \cdot \mathbf{B}(\bar{\boldsymbol{\lambda}}_2)) d\Omega \quad \forall \bar{\boldsymbol{\lambda}}_1,\ \bar{\boldsymbol{\lambda}}_2 \in \Phi \tag{4.2.16}$$

When (4.2.16) is compared with (4.2.14), the left-hand side of (4.2.16) is the same as the right-hand side of (4.2.14). Thus, (4.2.14) can be written as

$$\int_{\Omega_1} \nu_1 \mathbf{B}(\boldsymbol{\lambda}_1) \cdot \mathbf{B}(\bar{\boldsymbol{\lambda}}_1) d\Omega + \int_{\Omega_2} \nu_2 \mathbf{B}(\boldsymbol{\lambda}_2) \cdot \mathbf{B}(\bar{\boldsymbol{\lambda}}_2) d\Omega$$
$$= \int_{\Omega_1} (\mathbf{J}_1 \cdot \bar{\boldsymbol{\lambda}}_1 + \mathbf{M}_{o1} \cdot \mathbf{B}(\bar{\boldsymbol{\lambda}}_1)) d\Omega + \int_{\Omega_2} (\mathbf{J}_2 \cdot \bar{\boldsymbol{\lambda}}_2 + \mathbf{M}_{o2} \cdot \mathbf{B}(\bar{\boldsymbol{\lambda}}_2)) d\Omega \quad \forall \bar{\boldsymbol{\lambda}}_1,\ \bar{\boldsymbol{\lambda}}_2 \in \Phi \tag{4.2.17}$$

The adjoint Eq. (4.2.17) has the same form and the same source. Just as the variational state equation (4.2.15) is equivalent to the differential state equations of (4.1.4) and (4.1.5), the variational adjoint Eq. (4.2.17) is equivalent to differential adjoint equations:

$$\nabla \times \nu_1 \nabla \times \boldsymbol{\lambda}_1 = \mathbf{J}_1 + \nabla \times \mathbf{M}_{o1} \quad \text{in} \quad \Omega_1 \tag{4.2.18}$$

$$\nabla \times \nu_2 \nabla \times \boldsymbol{\lambda}_2 = \mathbf{J}_2 + \nabla \times \mathbf{M}_{o2} \quad \text{in} \quad \Omega_2 \tag{4.2.19}$$

If the boundary condition for this adjoint equation is the same as the original state equation, the adjoint variable is the same as the state variable in the whole region. Thus, the boundary conditions for the adjoint equation are gives as

$$\boldsymbol{\lambda}_1 = \mathbf{0} \quad \text{on} \quad \Gamma^0 \quad \text{(homogeneous Dirichlet boundary condition)} \tag{4.2.20}$$

$$\frac{\partial \lambda_1}{\partial n} = \mathbf{0} \quad \text{on} \quad \Gamma^1 \quad \text{(homogeneous Neumann boundary condition)} \tag{4.2.21}$$

That is, the adjoint variable, which is determined from the adjoint Eq. (4.2.17) with the boundary conditions of (4.2.20) and (4.2.21), is the same as the state variable;

$$\boldsymbol{\lambda} = \mathbf{A} \quad \text{in} \quad \Omega \quad \text{and} \quad \text{on}\,\Gamma \tag{4.2.22}$$

Consequently, solving the adjoint equation is not necessary for the sensitivity (4.2.13) of the system energy. By inserting (4.2.22) into (4.2.13), the sensitivity (4.2.13) becomes

$$\begin{aligned}\dot{G} = & \int_{\Omega_1} [\nu_1 \mathbf{B}(\mathbf{A}_1)\cdot\mathbf{B}(\mathbf{V}\cdot\nabla\mathbf{A}_1) - \mathbf{J}_1\cdot(\mathbf{V}\cdot\nabla\mathbf{A}_1) - \mathbf{M}_{o1}\cdot\mathbf{B}(\mathbf{V}\cdot\nabla\mathbf{A}_1)]\mathrm{d}\Omega \\ & + \int_{\Omega_2} [\nu_2 \mathbf{B}(\mathbf{A}_2)\cdot\mathbf{B}(\mathbf{V}\cdot\nabla\mathbf{A}_2) - \mathbf{J}_2\cdot(\mathbf{V}\cdot\nabla\mathbf{A}_2) - \mathbf{M}_{o2}\cdot\mathbf{B}(\mathbf{V}\cdot\nabla\mathbf{A}_2)]\mathrm{d}\Omega \\ & - \int_{\gamma} \left(\frac{1}{2}\nu_1\mathbf{B}(\mathbf{A}_1)\cdot\mathbf{B}(\mathbf{A}_1) - \frac{1}{2}\nu_2\mathbf{B}(\mathbf{A}_2)\cdot\mathbf{B}(\mathbf{A}_2)\right)V_n\mathrm{d}\Gamma \\ & + \int_{\gamma} (\mathbf{M}_{o1}\cdot\mathbf{B}(\mathbf{A}_1) - \mathbf{M}_{o2}\cdot\mathbf{B}(\mathbf{A}_2))V_n\mathrm{d}\Gamma + \int_{\gamma} (\mathbf{J}_1\cdot\mathbf{A}_1 - \mathbf{J}_2\cdot\mathbf{A}_2)V_n\mathrm{d}\Gamma\end{aligned} \tag{4.2.23}$$

4.2.4 Boundary Expression of Shape Sensitivity

The domain integrals in the design sensitivity Eq. (4.2.23) can be expressed in boundary integrals by using the variational identities. For this purpose, two variational identities for the state and the adjoint equations are needed.

First, the variational identities for the state equation were given as (4.1.8) and (4.1.9);

$$\begin{aligned}& \int_{\Omega_1} \nu_1\mathbf{B}(\boldsymbol{\lambda}_1)\cdot\mathbf{B}(\bar{\lambda}_1)\mathrm{d}\Omega + \int_{\Omega_2} \nu_2\mathbf{B}(\boldsymbol{\lambda}_2)\cdot\mathbf{B}(\bar{\lambda}_2)\mathrm{d}\Omega \\ & = \int_{\Omega_1} (\mathbf{g}_{\mathbf{A}_1}\cdot\bar{\lambda}_1 + \mathbf{g}_{\mathbf{B}_1}\cdot\mathbf{B}(\bar{\lambda}_1))m_p\mathrm{d}\Omega \\ & + \int_{\Omega_2} (\mathbf{g}_{\mathbf{A}_2}\cdot\bar{\lambda}_2 + \mathbf{g}_{\mathbf{B}_2}\cdot\mathbf{B}(\bar{\lambda}_2))m_p\mathrm{d}\Omega \quad \forall\bar{\lambda}_1,\bar{\lambda}_2 \in \Phi\forall\bar{\mathbf{A}}_1 \in \Phi \quad \forall\bar{\mathbf{A}}_1 \in \Phi\end{aligned} \tag{4.2.24}$$

$$\begin{aligned}& \int_{\Omega_2} (\nu_2\mathbf{B}(\mathbf{A}_2)\cdot\mathbf{B}(\bar{\mathbf{A}}_2) - \mathbf{J}_2\cdot\bar{\mathbf{A}}_2 - \mathbf{M}_{o2}\cdot\mathbf{B}(\bar{\mathbf{A}}_2))\mathrm{d}\Omega \\ & \qquad = \int_{\gamma} (\mathbf{n}\times\mathbf{H}(\mathbf{A}_2))\cdot\bar{\mathbf{A}}_2\mathrm{d}\Gamma \quad \forall\bar{\mathbf{A}}_2 \in \Phi\end{aligned} \tag{4.2.25}$$

These variational identities are used to express the domain integrals in (4.2.23) as boundary integrals.

We choose $\bar{\mathbf{A}}_1 = \mathbf{V}\cdot\nabla\mathbf{A}_1$ and $\bar{\mathbf{A}}_2 = \mathbf{V}\cdot\nabla\mathbf{A}_2$ in (4.2.24) and (4.2.25), respectively;

$$\int_{\Omega_1} [\nu_1 \mathbf{B}(\mathbf{A}_1)\cdot\mathbf{B}(\mathbf{V}\cdot\nabla\mathbf{A}_1) - \mathbf{J}_1\cdot(\mathbf{V}\cdot\nabla\mathbf{A}_1) - \mathbf{M}_{o1}\cdot\mathbf{B}(\mathbf{V}\cdot\nabla\mathbf{A}_1)]\mathrm{d}\Omega = -\int_{\gamma} (\mathbf{n}\times\mathbf{H}(\mathbf{A}_1))\cdot(\mathbf{V}\cdot\nabla\mathbf{A}_1)\mathrm{d}\Gamma \quad (4.2.26)$$

$$\int_{\Omega_2} [\nu_2 \mathbf{B}(\mathbf{A}_2)\cdot\mathbf{B}(\mathbf{V}\cdot\nabla\mathbf{A}_2) - \mathbf{J}_2\cdot(\mathbf{V}\cdot\nabla\mathbf{A}_2) - \mathbf{M}_{o2}\cdot\mathbf{B}(\mathbf{V}\cdot\nabla\mathbf{A}_2)]\mathrm{d}\Omega = \int_{\gamma} (\mathbf{n}\times\mathbf{H}(\mathbf{A}_2))\cdot(\mathbf{V}\cdot\nabla\mathbf{A}_2)\mathrm{d}\Gamma \quad (4.2.27)$$

By substituting these two relations into (4.2.23), the domain integrals in (4.2.23) become boundary integrals;

$$\begin{aligned}\dot{G} = &\int_{\gamma} [-(\mathbf{n}\times\mathbf{H}(\mathbf{A}_1))\cdot(\mathbf{V}\cdot\nabla\mathbf{A}_1) + (\mathbf{n}\times\mathbf{H}(\mathbf{A}_2))\cdot(\mathbf{V}\cdot\nabla\mathbf{A}_2)]\mathrm{d}\Gamma \\ &- \int_{\gamma} \left(\frac{1}{2}\nu_1\mathbf{B}(\mathbf{A}_1)\cdot\mathbf{B}(\mathbf{A}_1) - \frac{1}{2}\nu_2\mathbf{B}(\mathbf{A}_2)\cdot\mathbf{B}(\mathbf{A}_2)\right)V_n\mathrm{d}\Gamma \\ &+ \int_{\gamma} (\mathbf{M}_{o1}\cdot\mathbf{B}(\mathbf{A}_1) - \mathbf{M}_{o2}\cdot\mathbf{B}(\mathbf{A}_2))V_n\mathrm{d}\Gamma + \int_{\gamma} (\mathbf{J}_1\cdot\mathbf{A}_1 - \mathbf{J}_2\cdot\mathbf{A}_2)V_n\mathrm{d}\Gamma \quad (4.2.28)\end{aligned}$$

Since this sensitivity has no adjoint variable $\boldsymbol{\lambda}$, we use the following notation for the simpler expression:

$$\mathbf{B}_1 \equiv \mathbf{B}(\mathbf{A}_1),\ \mathbf{B}_2 \equiv \mathbf{B}(\mathbf{A}_2),\ \mathbf{H}_1 \equiv \mathbf{H}(\mathbf{A}_1),\ \mathbf{H}_2 \equiv \mathbf{H}(\mathbf{A}_2) \quad (4.2.29)$$

With this notation, (4.2.28) is written as

$$\begin{aligned}\dot{G} = &\int_{\gamma} [(\mathbf{n}\times\mathbf{H}_2)\cdot(\mathbf{V}\cdot\nabla\mathbf{A}_2) - (\mathbf{n}\times\mathbf{H}_1)\cdot(\mathbf{V}\cdot\nabla\mathbf{A}_1)]\mathrm{d}\Gamma \\ &+ \int_{g} \left(\frac{1}{2}\nu_2\mathbf{B}_2\cdot\mathbf{B}_2 - \frac{1}{2}\nu_1\mathbf{B}_1\cdot\mathbf{B}_1\right)V_n\mathrm{d}\Gamma \\ &- \int_{\gamma} (\mathbf{M}_{o2}\cdot\mathbf{B}_2 - \mathbf{M}_{o1}\cdot\mathbf{B}_1)V_n\mathrm{d}\Gamma - \int_{\gamma} (\mathbf{J}_2\cdot\mathbf{A}_2 - \mathbf{J}_1\cdot\mathbf{A}_1)V_n\mathrm{d}\Gamma \quad (4.2.30)\end{aligned}$$

The integrand of the first integral in (4.2.30) is arranged using the interface condition and a vector identity. The interface conditions for **H** and **B** imply

$$\mathbf{n} \times \mathbf{H}(\mathbf{A}_1) = \mathbf{n} \times \mathbf{H}(\mathbf{A}_2) \quad \text{on} \quad \gamma \quad (\text{interface condition}) \tag{4.2.31}$$

$$B_{1n} = B_{2n} \quad \text{on} \quad \gamma \quad (\text{interface condition}) \tag{4.2.32}$$

Using the interface condition (4.2.31) and the relation (4.1.56), the integrand of the first integral in (4.2.30) is written without V_n as

$$(\mathbf{n} \times \mathbf{H}_2) \cdot (\mathbf{V} \cdot \nabla \mathbf{A}_2) - (\mathbf{n} \times \mathbf{H}_1) \cdot (\mathbf{V} \cdot \nabla \mathbf{A}_1) = (\mathbf{n} \times \mathbf{H}_1) \cdot \left(\frac{\partial \mathbf{A}_2}{\partial n} - \frac{\partial \mathbf{A}_1}{\partial n} \right) \tag{4.2.33}$$

Using the relations of (4.1.61) and (4.1.65), the interface condition (4.2.32), (4.2.33) is rewritten as

$$\begin{aligned}(\mathbf{n} \times \mathbf{H}_1) \cdot (\mathbf{B}_2 \times \mathbf{n} - \mathbf{B}_1 \times \mathbf{n}) &= -H_{1t}(B_{2t} - B_{1t}) \\ &= -H_{1t}(B_{2t} - B_{1t}) - H_{1n}(B_{2n} - B_{1n}) \\ &= -\mathbf{H}_1 \cdot (\mathbf{B}_2 - \mathbf{B}_1)\end{aligned} \tag{4.2.34}$$

The integrand of the second integral in (4.2.30) is written without V_n as

$$\frac{1}{2} \nu_2 \mathbf{B}_2 \cdot \mathbf{B}_2 - \frac{1}{2} \nu_1 \mathbf{B}_1 \cdot \mathbf{B}_1 = \frac{1}{2} (\nu_2 - \nu_1) \mathbf{B}_1 \cdot \mathbf{B}_2 + \frac{1}{2} (\nu_2 \mathbf{B}_2 + \nu_1 \mathbf{B}_1) \cdot (\mathbf{B}_2 - \mathbf{B}_1) \tag{4.2.35}$$

Using (4.2.34), the sum of two integrand terms of the first and the third integrals in (4.2.30) is written without V_n as

$$\begin{aligned}-\mathbf{H}_1 \cdot (\mathbf{B}_2 - \mathbf{B}_1) - \mathbf{M}_{o2} \cdot \mathbf{B}_2 + \mathbf{M}_{o1} \cdot \mathbf{B}_1 &= -(\nu_1 \mathbf{B}_1 - \mathbf{M}_{o1}) \cdot (\mathbf{B}_2 - \mathbf{B}_1) - \mathbf{M}_{o2} \cdot \mathbf{B}_2 + \mathbf{M}_{o1} \cdot \mathbf{B}_1 \\ &= -\nu_1 \mathbf{B}_1 \cdot (\mathbf{B}_2 - \mathbf{B}_1) + \mathbf{M}_{o1} \cdot \mathbf{B}_2 - \mathbf{M}_{o2} \cdot \mathbf{B}_2 \\ &= -\nu_1 \mathbf{B}_1 \cdot (\mathbf{B}_2 - \mathbf{B}_1) - (\mathbf{M}_{o2} - \mathbf{M}_{o1}) \cdot \mathbf{B}_2\end{aligned} \tag{4.2.36}$$

The sum of (4.2.35) and (4.2.36) is the sum of integrand terms of the first three integrals in (4.2.30);

$$\begin{aligned}&\frac{1}{2} (\nu_2 - \nu_1) \mathbf{B}_1 \cdot \mathbf{B}_2 + \frac{1}{2} (\nu_2 \mathbf{B}_2 + \nu_1 \mathbf{B}_1) \cdot (\mathbf{B}_2 - \mathbf{B}_1) - \nu_1 \mathbf{B}_1 \cdot (\mathbf{B}_2 - \mathbf{B}_1) - (\mathbf{M}_{o2} - \mathbf{M}_{o1}) \cdot \mathbf{B}_2 \\ &= \frac{1}{2} (\nu_2 - \nu_1) \mathbf{B}_1 \cdot \mathbf{B}_2 - (\mathbf{M}_{o2} - \mathbf{M}_{o1}) \cdot \mathbf{B}_2 + \frac{1}{2} (\nu_2 \mathbf{B}_2 - \nu_1 \mathbf{B}_1) \cdot (\mathbf{B}_2 - \mathbf{B}_1)\end{aligned} \tag{4.2.37}$$

By the constitutive relations of (4.1.10) and (4.1.11), and the interface conditions of (4.2.31) and (4.2.32), the last term in (4.2.37) is arranged as

$$\begin{aligned}\frac{1}{2}(\nu_2\mathbf{B}_2 - \nu_1\mathbf{B}_1)\cdot(\mathbf{B}_2 - \mathbf{B}_1) &= \frac{1}{2}(\mathbf{H}_2 + \mathbf{M}_{o2} - \mathbf{H}_1 - \mathbf{M}_{o1})\cdot(\mathbf{B}_2 - \mathbf{B}_1)\\ &= \frac{1}{2}(\mathbf{H}_2 - \mathbf{H}_1)\cdot(\mathbf{B}_2 - \mathbf{B}_1) + \frac{1}{2}(\mathbf{M}_{o2} - \mathbf{M}_{o1})\cdot(\mathbf{B}_2 - \mathbf{B}_1)\\ &= \frac{1}{2}(H_{2n} - H_{1n})(B_{2n} - B_{1n}) + \frac{1}{2}(H_{2t} - H_{1t})(B_{2t} - B_{1t})\\ &+ \frac{1}{2}(\mathbf{M}_{o2} - \mathbf{M}_{o1})\cdot(\mathbf{B}_2 - \mathbf{B}_1)\\ &= \frac{1}{2}(\mathbf{M}_{o2} - \mathbf{M}_{o1})\cdot(\mathbf{B}_2 - \mathbf{B}_1)\end{aligned} \tag{4.2.38}$$

By inserting (4.2.38) into (4.2.37), the sum of integrand terms of the first three integrals in (4.2.30) is obtained without V_n as

$$\begin{aligned}&\frac{1}{2}(\nu_2 - \nu_1)\mathbf{B}_1\cdot\mathbf{B}_2 - (\mathbf{M}_{o2} - \mathbf{M}_{o1})\cdot\mathbf{B}_2 + \frac{1}{2}(\mathbf{M}_{o2} - \mathbf{M}_{o1})\cdot(\mathbf{B}_2 - \mathbf{B}_1)\\ &= \frac{1}{2}(\nu_2 - \nu_1)\mathbf{B}_1\cdot\mathbf{B}_2 - \frac{1}{2}(\mathbf{M}_{o2} - \mathbf{M}_{o1})\cdot(\mathbf{B}_2 + \mathbf{B}_1)\end{aligned} \tag{4.2.39}$$

Finally, by substituting (4.2.39) into (4.2.30), the general three-dimensional sensitivity formula of the system energy by the interface variation is obtained as

$$\dot{G} = \int_\gamma \left[\frac{1}{2}(\nu_2 - \nu_1)\mathbf{B}_1\cdot\mathbf{B}_2 - \frac{1}{2}(\mathbf{M}_{o2} - \mathbf{M}_{o1})\cdot(\mathbf{B}_2 + \mathbf{B}_1) - (\mathbf{J}_2 - \mathbf{J}_1)\cdot\mathbf{A}_2\right] V_n \mathrm{d}\Gamma \tag{4.2.40}$$

4.2.5 Interface Problems

The general sensitivity formula (4.2.40) for the interface variation can be expressed in various ways according to the characteristics of the design problems of the magnetostatic system. For example, the sensitivity formula for the magnetostatic system without the permanent magnet is written as

$$\dot{G} = \int_\gamma \left[\frac{1}{2}(\nu_2 - \nu_1)\mathbf{B}_1\cdot\mathbf{B}_2 - (\mathbf{J}_2 - \mathbf{J}_1)\cdot\mathbf{A}_2\right] V_n \mathrm{d}\Gamma \tag{4.2.41}$$

The general sensitivity formula (4.2.40) is applied to the following specific cases for practical design problems.

(1) interface of air and ferromagnetic material

When the air is Ω_1 and the linear ferromagnetic material is Ω_2,

$$\dot{G} = \int_\gamma \frac{v_0}{2}(v_r - 1)\mathbf{B}_1 \cdot \mathbf{B}_2 V_n \mathrm{d}\Gamma \tag{4.2.42}$$

where $v_1 = v_0 = 1/\mu_0$, $v_2 = v_r v_0 = 1/\mu_r \mu_0$, $v_r = 1/\mu_r$.

(2) interface of magnetic material and permanent magnet

When the linear ferromagnetic material is Ω_1 and the permanent magnet is Ω_2,

$$\dot{G} = \int_\gamma \frac{1}{2}[(v_2 - v_1)\mathbf{B}_1 \cdot \mathbf{B}_2 - \mathbf{M}_{o2} \cdot (\mathbf{B}_2 + \mathbf{B}_1)]V_n \mathrm{d}\Gamma \tag{4.2.43}$$

where v_1 is the reluctivity of the linear ferromagnetic material, and v_2 is the reluctivity of the permanent magnet. If the relative permeability of the permanent magnet is assumed to be 1,

$$\dot{G} = \int_\gamma \frac{1}{2}[v_0(1 - v_r)\mathbf{B}_1 \cdot \mathbf{B}_2 - \mathbf{M}_{o2} \cdot (\mathbf{B}_2 + \mathbf{B}_1)]V_n \mathrm{d}\Gamma \tag{4.2.44}$$

where $v_1 = v_r v_0 = 1/\mu_r \mu_0$, $v_2 = v_0 = 1/\mu_0$.

(3) interface of magnetic material and current region

When the linear ferromagnetic material is Ω_1 and the current region is Ω_2,

$$\dot{G} = \int_\gamma \left[\frac{1}{2}(v_2 - v_1)\mathbf{B}_1 \cdot \mathbf{B}_2 - \mathbf{J}_2 \cdot \mathbf{A}_2\right] V_n \mathrm{d}\Gamma \tag{4.2.45}$$

where v_1 is the reluctivity of the linear ferromagnetic material, and v_2 is the reluctivity of the current region. If the relative permeability of the current region is taken to be 1,

$$\dot{G} = \int_\gamma \left[\frac{v_0}{2}(1 - v_r)\mathbf{B}_1 \cdot \mathbf{B}_2 - \mathbf{J}_2 \cdot \mathbf{A}_2\right] V_n \mathrm{d}\Gamma \tag{4.2.46}$$

where $v_1 = v_r v_0 = 1/\mu_r \mu_0$, $v_2 = v_0 = 1/\mu_0$.

(4) interface of permanent magnet and current region

When the permanent magnet is Ω_1 and the current region is Ω_2,

$$\dot{G} = \int_{\gamma} \left[\frac{1}{2}(v_2 - v_1)\mathbf{B}_1 \cdot \mathbf{B}_2 + \frac{1}{2}\mathbf{M}_{o1} \cdot (\mathbf{B}_2 + \mathbf{B}_1) - \mathbf{J}_2 \cdot \mathbf{A}_2 \right] V_n \mathrm{d}\Gamma \tag{4.2.47}$$

where v_1 is the reluctivity of the permanent magnet, and v_2 is the reluctivity of the current region. If the relative permeabilities of the permanent magnet and the current region are taken to be 1,

$$\dot{G} = \int_{\gamma} \left[\frac{1}{2}\mathbf{M}_{o1} \cdot (\mathbf{B}_2 + \mathbf{B}_1) - \mathbf{J}_2 \cdot \mathbf{A}_2 \right] V_n \mathrm{d}\Gamma \tag{4.2.48}$$

(5) interface of two permanent magnets with different magnetization

When the permanent magnet 1 is Ω_1 and the permanent magnet 2 is Ω_2,

$$\dot{G} = -\int_{\gamma} \frac{1}{2}(\mathbf{M}_{o2} - \mathbf{M}_{o1}) \cdot (\mathbf{B}_2 + \mathbf{B}_1) V_n \mathrm{d}\Gamma \tag{4.2.49}$$

(6) interface of permanent magnet and air

When the permanent magnet 1 is Ω_1 and the air is Ω_2,

$$\dot{G} = \int_{\gamma} \frac{1}{2}[(v_2 - v_1)\mathbf{B}_1 \cdot \mathbf{B}_2 + \mathbf{M}_{o1} \cdot (\mathbf{B}_2 + \mathbf{B}_1)] V_n \mathrm{d}\Gamma \tag{4.2.50}$$

where v_1 is the reluctivity of the permanent magnet, and v_2 is the reluctivity of the air. If the relative permeability of the permanent magnet is taken to be 1,

$$\dot{G} = \int_{\gamma} \frac{1}{2}\mathbf{M}_{o1} \cdot (\mathbf{B}_2 + \mathbf{B}_1) V_n \mathrm{d}\Gamma \tag{4.2.51}$$

(7) interface of two current regions with different current density

When the current region 1 is Ω_1 and the current region 2 is Ω_2,

$$\dot{G} = -\int_{\gamma} (\mathbf{J}_2 - \mathbf{J}_1) \cdot \mathbf{A}_2 V_n \mathrm{d}\Gamma \tag{4.2.52}$$

(8) interface of current region and air

When the current region is Ω_1 and the air is Ω_2,

$$\dot{G} = \int_{\gamma} \mathbf{J}_1 \cdot \mathbf{A}_2 V_n \mathrm{d}\Gamma \tag{4.2.53}$$

(9) interface of two ferromagnetic materials with different reluctivity

When the ferromagnetic material 1 is Ω_1 and the other ferromagnetic material 2 is Ω_2,

$$\dot{G} = \int_{\gamma} \frac{1}{2}(\nu_2 - \nu_1)\mathbf{B}_1 \cdot \mathbf{B}_2 V_n \mathrm{d}\Gamma \tag{4.2.54}$$

4.2.6 Source Condition and Inductance Sensitivity

In this section, the sign of the sensitivity is examined in the electric-circuit point of view and the inductance sensitivity is derived using the energy sensitivity obtained in Sect. 4.2.4 [4–6]. In Sects. 3.3.5 and 3.5.5, we examined how the sign of the energy sensitivity in the electrostatic system changes according to the condition of source application. The same phenomenon occurs in the magnetic system. There are two conditions: the current source condition and the voltage source condition. The first source condition is that the coil terminals are connected to an external current source in Fig. 4.25. The second source condition is that the coil terminals are connected to an external voltage source. The fact that a voltage **V** for the terminals is given means that a flux linkage $\boldsymbol{\lambda}$ is given since the voltage is the time derivative of the flux linkage. Even while the shape of the magnetic system is changed, the first and second conditions are maintained. As mentioned as in Sect. 4.2.1, when the inductance is evaluated using the system energy, the permanent magnet is not included in calculation of the system energy.

Under the first condition of current source, the stored energy of the inductor can be written with the inductance L and the given current I as

$$W_m = \frac{1}{2} L I^2 \tag{4.2.55}$$

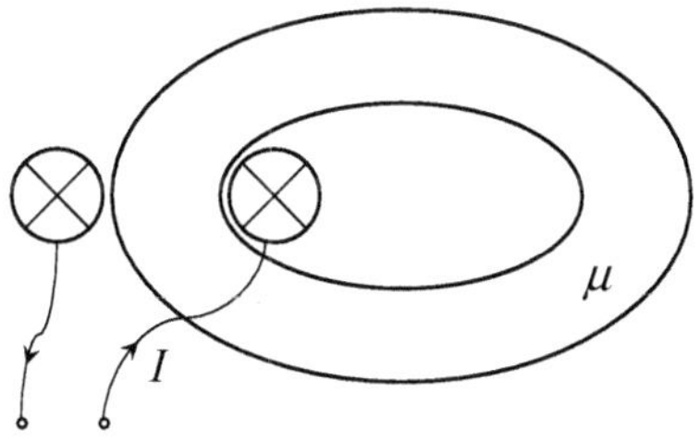

Fig. 4.25 Inductor model for system energy

A shape variation of the magnetic system causes a variation in the inductance, which is determined only by its geometry and material property. It results in a variation of the system energy (4.2.55). This energy variation can be expressed by taking the total derivative of (4.2.55) as

$$\dot{W}_m = \frac{1}{2}\dot{L}I^2 \tag{4.2.56}$$

where, with the current I given, the variation of the inductance L is proportional to the variation of the system energy. Since $\lambda = LI$, the variation of the inductance causes the variation of the flux linkage for the terminals, which means an induced voltage.

Under the second condition of voltage source, the stored energy of the inductor is written with the inductance L and a given flux linkage λ as

$$W_m = \frac{1}{2}\frac{\lambda^2}{L} \tag{4.2.57}$$

The inductance variation due to the shape change of the magnetic system results in the variation of the system energy (4.2.57). This energy variation can be expressed by taking the total derivative of (4.2.57) as

$$\dot{W}_m = -\frac{1}{2}\frac{\lambda^2}{L^2}\dot{L} \tag{4.2.58}$$

Since the flux linkage λ is given, the increase of the inductance L results in the decrease of the system energy, and vice versa. Using $\lambda = LI$, (4.2.58) can be rewritten as

$$\dot{W}_m = -\frac{1}{2}\dot{L}I^2 \tag{4.2.59}$$

Comparing the two energy sensitivities of (4.2.56) and (4.2.59), we see that they have the opposite sign. That is, the sign of the energy sensitivity depends on the condition of external source.

In Sect. 4.2.1, the objective function of system energy was defined with a given current and the sensitivity was also derived from the state equations with the given current distribution. That is, the sensitivity formula (4.2.41) was derived under the first condition of current source. When we deal with the magnetostatic systems with the current source, the sign of the sensitivity formula (4.2.41) does not change.

Hence, the sensitivity formula for the current-source magnetostatic system is written as the same as (4.2.41);

$$\dot{G} = \int_{\gamma} \left[\frac{1}{2}(\nu_2 - \nu_1)\mathbf{B}_1 \cdot \mathbf{B}_2 - (\mathbf{J}_2 - \mathbf{J}_1) \cdot \mathbf{A}_2 \right] V_n \mathrm{d}\Gamma \qquad (4.2.60)$$

This sensitivity formula can be used to obtain the inductance sensitivity.

The total derivative of the inductor-stored energy (4.2.56) is equal to the energy sensitivity with the current source condition (4.2.60);

$$\frac{1}{2}\dot{L}I^2 = \int_{\gamma} \left[\frac{1}{2}(\nu_2 - \nu_1)\mathbf{B}_1 \cdot \mathbf{B}_2 - (\mathbf{J}_2 - \mathbf{J}_1) \cdot \mathbf{A}_2 \right] V_n \mathrm{d}\Gamma \qquad (4.2.61)$$

From this relation, the inductance sensitivity $\dot{L}$ is obtained as

$$\dot{L} = \frac{1}{I^2} \int_{\gamma} [(\nu_2 - \nu_1)\mathbf{B}_1 \cdot \mathbf{B}_2 - 2(\mathbf{J}_2 - \mathbf{J}_1) \cdot \mathbf{A}_2] V_n \mathrm{d}\Gamma \qquad (4.2.62)$$

4.2.7 *Analytical Examples*

For the energy objective problem in Sect. 4.2.4, the sensitivity formula was derived as (4.2.40). To show that the sensitivity formula is correct, one-dimensional analytical examples, which have the analytic field solutions, are taken. The sensitivity results analytically calculated in the examples are compared with the results of the sensitivity formula. For this purpose, two analytical examples are employed to compare the sensitivity for the ferromagnetic interface and the current region interface, respectively.

4.2.7.1 Infinite Solenoid-Ferromagnetic Material Interface

As an example that can be analytically calculated, consider an infinite solenoid shown in Fig. 4.26, where two ferromagnetic materials are inside a solenoid coil. The thickness of the solenoid coil is w and the current of density J_ϕ flows in it. The inner ferromagnetic material has radius a and permeability μ_1, and the outer one has thickness $b - a$ and permeability μ_2. The objective is to obtain the sensitivity of the system energy with respect to the interface γ, where the two ferromagnetic materials meet; the design variable is the radius a of the inner ferromagnetic material. The design sensitivity of the systme energy with respect to the design variable a is calculated.

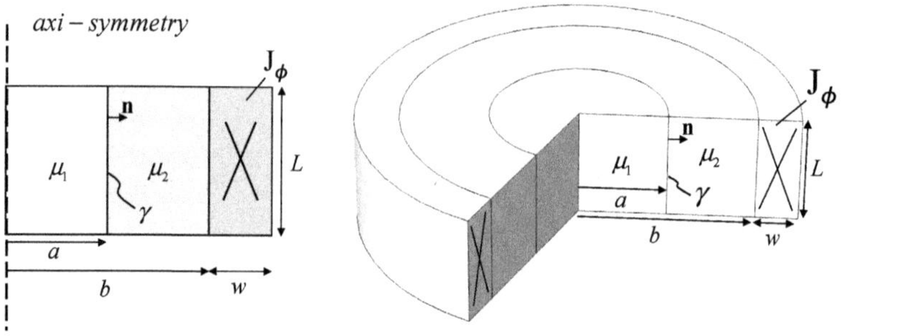

Fig. 4.26 Solenoid-ferromagnetic material design

The objective function is the system energy;

$$F = W_m = \frac{1}{2}\int_{\Omega} \frac{1}{\mu} B^2 \mathrm{d}\Omega \tag{4.2.63}$$

where

$$\mathbf{B}_1 = \mu_1 J_\phi w\,\mathbf{z} \quad \text{for} \quad 0 \le s < a \tag{4.2.64}$$

$$\mathbf{B}_2 = \mu_2 J_\phi w\,\mathbf{z} \quad \text{for} \quad a < s < b \tag{4.2.65}$$

$$\mathbf{B} = \mathbf{0} \quad \text{for} \quad b < s \tag{4.2.66}$$

The objective function is rewritten by using the magnetic flux density (4.2.64), (4.2.65), and (4.2.66);

$$\begin{aligned} F &= \frac{1}{2}\int_0^{2\pi} \mathrm{d}\phi \int_0^L \mathrm{d}z \left(\int_0^a \frac{1}{\mu_1} B(\mathbf{A}_1)^2 s \mathrm{d}s + \int_a^b \frac{1}{\mu_2} B(\mathbf{A}_2)^2 r \mathrm{d}r \right) \\ &= \frac{1}{2}\pi L J_\phi^2 w^2 \left(\mu_1 a^2 + \mu_2 (b^2 - a^2)\right) \end{aligned} \tag{4.2.67}$$

The analytical sensitivity per unit length is obtained by differentiating the objective function (4.2.67) with respect to a;

$$\frac{\mathrm{d}F}{\mathrm{d}a} = \pi a(\mu_1 - \mu_2) J_\phi^2 w^2 \tag{4.2.68}$$

Alternatively, the sensitivity can be calculated by using the sensitivity formula (4.2.40) in Sect. 4.2.4:

$$\dot{G} = \frac{1}{2} \int_{\gamma} \left(\frac{1}{\mu_2} - \frac{1}{\mu_1} \right) \mathbf{B}_1 \cdot \mathbf{B}_2 \, V_n \, \mathrm{d}\Gamma \qquad (4.2.69)$$

(4.2.64) and (4.2.65) are inserted into the sensitivity formula (4.2.69);

$$\dot{G} = \pi a(\mu_1 - \mu_2) J^2 w^2 V_n L \qquad (4.2.70)$$

Using $V_n = \frac{\mathrm{d}a}{\mathrm{d}_t}$, the design sensitivity per unit length is obtained as

$$\frac{\mathrm{d}G}{\mathrm{d}a} = \pi a(\mu_1 - \mu_2) J^2 w^2 \qquad (4.2.71)$$

which is the correct result when compared with the analytical result of (4.2.68).

4.2.7.2 Infinite Solenoid-Current Region Interface

For an analytical example, an infinite length solenoid is given in Fig. 4.27, where an inside of the solenoid coil is air and the current of density J_ϕ flows in the solenoid coil. The solenoid coil has thickness $b - a$ and the length L in the axial direction is infinite. The objective is to obtain the sensitivity of the system energy with respect to the interface γ, which is the outer surface of the solenoid coil. The design variable is the radius b of the outer suface of the coil. The design sensitivity of the system energy with respect to the design variable b is calculated.

The objective function is the system energy;

$$F = W_m = \frac{1}{2} \int_{\Omega} \frac{1}{\mu} B^2 \mathrm{d}\Omega \qquad (4.2.72)$$

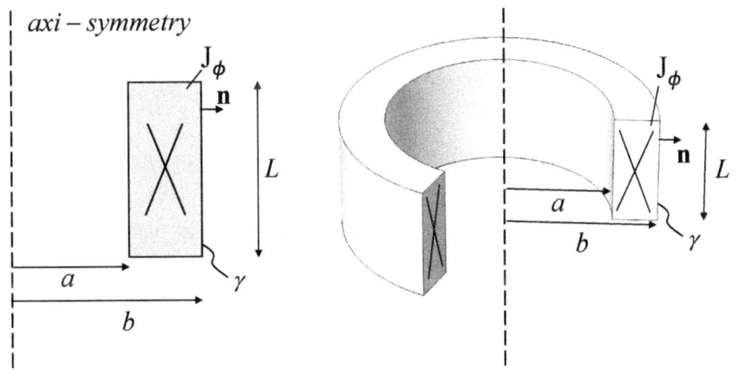

Fig. 4.27 Air core solenoid-current region interface

where

$$\mathbf{B} = \mu_0 J_\phi (b - a)\,\mathbf{z} \quad \text{for } s < a \tag{4.2.73}$$

$$\mathbf{B} = 0 \quad \text{for } a < s \tag{4.2.74}$$

The objective function is rewritten by using the field (4.2.73) and (4.2.74);

$$F = \frac{1}{2}\int_0^a \frac{1}{\mu_0}\left[\mu_0 J_\phi (b-a)\right]^2 r\mathrm{d}r \int_0^{2\pi} \mathrm{d}\phi \int_0^L \mathrm{d}z = \frac{1}{2}\pi a^2 L \frac{1}{\mu_0}\left[\mu_0 J_\phi (b-a)\right]^2 \tag{4.2.75}$$

The analytical sensitivity per unit length is obtained by differentiating the objective function (4.2.75) with respect to b;

$$\frac{\mathrm{d}F}{\mathrm{d}b} = \pi a^2 \mu_0 J_\phi^2 (b-a) \tag{4.2.76}$$

Alternatively, the sensitivity can be calculated by using the sensitivity formula (4.2.40) in Sect. 4.2.4:

$$\dot{G} = \int_\gamma J_\phi A_\phi \, V_n \, \mathrm{d}\Gamma \tag{4.2.77}$$

By using the notation $B(A) = \nabla \times A$, A_ϕ on the γ is obtained as

$$\int_0^a B\, r\mathrm{d}r \int_0^{2\pi} \mathrm{d}\phi = \int_\gamma A_\phi \, b \, \mathrm{d}\phi \tag{4.2.78}$$

$$A_\phi = \frac{a^2}{2b}\mu_0 J_\phi (b-a) \tag{4.2.79}$$

(4.2.79) is inserted into the sensitivity formula (4.2.77);

$$\dot{G} = \pi a^2 \mu_0 J_\phi^2 (b-a) V_n L \tag{4.2.80}$$

Using $V_n = \frac{\mathrm{d}b}{\mathrm{d}t}$, the design sensitivity per unit length is obtained as

$$\frac{\mathrm{d}G}{\mathrm{d}b} = \pi a^2 \mu_0 J_\phi^2 (b-a) \tag{4.2.81}$$

which is the correct result when compared with the analytical result in (4.2.76).

4.2.8 Numerical Examples

The sensitivity formula (4.2.40) for the energy objective function derived in Sect. 4.2.4 is applied to six shape optimization problems of two-dimensional design model, of which the analytic field solutions are not given. These design models are taken to show that the sensitivity formula for energy objective function is well applied to the shape design problem of the interfaces in the magnetostatic system. In the first two numerical examples, the sensitivity formula for the energy objective function is used to optimize the material interface in simple numerical models, whose optimal designs are known. The results of the other four examples show that this optimization method is useful for the design of the material interface. In these four examples, the sensitivity formula requires the state variable, which is numerically calculated by the finite element method. The sensitivity obtained is used for the optimization algorithm to provide the evolution of the material interface shape. For the optimization algorithm, the level set method is used to represent the shape evolution of the design model.

4.2.8.1 Coaxial Cable-Ferromagnetic Interface Design

As an example with a known optimal design, consider a coaxial cable in Fig. 4.28, where the air and the ferromagnetic material of permeability μ are between an inner conductor and an outer conductor. These two conductors have the opposite currents flows of a same magnitude. When the ferromagnetic material is attached to the inner cable and is formed to be a circle, the stored system energy becomes the maximum under the constraint of constant ferromagnetic material volume. The design objective is to obtain the maximum energy and the design variable is the shape of the interface γ between the ferromagnetic region and the air.

The objective function to be maximized is defined as the magnetic system energy;

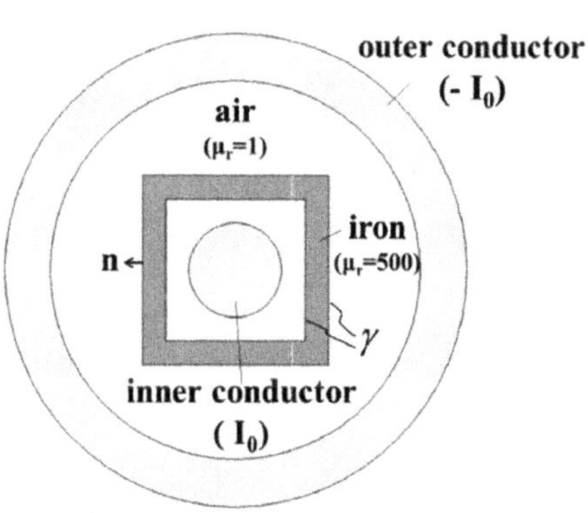

Fig. 4.28 Coaxial cable-ferromagnetic material design, system energy

$$F = W_m = \int_{\Omega} \frac{1}{2} \nu \mathbf{B}(\mathbf{A}) \cdot \mathbf{B}(\mathbf{A}) d\Omega \tag{4.2.82}$$

The shape sensitivity for this interface of the ferromagnetic material is the sensitivity formula (4.2.40):

$$\dot{G} = \int_{\gamma} \frac{\nu_0}{2} (1 - \nu_r) \mathbf{B}_1 \cdot \mathbf{B}_2 V_n d\Gamma \tag{4.2.83}$$

For this maximization problem, the design velocity is taken as

$$V_n = \frac{\nu_0}{2} (1 - \nu_r) \mathbf{B}_1 \cdot \mathbf{B}_2 \tag{4.2.84}$$

This problem has a constraint of constant volume of the ferromagnetic material;

$$\int_{\Omega_1} d\Omega = C \tag{4.2.85}$$

where C is a constant. The constant volume (4.2.85) is equivalent to the zero sum of the design velocity over the ferromagnetic interface, which is obtained by differentiating (4.2.85).

$$\int_{\gamma} V_n d\Gamma = 0 \tag{4.2.86}$$

The design velocity for this minimization problem with the volume constraint is taken as

$$U_n = V_n - V_{na} \tag{4.2.87}$$

where

$$V_{na} = \int_{\gamma} \frac{\nu_0}{2} (1 - \nu_r) \mathbf{B}_1 \cdot \mathbf{B}_2 d\Gamma / \int_{\gamma} d\Gamma \tag{4.2.88}$$

The design result is shown in Fig. 4.29, where the shape of the ferromagnetic material becomes gradually a circle with the iteration, and finally, the material is attached to the inner conductor as expected. The final design of the circular shape is obtained at the 70 s, when the system energy converges to the maximum value as in Fig. 4.30. The result shows that the shape sensitivity analysis is feasible for the

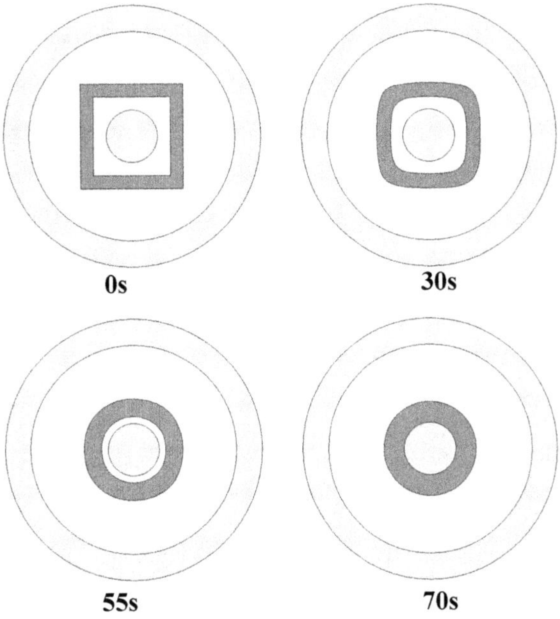

Fig. 4.29 Coaxial cable: shape variation

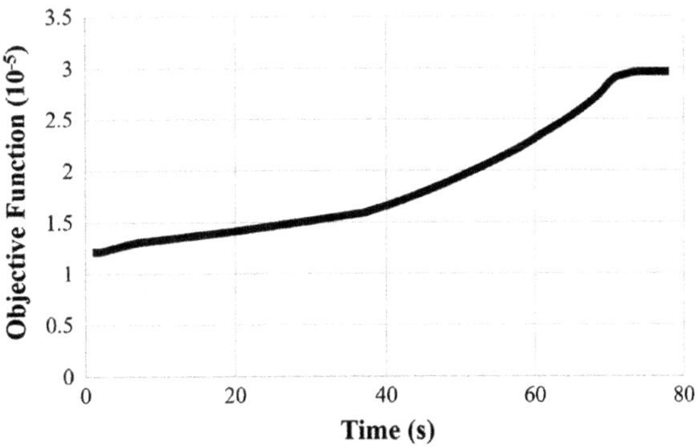

Fig. 4.30 Coaxial cable: evolution of objective function

ferromagnetic interface problem with the system energy objective function in the axi-symmetric magnetostatic system.

4.2.8.2 Coaxial Cable-Current Region Design

For another example with a known optimal design, consider a current-carrying cable, which consists of four inner conductors and an outer conductor as in Fig. 4.31. In the inner conductors and the outer conductor, the currents of a same magnitude flow into the opposite direction. When the shape of the inner conductors

Fig. 4.31 Coaxial cable-current region design, system energy maximization

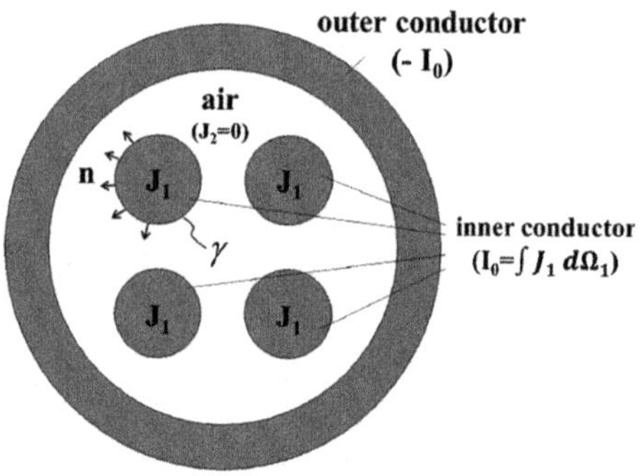

are changed to be an integrated circle, the stored system energy becomes a maximum under the constraint of constant conductor volume. The design objective is to obtain the maximum energy and the design variable is the shape of the interface γ between the inner cables and their surrounding air.

The objective function to be maximized is defined as the magnetic system energy;

$$F = W_m = \int_{\Omega} \frac{1}{2} \nu \mathbf{B}(\mathbf{A}) \cdot \mathbf{B}(\mathbf{A}) \mathrm{d}\Omega \tag{4.2.89}$$

The shape sensitivity for the current region interface is the sensitivity formula (4.2.40):

$$\dot{G} = \int_{\gamma} \mathbf{J}_1 \cdot \mathbf{A}_2 V_n \mathrm{d}\Gamma \tag{4.2.90}$$

This problem has a constraint of constant volume of the inner conductors. For this maximization problem with the volume constraint, the design velocity is taken as

$$U_n = V_n - V_{na} \tag{4.2.91}$$

where

$$V_n = \mathbf{J}_1 \cdot \mathbf{A}_2 \tag{4.2.92}$$

$$V_{na} = \int_{\gamma} \mathbf{J}_1 \cdot \mathbf{A}_2 \mathrm{d}\Gamma \Big/ \int_{\gamma} \mathrm{d}\Gamma \tag{4.2.93}$$

The design result is shown in Fig. 4.32, where the four current regions in the inner conductors are merged into one current region and it becomes gradually a circle as expected. The final design of the circular shape is obtained at the 80 s,

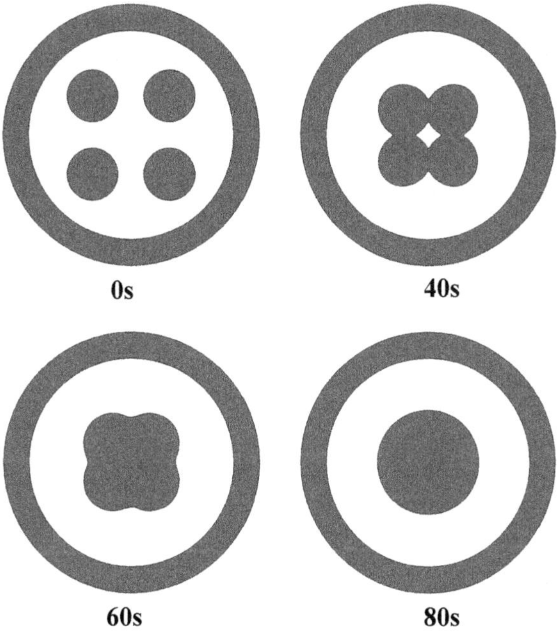

Fig. 4.32 Coaxial cable: shape variation

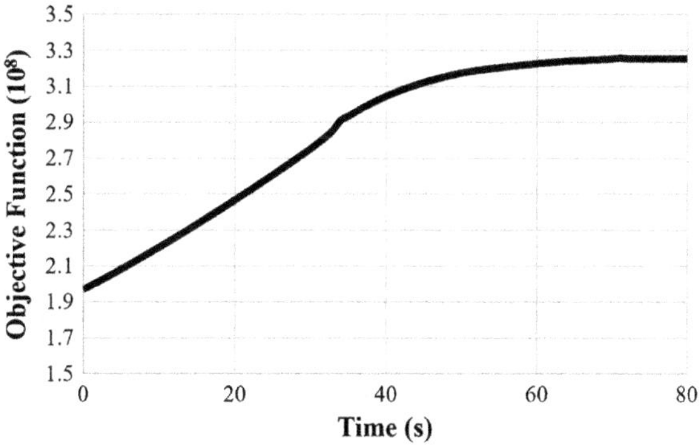

Fig. 4.33 Coaxial cable: evolution of objective function

when the system energy converges to the maximum value as in Fig. 4.33. The result shows that the shape sensitivity analysis is feasible for the interface of the current region problem with the system energy objective function in the axi-symmetric magnetostatic system.

4.2.8.3 Permanent Magnet Motor-Ferromagnetic Interface Design

The shape optimization using the interface sensitivity formula is applied to a shape design problem without the analytical field solution nor the known optimal shape.

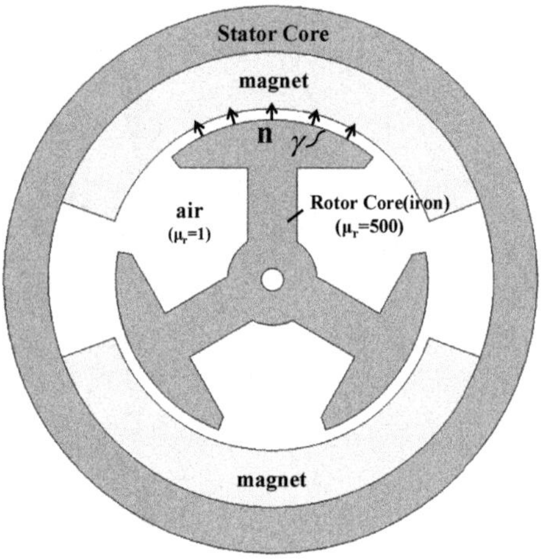

Fig. 4.34 Permanent magnet motor-ferromagnetic material design

As an example, consider a permanent magnet motor in Fig. 4.34, where the motor consists of two magnets with a stator core and a ferromagnetic rotor core. The design objective is to find a rotor shape, which minimize the cogging torque [7]. In this motor design, coil windings are not considered because it does not affect generation of the cogging torque.

The cogging torque on the ferromagnetic rotor can be expressed as the derivative of system energy W_m with respect to the angular position θ of the rotor;

$$T = \frac{dW_m}{d\theta} \tag{4.2.94}$$

As the energy variation between with respect to the angular position decreases, the generated cogging torque decreases. To minimize the energy variation with respect to the angular position, the objective function is defined as

$$F = \sum_{\theta=\theta_1}^{\theta_{10}} \left(W_{m,\theta} - W_0\right)^2 \tag{4.2.95}$$

where W_0 is a constant target energy and the energy at each rotor position θ is

$$W_{m,\theta} = \int_{\Omega} \frac{1}{2} \nu \mathbf{B}_\theta(\mathbf{A}) \cdot \mathbf{B}_\theta(\mathbf{A}) d\Omega \tag{4.2.96}$$

At each position, the shape sensitivity for the system energy is written as

$$\dot{G}_\theta = \int_\gamma \frac{v_0}{2}(1 - v_r)\mathbf{B}_{1,\theta} \cdot \mathbf{B}_{2,\theta} V_n \mathrm{d}\Gamma \qquad (4.2.97)$$

Using (4.2.96) and (4.2.97), the shape sensitivity for the objective function (4.2.95) is rewritten as

$$\dot{G} = \sum_{\theta=\theta_1}^{\theta_{10}} 2(W_{m,\theta} - W_0)\dot{G}_\theta = \int_\gamma \left[\sum_{\theta=\theta_1}^{\theta_{10}} 2(W_{m,\theta} - W_0)\frac{v_0}{2}(1 - v_r)\mathbf{B}_{1,\theta} \cdot \mathbf{B}_{2,\theta}\right] V_n \mathrm{d}\Gamma \qquad (4.2.98)$$

For this minimization problem, the design velocity is taken as

$$V_n = -\left[\sum_{\theta=\theta_1}^{\theta_{10}} 2(W_{m,\theta} - W_0)\frac{v_0}{2}(1 - v_r)\mathbf{B}_{1,\theta} \cdot \mathbf{B}_{2,\theta}\right] \qquad (4.2.99)$$

Figure 4.35 shows the initial and the final designs of the ferromagnetic rotor. In the optimization process, the edge sides of the rotor pole move down and the final shape of rotor becomes salient pole. According to this shape evolution, the energy

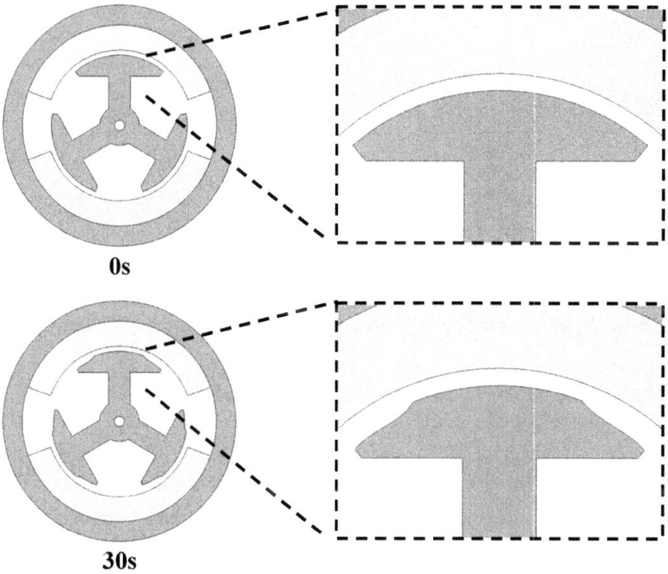

Fig. 4.35 Permanent magnet motor: initial and final shapes

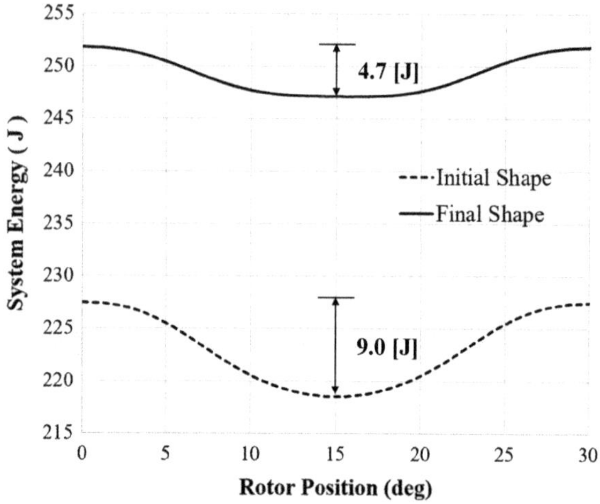

Fig. 4.36 Permanent magnet motor: system energy to rotor position

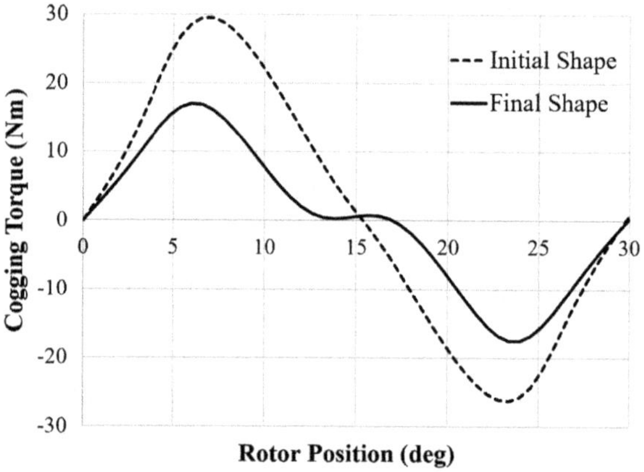

Fig. 4.37 Permanent magnet motor: cogging torque to rotor position

variation with respect to the rotor position decreases as in Fig. 4.36; the cogging torque of the final design decreases by 48% as in Fig. 4.37.

4.2.8.4 Shell-Type Transformer-Ferromagnetic Interface Design

The shape optimization using the interface sensitivity formula is applied to a shape design problem without the analytical field solution nor the known optimal shape. As an example, consider a shell-type transformer in Fig. 4.38, where the primary and the secondary coils are wound around the shell-type ferromagnetic core [8]. The design objective is to find a shape of ferromagnetic core that maximizes the self inductance and the mutual inductance of the windings. Here, the design variable is the shape of the interface γ between the outer surface of the ferromagnetic core and the air.

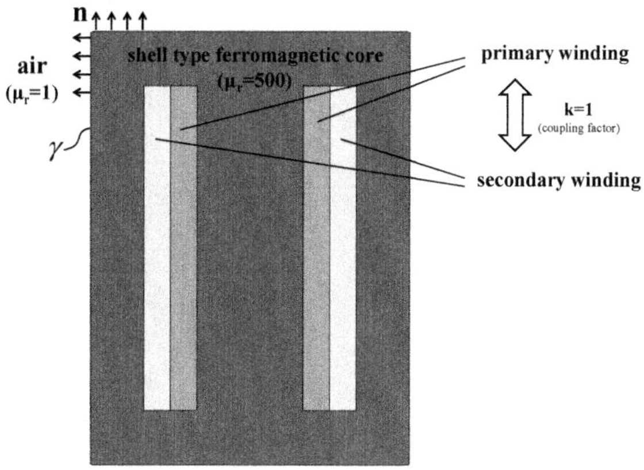

Fig. 4.38 Shell type transformer-ferromagnetic material design, system energy

Since the coupling coefficient of the shell type transformer is close to 1, the self inductance and the mutual inductance are assumed to be proportional to the magnetic system energy. To obtain the maximum inductances, the objective function is defined as the magnetic system energy;

$$F = W_m = \int_{\Omega} \frac{1}{2} \nu \mathbf{B}(\mathbf{A}) \cdot \mathbf{B}(\mathbf{A}) \mathrm{d}\Omega \tag{4.2.100}$$

The shape sensitivity for this ferromagnetic material interface is the sensitivity formula (4.2.40):

$$\dot{G} = \int_{\gamma} \frac{\nu_0}{2} (1 - \nu_r) \mathbf{B}_1 \cdot \mathbf{B}_2 V_n \mathrm{d}\Gamma \tag{4.2.101}$$

This problem has a constraint of constant volume of the ferromagnetic material. For this maximization problem with the volume constraint, the design velocity is taken as

$$U_n = V_n - V_{na} \tag{4.2.102}$$

where

$$V_n = \frac{\nu_0}{2} (1 - \nu_r) \mathbf{B}_1 \cdot \mathbf{B}_2 \tag{4.2.103}$$

$$V_{na} = \int_{\gamma} \frac{\nu_0}{2} (1 - \nu_r) \mathbf{B}_1 \cdot \mathbf{B}_2 \mathrm{d}\Gamma \Big/ \int_{\gamma} \mathrm{d}\Gamma \tag{4.2.104}$$

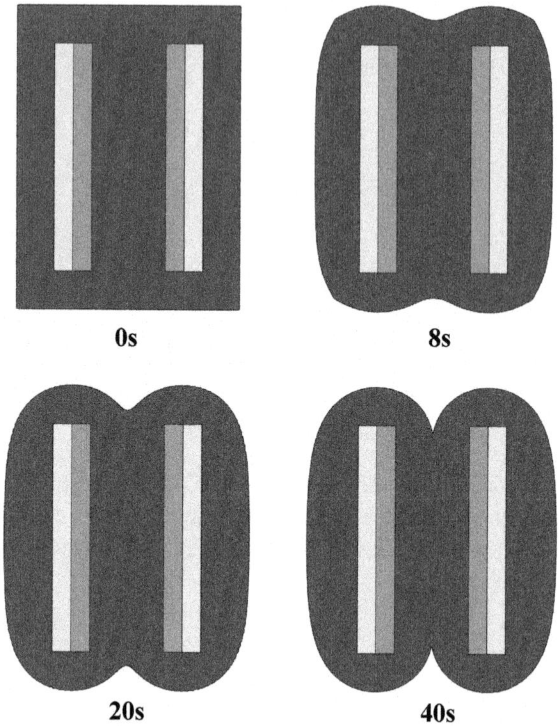

Fig. 4.39 Shell type transformer: shape variation

Figure 4.39 shows the shape variation of the ferromagnetic core. The shape of the outer interface becomes gradually curved with the increase of the iteration number. According to this shape variation, the width of the magnetic path becomes constant in the ferromagnetic core; this results in the minimum magnetic reluctance and the maximum system energy. The final design is obtained at the 40 s, when the system energy converges to the maximum value as in Fig. 4.40.

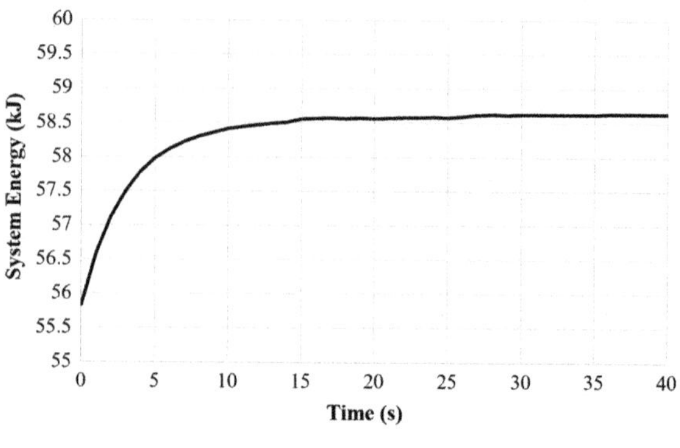

Fig. 4.40 Shell type transformer: evolution of objective function

4.2.8.5 Shell Type Transformer-Ferromagnetic Material, Current Region Interface Design

The shell type transformer, which has the same structure as in the previous example, is considered again in Fig. 4.41. Unlike the previous example, the interface between the windings and the inner surface of the ferromagnetic core is designed in this design problem [8].

The design objective is to obtain a maximum system energy and the objective function is the system energy;

$$F = W_m = \int_{\Omega} \frac{1}{2} \nu \mathbf{B}(\mathbf{A}) \cdot \mathbf{B}(\mathbf{A}) \mathrm{d}\Omega \tag{4.2.105}$$

The shape sensitivity for the interface γ is the sensitivity formula (4.2.40):

$$\dot{G} = \int_{\gamma} \left[\frac{\nu_0}{2} (1 - \nu_r) \mathbf{B}_1 \cdot \mathbf{B}_2 - \mathbf{J}_2 \cdot \mathbf{A}_2 \right] V_n \mathrm{d}\Gamma \tag{4.2.106}$$

This problem has a constraint of constant volume of the ferromagnetic material. For this maximization problem with the volume constraint, the design velocity is taken as

$$U_n = V_n - V_{na} \tag{4.2.107}$$

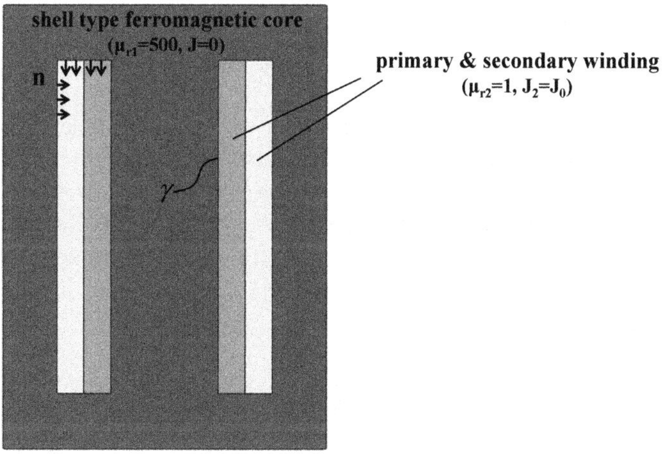

Fig. 4.41 Shell type transformer-ferromagnetic material and current region design, system energy

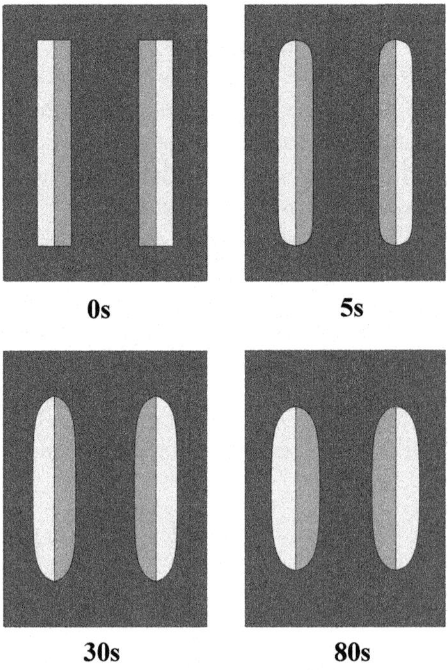

Fig. 4.42 Shell type transformer: shape variation

where

$$V_n = \frac{\nu_0}{2}(1 - \nu_r)\mathbf{B}_1 \cdot \mathbf{B}_2 - \mathbf{J}_2 \cdot \mathbf{A}_2 \tag{4.2.108}$$

$$V_{na} = \int_{\gamma} \left[\frac{\nu_0}{2}(1 - \nu_r)\mathbf{B}_1 \cdot \mathbf{B}_2 - \mathbf{J}_2 \cdot \mathbf{A}_2 d\Gamma\right] / \int_{\gamma} \mathrm{d}\Gamma \tag{4.2.109}$$

Figure 4.42 shows the shape evolution of the ferromagnetic core and the winding. The shape of the inner interface becomes gradually curved; so this variation makes the width of the magnetic path constant. This results in the minimum magnetic reluctance and the maximum system energy. The final design is obtained at the 80s, when the system energy converges to the maximum value as in Fig. 4.43.

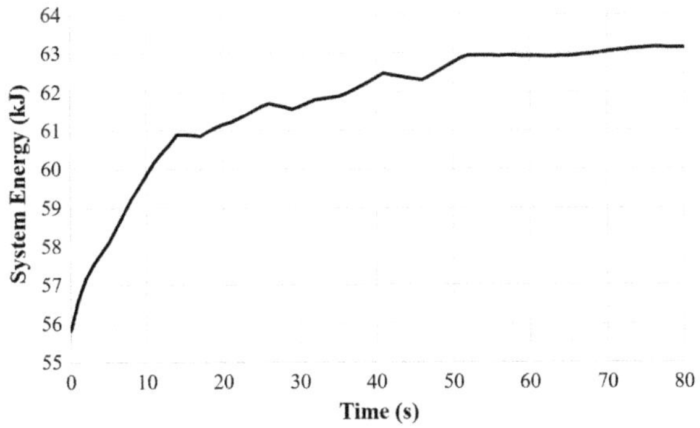

Fig. 4.43 Shell type transformer: evolution of objective function

4.2.8.6 Shell Type Transformer-Multi Interface Design

The shell type transformer, which is the same as in the previous examples, is considered once more in Fig. 4.44. In this example, both the outer interface γ_o and the inner interface γ_i of the magnetic core, of which each was treated in the previous two examples, are designed simultaneously [8].

The design objective is to obtain maximum system energy and the objective function is the system energy;

$$F = W_m = \int_{\Omega} \frac{1}{2} \nu \mathbf{B}(\mathbf{A}) \cdot \mathbf{B}(\mathbf{A}) \mathrm{d}\Omega \tag{4.2.110}$$

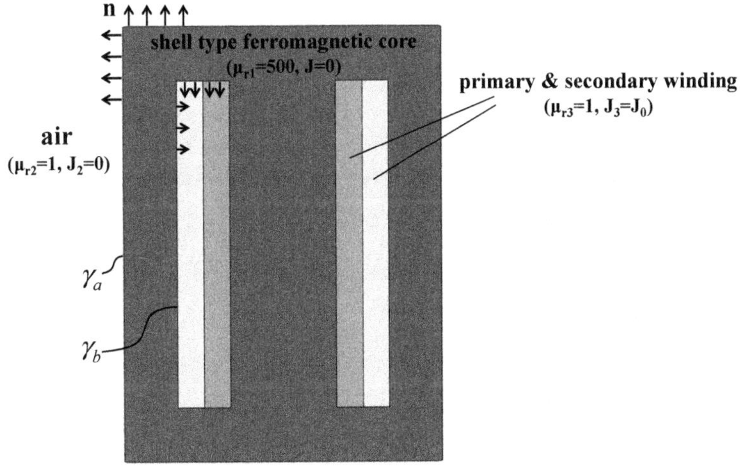

Fig. 4.44 Shell type transformer-multi-interface design, system energy

The shape sensitivity for the two interfaces γ_i and γ_o is expressed as (4.2.111), respectively:

$$\dot{G} = \int_{\gamma_o} \frac{\nu_0}{2}(1 - \nu_r)\mathbf{B}_1 \cdot \mathbf{B}_2 V_{n,o} \mathrm{d}\Gamma + \int_{\gamma_i} \left[\frac{\nu_0}{2}(1 - \nu_r)\mathbf{B}_1 \cdot \mathbf{B}_3 - \mathbf{J}_3 \cdot \mathbf{A}_3\right] V_{n,i} \mathrm{d}\Gamma \tag{4.2.111}$$

This problem has a constraint of constant volume of the ferromagnetic core. For this maximization problem with the volume constraint, the design velocities for the outer and inner interfaces are taken as

$$U_{n,o} = V_{n,o} - V_{na,o} \quad \text{for the interface } \gamma_o \tag{4.2.112}$$

$$U_{n,i} = V_{n,i} - V_{na,i} \quad \text{for the interface } \gamma_i \tag{4.2.113}$$

where

$$V_{n,o} = \frac{\nu_0}{2}(1 - \nu_{r1})\mathbf{B}_1 \cdot \mathbf{B}_2 \quad \text{for the interface} \gamma_o \tag{4.2.114}$$

$$V_{n,i} = \frac{\nu_0}{2}(1 - \nu_{r1})\mathbf{B}_1 \cdot \mathbf{B}_2 - \mathbf{J}_3 \cdot \mathbf{A}_3 \quad \text{for the interface} \gamma_o \tag{4.2.115}$$

$$V_{na,o} = \int_{\gamma_o} \frac{\nu_0}{2}(1 - \nu_{r1})\mathbf{B}_1 \cdot \mathbf{B}_2 d\Gamma \Big/ \int_{\gamma_o} \mathrm{d}\Gamma \tag{4.2.116}$$

$$V_{na} = \int_{\gamma_i} \left[\frac{\nu_0}{2}(1 - \nu_{r1})\mathbf{B}_1 \cdot \mathbf{B}_2 - \mathbf{J}_3 \cdot \mathbf{A}_3 \mathrm{d}\Gamma\right] \Big/ \int_{\gamma_i} \mathrm{d}\Gamma \tag{4.2.117}$$

The shape variation of this multi-interface problem is shown in Fig. 4.45c, where both the core and the winding become circular. For comparison of the results of the previous two examples, their shape variations are also shown in Fig. 4.45a, b. Since the design space of the multi-interface design is much larger than the single-interface design, the possible design is more various. The variation of the objective function of the multi-interface design is compared with the ones of the single-interface design in Fig. 4.46, where we see the final objective function by the multi-interface design increases by over 70% compared with the ones of the single-interface designs.

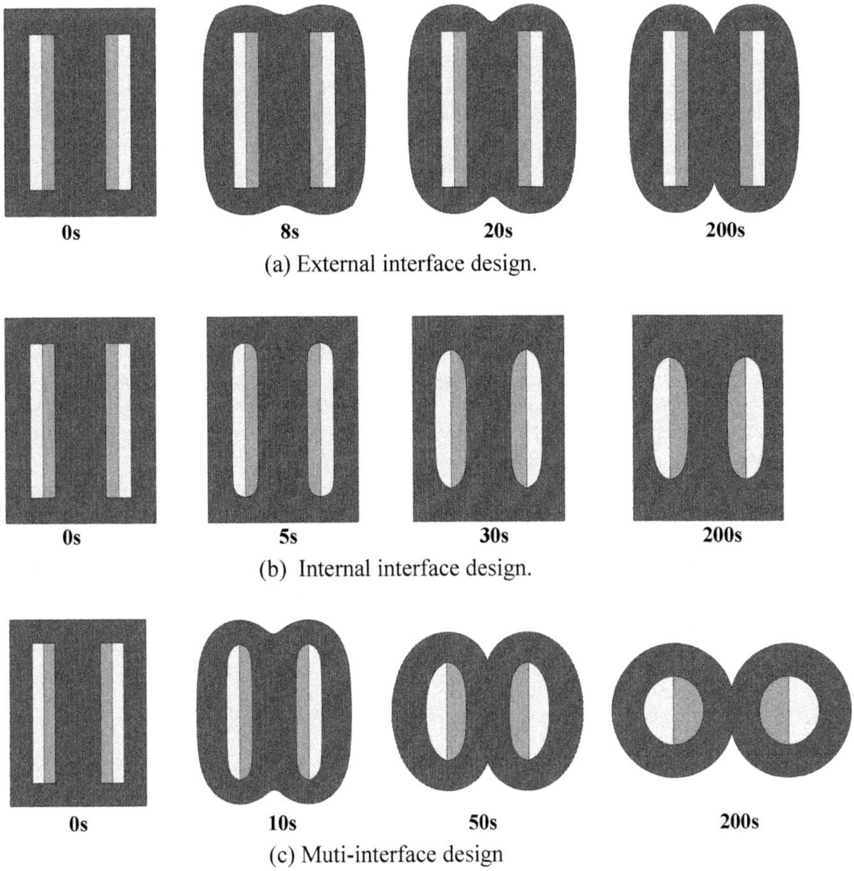

Fig. 4.45 Shell-type transformer: shape variation

Fig. 4.46 Shell type transformer: evolution of objective function

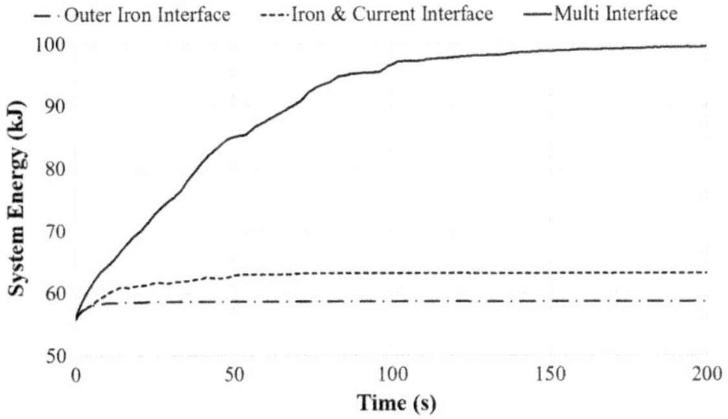

References

1. Park, I.H., Lee, B.T., Hahn, S.Y.: Sensitivity analysis based on analytic approach for shape optimization of electromagnetic devices: interface problem of iron and air. IEEE Trans. Magn. **27**, 4142–4145 (1991)
2. Park, I.H., Coulomb, J.L., Hahn, S.Y.: Design sensitivity analysis for nonlinear magnetostatic problems by continuum approach. J. Phys. III (France) **2**, 2045–2053 (1992)
3. Choi, K.K., Seong, H.G.: Design component method for sensitivity analysis of built-up structures. J. Struct. Mech. **14**, 379–399 (1986)
4. Coulomb, J.L.: A methodology for the determination of global electromechanical quantities from a finite element analysis and its application to the evaluation of magnetic forces, torques and stiffness. IEEE Trans. Magn. Mag-19, 2514–2519 (1983)
5. Coulomb, J.L., Meunier, G.: Finite element implementation of virtual work principle for magnetic or electric force and torque computation. IEEE Trans. Magn. **20**, 1894–1896 (1984)
6. Coulomb, J.L., Meunier, G., Sabonnadière, J.C.: Energy methods for the evaluation of global quantities and integral parameters in a finite elements analysis of electromagnetic devices. IEEE Trans. Magn. **21**, 1817–1822 (1985)
7. Park, I.H., Lee, B.T., Hahn, S.Y.: Pole shape optimization for reduction of cogging torque by sensitivity analysis. Compel 9, Supplement A, 111–114 (1990)
8 Lee, K.H., Seo, K.S., Choi, H.S., Park, I.H.: Multiple level set method for multi-material shape optimization in electromagnetic system. Int. J. Appl. Electromagn. Mech. **56**, 183–193 (2018)

Chapter 5
Continuum Shape Design Sensitivity of Eddy Current System

The eddy current system is composed of ferromagnetic material, conductive material, and source current. We have examples of the design objective such as improving device performance, obtaining a desired distribution of magnetic field, obtaining a desired inductance. The eddy current system has various examples such as induction motor, induction heating, magnetic launcher, NDT/NDE, induction shielding, induction cooker [1–5].

In this chapter, the continuum shape sensitivity for the eddy current system is derived, just like in the previous Chaps. 3 and 4. As mentioned in Sect. 2.3.1, we deal with only the linear eddy current system of the complex state variable without the gradient phi term. The derivation of a general three-dimensional sensitivity for the eddy current system is still an open problem.

The design variable of the eddy current system, like in the case of the magnetostatic system, is only the interface. The shape sensitivity for the eddy current system is classified into two categories: the domain integral objective function and the system power objective function. The system power is defined as the input power supplied by an external current source. Since the input power objective function does not require solving for the adjoint variable and is used for the equivalent resistance and inductance, it is taken as another objective function.

In Sect. 5.1, a three-dimensional sensitivity of the interface design for the domain integral objective function is derived. Since the eddy current system is expressed with three material properties: permeability, conductivity, and current density, the derived sensitivity formula has the interface integral of three terms. If the air is treated as another material, the eddy current system has four material properties. Thus, the eddy current system has nine kinds of interfaces, each sensitivity of which are arranged in Sect. 5.1.5.

In Sect. 5.2, a three-dimensional sensitivity of the interface design for the system power objective function is derived. It also has the nine interfaces, each sensitivity of which are arranged in Sect. 5.2.5. The power sensitivity is examined in the electric-circuit point of view, and then, the inductance sensitivity and the resistance sensitivity are derived using the power sensitivity.

I. H. Park, *Design Sensitivity Analysis and Optimization of Electromagnetic Systems*, Mathematical and Analytical Techniques with Applications to Engineering,
https://doi.org/10.1007/978-981-13-0230-5_5

At the end of Sects. 5.1–5.2, the two sensitivity formulas derived are applied to numerical examples to be validated. Since analytical models with an analytical solution in the eddy current system are not easy to find, two-dimensional models with known or unknown optimal designs are tested by the two-dimensional finite element method.

5.1 Interface Shape Sensitivity

The eddy current system is another typical open boundary problem like the magnetostatic system. In the eddy current system, since the boundary conditions (homogeneous Dirichlet and homogeneous Neumann condition) are applied to the symmetry surface or the infinite boundary, its outer boundary is not taken as the design variable. The design variable for the eddy current problem is, therefore, the interface where two different materials meet. The shape variation of the interface results in the variation in the magnetic and the electric fields. In this section, a three-dimensional sensitivity for the interface variation is derived [6]. First, the eddy current system for the interface design is depicted and a general objective function is defined as a domain integral. Second, the Lagrange multiplier method is introduced for the constraint of the variational state equation. Third, the adjoint variable method is used to express the sensitivity in terms of design variation. Fourth, the variational identities are used to transform the domain integral of the sensitivity into the interface integral, which provides the three-dimensional sensitivity formula for the interface design. Finally, the obtained sensitivity formula is tested and validated with numerical examples.

5.1.1 Problem Definition and Objective Function

An eddy current system for interface design is given as in Fig. 5.1, where the whole domain Ω comprises two domains Ω_1 and Ω_2 divided by an interface γ. The domain Ω_1 has a distribution of ν_1, σ_1, and $\mathbf{J}_1$, and the domain Ω_2 has a distribution of ν_2, σ_2, and $\mathbf{J}_2$. The domain Ω_1 has the outer boundary Γ where $\mathbf{n}$ is defined as the outward normal vector on the outer boundary. The outer boundary consists of the

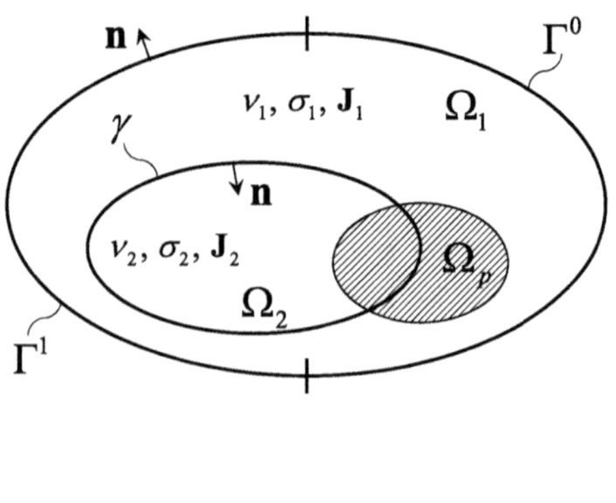

Fig. 5.1 Interface design of eddy current system

Dirichlet boundary Γ^0 and the Neumann boundary Γ^1. In this shape sensitivity analysis, the interface is taken as design variable.

A general objective function is defined in the integral form as

$$\begin{aligned} F &= \int_{\Omega} g(A, B(A)) m_p \mathrm{d}\Omega \\ &= \int_{\Omega_1} g_1 m_p \mathrm{d}\Omega + \int_{\Omega_2} g_2 m_p \mathrm{d}\Omega \end{aligned} \tag{5.1.1}$$

where g_1 and g_2 are any functions that are continuously differentiable with respect to their arguments;

$$g_1 \equiv g(\mathbf{A}_1, \mathbf{B}(\mathbf{A}_1)) \text{ and } g_2 \equiv g(\mathbf{A}_2, \mathbf{B}(\mathbf{A}_2)) \tag{5.1.2}$$

and m_p is a characteristic function that is defined as

$$m_p = \begin{pmatrix} 1 & \mathbf{x} \in \Omega_p \\ 0 & \mathbf{x} \notin \Omega_p \end{pmatrix} \tag{5.1.3}$$

The region Ω_p, the integral domain for the objective function, can include the interface as in Fig. 5.1.

The governing differential equations for the state variables of the magnetic vector potential $\mathbf{A}_1$ and $\mathbf{A}_2$ are given as;

$$\nabla \times \nu_1 \nabla \times \mathbf{A}_1 + j\omega\sigma_1 \mathbf{A}_1 = \mathbf{J}_1 \quad \text{in } \Omega_1 \tag{5.1.4}$$

$$\nabla \times \nu_2 \nabla \times \mathbf{A}_2 + j\omega\sigma_2 \mathbf{A}_2 = \mathbf{J}_2 \quad \text{in } \Omega_2 \tag{5.1.5}$$

These equations have a unique solution with the boundary condition:

$$\mathbf{A}_1 = \mathbf{0} \quad \text{on } \Gamma^0 \text{ (homogeneous Dirichlet boundary condition)} \tag{5.1.6}$$

$$\frac{\partial \mathbf{A}_1}{\partial n} = \mathbf{0} \quad \text{on } \Gamma^1 \text{ (homogeneous Neumann boundary condition)} \tag{5.1.7}$$

The variational identities for the state equations are obtained by multiplying both sides of (5.1.4) and (5.1.5) by an arbitrary virtual vector potential $\bar{\mathbf{A}}_1$ and $\bar{\mathbf{A}}_2$, respectively, and by using the vector identity and the divergence theorem;

$$\begin{aligned} &\int_{\Omega_1} \left(\nu_1 \mathbf{B}(\mathbf{A}_1) \cdot \mathbf{B}(\overline{\mathbf{A}}_1) + j\omega\sigma_1 \mathbf{A}_1 \cdot \overline{\mathbf{A}}_1 - \mathbf{J}_1 \cdot \overline{\mathbf{A}}_1 \right) \mathrm{d}\Omega \\ &\quad = -\int_{\gamma + \Gamma} (\mathbf{n} \times \mathbf{H}(\mathbf{A}_1)) \cdot \overline{\mathbf{A}}_1 \mathrm{d}\Gamma \quad \forall \overline{\mathbf{A}}_1 \in \Phi \end{aligned} \tag{5.1.8}$$

$$\begin{aligned}&\int\limits_{\Omega_2}\left(\nu_2\mathbf{B}(\mathbf{A}_2)\cdot\mathbf{B}(\overline{\mathbf{A}}_2)+j\omega\sigma_2\mathbf{A}_2\cdot\overline{\mathbf{A}}_2-\mathbf{J}_2\cdot\overline{\mathbf{A}}_2\right)\mathrm{d}\Omega\\&=\int\limits_{\gamma}(\mathbf{n}\times\mathbf{H}(\mathbf{A}_2))\cdot\overline{\mathbf{A}}_2\mathrm{d}\Gamma\quad\forall\overline{\mathbf{A}}_2\in\Phi\end{aligned}\tag{5.1.9}$$

where

$$\mathbf{H}(\mathbf{A}_1)=\nu_1\mathbf{B}(\mathbf{A}_1)\tag{5.1.10}$$

$$\mathbf{H}(\mathbf{A}_2)=\nu_2\mathbf{B}(\mathbf{A}_2)\tag{5.1.11}$$

and Φ is the space of the admissible complex vector potential defined in Sect. 2.3.2 as

$$\Phi=\left\{\bar{\mathbf{A}}\in\left[H^1(\Omega)\right]^3|\bar{\mathbf{A}}=\mathbf{0}\,\text{on}\,\mathbf{x}\in\Gamma^0\right\}\tag{5.1.12}$$

The boundary conditions of (5.1.6) and (5.1.7) can be rewritten for the variational equation;

$$\bar{\mathbf{A}}_1=\mathbf{0}\quad\text{on}\quad\Gamma^0\quad\text{(homogeneous Dirichlet boundary condition)}\tag{5.1.13}$$

$$\mathbf{n}\times\mathbf{H}(\mathbf{A}_1)=\mathbf{0}\quad\text{on}\quad\Gamma^1\quad\text{(homogeneous Neumann boundary condition)}\tag{5.1.14}$$

and the interface condition is

$$\mathbf{n}\times\mathbf{H}(\mathbf{A}_1)=\mathbf{n}\times\mathbf{H}(\mathbf{A}_2)\quad\text{on}\quad\gamma\tag{5.1.15}$$

By summing (5.1.8) and (5.1.9) and by imposing the boundary conditions and the interface condition, the variational state equation reduced by the differential Eqs. (5.1.4) and (5.1.5) is obtained as

$$\begin{aligned}&\int\limits_{\Omega_1}\left(\nu_1\mathbf{B}(\mathbf{A}_1)\cdot\mathbf{B}(\bar{\mathbf{A}}_1)+j\omega\sigma_1\mathbf{A}_1\cdot\bar{\mathbf{A}}_1\right)\mathrm{d}\Omega\\&+\int\limits_{\Omega_2}\left(\nu_2\mathbf{B}(\mathbf{A}_2)\cdot\mathbf{B}(\bar{\mathbf{A}}_2)+j\omega\sigma_2\mathbf{A}_2\cdot\bar{\mathbf{A}}_2\right)\mathrm{d}\Omega\\&=\int\limits_{\Omega_1}\mathbf{J}_1\cdot\bar{\mathbf{A}}_1\mathrm{d}\Omega+\int\limits_{\Omega_2}\mathbf{J}_2\cdot\bar{\mathbf{A}}_2\mathrm{d}\Omega\quad\forall\bar{\mathbf{A}}_1,\bar{\mathbf{A}}_2\in\Phi\end{aligned}\tag{5.1.16}$$

The bilinear form $a(\mathbf{A}, \bar{\mathbf{A}})$ and the source linear form $l(\bar{\mathbf{A}})$ are defined as

$$a(\mathbf{A}, \bar{\mathbf{A}}) \equiv \int_{\Omega_1} (v_1 \mathbf{B}(\mathbf{A}_1) \cdot \mathbf{B}(\bar{\mathbf{A}}_1) + j\omega\sigma_1 \mathbf{A}_1 \cdot \bar{\mathbf{A}}_1) \mathrm{d}\Omega + \int_{\Omega_2} (v_2 \mathbf{B}(\mathbf{A}_2) \cdot \mathbf{B}(\bar{\mathbf{A}}_2) + j\omega\sigma_2 \mathbf{A}_2 \cdot \bar{\mathbf{A}}_2) \mathrm{d}\Omega \tag{5.1.17}$$

$$l(\bar{\mathbf{A}}) \equiv \int_{\Omega_1} \mathbf{J}_1 \cdot \bar{\mathbf{A}}_1 \mathrm{d}\Omega + \int_{\Omega_2} \mathbf{J}_2 \cdot \bar{\mathbf{A}}_2 \mathrm{d}\Omega \tag{5.1.18}$$

The variational Eq. (5.1.16) is rewritten with the bilinear form and the source linear form as

$$a(\mathbf{A}, \bar{\mathbf{A}}) = l(\bar{\mathbf{A}}) \quad \forall\, \bar{\mathbf{A}} \in \Phi \tag{5.1.19}$$

where

$$\mathbf{A} = \mathbf{A}_1 \cup \mathbf{A}_2 \text{ and } \bar{\mathbf{A}} = \bar{A}_1 \cup \bar{\mathbf{A}}_2 \tag{5.1.20}$$

5.1.2 Lagrange Multiplier Method for Sensitivity Derivation

Since the variational state Eq. (5.1.19) holds regardless of the change of the interface shape, it is treated as an equality constraint in the shape sensitivity analysis. The method of Lagrange multipliers is employed for the implicit constraint of the variational state equation, and it provides an augmented objective function G as

$$G = F + l(\bar{\mathbf{A}}) - a(\mathbf{A}, \bar{\mathbf{A}}) \quad \forall\, \bar{\mathbf{A}} \in \Phi \tag{5.1.21}$$

where the arbitrary virtual potential $\bar{\mathbf{A}}$ plays the role of Lagrange multipliers and Φ is the space of admissible complex vector potential: $\Phi = \left\{ \bar{\mathbf{A}} \in [H^1(\Omega)]^3 | \bar{\mathbf{A}} = \mathbf{0} \right.$ on $\mathbf{x} \in \Gamma^0.\}$.

The sensitivity, the material derivative of the augmented objective function, is written as

$$\dot{G} = -\dot{a}(\mathbf{A}, \bar{\mathbf{A}}) + \dot{l}(\bar{\mathbf{A}}) + \dot{F} \quad \forall\, \bar{\mathbf{A}} \in \Phi \tag{5.1.22}$$

By applying the material derivative formula (3.1.22) to the variational state Eq. (5.1.16) and the objective function (5.1.1) and using the relation (3.1.17), each term in (5.1.22) is obtained below as (5.1.23), (5.1.24) and (5.1.25). In this interface sensitivity problem, when the material derivative formula is applied, only the integrals on the interface remain since the outer boundary is not taken as design variable ($V_n = 0$ on Γ).

$$\begin{aligned}
\dot{a}(\mathbf{A},\bar{\mathbf{A}}) &= \int_{\Omega_1} \left(v_1\mathbf{B}(\mathbf{A}'_1)\cdot\mathbf{B}(\bar{\mathbf{A}}_1) + v_1\mathbf{B}(\mathbf{A}_1)\cdot\mathbf{B}(\bar{\mathbf{A}}'_1) + j\omega\sigma_1\mathbf{A}'_1\cdot\bar{\mathbf{A}}_1 + j\omega\sigma_1\mathbf{A}_1\cdot\bar{\mathbf{A}}'_1\right)\mathrm{d}\Omega \\
&+ \int_{\gamma} \left(v_1\mathbf{B}(\mathbf{A}_1)\cdot\mathbf{B}(\bar{\mathbf{A}}_1) + j\omega\sigma_1\mathbf{A}_1\cdot\bar{\mathbf{A}}_1\right)V_n\mathrm{d}\Gamma \\
&+ \int_{\Omega_2} \left(v_2\mathbf{B}(\mathbf{A}'_2)\cdot\mathbf{B}(\bar{\mathbf{A}}_2) + v_2\mathbf{B}(\mathbf{A}_2)\cdot\mathbf{B}(\bar{\mathbf{A}}'_2) + j\omega\sigma_2\mathbf{A}'_2\cdot\bar{\mathbf{A}}_2 + j\omega\sigma_2\mathbf{A}_2\cdot\bar{\mathbf{A}}'_2\right)\mathrm{d}\Omega \\
&- \int_{\gamma} \left(v_2\mathbf{B}(\mathbf{A}_2)\cdot\mathbf{B}(\bar{\mathbf{A}}_2) + j\omega\sigma_2\mathbf{A}_2\cdot\bar{\mathbf{A}}_2\right)V_n\mathrm{d}\Gamma \\
&= \int_{\Omega_1} \Big(v_1\mathbf{B}(\dot{\mathbf{A}}_1)\cdot\mathbf{B}(\bar{\mathbf{A}}_1) - v_1\mathbf{B}(\mathbf{V}\cdot\nabla\mathbf{A}_1)\cdot\mathbf{B}(\bar{\mathbf{A}}_1) + v_1\mathbf{B}(\mathbf{A}_1)\cdot\mathbf{B}(\dot{\bar{\mathbf{A}}}_1) \\
&\quad - v_1\mathbf{B}(\mathbf{A}_1)\cdot\mathbf{B}(\mathbf{V}\cdot\nabla\bar{\mathbf{A}}_1)\Big)\mathrm{d}\Omega + \int_{\Omega_1}\Big[j\omega\sigma_1\dot{\mathbf{A}}_1\cdot\bar{\mathbf{A}}_1 - j\omega\sigma_1(\mathbf{V}\cdot\nabla\mathbf{A}_1)\cdot\bar{\mathbf{A}}_1 \\
&\quad + j\omega\sigma_1\mathbf{A}_1\cdot\dot{\bar{\mathbf{A}}}_1 - j\omega\sigma_1\mathbf{A}_1\cdot\left(\mathbf{V}\cdot\nabla\bar{\mathbf{A}}_1\right)\Big]\mathrm{d}\Omega + \int_{\Omega_2}\Big(v_2\mathbf{B}(\dot{\mathbf{A}}_2)\cdot\mathbf{B}(\bar{\mathbf{A}}_2) \\
&\quad - v_2\mathbf{B}(\mathbf{V}\cdot\nabla\mathbf{A}_2)\cdot\mathbf{B}(\bar{\mathbf{A}}_2) + v_2\mathbf{B}(\mathbf{A}_2)\cdot\mathbf{B}(\dot{\bar{\mathbf{A}}}_2) - v_2\mathbf{B}(\mathbf{A}_2)\cdot\mathbf{B}(\mathbf{V}\cdot\nabla\bar{\mathbf{A}}_2)\Big)\mathrm{d}\Omega \\
&+ \int_{\Omega_2}\Big[j\omega\sigma_2\dot{\mathbf{A}}_2\cdot\bar{\mathbf{A}}_2 - j\omega\sigma_2(\mathbf{V}\cdot\nabla\mathbf{A}_2)\cdot\bar{\mathbf{A}}_2 + j\omega\sigma_2\mathbf{A}_2\cdot\dot{\bar{\mathbf{A}}}_2 - j\omega\sigma_2\mathbf{A}_2\cdot\left(\mathbf{V}\cdot\nabla\bar{\mathbf{A}}_2\right)\Big]\mathrm{d}\Omega \\
&+ \int_{\gamma}\left(v_1\mathbf{B}(\mathbf{A}_1)\cdot\mathbf{B}(\bar{\mathbf{A}}_1) - v_2\mathbf{B}(\mathbf{A}_2)\cdot\mathbf{B}(\bar{\mathbf{A}}_2) + j\omega\sigma_1\mathbf{A}_1\cdot\bar{\mathbf{A}}_1 - j\omega\sigma_2\mathbf{A}_2\cdot\bar{\mathbf{A}}_2\right)V_n\mathrm{d}\Gamma
\end{aligned} \tag{5.1.23}$$

$$\begin{aligned}
\dot{l}(\bar{\mathbf{A}}) &= \int_{\Omega_1}\mathbf{J}_1\cdot\bar{\mathbf{A}}'_1\mathrm{d}\Omega + \int_{\gamma}\mathbf{J}_1\cdot\bar{\mathbf{A}}_1V_n\mathrm{d}\Gamma + \int_{\Omega_2}\mathbf{J}_2\cdot\bar{\mathbf{A}}'_2\mathrm{d}\Omega - \int_{\gamma}\mathbf{J}_2\cdot\bar{\mathbf{A}}_2V_n\mathrm{d}\Gamma \\
&= \int_{\Omega_1}\Big[\mathbf{J}_1\cdot\dot{\bar{\mathbf{A}}}_1 - \mathbf{J}_1\cdot\left(\mathbf{V}\cdot\nabla\bar{\mathbf{A}}_1\right)\Big]\mathrm{d}\Omega + \int_{\Omega_2}\Big[\mathbf{J}_2\cdot\dot{\bar{\mathbf{A}}}_2 - \mathbf{J}_2\cdot\left(\mathbf{V}\cdot\nabla\bar{\mathbf{A}}_2\right)\Big]\mathrm{d}\Omega \\
&+ \int_{\gamma}\left(\mathbf{J}_1\cdot\bar{\mathbf{A}}_1 - \mathbf{J}_2\cdot\bar{\mathbf{A}}_2\right)V_n\mathrm{d}\Gamma
\end{aligned} \tag{5.1.24}$$

$$
\begin{aligned}
\dot{F} &= \int_{\Omega_1} g_1' m_p \mathrm{d}\Omega + \int_{\gamma} g_1 m_p V_n \mathrm{d}\Gamma + \int_{\Omega_2} g_1' m_p \mathrm{d}\Omega - \int_{\gamma} g_2 m_p V_n \mathrm{d}\Gamma \\
&= \int_{\Omega_1} \left(\mathbf{g}_{\mathbf{A}_1} \cdot \mathbf{A}'_1 + \mathbf{g}_{\mathbf{B}_1} \cdot \mathbf{B}(\mathbf{A}'_1)\right) m_p \mathrm{d}\Omega + \int_{\gamma} g_1 m_p V_n \mathrm{d}\Gamma \\
&\quad + \int_{\Omega_2} \left(\mathbf{g}_{\mathbf{A}_2} \cdot \mathbf{A}'_2 + \mathbf{g}_{\mathbf{B}_2} \cdot \mathbf{B}(\mathbf{A}'_2)\right) m_p \mathrm{d}\Omega - \int_{\gamma} g_2 m_p V_n \mathrm{d}\Gamma \\
&= \int_{\Omega_1} \left[\mathbf{g}_{\mathbf{A}_1} \cdot \dot{\mathbf{A}}_1 - \mathbf{g}_{\mathbf{A}_1} \cdot (\mathbf{V} \cdot \nabla \mathbf{A}_1) + \mathbf{g}_{\mathbf{B}_1} \cdot \mathbf{B}(\dot{\mathbf{A}}_1) - \mathbf{g}_{\mathbf{B}_1} \cdot \mathbf{B}(\mathbf{V} \cdot \nabla \mathbf{A}_1)\right] m_p \mathrm{d}\Omega \\
&\quad + \int_{\Omega_2} \left[\mathbf{g}_{\mathbf{A}_2} \cdot \dot{\mathbf{A}}_2 - \mathbf{g}_{\mathbf{A}_2} \cdot (\mathbf{V} \cdot \nabla \mathbf{A}_2) + \mathbf{g}_{\mathbf{B}_2} \cdot \mathbf{B}(\dot{\mathbf{A}}_2) - \mathbf{g}_{\mathbf{B}_2} \cdot \mathbf{B}(\mathbf{V} \cdot \nabla \mathbf{A}_2)\right] m_p \mathrm{d}\Omega \\
&\quad + \int_{\gamma} (g_1 - g_2) m_p V_n \mathrm{d}\Gamma
\end{aligned}
\tag{5.1.25}
$$

For g_1' and g_2' in (5.1.25), we used the relation (3.1.27):

$$
g' = \frac{\partial g}{\partial \mathbf{A}} \cdot \mathbf{A}' + \frac{\partial g}{\partial \mathrm{B}} \cdot \mathbf{B}(\mathbf{A}') = \mathbf{g}_{\mathbf{A}} \cdot \mathbf{A}' + \mathbf{g}_{\mathbf{B}} \cdot \mathbf{B}(\mathbf{A}') \tag{5.1.26}
$$

where

$$
\mathbf{g}_{\mathbf{A}} \equiv \frac{\partial g}{\partial \mathbf{A}} = \left[\frac{\partial g}{\partial A_x}, \frac{\partial g}{\partial A_y}, \frac{\partial g}{\partial A_z}\right]^T \text{ and } \mathbf{g}_{\mathbf{B}} \equiv \frac{\partial g}{\partial \mathbf{B}} = \left[\frac{\partial g}{\partial B_x}, \frac{\partial g}{\partial B_y}, \frac{\partial g}{\partial B_z}\right]^T \tag{5.1.27}
$$

In derivation of (5.1.23) and (5.1.24), v_1', $v_2' = 0$, $\sigma_1', \sigma_2' = 0$ and $\mathbf{J}'_1, \mathbf{J}'_2 = 0$ were used. (5.1.23), (5.1.24), and (5.1.25) are inserted into (5.1.22) to provide

$$
\begin{aligned}
\dot{G} = & -\int_{\Omega_1} \left(\nu_1 \mathbf{B}(\dot{\mathbf{A}}_1)\cdot \mathbf{B}(\bar{\mathbf{A}}_1) + j\omega\sigma_1 \dot{\mathbf{A}}_1 \cdot \bar{\mathbf{A}}_1\right)\mathrm{d}\Omega - \int_{\Omega_2} \left(\nu_2 \mathbf{B}(\dot{\mathbf{A}}_2)\cdot \mathbf{B}(\bar{\mathbf{A}}_2) + j\omega\sigma_2 \dot{\mathbf{A}}_2 \cdot \bar{\mathbf{A}}_2\right)\mathrm{d}\Omega \\
& -\int_{\Omega_1} \left(\nu_1 \mathbf{B}(\mathbf{A}_1)\cdot \mathbf{B}(\dot{\bar{\mathbf{A}}}_1) + j\omega\sigma_1 \mathbf{A}_1 \cdot \dot{\bar{\mathbf{A}}}_1\right)\mathrm{d}\Omega - \int_{\Omega_2} \left(\nu_2 \mathbf{B}(\mathbf{A}_2)\cdot \mathbf{B}(\dot{\bar{\mathbf{A}}}_2) + j\omega\sigma_2 \mathbf{A}_2 \cdot \dot{\bar{\mathbf{A}}}_2\right)\mathrm{d}\Omega \\
& + \int_{\Omega_1} \mathbf{J}_1 \cdot \dot{\bar{\mathbf{A}}}_1 \mathrm{d}\Omega + \int_{\Omega_2} \mathbf{J}_2 \cdot \dot{\bar{\mathbf{A}}}_2 \mathrm{d}\Omega + \int_{\Omega_1} \left(\mathbf{g}_{\mathbf{A}_1} \cdot \dot{\mathbf{A}}_1 + \mathbf{g}_{\mathbf{B}_1} \cdot \mathbf{B}(\dot{\mathbf{A}}_1)\right) m_p \mathrm{d}\Omega \\
& + \int_{\Omega_2} \left(\mathbf{g}_{\mathbf{A}_2} \cdot \dot{\mathbf{A}}_2 + \mathbf{g}_{\mathbf{B}_2} \cdot \mathbf{B}(\dot{\mathbf{A}}_2)\right) m_p \mathrm{d}\Omega + \int_{\Omega_1} \left[\nu_1 \mathbf{B}(\mathbf{V}\cdot\nabla\mathbf{A}_1)\cdot \mathbf{B}(\bar{\mathbf{A}}_1) + j\omega\sigma_1 (\mathbf{V}\cdot\nabla\mathbf{A}_1)\cdot \bar{\mathbf{A}}_1\right]\mathrm{d}\Omega \\
& + \int_{\Omega_2} \left[\nu_2 \mathbf{B}(\mathbf{V}\cdot\nabla\mathbf{A}_2)\cdot \mathbf{B}(\bar{\mathbf{A}}_2) + j\omega\sigma_2 (\mathbf{V}\cdot\nabla\mathbf{A}_2)\cdot \bar{\mathbf{A}}_2\right]\mathrm{d}\Omega + \int_{\Omega_1} \left[\nu_1 \mathbf{B}(\mathbf{A}_1)\cdot \mathbf{B}(\mathbf{V}\cdot\nabla\bar{\mathbf{A}}_1)\right. \\
& \left. + j\omega\sigma_1 \mathbf{A}_1 \cdot (\mathbf{V}\cdot\nabla\bar{\mathbf{A}}_1)\right]\mathrm{d}\Omega + \int_{\Omega_2} \left[\nu_2 \mathbf{B}(\mathbf{A}_2)\cdot \mathbf{B}(\mathbf{V}\cdot\nabla\bar{\mathbf{A}}_2) + j\omega\sigma_2 \mathbf{A}_2 \cdot (\mathbf{V}\cdot\nabla\bar{\mathbf{A}}_2)\right]\mathrm{d}\Omega \\
& - \int_{\Omega_1} \mathbf{J}_1 \cdot (\mathbf{V}\cdot\nabla\bar{\mathbf{A}}_1)\mathrm{d}\Omega - \int_{\Omega_2} \mathbf{J}_2 \cdot (\mathbf{V}\cdot\nabla\bar{\mathbf{A}}_2)\mathrm{d}\Omega \\
& - \int_{\Omega_1} \left[\mathbf{g}_{\mathbf{A}_1} \cdot (\mathbf{V}\cdot\nabla\mathbf{A}_1) + \mathbf{g}_{\mathbf{B}_1} \cdot \mathbf{B}(\mathbf{V}\cdot\nabla\mathbf{A}_1)\right] m_p \mathrm{d}\Omega - \int_{\Omega_2} \left[\mathbf{g}_{\mathbf{A}_2} \cdot (\mathbf{V}\cdot\nabla\mathbf{A}_2)\right. \\
& \left. + \mathbf{g}_{\mathbf{B}_2} \cdot \mathbf{B}(\mathbf{V}\cdot\nabla\mathbf{A}_2)\right] m_p \mathrm{d}\Omega - \int_{\gamma} \left(\nu_1 \mathbf{B}(\mathbf{A}_1)\cdot \mathbf{B}(\bar{\mathbf{A}}_1) - \nu_2 \mathbf{B}(\mathbf{A}_2)\cdot \mathbf{B}(\bar{\mathbf{A}}_2) + j\omega\sigma_1 \mathbf{A}_1 \cdot \bar{\mathbf{A}}_1\right. \\
& \left. - j\omega\sigma_2 \mathbf{A}_2 \cdot \bar{\mathbf{A}}_2\right) V_n \mathrm{d}\Gamma + \int_{\gamma} \left[\mathbf{J}_1 \cdot \bar{\mathbf{A}}_1 - \mathbf{J}_2 \cdot \bar{\mathbf{A}}_2 + (g_1 - g_2) m_p\right] V_n \mathrm{d}\Gamma \quad \forall \bar{\mathbf{A}}_1, \bar{\mathbf{A}}_2 \in \Phi
\end{aligned}
\tag{5.1.28}
$$

Since $\dot{\bar{\mathbf{A}}}_1$ and $\dot{\bar{\mathbf{A}}}_2$ belong to Φ, the variational state equation of (5.1.16) provides

$$
\begin{aligned}
& \int_{\Omega_1} \left(\nu_1 \mathbf{B}(\mathbf{A}_1)\cdot \mathbf{B}(\dot{\bar{\mathbf{A}}}_1) + j\omega\sigma_1 \mathbf{A}_1 \cdot \dot{\bar{\mathbf{A}}}_1\right)\mathrm{d}\Omega + \int_{\Omega_2} \left(\nu_2 \mathbf{B}(\mathbf{A}_2)\cdot \mathbf{B}(\dot{\bar{\mathbf{A}}}_2) + j\omega\sigma_2 \mathbf{A}_2 \cdot \dot{\bar{\mathbf{A}}}_2\right)\mathrm{d}\Omega \\
& = \int_{\Omega_1} \mathbf{J}_1 \cdot \dot{\bar{\mathbf{A}}}_1 \mathrm{d}\Omega + \int_{\Omega_2} \mathbf{J}_2 \cdot \dot{\bar{\mathbf{A}}}_2 \mathrm{d}\Omega
\end{aligned}
\tag{5.1.29}
$$

Hence, all terms with $\underset{1}{\dot{\bar{\mathbf{A}}}}$ and $\underset{2}{\dot{\bar{\mathbf{A}}}}$ in (5.1.28) are canceled out;

$$
\begin{aligned}
\dot{G} = & -\int_{\Omega_1} \left(v_1\mathbf{B}(\dot{\mathbf{A}}_1)\cdot\mathbf{B}(\bar{\mathbf{A}}_1) + j\omega\sigma_1\dot{\mathbf{A}}_1\cdot\bar{\mathbf{A}}_1\right)\mathrm{d}\Omega - \int_{\Omega_2} \left(v_2\mathbf{B}(\dot{\mathbf{A}}_2)\cdot\mathbf{B}(\bar{\mathbf{A}}_2) + j\omega\sigma_2\dot{\mathbf{A}}_2\cdot\bar{\mathbf{A}}_2\right)\mathrm{d}\Omega \\
& + \int_{\Omega_1} \left(\mathbf{g}_{\mathbf{A}_1}\cdot\dot{\mathbf{A}}_1 + \mathbf{g}_{\mathbf{B}_1}\cdot\mathbf{B}(\dot{\mathbf{A}}_1)\right)m_p\mathrm{d}\Omega + \int_{\Omega_2} \left(\mathbf{g}_{\mathbf{A}_2}\cdot\dot{\mathbf{A}}_2 + \mathbf{g}_{\mathbf{B}_2}\cdot\mathbf{B}(\dot{\mathbf{A}}_2)\right)m_p\mathrm{d}\Omega \\
& + \int_{\Omega_1} \left[v_1\mathbf{B}(\mathbf{V}\cdot\nabla\mathbf{A}_1)\cdot\mathbf{B}(\bar{\mathbf{A}}_1) + j\omega\sigma_1(\mathbf{V}\cdot\nabla\mathbf{A}_1)\cdot\bar{\mathbf{A}}_1\right]\mathrm{d}\Omega + \int_{\Omega_2} \left[v_2\mathbf{B}(\mathbf{V}\cdot\nabla\mathbf{A}_2)\cdot\mathbf{B}(\bar{\mathbf{A}}_2)\right. \\
& \left. + j\omega\sigma_2(\mathbf{V}\cdot\nabla\mathbf{A}_2)\cdot\bar{\mathbf{A}}_2\right]\mathrm{d}\Omega + \int_{\Omega_1} \left[v_1\mathbf{B}(\mathbf{A}_1)\cdot\mathbf{B}(\mathbf{V}\cdot\nabla\bar{\mathbf{A}}_1) + j\omega\sigma_1\mathbf{A}_1\cdot\left(\mathbf{V}\cdot\nabla\bar{\mathbf{A}}_1\right)\right]\mathrm{d}\Omega \\
& + \int_{\Omega_2} \left[v_2\mathbf{B}(\mathbf{A}_2)\cdot\mathbf{B}(\mathbf{V}\cdot\nabla\bar{\mathbf{A}}_2) + j\omega\sigma_2\mathbf{A}_2\cdot\left(\mathbf{V}\cdot\nabla\bar{\mathbf{A}}_2\right)\right]\mathrm{d}\Omega \\
& - \int_{\Omega_1} \mathbf{J}_1\cdot\left(\mathbf{V}\cdot\nabla\bar{\mathbf{A}}_1\right)\mathrm{d}\Omega - \int_{\Omega_2} \mathbf{J}_2\cdot\left(\mathbf{V}\cdot\nabla\bar{\mathbf{A}}_2\right)\mathrm{d}\Omega \\
& - \int_{\Omega_1} \left[\mathbf{g}_{\mathbf{A}_1}\cdot(\mathbf{V}\cdot\nabla\mathbf{A}_1) + \mathbf{g}_{\mathbf{B}_1}\cdot\mathbf{B}(\mathbf{V}\cdot\nabla\mathbf{A}_1)\right]m_p\mathrm{d}\Omega \\
& - \int_{\Omega_2} \left[\mathbf{g}_{\mathbf{A}_2}\cdot(\mathbf{V}\cdot\nabla\mathbf{A}_2) + \mathbf{g}_{\mathbf{B}_2}\cdot\mathbf{B}(\mathbf{V}\cdot\nabla\mathbf{A}_2)\right]m_p\mathrm{d}\Omega \\
& - \int_{\gamma} \left(v_1\mathbf{B}(\mathbf{A}_1)\cdot\mathbf{B}(\bar{\mathbf{A}}_1) - v_2\mathbf{B}(\mathbf{A}_2)\cdot\mathbf{B}(\bar{\mathbf{A}}_2) + j\omega\sigma_1\mathbf{A}_1\cdot\bar{\mathbf{A}}_1 - j\omega\sigma_2\mathbf{A}_2\cdot\bar{\mathbf{A}}_2\right)V_n\mathrm{d}\Gamma \\
& + \int_{\gamma} \left[\mathbf{J}_1\cdot\bar{\mathbf{A}}_1 - \mathbf{J}_2\cdot\bar{\mathbf{A}}_2 + (g_1 - g_2)m_p\right]V_n\mathrm{d}\Gamma \quad \forall\bar{\mathbf{A}}_1, \bar{\mathbf{A}}_2 \in \Phi
\end{aligned} \tag{5.1.30}
$$

5.1.3 Adjoint Variable Method for Sensitivity Analysis

To explicitly express (5.1.30) in terms of the velocity field $\mathbf{V}$, an adjoint equation is introduced. The adjoint equation is obtained by replacing $\dot{\mathbf{A}}_1$ and $\dot{\mathbf{A}}_2$ with the g-related terms of (5.1.30) with a virtual potential $\bar{\boldsymbol{\lambda}}_1$ and $\bar{\boldsymbol{\lambda}}_2$, respectively, and by equating the integrals to the bilinear form $a(\boldsymbol{\lambda}, \bar{\boldsymbol{\lambda}})$. The adjoint equation so obtained is written as

$$\int_{\Omega_1} (v_1 \mathbf{B}(\boldsymbol{\lambda}_1) \cdot \mathbf{B}(\bar{\lambda}_1) + j\omega\sigma_1 \boldsymbol{\lambda}_1 \cdot \bar{\lambda}_1) \mathrm{d}\Omega + \int_{\Omega_2} (v_2 \mathbf{B}(\boldsymbol{\lambda}_2) \cdot \mathbf{B}(\bar{\lambda}_2) + j\omega\sigma_2 \boldsymbol{\lambda}_2 \cdot \bar{\lambda}_2) \mathrm{d}\Omega$$
$$= \int_{\Omega_1} (\mathbf{g}_{\mathbf{A}_1} \cdot \bar{\lambda}_1 + \mathbf{g}_{\mathbf{B}_1} \cdot \mathbf{B}(\bar{\lambda}_1)) m_p \mathrm{d}\Omega + \int_{\Omega_2} (\mathbf{g}_{\mathbf{A}_2} \cdot \bar{\lambda}_2 + \mathbf{g}_{\mathbf{B}_2} \cdot \mathbf{B}(\bar{\lambda}_2)) m_p \mathrm{d}\Omega \quad \forall \bar{\lambda}_1,\ \bar{\lambda}_2 \in \Phi \tag{5.1.31}$$

where $\boldsymbol{\lambda}_1$ and $\boldsymbol{\lambda}_2$ are the adjoint variables, and their solutions are desired, and Φ is the space of admissible complex vector potential: $\Phi = \left\{ \bar{\lambda} \in [H^1(\Omega)]^3 | \bar{\lambda} = \mathbf{0} \right.$ on $\mathbf{x} \in \Gamma^0 . \}$.

(5.1.31) is evaluated at specific $\bar{\lambda}_1 = \dot{\mathbf{A}}_1$ and $\bar{\lambda}_2 = \dot{\mathbf{A}}_2$ since (5.1.31) holds for all $\forall \bar{\lambda}_1, \bar{\lambda}_2 \in \Phi$, to yield

$$\int_{\Omega_1} (v_1 \mathbf{B}(\boldsymbol{\lambda}_1) \cdot \mathbf{B}(\dot{\mathbf{A}}_1) + j\omega\sigma_1 \boldsymbol{\lambda}_1 \cdot \dot{\mathbf{A}}_1) \mathrm{d}\Omega + \int_{\Omega_2} (v_2 \mathbf{B}(\boldsymbol{\lambda}_2) \cdot \mathbf{B}(\dot{\mathbf{A}}_2) + j\omega\sigma_2 \boldsymbol{\lambda}_2 \cdot \dot{\mathbf{A}}_2) \mathrm{d}\Omega$$
$$= \int_{\Omega_1} (\mathbf{g}_{\mathbf{A}_1} \cdot \dot{\mathbf{A}}_1 + \mathbf{g}_{\mathbf{B}_1} \cdot \mathbf{B}(\dot{\mathbf{A}}_1)) m_p \mathrm{d}\Omega + \int_{\Omega_2} (\mathbf{g}_{\mathbf{A}_2} \cdot \dot{\mathbf{A}}_2 + \mathbf{g}_{\mathbf{B}_2} \cdot \mathbf{B}(\dot{\mathbf{A}}_2)) m_p \mathrm{d}\Omega \tag{5.1.32}$$

Similarly, the sensitivity Eq. (5.1.30) is evaluated at the specific $\bar{\mathbf{A}}_1 = \boldsymbol{\lambda}_1$, $\bar{\mathbf{A}}_2 = \boldsymbol{\lambda}_2$ since the $\boldsymbol{\lambda}_1$ and $\boldsymbol{\lambda}_2$ belong to the admissible space Φ, to yield

$$\begin{aligned}
\dot{G} = & -\int_{\Omega_1} \left(\nu_1 \mathbf{B}(\dot{\mathbf{A}}_1)\cdot\mathbf{B}(\boldsymbol{\lambda}_1) + j\omega\sigma_1 \dot{\mathbf{A}}_1\cdot\boldsymbol{\lambda}_1\right) \mathrm{d}\Omega - \int_{\Omega_2} \left(\nu_2 \mathbf{B}(\dot{\mathbf{A}}_2)\cdot\mathbf{B}(\boldsymbol{\lambda}_2) + j\omega\sigma_2 \dot{\mathbf{A}}_2\cdot\boldsymbol{\lambda}_2\right) \mathrm{d}\Omega \\
& + \int_{\Omega_1} \left(\mathbf{g}_{\mathbf{A}_1}\cdot\dot{\mathbf{A}}_1 + \mathbf{g}_{\mathbf{B}_1}\cdot\mathbf{B}(\dot{\mathbf{A}}_1)\right) m_p \mathrm{d}\Omega + \int_{\Omega_2} \left(\mathbf{g}_{\mathbf{A}_2}\cdot\dot{\mathbf{A}}_2 + \mathbf{g}_{\mathbf{B}_2}\cdot\mathbf{B}(\dot{\mathbf{A}}_2)\right) m_p \mathrm{d}\Omega \\
& + \int_{\Omega_1} \left[\nu_1 \mathbf{B}(\mathbf{V}\cdot\nabla\mathbf{A}_1)\cdot\mathbf{B}(\boldsymbol{\lambda}_1) + j\omega\sigma_1 (\mathbf{V}\cdot\nabla\mathbf{A}_1)\cdot\boldsymbol{\lambda}_1\right] \mathrm{d}\Omega \\
& + \int_{\Omega_2} \left[\nu_2 \mathbf{B}(\mathbf{V}\cdot\nabla\mathbf{A}_2)\cdot\mathbf{B}(\boldsymbol{\lambda}_2) + j\omega\sigma_2 (\mathbf{V}\cdot\nabla\mathbf{A}_2)\cdot\boldsymbol{\lambda}_2\right] \mathrm{d}\Omega \\
& + \int_{\Omega_1} \left[\nu_1 \mathbf{B}(\mathbf{A}_1)\cdot\mathbf{B}(\mathbf{V}\cdot\nabla\boldsymbol{\lambda}_1) + j\omega\sigma_1 \mathbf{A}_1\cdot(\mathbf{V}\cdot\nabla\boldsymbol{\lambda}_1)\right] \mathrm{d}\Omega \\
& + \int_{\Omega_2} \left[\nu_2 \mathbf{B}(\mathbf{A}_2)\cdot\mathbf{B}(\mathbf{V}\cdot\nabla\boldsymbol{\lambda}_2) + j\omega\sigma_2 \mathbf{A}_2\cdot(\mathbf{V}\cdot\nabla\boldsymbol{\lambda}_2)\right] \mathrm{d}\Omega \\
& - \int_{\Omega_1} \mathbf{J}_1\cdot(\mathbf{V}\cdot\nabla\boldsymbol{\lambda}_1) \mathrm{d}\Omega - \int_{\Omega_2} \mathbf{J}_2\cdot(\mathbf{V}\cdot\nabla\boldsymbol{\lambda}_2) \mathrm{d}\Omega \\
& - \int_{\Omega_1} \left[\mathbf{g}_{\mathbf{A}_1}\cdot(\mathbf{V}\cdot\nabla\mathbf{A}_1) + \mathbf{g}_{\mathbf{B}_1}\cdot\mathbf{B}(\mathbf{V}\cdot\nabla\mathbf{A}_1)\right] m_p \mathrm{d}\Omega \\
& - \int_{\Omega_2} \left[\mathbf{g}_{\mathbf{A}_2}\cdot(\mathbf{V}\cdot\nabla\mathbf{A}_2) + \mathbf{g}_{\mathbf{B}_2}\cdot\mathbf{B}(\mathbf{V}\cdot\nabla\mathbf{A}_2)\right] m_p \mathrm{d}\Omega \\
& - \int_{\gamma} \left(\nu_1 \mathbf{B}(\mathbf{A}_1)\cdot\mathbf{B}(\boldsymbol{\lambda}_1) - \nu_2 \mathbf{B}(\mathbf{A}_2)\cdot\mathbf{B}(\boldsymbol{\lambda}_2) + j\omega\sigma_1 \mathbf{A}_1\cdot\boldsymbol{\lambda}_1 - j\omega\sigma_2 \mathbf{A}_2\cdot\boldsymbol{\lambda}_2\right) V_n \mathrm{d}\Gamma \\
& + \int_{\gamma} \left[\mathbf{J}_1\cdot\boldsymbol{\lambda}_1 - \mathbf{J}_2\cdot\boldsymbol{\lambda}_2 + (g_1 - g_2) m_p\right] V_n \mathrm{d}\Gamma
\end{aligned} \tag{5.1.33}$$

The bilinear form is symmetric in its arguments;

$$\int_{\Omega_1} \left(\nu_1 \mathbf{B}(\dot{\mathbf{A}}_1)\cdot\mathbf{B}(\boldsymbol{\lambda}_1) + j\omega\sigma_1 \dot{\mathbf{A}}_1\cdot\boldsymbol{\lambda}_1\right) \mathrm{d}\Omega = \int_{\Omega_1} \left(\nu_1 \mathbf{B}(\boldsymbol{\lambda}_1)\cdot\mathbf{B}(\dot{\mathbf{A}}_1) + j\omega\sigma_1 \boldsymbol{\lambda}_1\cdot\dot{\mathbf{A}}_1\right) \mathrm{d}\Omega \tag{5.1.34}$$

$$\int_{\Omega_2} \left(\nu_2 \mathbf{B}(\dot{\mathbf{A}}_2)\cdot\mathbf{B}(\boldsymbol{\lambda}_2) + j\omega\sigma_2 \dot{\mathbf{A}}_2\cdot\boldsymbol{\lambda}_2\right) \mathrm{d}\Omega = \int_{\Omega_2} \left(\nu_2 \mathbf{B}(\boldsymbol{\lambda}_2)\cdot\mathbf{B}(\dot{\mathbf{A}}_2) + j\omega\sigma_2 \boldsymbol{\lambda}_2\cdot\dot{\mathbf{A}}_2\right) \mathrm{d}\Omega \tag{5.1.35}$$

By using the relations (5.1.32), (5.1.34), and (5.1.35), all terms containing $\dot{\mathbf{A}}_1$, $\dot{\mathbf{A}}_2$ in (5.1.33) are canceled out and all terms are expressed with the velocity field $\mathbf{V}$;

$$\begin{aligned}
\dot{G} = & \int_{\Omega_1} [\nu_1 \mathbf{B}(\mathbf{A}_1) \cdot \mathbf{B}(\mathbf{V} \cdot \nabla \lambda_1) + j\omega\sigma_1 \mathbf{A}_1 \cdot (\mathbf{V} \cdot \nabla \lambda_1) - \mathbf{J}_1 \cdot (\mathbf{V} \cdot \nabla \boldsymbol{\lambda}_1)] d\Omega \\
& + \int_{\Omega_2} [\nu_2 \mathbf{B}(\mathbf{A}_2) \cdot \mathbf{B}(\mathbf{V} \cdot \nabla \boldsymbol{\lambda}_2) + j\omega\sigma_2 \mathbf{A}_2 \cdot (\mathbf{V} \cdot \nabla \boldsymbol{\lambda}_2) - \mathbf{J}_2 \cdot (\mathbf{V} \cdot \nabla \boldsymbol{\lambda}_2)] d\Omega \\
& + \int_{\Omega_1} \big[\nu_1 \mathbf{B}(\boldsymbol{\lambda}_1) \cdot \mathbf{B}(\mathbf{V} \cdot \nabla \mathbf{A}_1) + j\omega\sigma_1 \boldsymbol{\lambda}_1 \cdot (\mathbf{V} \cdot \nabla \mathbf{A}_1) - \mathbf{g}_{\mathbf{A}_1} \cdot (\mathbf{V} \cdot \nabla \mathbf{A}_1) m_p \\
& - \mathbf{g}_{\mathbf{B}_1} \cdot \mathbf{B}(\mathbf{V} \cdot \nabla \mathbf{A}_1) m_p\big] d\Omega + \int_{\Omega_2} [\nu_2 \mathbf{B}(\boldsymbol{\lambda}_2) \cdot \mathbf{B}(\mathbf{V} \cdot \nabla \mathbf{A}_2) + j\omega\sigma_2 \boldsymbol{\lambda}_2 \cdot (\mathbf{V} \cdot \nabla \mathbf{A}_2) \\
& - \mathbf{g}_{\mathbf{A}_2} \cdot (\mathbf{V} \cdot \nabla \mathbf{A}_2) m_p - \mathbf{g}_{\mathbf{B}_2} \cdot \mathbf{B}(\mathbf{V} \cdot \nabla \mathbf{A}_2) m_p\big] d\Omega \\
& - \int_{\gamma} (\nu_1 \mathbf{B}(\mathbf{A}_1) \cdot \mathbf{B}(\boldsymbol{\lambda}_1) - \nu_2 \mathbf{B}(\mathbf{A}_2) \cdot \mathbf{B}(\boldsymbol{\lambda}_2) + j\omega\sigma_1 \mathbf{A}_1 \cdot \boldsymbol{\lambda}_1 - j\omega\sigma_2 \mathbf{A}_2 \cdot \boldsymbol{\lambda}_2) V_n d\Gamma \\
& + \int_{\gamma} \big[\mathbf{J}_1 \cdot \boldsymbol{\lambda}_1 - \mathbf{J}_2 \cdot \boldsymbol{\lambda}_2 + (g_1 - g_2) m_p\big] V_n d\Gamma
\end{aligned} \tag{5.1.36}$$

5.1.4 Boundary Expression of Shape Sensitivity

The domain integrals in the design sensitivity Eq. (5.1.36) are expressed in boundary integrals by using the variational identities. The fact that the sensitivity is expressed as a boundary integral provides an advantage in numerical implementation. For this purpose, two variational identities for the state and adjoint equations are required.

First, the variational identities for the state equation were given as (5.1.8) and (5.1.9);

$$\begin{aligned}
& \int_{\Omega_1} \left(\nu_1 \mathbf{B}(\mathbf{A}_1) \cdot \mathbf{B}(\overline{\mathbf{A}}_1) + j\omega\sigma_1 \mathbf{A}_1 \cdot \overline{\mathbf{A}}_1 - \mathbf{J}_1 \cdot \overline{\mathbf{A}}_1\right) d\Omega \\
& = - \int_{\gamma+\Gamma} (\mathbf{n} \times \mathbf{H}(\mathbf{A}_1)) \cdot \overline{\mathbf{A}}_1 d\Gamma \quad \forall \overline{\mathbf{A}}_1 \in \Phi
\end{aligned} \tag{5.1.37}$$

$$\int_{\Omega_2} \left(v_2 \mathbf{B}(\mathbf{A}_2) \cdot \mathbf{B}(\overline{\mathbf{A}}_2) + j\omega\sigma_2 \mathbf{A}_2 \cdot \overline{\mathbf{A}}_2 - \mathbf{J}_2 \cdot \overline{\mathbf{A}}_2 \right) d\Omega$$

$$= \int_{\gamma} (\mathbf{n} \times \mathbf{H}(\mathbf{A}_2)) \cdot \overline{\mathbf{A}}_2 d\Gamma \quad \forall \overline{\mathbf{A}}_2 \in \Phi \tag{5.1.38}$$

Next, the variational identity for the adjoint equation can be derived from a differential adjoint equation, which is obtained from (5.1.31) by the same procedure in Sect. 4.1. The variational adjoint equation (4.1.39) is equivalent to differential adjoint equations of (4.1.41) and (4.1.42). What is different is that the eddy current terms of $j\omega\sigma_1\boldsymbol{\lambda}_1$ and $j\omega\sigma_2\boldsymbol{\lambda}_2$ are added. The differential adjoint equations so obtained are written as

$$\nabla \times v_1 \nabla \times \boldsymbol{\lambda}_1 + j\omega\sigma_1\boldsymbol{\lambda}_1 = \mathbf{g}_{\mathbf{A}_1} m_p + \nabla \times \mathbf{g}_{\mathbf{B}_1} m_p \text{ in } \Omega_1 \tag{5.1.39}$$

$$\nabla \times v_2 \nabla \times \boldsymbol{\lambda}_2 + j\omega\sigma_2\boldsymbol{\lambda}_2 = \mathbf{g}_{\mathbf{A}_2} m_p + \nabla \times \mathbf{g}_{\mathbf{B}_2} m_p \text{ in } \Omega_2 \tag{5.1.40}$$

with the boundary condition:

$$\boldsymbol{\lambda}_1 = \mathbf{0} \quad \text{on } \Gamma^0 \text{ (homogeneous Dirichlet boundary condition)} \tag{5.1.41}$$

$$\frac{\partial \boldsymbol{\lambda}_1}{\partial n} = \mathbf{0} \quad \text{on } \Gamma^1 \text{ (homogeneous Neumann boundary condition)} \tag{5.1.42}$$

The adjoint sources $\mathbf{g}_{\mathbf{A}_1} m_p$, $\mathbf{g}_{\mathbf{A}_2} m_p$ and $\mathbf{g}_{\mathbf{B}_1} m_p$, $\mathbf{g}_{\mathbf{B}_2} m_p$ exist only in the domain Ω as the original sources $\mathbf{J}_1$, $\mathbf{J}_2$. Thus, there is no adjoint surface source equivalent to Dirichlet boundary condition. That is, while the original state equation may have surface sources equivalent to its Dirichlet boundary condition, the adjoint equation has no such surface source. Hence, the Dirichlet boundary condition for the adjoint equation is given as zero. In addition, since the structure symmetry is maintained in the adjoint system, the homogeneous Neumann condition is imposed on Γ^1 of the adjoint system.

Just as the variational identity of (5.1.37) and (5.1.38) for the state equation is obtained from the differential state Eqs. (5.1.4) and (5.1.5), the variational identities for the adjoint equation are obtained as

$$\int_{\Omega_1} \left(v_1 \mathbf{B}(\boldsymbol{\lambda}_1) \cdot \mathbf{B}(\bar{\lambda}_1) + j\omega\sigma_1\boldsymbol{\lambda}_1 \cdot \bar{\lambda}_1 - \mathbf{g}_{\mathbf{A}_1} \cdot \bar{\lambda}_1 m_p - \mathbf{g}_{\mathbf{B}_1} \cdot \mathbf{B}(\bar{\lambda}_1) m_p \right) d\Omega$$

$$= -\int_{\gamma+\Gamma} (\mathbf{n} \times \mathbf{H}(\boldsymbol{\lambda}_1)) \cdot \bar{\lambda}_1 d\Gamma \quad \forall \bar{\lambda}_1 \in \Phi \tag{5.1.43}$$

$$\int_{\Omega_2} \left(v_2\mathbf{B}(\boldsymbol{\lambda}_2)\cdot\mathbf{B}(\bar{\lambda}_2) + j\omega\sigma_2\boldsymbol{\lambda}_2\cdot\bar{\lambda}_2 - \mathbf{g}_{\mathbf{A}_2}\cdot\bar{\lambda}_2 m_p - \mathbf{g}_{\mathbf{B}_2}\cdot\mathbf{B}(\bar{\lambda}_2)m_p\right)d\Omega = \int_{\gamma}(\mathbf{n}\times\mathbf{H}(\boldsymbol{\lambda}_2))\cdot\bar{\lambda}_2 d\Gamma \quad \forall\bar{\lambda}_2\in\Phi \tag{5.1.44}$$

where

$$\mathbf{H}(\boldsymbol{\lambda}_1) = v_1\mathbf{B}(\boldsymbol{\lambda}_1) - \mathbf{g}_{\mathbf{B}_1}m_p \tag{5.1.45}$$

$$\mathbf{H}(\boldsymbol{\lambda}_2) = v_2\mathbf{B}(\boldsymbol{\lambda}_2) - \mathbf{g}_{\mathbf{B}_2}m_p \tag{5.1.46}$$

The variational identities of (5.1.37), (5.1.38) and (5.1.43), (5.1.44) are used to transform the domain integrals of (5.1.36) into boundary integrals. First, (5.1.37) and (5.1.38) are evaluated at $\bar{\mathbf{A}}_1 = \mathbf{V}\cdot\nabla\boldsymbol{\lambda}_1$ and $\bar{\mathbf{A}}_2 = \mathbf{V}\cdot\nabla\boldsymbol{\lambda}_2$, respectively;

$$\int_{\Omega_1}[v_1\mathbf{B}(\mathbf{A}_1)\cdot\mathbf{B}(\mathbf{V}\cdot\nabla\boldsymbol{\lambda}_1) + j\omega\sigma_1\mathbf{A}_1\cdot(\mathbf{V}\cdot\nabla\boldsymbol{\lambda}_1) - \mathbf{J}_1\cdot(\mathbf{V}\cdot\nabla\boldsymbol{\lambda}_1)]d\Omega = -\int_{\gamma}(\mathbf{n}\times\mathbf{H}(\mathbf{A}_1))\cdot(\mathbf{V}\cdot\nabla\boldsymbol{\lambda}_1)d\Gamma \tag{5.1.47}$$

$$\int_{\Omega_2}[v_2\mathbf{B}(\mathbf{A}_2)\cdot\mathbf{B}(\mathbf{V}\cdot\nabla\boldsymbol{\lambda}_2) + j\omega\sigma_2\mathbf{A}_2\cdot(\mathbf{V}\cdot\nabla\boldsymbol{\lambda}_2) - \mathbf{J}_2\cdot(\mathbf{V}\cdot\nabla\boldsymbol{\lambda}_2)]d\Omega = \int_{\gamma}(\mathbf{n}\times\mathbf{H}(\mathbf{A}_2))\cdot(\mathbf{V}\cdot\nabla\boldsymbol{\lambda}_2)d\Gamma \tag{5.1.48}$$

Second, (5.1.43) and (5.1.44) are evaluated at $\bar{\lambda}_1 = \mathbf{V}\cdot\nabla\mathrm{A}_1$ and $\bar{\lambda}_2 = \mathbf{V}\cdot\nabla\mathrm{A}_2$, respectively;

$$\int_{\Omega_1}\left[v_1\mathbf{B}(\boldsymbol{\lambda}_1)\cdot\mathbf{B}(\mathbf{V}\cdot\nabla\mathbf{A}_1) + j\omega\sigma_1\boldsymbol{\lambda}_1\cdot(\mathbf{V}\cdot\nabla\mathbf{A}_1) - \mathbf{g}_{\mathbf{A}_1}\cdot(\mathbf{V}\cdot\nabla\mathbf{A}_1)m_p - \mathbf{g}_{\mathbf{B}_1}\cdot\mathbf{B}(\mathbf{V}\cdot\nabla\mathbf{A}_1)m_p\right]d\Omega = \int_{\gamma}(\mathbf{n}\times\mathbf{H}(\boldsymbol{\lambda}_1))\cdot(\mathbf{V}\cdot\nabla\mathbf{A}_1)d\Gamma \tag{5.1.49}$$

$$\int_{\Omega_2}\left[v_2\mathbf{B}(\boldsymbol{\lambda}_2)\cdot\mathbf{B}(\mathbf{V}\cdot\nabla\mathbf{A}_2) + j\omega\sigma_2\boldsymbol{\lambda}_2\cdot(\mathbf{V}\cdot\nabla\mathbf{A}_2) - \mathbf{g}_{\mathbf{A}_2}\cdot(\mathbf{V}\cdot\nabla\mathbf{A}_2)m_p - \mathbf{g}_{\mathbf{B}_2}\cdot\mathbf{B}(\mathbf{V}\cdot\nabla\mathbf{A}_2)m_p\right]d\Omega = \int_{\gamma}(\mathbf{n}\times\mathbf{H}(\boldsymbol{\lambda}_2))\cdot(\mathbf{V}\cdot\nabla\mathbf{A}_2)d\Gamma \tag{5.1.50}$$

By inserting (5.1.47)–(5.1.50) into (5.1.36), the domain integrals in (5.1.36) are transformed in the boundary integrals;

$$\begin{aligned}\dot{G} = &\int_{\gamma} [-(\mathbf{n} \times \mathbf{H}(\mathbf{A}_1)) \cdot (\mathbf{V} \cdot \nabla\boldsymbol{\lambda}_1) + (\mathbf{n} \times \mathbf{H}(\mathbf{A}_2)) \cdot (\mathbf{V} \cdot \nabla\boldsymbol{\lambda}_2) \\ &-(\mathbf{n} \times \mathbf{H}(\boldsymbol{\lambda}_1)) \cdot (\mathbf{V} \cdot \nabla\mathbf{A}_1) + (\mathbf{n} \times \mathbf{H}(\boldsymbol{\lambda}_2)) \cdot (\mathbf{V} \cdot \nabla\mathbf{A}_2)]\mathrm{d}\Gamma \\ &- \int_{\gamma} (v_1\mathbf{B}(\mathbf{A}_1) \cdot \mathbf{B}(\boldsymbol{\lambda}_1) - v_2\mathbf{B}(\mathbf{A}_2) \cdot \mathbf{B}(\boldsymbol{\lambda}_2))V_n\mathrm{d}\Gamma \\ &- \int_{\gamma} (j\omega\sigma_1\mathbf{A}_1 \cdot \boldsymbol{\lambda}_1 - j\omega\sigma_2\mathbf{A}_2 \cdot \boldsymbol{\lambda}_2)V_n\mathrm{d}\Gamma \\ &+ \int_{\gamma} \left[\mathbf{J}_1 \cdot \boldsymbol{\lambda}_1 - \mathbf{J}_2 \cdot \boldsymbol{\lambda}_2 + (g_1 - g_2)m_p\right] V_n\mathrm{d}\Gamma \end{aligned} \tag{5.1.51}$$

This sensitivity formula is the same as the sensitivity formula (4.1.53) in Sect. 4.1 except the third term. Thus, all the terms except the third one are arranged in the same manner in Sect. 4.1. The integrand of the first integral in (5.1.51) is obtained as (5.1.52), which is the same as (4.1.66) except for that the permanent magnetization term is eliminated.

$$\begin{aligned}&-(\mathbf{n} \times \mathbf{H}(\mathbf{A}_1)) \cdot (\mathbf{V} \cdot \nabla\boldsymbol{\lambda}_1) + (\mathbf{n} \times \mathbf{H}(\mathbf{A}_2)) \cdot (\mathbf{V} \cdot \nabla\boldsymbol{\lambda}_2) - (\mathbf{n} \times \mathbf{H}(\boldsymbol{\lambda}_1)) \cdot (\mathbf{V} \cdot \nabla\mathbf{A}_1) \\ &+ (\mathbf{n} \times \mathbf{H}(\boldsymbol{\lambda}_2)) \cdot (\mathbf{V} \cdot \nabla\mathbf{A}_2) = -v_1B_t(\mathbf{A}_1)(B_t(\boldsymbol{\lambda}_2) - B_t(\boldsymbol{\lambda}_1)) \\ &- v_2B_t(\boldsymbol{\lambda}_2)(B_t(\mathbf{A}_2) - B_t(\mathbf{A}_1)) + g_{\mathbf{B}_{2}t}(B_t(\mathbf{A}_2) - B_t(\mathbf{A}_1))m_p\end{aligned} \tag{5.1.52}$$

The integrand of the second integral in (5.1.51) is the same as (4.1.67);

$$\begin{aligned}&- v_1\mathbf{B}(\mathbf{A}_1) \cdot \mathbf{B}(\boldsymbol{\lambda}_1) + v_2\mathbf{B}(\mathbf{A}_2) \cdot \mathbf{B}(\boldsymbol{\lambda}_2) = -v_1B_n(\mathbf{A}_1)B_n(\boldsymbol{\lambda}_1) \\ &- v_1B_t(\mathbf{A}_1)B_t(\boldsymbol{\lambda}_1) + v_2B_n(\mathbf{A}_2)B_n(\boldsymbol{\lambda}_2) + v_2B_t(\mathbf{A}_2)B_t(\boldsymbol{\lambda}_2)\end{aligned} \tag{5.1.53}$$

Summing (5.1.52) and (5.1.53) results in

$$(v_2 - v_1)\mathbf{B}(\mathbf{A}_1) \cdot \mathbf{B}(\boldsymbol{\lambda}_2) + g_{\mathbf{B}_{2}t}(B_t(\mathbf{A}_2) - B_t(\mathbf{A}_1))m_p \tag{5.1.54}$$

which is the sum of two integrand terms without V_n of the first two integrals in (5.1.51). Inserting (5.1.54) into (5.1.51) provides

$$\dot{G} = \int\limits_{\gamma} \left[(\nu_2 - \nu_1)\mathbf{B}(\mathbf{A}_1)\cdot\mathbf{B}(\boldsymbol{\lambda}_2) + g_{\mathbf{B}_{2t}}(B_t(\mathbf{A}_2) - B_t(\mathbf{A}_1))m_p\right]V_n \mathrm{d}\Gamma$$
$$- \int\limits_{\gamma} (j\omega\sigma_1\mathbf{A}_1\cdot\boldsymbol{\lambda}_1 - j\omega\sigma_2\mathbf{A}_2\cdot\boldsymbol{\lambda}_2)V_n \mathrm{d}\Gamma + \int\limits_{\gamma} \left[\mathbf{J}_1\cdot\boldsymbol{\lambda}_1 - \mathbf{J}_2\cdot\boldsymbol{\lambda}_2 + (g_1 - g_2)m_p\right]V_n \mathrm{d}\Gamma \tag{5.1.55}$$

Finally, by the continuity conditions of $\mathbf{A}$ and $\boldsymbol{\lambda}$, the general sensitivity formula for the interface variation in the eddy current system is obtained as

$$\dot{G} = \int\limits_{\gamma} \left[(\nu_2 - \nu_1)\mathbf{B}(\mathbf{A}_1)\cdot\mathbf{B}(\boldsymbol{\lambda}_2) + j\omega(\sigma_2 - \sigma_1)\mathbf{A}_1\cdot\boldsymbol{\lambda}_2 - (\mathbf{J}_2 - \mathbf{J}_1)\cdot\boldsymbol{\lambda}_2\right.$$
$$\left. -(g_2 - g_1)m_p + g_{\mathbf{B}_{2t}}(B_t(\mathbf{A}_2) - B_t(\mathbf{A}_1))m_p\right]V_n \mathrm{d}\Gamma \tag{5.1.56}$$

When the objective function is defined on the inner area in the domain Ω that does not intersect with the design boundary, $m_p = 0$ on γ. Hence, (5.1.56) becomes

$$\dot{G} = \int\limits_{\gamma} [(\nu_2 - \nu_1)\mathbf{B}(\mathbf{A}_1)\cdot\mathbf{B}(\boldsymbol{\lambda}_2) + j\omega(\sigma_2 - \sigma_1)\mathbf{A}_1\cdot\boldsymbol{\lambda}_2 - (\mathbf{J}_2 - \mathbf{J}_1)\cdot\boldsymbol{\lambda}_2]V_n \mathrm{d}\Gamma \tag{5.1.57}$$

5.1.5 *Interface Problems*

The sensitivity formula (5.1.57) for the interface variation can be expressed in various ways according to the characteristics of the design problems of the eddy current system. It is reduced to the following specific cases that are frequently found in practical design problems.

(1) interface of air and ferromagnetic material
When the air is Ω_1 and the linear ferromagnetic material is Ω_2,

$$\dot{G} = \int\limits_{\gamma} \nu_0(\nu_r - 1)\mathbf{B}(\mathbf{A}_1)\cdot\mathbf{B}(\boldsymbol{\lambda}_2)V_n \mathrm{d}\Gamma \tag{5.1.58}$$

where $\nu_1 = \nu_0 = 1/\mu_0$ $\nu_2 = \nu_r\nu_0 = 1/\mu_r\mu_0$ $\nu_r = 1/\mu_r$, $\nu_2 = \nu_r\nu_0 = 1/\mu_r\mu_0$, $\nu_r = 1/\mu_r$.

(2) interface of magnetic material and conductor
When the ferromagnetic material is Ω_1 and the conductor is Ω_2 where the eddy current is induced,

$$\dot{G} = \int_{\gamma} [(\nu_2 - \nu_1)\mathbf{B}(\mathbf{A}_1) \cdot \mathbf{B}(\boldsymbol{\lambda}_2) + j\omega\sigma_2\mathbf{A}_1 \cdot \boldsymbol{\lambda}_2] V_n \mathrm{d}\Gamma \tag{5.1.59}$$

where ν_1 is the reluctivity of the ferromagnetic material and ν_2 is the reluctivity of the conductor. If the relative permeability of the conductor is assumed to be 1,

$$\dot{G} = \int_{\gamma} [\nu_0(1 - \nu_r)\mathbf{B}(\mathbf{A}_1) \cdot \mathbf{B}(\boldsymbol{\lambda}_2) + j\omega\sigma_2\mathbf{A}_1 \cdot \boldsymbol{\lambda}_2] V_n \mathrm{d}\Gamma \tag{5.1.60}$$

where $\nu_1 = \nu_r\nu_0 = 1/\mu_r\mu_0$, $\nu_2 = \nu_0 = 1/\mu_0$

(3) interface of magnetic material and current region

When a linear ferromagnetic material is Ω_1 and the current region is Ω_2,

$$\dot{G} = \int_{\gamma} [(\nu_2 - \nu_1)\mathbf{B}(\mathbf{A}_1) \cdot \mathbf{B}(\boldsymbol{\lambda}_2) - \mathbf{J}_2 \cdot \boldsymbol{\lambda}_2] V_n \mathrm{d}\Gamma \tag{5.1.61}$$

where ν_1 is the reluctivity of the ferromagnetic material and ν_2 is the reluctivity of the conductor. If the relative permeability of the current region is taken to be 1,

$$\dot{G} = \int_{\gamma} [\nu_0(1 - \nu_r)\mathbf{B}(\mathbf{A}_1) \cdot \mathbf{B}(\boldsymbol{\lambda}_2) - \mathbf{J}_2 \cdot \boldsymbol{\lambda}_2] V_n \mathrm{d}\Gamma \tag{5.1.62}$$

where $\nu_1 = \nu_r\nu_0 = 1/\mu_r\mu_0$, $\nu_2 = \nu_0 = 1/\mu_0$.

(4) interface of conductor and current region

When the conductor is Ω_1 and the current region is Ω_2,

$$\dot{G} = \int_{\gamma} [(\nu_2 - \nu_1)\mathbf{B}(\mathbf{A}_1) \cdot \mathbf{B}(\boldsymbol{\lambda}_2) - j\omega\sigma_1\mathbf{A}_1 \cdot \boldsymbol{\lambda}_2 - \mathbf{J}_2 \cdot \boldsymbol{\lambda}_2] V_n \mathrm{d}\Gamma \tag{5.1.63}$$

If the conductor is a conductive ferromagnetic material and the relative permeability of the current region is taken to be 1,

$$\dot{G} = \int_{\gamma} [\nu_0(1 - \nu_r)\mathbf{B}(\mathbf{A}_1) \cdot \mathbf{B}(\boldsymbol{\lambda}_2) - j\omega\sigma_1\mathbf{A}_1 \cdot \boldsymbol{\lambda}_2 - \mathbf{J}_2 \cdot \boldsymbol{\lambda}_2] V_n \mathrm{d}\Gamma \tag{5.1.64}$$

where $\nu_1 = \nu_r\nu_0 = 1/\mu_r\mu_0$, $\nu_2 = \nu_0 = 1/\mu_0$. If the conductor is non-magnetic,

$$\dot{G} = -\int_{\gamma} (j\omega\sigma_1\mathbf{A}_1 \cdot \boldsymbol{\lambda}_2 + \mathbf{J}_2 \cdot \boldsymbol{\lambda}_2) V_n \mathrm{d}\Gamma \tag{5.1.65}$$

(5) interface of conductor and air
When the conductor is Ω_1 and the air is Ω_2

$$\dot{G} = \int_{\gamma} [\nu_0(1 - \nu_r)\mathbf{B}(\mathbf{A}_1) \cdot \mathbf{B}(\boldsymbol{\lambda}_2) - j\omega\sigma_1\mathbf{A}_1 \cdot \boldsymbol{\lambda}_2] V_n \mathrm{d}\Gamma \tag{5.1.66}$$

If the conductor is non-magnetic,

$$\dot{G} = -\int_{\gamma} j\omega\sigma_1\mathbf{A}_1 \cdot \boldsymbol{\lambda}_2 V_n \mathrm{d}\Gamma \tag{5.1.67}$$

(6) interface of two current regions with different current density
When the current region 1 is Ω_1 and the current region 2 is Ω_2,

$$\dot{G} = -\int_{\gamma} (\mathbf{J}_2 - \mathbf{J}_1) \cdot \boldsymbol{\lambda}_2 V_n \mathrm{d}\Gamma \tag{5.1.68}$$

(7) interface of current region and air
When the current region is Ω_1 and the air is Ω_2,

$$\dot{G} = \int_{\gamma} \mathbf{J}_1 \cdot \boldsymbol{\lambda}_2 V_n \mathrm{d}\Gamma \tag{5.1.69}$$

(8) interface of two ferromagnetic regions with different reluctivity
When the ferromagnetic material 1 is Ω_1 and the ferromagnetic material 2 is Ω_2,

$$\dot{G} = \int_{\gamma} (\nu_2 - \nu_1)\mathbf{B}(\mathbf{A}_1) \cdot \mathbf{B}(\boldsymbol{\lambda}_2) V_n \mathrm{d}\Gamma \tag{5.1.70}$$

(9) interface of two current regions with different current density
When the current conductor 1 is Ω_1 and the current conductor 2 is Ω_2,

$$\dot{G} = \int_{\gamma} [j\omega(\sigma_2 - \sigma_1)\mathbf{A}_1 \cdot \boldsymbol{\lambda}_2] V_n d\Gamma \tag{5.1.71}$$

5.1.6 Numerical Examples

The sensitivity formula (5.1.57) derived in the Sect. 5.1.4 is applied to shape optimization problems of two-dimensional and axi-symmetric design models, of which the analytic field solutions are not given. These design models are taken to illustrate how well the sensitivity formula is applied to the shape design problem in the eddy current system. The numerical examples are two magnetic shielding problems. The results of the examples show that the sensitivity formula is useful for the design of the eddy current system. In these examples, the sensitivity formula requires the state and the adjoint variables, which are numerically calculated by the finite element method. The sensitivity obtained is used for the optimization algorithm to evolve the shape of the design model. For the optimization algorithm, the level set method is used to represent the shape variations of the design model. The level set method is described in the Chap. 7, where the shape evolution is expressed by the parameter t of unit s for the amount of shape change.

5.1.7 Magnetic shielding problem I

The sensitivity formula of the eddy current system is applied to a two-dimensional shape design problem. Consider a current coil enclosed by a conductor as shown in Fig. 5.1.2. The coil, which carries a sinusoidal current, produces a time-varying magnetic field in the surrounding space. If a conductor is placed around the current coil, the changing magnetic field induces an eddy current in the conductor, which decreases the field intensity in the surrounding space. The conductive enclosure, which acts as a barrier to the magnetic field, is designed to effectively reduce the leakage magnetic field.

The design objective is to obtain the minimum magnetic field $\mathbf{B}$ in the outside domain Ω_p by deforming the shape of the conductor. In this design model where the zero magnetic vector potential $\mathbf{A}$ is applied on the outer and symmetry boundaries, the objective function to be minimized is defined as

$$F = \int_{\Omega} A_z m_p \mathrm{d}\Omega \tag{5.1.72}$$

where A_z is the z component of the magnetic vector potential $\mathbf{A}$. The design variable is the shape of the conductor-air interface γ in the design domain as shown in Fig. 5.2. The shape sensitivity for this interface is the sensitivity formula (5.1.67);

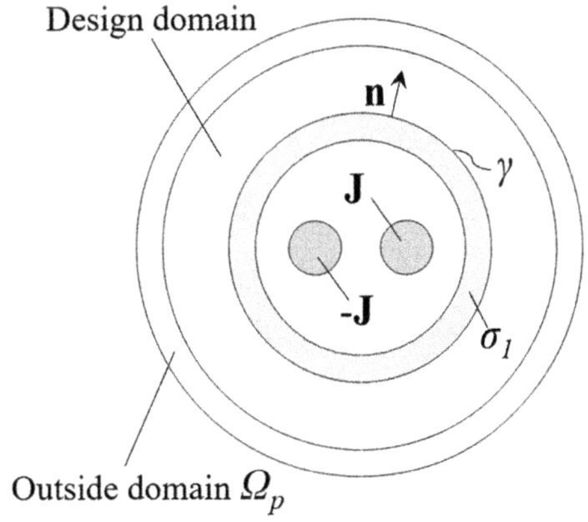

Fig. 5.2 Magnetic shielding problem I

$$\dot{G} = -\int_{\gamma} j\omega\sigma_1 \mathbf{A}_1 \cdot \boldsymbol{\lambda}_2 V_n \mathrm{d}\Gamma \tag{5.1.73}$$

The velocity field $V_n$in (5.1.73) is taken for the value of the shape sensitivity $\dot{G}$ to be negative as

$$V_n = \left[\mathrm{Re}(j\omega\sigma_1 \mathbf{A}_1 \cdot \boldsymbol{\lambda}_2)^2 + \mathrm{Im}(j\omega\sigma_1 \mathbf{A}_1 \cdot \boldsymbol{\lambda}_2)^2\right]^{1/2} \tag{5.1.74}$$

This design problem has a constraint of constant conductor volume;

$$\int_{\Omega_c} \mathrm{d}\Omega = C \tag{5.1.75}$$

where Ω_c is the conductor region and C is a given value of the constraint. The material derivative of the constraint (5.1.75) is obtained as

$$\int_{\Gamma} V_n \mathrm{d}\Gamma = 0 \tag{5.1.76}$$

which is a different form of the constraint (5.1.75) expressed with the design velocity field V_n. In order to satisfy the constraint (5.1.76), the design velocity (5.1.74) is modified by subtracting its average V_{na} to become U_n as

$$U_n = V_n - V_{na} \tag{5.1.77}$$

where

$$V_{na} = \left[\mathrm{Re}(j\omega\sigma_1 \mathbf{A}_1 \cdot \boldsymbol{\lambda}_2)^2 + \mathrm{Im}(j\omega\sigma_1 \mathbf{A}_1 \cdot \boldsymbol{\lambda}_2)^2\right]^{1/2} / \int_{\gamma} \mathrm{d}\Gamma \tag{5.1.78}$$

The shape design result is shown in Fig. 5.3, where the shape of the conductor–air interface is gradually deformed and the conductor is separated into two ones. In Fig. 5.4, the magnetic field of the final design, which is obtained at 6000 s, is 28.4% lower in the outside domain than that of the initial design. The objective function converges around 0.072 mT. The result of this example shows the feasibility of the shape sensitivity analysis in the two-dimensional eddy current system.

5.1.8 Magnetic shielding problem II

The sensitivity formula is applied to an axi-symmetric shape design problem. Consider a solenoid carrying a sinusoidal current and a conductor ring inside the

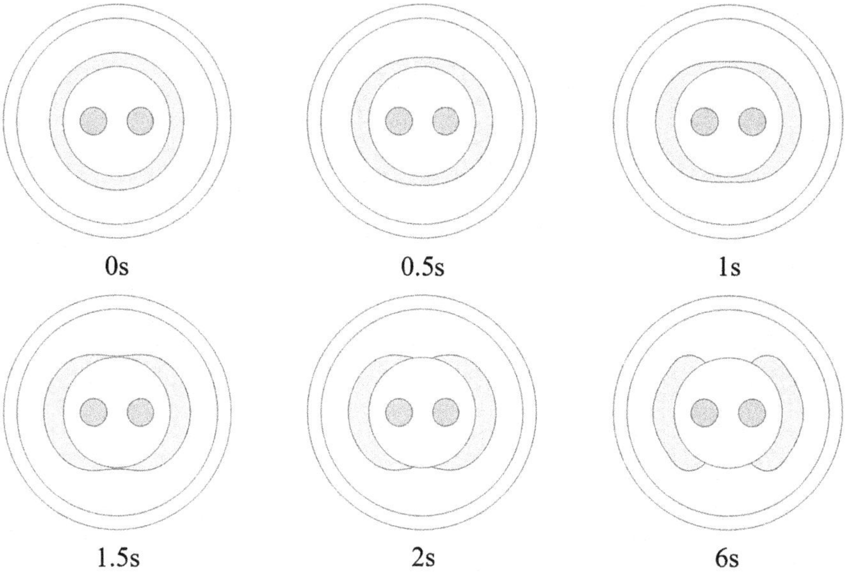

Fig. 5.3 Magnetic shielding problem I: shape variation

Fig. 5.4 Magnetic shielding problem I: evolution of objective function

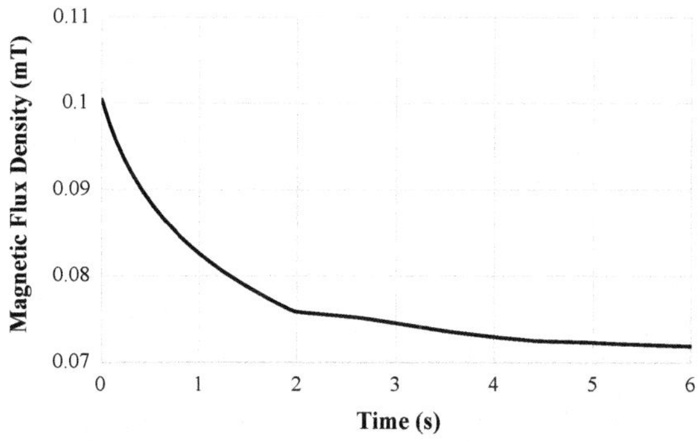

solenoid as shown in Fig. 5.5. The time-varying magnetic field produced by the current-carrying solenoid causes the eddy current in the conductor ring, which acts to cancel the magnetic field in the inside domain Ω_p. The amount of the magnetic field reduction depends on the shape of the conductor ring with a given volume.

The design objective is to obtain the minimum magnetic field $\mathbf{B}$ in the inside domain Ω_p by deforming the conductor shape. The objective function to be minimized is defined as the integral of the magnetic vector potential in Ω_p;

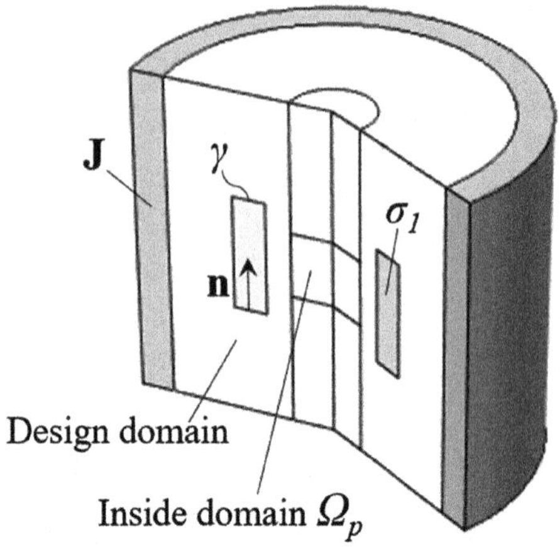

Fig. 5.5 Magnetic shielding problem II

$$F = \int_{\Omega} A_{\phi} m_p \mathrm{d}\Omega \tag{5.1.79}$$

where A_{ϕ} is the azimuthal component of the magnetic vector potential $\mathbf{A}$. The shape sensitivity formula $\dot{G}$ and the velocity field U_n are the same as (5.1.73) and (5.1.77), respectively.

The design result is shown in Fig. 5.6, where the shape of the conductor ring is gradually deformed and the conductor approaches the inside domain to cancel the magnetic field effectively. The final design is obtained at 6 *s*. Figure 5.7 shows that the magnetic field in the inside domain of the final design is 6.6% lower than that of the initial design. The objective function value converges around 6.42 mT.

5.2 Interface Shape Sensitivity for System Power

In this section, the shape sensitivity for the interface design is also developed in the eddy current system as in the Sect. 5.1; but the objective function is the system power. The system power of the eddy current system is related to the resistance and inductance of its equivalent electric circuit, and it can be applied to design problems for reduction of eddy current loss and shielding of AC magnetic field, etc. The derivation procedure is similar to Sect. 4.1. The difference is that the adjoint variable for the system energy is obtained as $-j\omega$ times the state variable. Thus, solving the adjoint variable equation is not necessary. The derived sensitivity formula is tested and validated with numerical examples.

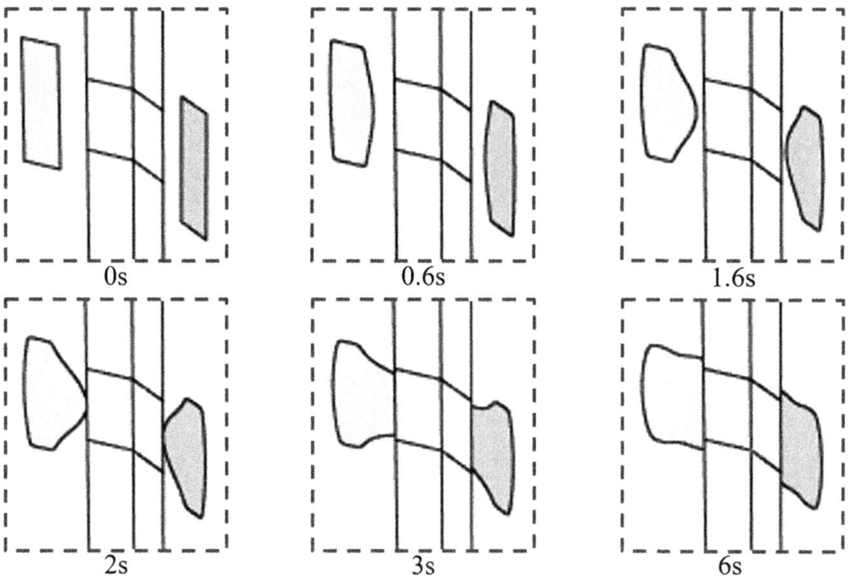

Fig. 5.6 Magnetic shielding problem II: shape variation

Fig. 5.7 Magnetic shielding problem II: evolution of objective function

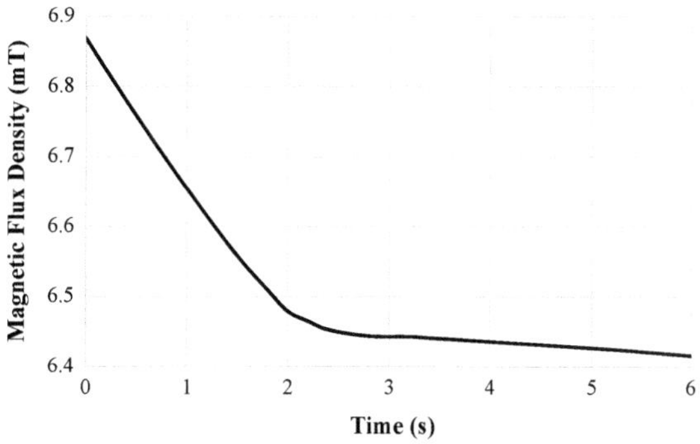

5.2.1 Problem Definition

An eddy current system for interface design is given as in Fig. 5.8, where the whole domain Ω comprises two domains Ω_1 and Ω_2 divided by an interface γ. The domain Ω_1 has a distribution of ν_1 and σ_1, and the domain Ω_2 has a distribution of ν_2 and σ_2. The current density $\mathbf{J}_1$ is given by a single-phase current source in the domain $\Omega_p \subset \Omega_1$.

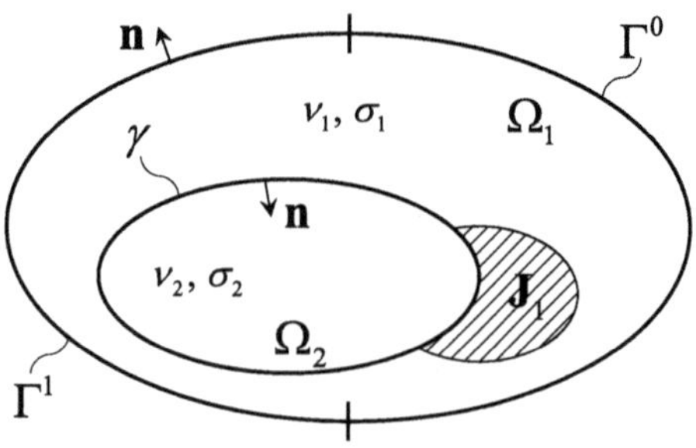

Fig. 5.8 Interface design of eddy current system for system power

The objective function is the complex system power of the eddy current system. The system power is expressed as the input power by the current source;

$$P = \int_{\Omega_1} \mathbf{J}_1 \cdot \mathbf{E}_1^* m_p \mathrm{d}\Omega \tag{5.2.1}$$

where $\mathbf{J}_1 \cdot \mathbf{E}_1^*$ is the power density in the domains Ω_1 and m_p is a characteristic function that is defined as

$$m_p = \begin{pmatrix} 1 & \mathbf{x} \in \Omega_p \\ 0 & \mathbf{x} \notin \Omega_p \end{pmatrix} \tag{5.2.2}$$

The region Ω_p, the integral domain for the objective function, is the current-source domain, and it can intersect with the interface as shown in Fig. 5.8.

However, this definition of the objective function (5.2.1) causes some difficulty for the adjoint equation due to the adjoint sources with $\mathbf{J}_1^*$. When this source is equated to the the bilinear form, it is not consistent with the bilinear form and the adjoint equation is not solvable. Hence, an alternative objective function Q for the complex system power is defined as

$$Q = \int_{\Omega_1} \mathbf{J}_1 \cdot \mathbf{E}_1 m_p \mathrm{d}\Omega \tag{5.2.3}$$

It is examined how this objective function is related to the one (5.2.1). In the term $\mathbf{J}_1 \cdot \mathbf{E}_1$ of (5.2.3), the complex variable $\mathbf{J}_1$ and $\mathbf{E}_1$ can be written as

$$\mathbf{J}_1 = \mathbf{J}_0 \angle 0 \text{ and } \mathbf{E}_1 = \mathbf{E}_0 \angle \alpha \tag{5.2.4}$$

where 0 and α are their phase angles and $\mathbf{J}_0$ and $\mathbf{E}_0$ are the magnitudes for $\mathbf{J}_1$ and $\mathbf{E}_1$, respectively. For this current source system, since the phase angle of $\mathbf{J}_1$ can be taken as zero, $\mathbf{J}_1$ is written only with the real part $\mathbf{J}_r$ as

$$\mathbf{J}_1 = \mathbf{J}_r \tag{5.2.5}$$

But since the phase angle of $\mathbf{E}_1$ is determined by the system, $\mathbf{E}_1$ can be written as

$$\mathbf{E}_1 = \mathbf{E}_r + j\mathbf{E}_i \tag{5.2.6}$$

where $\mathbf{E}_r$ and $\mathbf{E}_i$ are the real and imaginary parts, respectively. Using (5.2.5) and (5.2.6), $\mathbf{J}_1 \cdot \mathbf{E}_1$ is expressed as

$$\mathbf{J}_1 \cdot \mathbf{E}_1 = \mathbf{J}_r \cdot \mathbf{E}_r + j\mathbf{J}_r \cdot \mathbf{E}_i \tag{5.2.7}$$

On the other hand, the term $\mathbf{J}_1 \cdot \mathbf{E}_1^*$ of (5.2.1) is written as

$$\mathbf{J}_1 \cdot \mathbf{E}_1^* = \mathbf{J}_r \cdot \mathbf{E}_r - j\mathbf{J}_r \cdot \mathbf{E}_i \tag{5.2.8}$$

Comparing (5.2.7) and (5.2.8) provides the relation:

$$\mathrm{Re}(\mathbf{J}_1 \cdot \mathbf{E}_1) = \mathrm{Re}\left(\mathbf{J}_1 \cdot \mathbf{E}_1^*\right) \tag{5.2.9}$$

$$\mathrm{Im}(\mathbf{J}_1 \cdot \mathbf{E}_1) = -\mathrm{Im}\left(\mathbf{J}_1 \cdot \mathbf{E}_1^*\right) \tag{5.2.10}$$

By introduction of the alternative objective function (5.2.3), the resistive power is unchanged, but the sign of the reactive power is changed. For this power sensitivity problem, the new objective function (5.2.3) can be employed and its sensitivity can be developed by taking into account the relations of (5.2.9) and (5.2.10).

As mentioned in Sect. 2.3, since the eddy current system does not have the term $\nabla\phi$, the electric field intensity in (2.3.9) is written with the magnetic vector potential;

$$\mathbf{E} = -j\omega\mathbf{A} \tag{5.2.11}$$

The system power of (5.2.3) is rewritten with (5.2.11);

$$Q = -j\omega \int_{\Omega_1} \mathbf{J}_1 \cdot \mathbf{A}_1 m_p \mathrm{d}\Omega \tag{5.2.12}$$

This objective function is a specific case of the general objective function of (5.1.1);

$$g_1 = -j\omega\mathbf{J}_1 \cdot \mathbf{A}_1 \text{ and } g_2 = 0 \tag{5.2.13}$$

5.2.2 Adjoint Variable Method for Power Sensitivity

To explicitly express the sensitivity in terms of the velocity field $\mathbf{V}$, an adjoint equation is introduced. The results obtained in Sect. 5.1 can be easily applied to this problem. The overall procedure for developing the sensitivity formula for the system power is almost the same as in Sect. 5.1. Only the different things are described.

Since the g_1 in the general objective function of (5.1.1) was defined as a function of the magnetic vector potential $\mathbf{A}$ and the magnetic flux density $\mathbf{B}$,

$$g_1 \equiv g(\mathbf{A}_1, \mathbf{B}(\mathbf{A}_1)) \tag{5.2.14}$$

But the g_1 in this problem is only a function of the magnetic vector potential $\mathbf{A}$ as in (5.2.12). Thus, the relation (5.1.26) is written as

$$g' = \frac{\partial g}{\partial \mathbf{A}} \cdot \mathbf{A}' = \mathbf{g}_{\mathbf{A}} \cdot \mathbf{A}' \tag{5.2.15}$$

where

$$\mathbf{g}_{\mathbf{A}} \equiv \frac{\partial g}{\partial \mathbf{A}} = \left[\frac{\partial g}{\partial A_x}, \frac{\partial g}{\partial A_y}, \frac{\partial g}{\partial A_z}\right]^T \tag{5.2.16}$$

That is, the second term of (5.1.26) is zero;

$$\mathbf{g}_{\mathbf{B}} = \mathbf{0} \tag{5.2.17}$$

g_1 and g_2 of (5.2.13) in this problem provide

$$\mathbf{g}_{\mathbf{A}_1} = -j\omega \mathbf{J}_1 \text{ and } \mathbf{g}_{\mathbf{A}_2} = \mathbf{0} \tag{5.2.18}$$

With these ones, the general adjoint Eq. (5.1.31) is reduced to the variational adjoint equation as

$$\int_{\Omega_1} \left(\nu_1 \mathbf{B}(\boldsymbol{\lambda}_1) \cdot \mathbf{B}(\bar{\lambda}_1) + j\omega\sigma_1 \boldsymbol{\lambda}_1 \cdot \bar{\lambda}_1\right) \mathrm{d}\Omega + \int_{\Omega_2} \left(\nu_2 \mathbf{B}(\boldsymbol{\lambda}_2) \cdot \mathbf{B}(\bar{\lambda}_2) + j\omega\sigma_2 \boldsymbol{\lambda}_2 \cdot \bar{\lambda}_2\right) \mathrm{d}\Omega$$
$$= -j\omega \int_{\Omega_1} \mathbf{J}_1 \cdot \bar{\lambda}_1 m_p \mathrm{d}\Omega \quad \forall \bar{\lambda}_1, \bar{\lambda}_2 \in \Phi \tag{5.2.19}$$

where Φ is the space of admissible complex vector potential: $\Phi = \left\{ \bar{\mathbf{A}} \in [H^1(\Omega)]^3 \right|$ $\bar{\mathbf{A}} = \mathbf{0} \quad \text{on } \mathbf{x} \in \Gamma^0.\}$.

By using (5.2.17) and (5.2.18), the general differential adjoint equations of (5.1.39) and (5.1.40) are also reduced to the differential adjoint equation:

$$\nabla \times \nu_1 \nabla \times \boldsymbol{\lambda}_1 + j\omega\sigma_1 \boldsymbol{\lambda}_1 = -j\omega \mathbf{J}_1 m_p \quad \text{in } \Omega_1 \tag{5.2.20}$$

$$\nabla \times \nu_2 \nabla \times \boldsymbol{\lambda}_2 + j\omega\sigma_2 \boldsymbol{\lambda}_2 = \mathbf{0} \quad \text{in } \Omega_2 \tag{5.2.21}$$

Since the current source is given only in the domain Ω_1 as in Fig. 5.1.8, the variational and differential state equations of (5.1.16), (5.1.4) and (5.1.5) are written as

$$\int_{\Omega_1} \left(\nu_1 \mathbf{B}(\mathbf{A}_1) \cdot \mathbf{B}(\bar{\mathbf{A}}_1) + j\omega\sigma_1 \mathbf{A}_1 \cdot \bar{\mathbf{A}}_1\right) d\Omega + \int_{\Omega_2} \left(\nu_2 \mathbf{B}(\mathbf{A}_2) \cdot \mathbf{B}(\bar{\mathbf{A}}_2) + j\omega\sigma_2 \mathbf{A}_2 \cdot \bar{\mathbf{A}}_2\right) d\Omega$$
$$= \int_{\Omega_1} \mathbf{J}_1 \cdot \bar{\mathbf{A}}_1 d\Omega \quad \forall\, \bar{\mathbf{A}}_1, \bar{\mathbf{A}}_2 \in \Phi \tag{5.2.22}$$

$$\nabla \times \nu_1 \nabla \times \mathbf{A}_1 + j\,\omega\sigma_1 \mathbf{A}_1 = \mathbf{J}_1 \quad \text{in } \Omega_1 \tag{5.2.23}$$

$$\nabla \times \nu_2 \nabla \times \mathbf{A}_2 + j\omega\sigma_2 \mathbf{A}_2 = \mathbf{0} \quad \text{in } \Omega_2 \tag{5.2.24}$$

Comparing these state equations with the adjoint equations of (5.2.19)–(5.2.21) shows that they have only the different sources of $\mathbf{J}_1$ and $-j\omega\mathbf{J}_1$ in Ω_1. Therefore, the adjoint variable, which is determined from the adjoint Eq. (5.2.22), is obtained as

$$\boldsymbol{\lambda} = -j\omega \mathbf{A} \quad \text{in} \quad \Omega \text{ and on } \Gamma \tag{5.2.25}$$

Since the adjoint variable is obtained from the state variable $\mathbf{A}$ in (5.2.25), solving the adjoint equation for the adjoint variable is not necessary.

5.2.3 *Boundary Expression of Shape Sensitivity*

Inserting the relations of (5.2.13), (5.2.17), and (5.2.25) into the general sensitivity formula (5.1.56), the sensitivity formula for the objective function Q of (5.2.3) is obtained as

$$\dot{G} = \int_{\gamma} \left[-j\omega(\nu_2 - \nu_1)\mathbf{B}_1 \cdot \mathbf{B}_2 + \omega^2(\sigma_2 - \sigma_1)\mathbf{A}_1 \cdot \mathbf{A}_2 - 2j\omega \mathbf{J}_1 \cdot \mathbf{A}_2\right] V_n d\Gamma \tag{5.2.26}$$

where the continuity condition of $\mathbf{A}$ and the notation $\mathbf{B}_1 = \mathbf{B}(\mathbf{A}_1)$ and $\mathbf{B}_2 = \mathbf{B}(\mathbf{A}_2)$ were used.

The system power of the original objective function (5.2.1) is the apparent power, which is expressed with the resistance loss power and the reactive power;

$$P = P_{\mathrm{r}} + jP_x \tag{5.2.27}$$

where P_r and P_x are the real and the imaginary parts of P, respectively. By the relation of (5.2.9) and (5.2.10), the objective function Q of (5.2.3) is expressed as

$$Q = P_{\mathrm{r}} - jP_x \tag{5.2.28}$$

where

$$P_{\mathrm{r}} = \mathrm{Re}(Q) \tag{5.2.29}$$

$$P_x = -\mathrm{Im}(Q) \tag{5.2.30}$$

Since the sensitivity formula of (5.2.3) was developed from the objective function Q, it can be rewritten as

$$\dot{Q} = \int_{\gamma} \left[-j\omega(v_2 - v_1)\mathbf{B}_1 \cdot \mathbf{B}_2 + \omega^2(\sigma_2 - \sigma_1)\mathbf{A}_1 \cdot \mathbf{A}_2 - 2j\omega\mathbf{J}_1 \cdot \mathbf{A}_2\right] V_n d\Gamma \tag{5.2.31}$$

The material derivative of (5.2.28) is also written as

$$\dot{Q} = \dot{P}_{\mathrm{r}} - j\dot{P}_x \tag{5.2.32}$$

Thus, the following relations are obtained;

$$\dot{P}_{\mathrm{r}} = \mathrm{Re}\left(\dot{Q}\right) \tag{5.2.33}$$

$$\dot{P}_x = -\mathrm{Im}\left(\dot{Q}\right) \tag{5.2.34}$$

By using the relations of (5.2.31), (5.2.32), and (5.2.33), the sensitivity of the resistance loss power is obtained as

$$\dot{P}_{\mathrm{r}} = \mathrm{Re}\left(\int_{\gamma} \left[-j\omega(v_2 - v_1)\mathbf{B}_1 \cdot \mathbf{B}_2 + \omega^2(\sigma_2 - \sigma_1)\mathbf{A}_1 \cdot \mathbf{A}_2 - 2j\omega\mathbf{J}_1 \cdot \mathbf{A}_2\right] V_n d\Gamma\right) \tag{5.2.35}$$

And the sensitivity of the reactive power is obtained as

$$\dot{P}_x = -\mathrm{Im}\left(\int_{\gamma} \left[-j\omega(\nu_2 - \nu_1)\mathbf{B}_1 \cdot \mathbf{B}_2 + \omega^2(\sigma_2 - \sigma_1)\mathbf{A}_1 \cdot \mathbf{A}_2 - 2j\omega\mathbf{J}_1 \cdot \mathbf{A}_2 \right] V_n d\Gamma \right) \tag{5.2.36}$$

5.2.4 Sensitivities of Resistance and Inductance

In this section, the resistance sensitivity and the inductance sensitivity are derived in the electric-circuit point of view by using the system power sensitivity obtained in Sect. 5.2.3. The eddy current system is given as in Fig. 5.8, where a single-phase current source supplies the system power into the eddy current system through a coil-winding terminal. The terminal current–voltage relation can be modeled with its equivalent circuit of a serial connection of a resistance and an inductance. The resistance represents the Joule loss, and the inductance represents the stored energy of magnetic field. This resistance represents only the Joule loss by the eddy current, but it does not include the resistance of coil winding.

The system power supplied by the current source is written as

$$P = V^* I \tag{5.2.37}$$

The voltage at the coil terminal is written with the resistance and the inductance as

$$V = (R + j\omega L)I \tag{5.2.38}$$

Inserting this relation into (5.2.37), the system power is rewritten as

$$P = (RI^* - j\omega LI^*)I \tag{5.2.39}$$

Hence, the Joule loss power P_r and the reactive power P_x are written, respectively, as

$$P_{\mathrm{r}} = RII^* \tag{5.2.40}$$

$$P_x = -\omega LII^* \tag{5.2.41}$$

The shape variation of the eddy current system causes the variation of the field distribution, which again results in the variation of the resistance R and the inductance L. With the current I given, the sensitivity of the Joule loss power is expressed by taking the total derivative of (5.2.40);

$$\dot{P}_r = \dot{R}II^* \tag{5.2.42}$$

where $\dot{R}$ is the sensitivity of the resistance. The variation of the reactive power is expressed by taking the total derivative of (5.2.41);

$$\dot{P}_x = -\omega\dot{L}II^* \tag{5.2.43}$$

where $\dot{L}$ is the sensitivity of the inductance.

In the previous Sect. 5.2.3, the sensitivity of the resistance loss power (5.2.35) was obtained as

$$\dot{P}_{\mathrm{r}} = \mathrm{Re}\left(\int_\gamma \left[-j\omega(v_2 - v_1)\mathbf{B}_1 \cdot \mathbf{B}_2 + \omega^2(\sigma_2 - \sigma_1)\mathbf{A}_1 \cdot \mathbf{A}_2 - 2j\omega\mathbf{J}_1 \cdot \mathbf{A}_2\right] V_n d\Gamma\right) \tag{5.2.44}$$

and the sensitivity of the reactive power (5.2.36) was obtained as

$$\dot{P}_x = -\mathrm{Im}\left(\int_\gamma \left[-j\omega(v_2 - v_1)\mathbf{B}_1 \cdot \mathbf{B}_2 + \omega^2(\sigma_2 - \sigma_1)\mathbf{A}_1 \cdot \mathbf{A}_2 - 2j\omega\mathbf{J}_1 \cdot \mathbf{A}_2\right] V_n d\Gamma\right) \tag{5.2.45}$$

From the two expressions of (5.2.42) and (5.2.44), the sensitivity of the resistance is obtained as

$$\dot{R} = \frac{1}{II^*}\mathrm{Re}\left(\int_\gamma \left[-j\omega(v_2 - v_1)\mathbf{B}_1 \cdot \mathbf{B}_2 + \omega^2(\sigma_2 - \sigma_1)\mathbf{A}_1 \cdot \mathbf{A}_2 - 2j\omega\mathbf{J}_1 \cdot \mathbf{A}_2\right] V_n d\Gamma\right) \tag{5.2.46}$$

From the two expressions of (5.2.43) and (5.2.45), the sensitivity of the inductance is obtained as

$$\dot{L} = \frac{1}{\omega II^*}\mathrm{Im}\left(\int_\gamma \left[-j\omega(v_2 - v_1)\mathbf{B}_1 \cdot \mathbf{B}_2 + \omega^2(\sigma_2 - \sigma_1)\mathbf{A}_1 \cdot \mathbf{A}_2 - 2j\omega\mathbf{J}_1 \cdot \mathbf{A}_2\right] V_n d\Gamma\right) \tag{5.2.47}$$

5.2.5 Numerical Examples

The sensitivity formula (5.2.35) in the Sect. 5.2.3 is applied to three shape optimization problems of axi-symmetric design model, where the analytic field solutions are unknown. These design models show that the sensitivity formula is well applied to the shape design for the Joule loss by the eddy current. The numerical examples are the design problems of three interfaces: conductor–air, current region–air, and ferromagnetic material–air. These examples have the known optimal designs. If the results of the examples are obtained as the expected optimal designs, it can be said that the shape optimization using the sensitivity formula for the Joule loss is feasible for the design of the eddy current system. In these examples, the state variable, which is required to evaluate the sensitivity formula, is numerically calculated by the finite element method. The sensitivity obtained is used for the optimization algorithm to evolve the interface shape. The level set method is used to represent the shape evolution of the design model.

5.2.6 Conductor–Air Interface Design

As an example that has a known optimal design, consider an axi-symmetric eddy current system consisting of a conductor, a sinusoidal current source, and the air as shown in Fig. 5.9. The source current produces a time-varying magnetic field, which induces an eddy current in the conductor. If the shape of the conductor is changed to a cylinder, the system has the minimum Joule loss under the constraint of the constant conductor volume.

The design objective is to obtain the minimum eddy loss power in the conductor. The objective function F to be minimized is defined as

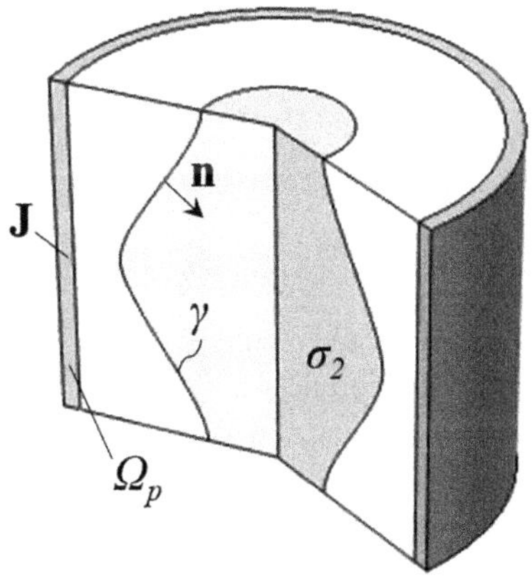

Fig. 5.9 Conductor–air interface design

$$F = \mathrm{Re}\left(\int_{\Omega} \mathbf{J} \cdot \mathbf{E}^{*} m_p \mathrm{d}\Omega\right) \tag{5.2.48}$$

The design variable is the shape of the conductor–air interface γ as shown in Fig. 5.9. The shape sensitivity of this interface for the resistance loss power is obtained from (5.2.35);

$$\dot{P}_{\mathrm{r}} = \mathrm{Re}\left(\int_{\gamma} \omega^2 \sigma_2 \mathbf{A}_1 \cdot \mathbf{A}_2 V_n \mathrm{d}\Gamma\right) \tag{5.2.49}$$

The velocity field V_n is taken for the negative shape sensitivity as

$$V_n = -\mathrm{Re}\left(\omega^2 \sigma_2 \mathbf{A}_1 \cdot \mathbf{A}_2\right) \tag{5.2.50}$$

This design problem has a constraint of constant conductor volume;

$$\int_{\Omega_{\mathrm{c}}} \mathrm{d}\Omega = C \tag{5.2.51}$$

where Ω_{c} is the conductor region and C is a given value of the constraint.

The material derivative of the constraint (5.2.51) is obtained as

$$\int_{\Gamma} V_n \mathrm{d}\Gamma = 0 \tag{5.2.52}$$

which is a different form of the constraint (5.2.51) expressed with the design velocity field V_n. In order to satisfy the constraint (5.2.52), the design velocity (5.2.50) is modified by subtracting its average V_{na} to become U_n as

$$U_n = V_n - V_{na} \tag{5.2.53}$$

where

$$V_{na} = -\mathrm{Re}\left(\omega^2 \sigma_2 \mathbf{A}_1 \cdot \mathbf{A}_2\right) \Big/ \int_{\gamma} d\Gamma \tag{5.2.54}$$

The design result is shown in Fig. 5.1.10, where the shape of the conductor becomes gradually a cylinder with the increase of the iteration number as expected. The final design of the cylinder shape is obtained at the 20 s, when the objective function value converges to 90 kW/m^3 as in Fig. 5.11. The result of this example shows the feasibility of the shape sensitivity for conductor–air interface in the axi-symmetric eddy current system.

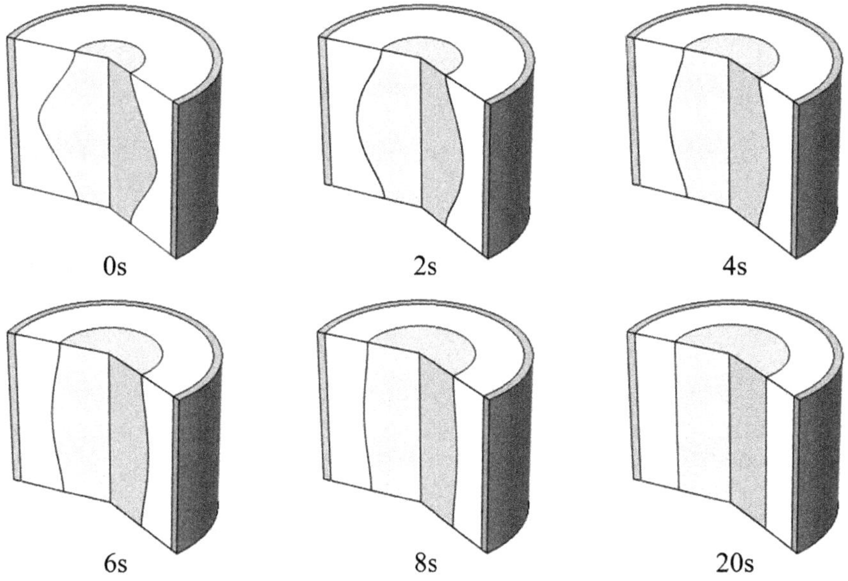

Fig. 5.10 Conductor–air interface design: shape variation

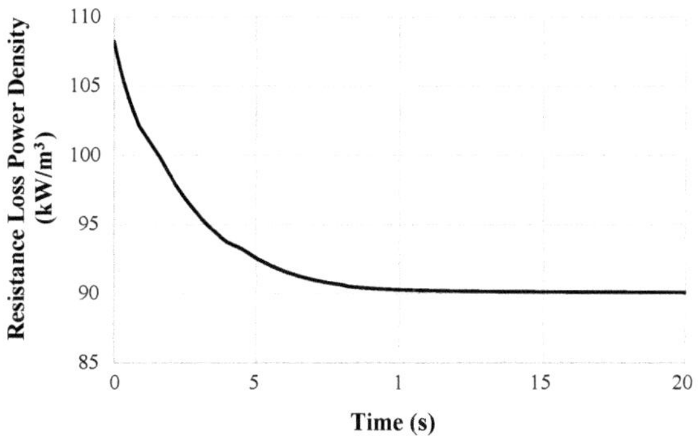

Fig. 5.11 Conductor–air interface design: evolution of objective function

5.2.7 *Current Region–Air Interface Design*

Consider an axi-symmetric eddy current problem, which consists of a conductor, a sinusoidal current source, and the air as shown in Fig. 5.12. The time-varying magnetic field by the source current induces an eddy current in the conductor. If the shape of the source current region is changed to a hollow cylinder, the eddy loss becomes the minimum value under the constant volume constraint of the source current region.

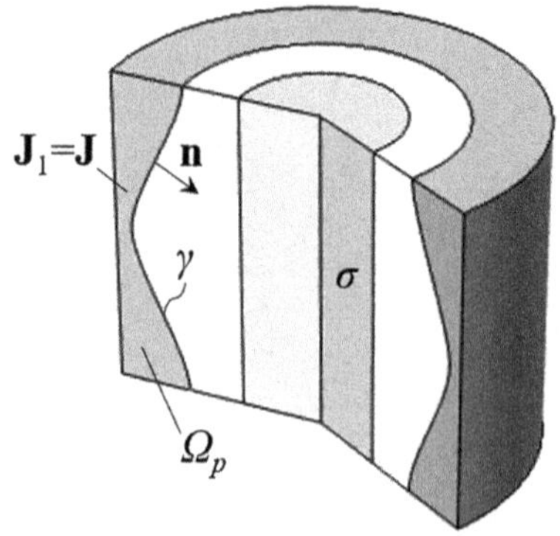

Fig. 5.12 Current region–air interface design

The design objective is to obtain the minimum eddy loss power of the system. The objective function F is defined as

$$F = \mathrm{Re}\left(\int_{\Omega} \mathbf{J} \cdot \mathbf{E}^{*} m_p \mathrm{d}\Omega\right) \tag{5.2.55}$$

The design variable is the shape of the current region–air interface γ as shown in Fig. 5.12. The shape sensitivity of this interface for the eddy loss power is obtained from (5.2.35);

$$\dot{P}_r = \mathrm{Re}\left(\int_{\gamma} -2j\omega \mathbf{J}_1 \cdot \mathbf{A}_2 V_n \mathrm{d}\Gamma\right) \tag{5.2.56}$$

The velocity field V_n for the negative shape sensitivity is taken as

$$V_n = -\mathrm{Re}(-2j\omega \mathbf{J}_1 \cdot \mathbf{A}_2) \tag{5.2.57}$$

In this problem, the modified velocity U_n for the volume constraint of the current region is taken as

$$U_n = V_n - V_{na} \tag{5.2.58}$$

where

$$V_{na} = -\mathrm{Re}(-2j\omega \mathbf{J}_1 \cdot \mathbf{A}_2) / \int_{\gamma} \mathrm{d}\Gamma \tag{5.2.59}$$

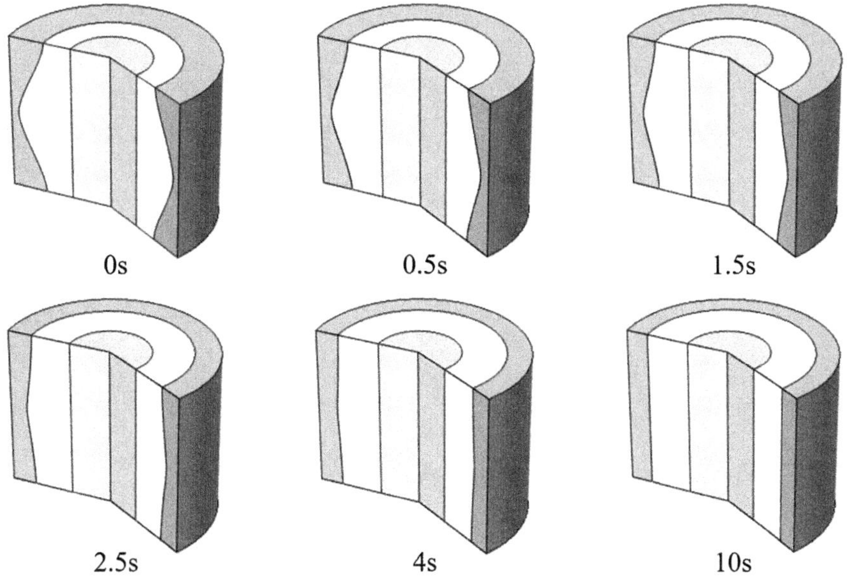

Fig. 5.13 Current region–air interface design: shape variation

Fig. 5.14 Current region–air interface design: evolution of objective function

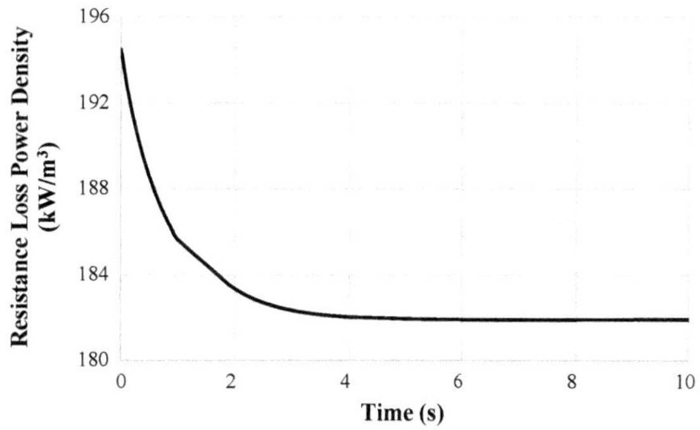

The design result is shown in Fig. 5.1.13, where the shape of the current region becomes gradually a hollow cylinder with the increase of the iteration number as expected. The final design of the hollow cylinder shape is obtained at the 10 *s*, when the objective function value converges to 182 kW/m^3 as in Fig. 5.14. The result of this example shows that the shape sensitivity for current region–air interface works well for the axi-symmetric eddy current system.

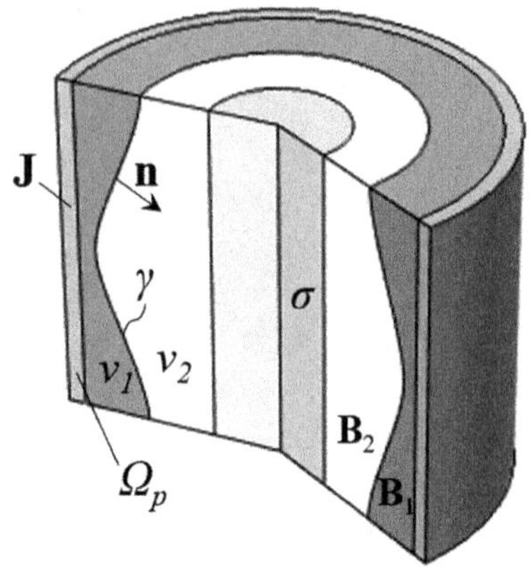

Fig. 5.15 Ferromagnetic material–air interface design

5.2.8 Ferromagnetic Material–Air Interface Design

Figure 5.15 shows an axi-symmetric eddy current problem, which consists of a conductor, a sinusoidal source current, a ferromagnetic material, and the air. An eddy current is induced in the conductor by the time-varying magnetic field by the source current. If the shape of the ferromagnetic material is changed to be a hollow cylinder, the eddy loss becomes the minimum value under the constant volume constraint of the current region.

The design objective is to obtain the minimum eddy loss power. The objective function F is defined as

$$F = \mathrm{Re}\left(\int_{\Omega} \mathbf{J}\cdot\mathbf{E}^{*} m_p \mathrm{d}\Omega\right) \tag{5.2.60}$$

The design variable is the shape of the ferromagnetic material–air interface γ as shown in Fig. 5.15. The shape sensitivity of this interface for the eddy loss power is obtained from (5.2.35);

$$\dot{P}_{\mathrm{r}} = \mathrm{Re}\left(\int_{\gamma} -j\omega(\nu_2-\nu_1)\mathbf{B}_1\cdot\mathbf{B}_2 V_n \mathrm{d}\Gamma\right) \tag{5.2.61}$$

The velocity field V_n can be taken for the negative value of the shape sensitivity (5.2.61) as

$$V_n = -\mathrm{Re}[-j\omega(\nu_2-\nu_1)\mathbf{B}_1\cdot\mathbf{B}_2] \tag{5.2.62}$$

In this problem, to satisfy a given volume constraint of the ferromagnetic material, the modified velocity field U_n is taken as

$$U_n = V_n - V_{na} \tag{5.2.63}$$

where

$$V_{na} = -\mathrm{Re}[-j\omega(v_2 - v_1)\mathbf{B}_1 \cdot \mathbf{B}_2] / \int_{\gamma} \mathrm{d}\Gamma \tag{5.2.64}$$

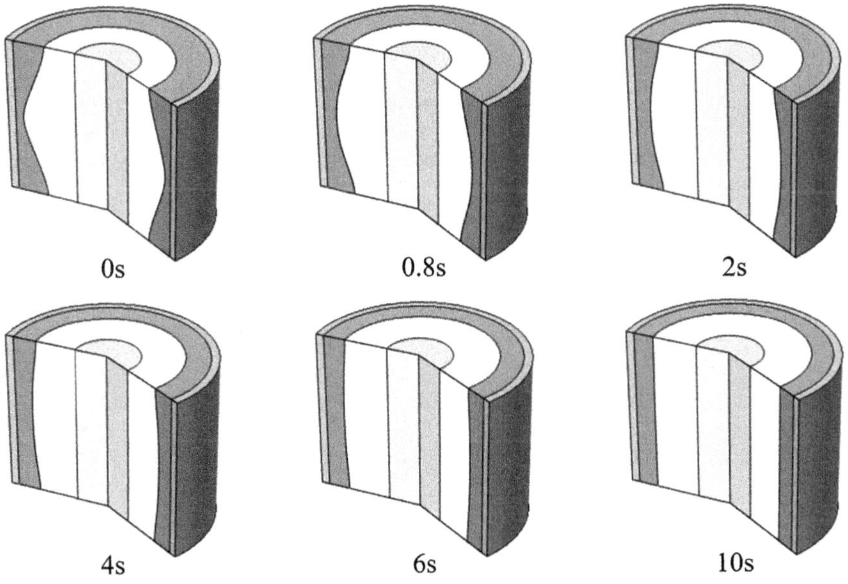

Fig. 5.16 Ferromagnetic material–air interface design: shape variation

Fig. 5.17 Ferromagnetic material–air interface design: evolution of objective function

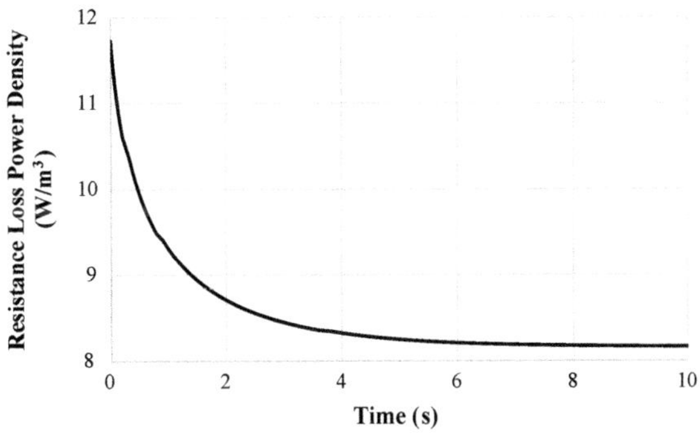

The design result is obtained as in Fig. 5.16, where the shape of the ferromagnetic material becomes gradually a hollow cylinder as expected. The final design of the hollow cylinder shape is obtained at the 10 *s*, when the objective function value converges to 8.15 W/m^3 as in Fig. 5.17. The result of this example shows that the shape sensitivity for the ferromagnetic material–air interface is well applied to the axi-symmetric eddy current system.

References

1. Krawczyk, A., Tegopoulos, A.J.A.: Numerical Modelling of Eddy Currents. Clarendon Press, Oxford (1993)
2. McCary, R.: Optimization of eddy current transducers surround coil. IEEE Trans. Magn. **15**, 1677–1679 (1979)
3. Davies, J., Simpson, P.: Induction Heating Handbook. McGraw-hill, London (1979)
4. Byun, J.K., Jung, H.K., Hahn, S.Y., Choi, K., Park, I.H.: Optimal temperature control for induction heating devices using physical and geometrical design sensitivity. IEEE Trans. Magn. **34**, 3114–3117 (1998)
5. Kwak, I.G., Byun, J.K., Park, I.H., Hahn, S.Y.: Design sensitivity of electro-thermal systems for exciting-coil positioning. Int. J. Appl. Electromag. Mech. **9**, 249–261 (1998)
6. Park, I.H., Lee, H.B., Kwak, I.G., Hahn, S.Y.: Design sensitivity analysis for steady state eddy current problems by continuum approach. IEEE Trans. Magn. **30**, 3411–3414 (1994)

Chapter 6
Continuum Shape Design Sensitivity of DC Conductor System

The DC conductor system is composed of conductive material and electrodes [1–4]. There are examples of the design objectives for the DC conductor system such as reducing Joule loss, obtaining a desired resistance, obtaining a desired current distribution, reducing leakage current loss of insulator. The DC conductor system includes circuit system, conductor connection, insulator, typical examples of which are semiconductor circuit layout, PCB layout, current distributor, connector, high voltage insulator, etc. [5, 6].

In this chapter, the continuum shape sensitivity for the DC conductor system is derived just like in the previous chapters. The design problem of the DC conductor system, although similar to that of the electrostatic system, has only the design variable of the outer boundary. The shape sensitivity for the DC conductor system is classified into two categories: the domain integral objective function and the objective function of system loss power. The objective function of the system loss power does not require solving for the state variable, and it is used to derive the resistance sensitivity.

In Sect. 6.1, a general three-dimensional sensitivity of the outer boundary for the domain integral objective function is derived. For the DC conductor system of a single conductive medium, the sensitivity formula is expressed as a boundary integral of only with only one term.

In Sect. 6.2, by using the relation of the system loss power and the system energy of the electrostatic system in Sect. 3.3, a general three-dimensional sensitivity for the system loss power is simply derived. The loss power sensitivity, which is related to the system resistance in the electric circuit, is used to derive the resistance sensitivity.

At the end of the Sects. 6.1–6.2, the two general sensitivity formulas are applied to design examples to be validated in the same way as in the previous chapters.

I. H. Park, *Design Sensitivity Analysis and Optimization of Electromagnetic Systems*, Mathematical and Analytical Techniques with Applications to Engineering,
https://doi.org/10.1007/978-981-13-0230-5_6

6.1 Shape Sensitivity of Outer Boundary

The design variable of the DC current-carrying conductor is the outer boundary, which consists of the Dirichlet boundary Γ^0 and the Neumann boundary Γ^1. The Dirichlet boundary condition is imposed on the electrode surfaces by the external voltage source. The homogeneous Neumann boundary condition is applied on all the conductor surfaces, where the current has only the tangential component. The shape variation of the outer boundary causes the variation in the electric field distribution in the domain. In this section, a general three-dimensional sensitivity for the outer boundary design is derived. The derivation procedure is the same as in Sect. 3.2. The derived sensitivity is expressed in a boundary integral [7]. The obtained sensitivity formula is tested and validated with analytical and numerical examples.

6.1.1 Problem Definition and Objective Function

A DC current-carrying conductor is given as in Fig. 6.1, where the domain Ω has the homogeneous distribution of conductivity σ. The domain Ω has the outer boundary Γ, where $\mathbf{n}$ is the outward normal vector. The electrode surfaces, where the Dirichlet boundary condition is imposed, is connected to the external voltage source. In this shape sensitivity analysis for the outer boundary design, the two boundaries of Dirichlet and Neumann boundaries are both taken as the design variable.

Consider a general objective function in the integral form;

$$F = \int_{\Omega} g(\phi, \nabla\phi) m_p \mathrm{d}\Omega \tag{6.1.1}$$

where g can be any function that is continuously differentiable with respect to the arguments of ϕ and $\nabla\phi$, and m_p is a characteristic function that is defined as

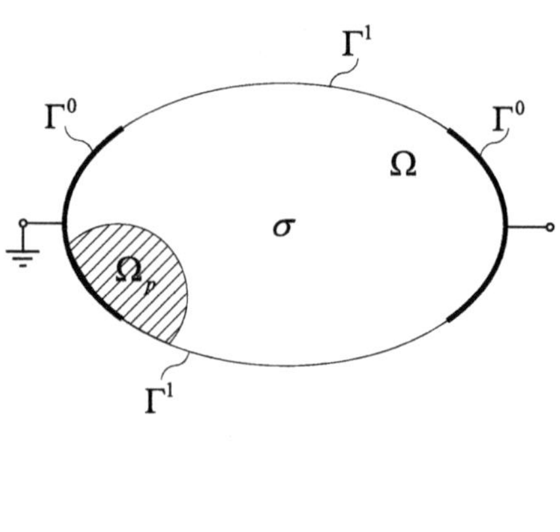

Fig. 6.1 Outer boundary design of DC conductor system

$$m_p = \begin{cases} 1 & \mathbf{x} \in \Omega_p \\ 0 & \mathbf{x} \notin \Omega_p \end{cases} \tag{6.1.2}$$

The region $\Omega_p \subset \Omega$, which is the integral domain for the objective function, can intersect with the outer boundary of the Dirichlet boundary Γ^0 or the Neumann boundary Γ^1 as shown schematically in Fig. 6.1.

6.1.2 Lagrange Multiplier Method for Sensitivity Derivation

The variational state equation Eq. (2.4.14), which holds regardless of the change of the boundary shape, is treated as an equality constraint in this shape sensitivity analysis. For the sensitivity of the objective function F (6.1.1) subject to the constraint (2.4.14), the method of Lagrange multipliers is employed. The method of Lagrange multipliers provides an augmented objective function G as

$$G = F - a(\phi, \bar{\phi}) \quad \forall \bar{\phi} \in \Phi \tag{6.1.3}$$

where the arbitrary virtual potential $\bar{\phi}$ plays the role of Lagrange multipliers and Φ is the space of admissible potential defined in Sect. 2.4.2;

$$\Phi = \left\{ \bar{\phi} \in H^1(\Omega) \middle| \bar{\phi} = 0 \quad \text{on} \quad \mathbf{x} \in \Gamma^0 \right\} \tag{6.1.4}$$

This augmented objective function is differentiated by using the concept of material derivative;

$$\dot{G} = -\dot{a}(\phi, \bar{\phi}) + \dot{F} \quad \forall \bar{\phi} \in \Phi \tag{6.1.5}$$

By applying the material derivative formula (3.1.22) to (2.4.13) and (6.1.1), and by using the relation (3.1.15), each term in (6.1.5) is obtained as

$$\begin{aligned} \dot{a}(\phi, \bar{\phi}) &= \int_{\Omega} (\sigma \nabla \phi' \cdot \nabla \bar{\phi} + \sigma \nabla \phi \cdot \nabla \bar{\phi}') \mathrm{d}\Omega + \int_{\Gamma} \sigma \nabla \phi \cdot \nabla \bar{\phi} V_n \mathrm{d}\Gamma \\ &= \int_{\Omega} \left(\sigma \nabla \dot{\phi} \cdot \nabla \bar{\phi} - \sigma \nabla (\mathbf{V} \cdot \nabla \phi) \cdot \nabla \bar{\phi} + \sigma \nabla \phi \cdot \nabla \dot{\bar{\phi}} - \sigma \nabla \phi \cdot \nabla (\mathbf{V} \cdot \nabla \bar{\phi}) \right) \mathrm{d}\Omega \\ &\quad + \int_{\Gamma} \sigma \nabla \phi \cdot \nabla \bar{\phi} V_n \mathrm{d}\Gamma \end{aligned} \tag{6.1.6}$$

$$\begin{aligned}\dot{F} &= \int_{\Omega} g' m_p \mathrm{d}\Omega + \int_{\Gamma} g m_p V_n \mathrm{d}\Gamma \\ &= \int_{\Omega} (g_\phi \phi' + \mathbf{g_E} \cdot \nabla \phi') m_p \mathrm{d}\Omega + \int_{\Gamma} g m_p V_n \mathrm{d}\Gamma \\ &= \int_{\Omega} \left[g_\phi \dot{\phi} - g_\phi (\mathbf{V} \cdot \nabla \phi) + \mathbf{g_E} \cdot \nabla \dot{\phi} - \mathbf{g_E} \cdot \nabla (\mathbf{V} \cdot \nabla \phi) \right] m_p \mathrm{d}\Omega + \int_{\Gamma} g m_p V_n \mathrm{d}\Gamma\end{aligned} \tag{6.1.7}$$

In (6.1.7), the relation (3.1.23) was used:

$$g' = \frac{\partial g}{\partial \phi} \phi' + \frac{\partial g}{\partial \nabla \phi} \cdot \nabla \phi' = g_\phi \phi' + \mathbf{g_E} \cdot \nabla \phi' \tag{6.1.8}$$

where

$$g_\phi \equiv \frac{\partial g}{\partial \phi} \text{ and } \mathbf{g_E} \equiv \frac{\partial g}{\partial \nabla \phi} = \left[\frac{\partial g}{\partial (\nabla \phi)_x}, \frac{\partial g}{\partial (\nabla \phi)_y}, \frac{\partial g}{\partial (\nabla \phi)_z} \right]^T \tag{6.1.9}$$

In derivation of (6.1.6), $\sigma' = 0$ was used. (6.1.6) and (6.1.7) are inserted into (6.1.5) to provide

$$\begin{aligned}\dot{G} = &- \int_{\Omega} \sigma \nabla \dot{\phi} \cdot \nabla \bar{\phi} \mathrm{d}\Omega - \int_{\Omega} \sigma \nabla \phi \cdot \nabla \dot{\bar{\phi}} \mathrm{d}\Omega + \int_{\Omega} \left(g_\phi \dot{\phi} + \mathbf{g_E} \cdot \nabla \dot{\phi} \right) m_p \mathrm{d}\Omega \\ &+ \int_{\Omega} \sigma \nabla (\mathbf{V} \cdot \nabla \phi) \cdot \nabla \bar{\phi} \mathrm{d}\Omega + \int_{\Omega} \sigma \nabla \phi \cdot \nabla (\mathbf{V} \cdot \nabla \bar{\phi}) \mathrm{d}\Omega \\ &- \int_{\Omega} \left[g_\phi (\mathbf{V} \cdot \nabla \phi) + \mathbf{g_E} \cdot \nabla (\mathbf{V} \cdot \nabla \phi) \right] m_p \mathrm{d}\Omega \\ &- \int_{\Gamma} \left(\sigma \nabla \phi \cdot \nabla \bar{\phi} - g m_p \right) V_n \mathrm{d}\Gamma \qquad \forall \bar{\phi} \in \Phi\end{aligned} \tag{6.1.10}$$

Since $\dot{\bar{\phi}}$ belongs to Φ, the variational state equation of (2.4.12) gives the following relation:

$$\int_{\Omega} \sigma \nabla \phi \cdot \nabla \dot{\bar{\phi}} \mathrm{d}\Omega = 0 \tag{6.1.11}$$

Thus, the term with $\dot{\bar{\phi}}$ in (6.1.10) vanishes and (6.1.10) becomes

$$\begin{aligned}\dot{G} = &-\int_{\Omega} \sigma\nabla\dot{\phi}\cdot\nabla\bar{\phi}\mathrm{d}\Omega + \int_{\Omega}\left(g_{\phi}\dot{\phi} + \mathbf{g_E}\cdot\nabla\dot{\phi}\right)m_p\mathrm{d}\Omega\\ &+\int_{\Omega}\sigma\nabla(\mathbf{V}\cdot\nabla\phi)\cdot\nabla\bar{\phi}\mathrm{d}\Omega + \int_{\Omega}\sigma\nabla\phi\cdot\nabla(\mathbf{V}\cdot\nabla\bar{\phi})\mathrm{d}\Omega\\ &-\int_{\Omega}\left[g_{\phi}(\mathbf{V}\cdot\nabla\phi) + \mathbf{g_E}\cdot\nabla(\mathbf{V}\cdot\nabla\phi)\right]m_p\mathrm{d}\Omega\\ &-\int_{\Gamma}\left(\sigma\nabla\phi\cdot\nabla\bar{\phi} - gm_p\right)V_n d\Gamma \qquad \forall\bar{\phi}\in\Phi\end{aligned} \tag{6.1.12}$$

6.1.3 Adjoint Variable Method for Sensitivity Analysis

In order to express this $\dot{G}$ explicitly in terms of the velocity field $\mathbf{V}$, an adjoint equation is introduced, which is obtained by replacing $\dot{\phi}$ in the g-related terms ($g_{\phi}\dot{\phi}$ and $\mathbf{g_E}\cdot\nabla\dot{\phi}$) of (6.1.12) with a virtual potential $\bar{\lambda}$ and by equating the terms to the energy bilinear form $a(\lambda,\bar{\lambda})$. The adjoint equation so obtained is written as

$$\int_{\Omega}\sigma\nabla\lambda\cdot\nabla\bar{\lambda}\mathrm{d}\Omega = \int_{\Omega}\left(g_{\phi}\bar{\lambda} + \mathbf{g_E}\cdot\nabla\bar{\lambda}\right)m_p\mathrm{d}\Omega \quad \forall\bar{\lambda}\in\Phi \tag{6.1.13}$$

where Φ is the space of admissible potential: $\Phi = \left\{\bar{\phi}\in H^1(\Omega)\middle|\ \bar{\phi} = 0\right.$ on $\left.\mathbf{x}\in\Gamma^0\right\}$.

This adjoint equation is evaluated at the specific $\bar{\lambda} = \dot{\phi}$;

$$\int_{\Omega}\sigma\nabla\lambda\cdot\nabla\dot{\phi}\mathrm{d}\Omega = \int_{\Omega}\left(g_{\phi}\dot{\phi} + \mathbf{g_E}\cdot\nabla\dot{\phi}\right)m_p\mathrm{d}\Omega \tag{6.1.14}$$

Similarly, the sensitivity Eq. (6.1.12) is evaluated at the specific $\bar{\phi} = \lambda$ to yield

$$\begin{aligned}\dot{G} = &-\int_{\Omega} \sigma\nabla\dot{\phi}\cdot\nabla\lambda \mathrm{d}\Omega + \int_{\Omega} \left(g_{\phi}\dot{\phi} + \mathbf{g_E}\cdot\nabla\dot{\phi}\right) m_p \mathrm{d}\Omega \\ &+ \int_{\Omega} \sigma\nabla(\mathbf{V}\cdot\nabla\phi)\cdot\nabla\lambda \mathrm{d}\Omega + \int_{\Omega} \sigma\nabla\phi\cdot\nabla(\mathbf{V}\cdot\nabla\lambda)\mathrm{d}\Omega \\ &- \int_{\Omega} \left[g_{\phi}(\mathbf{V}\cdot\nabla\phi) + \mathbf{g_E}\cdot\nabla(\mathbf{V}\cdot\nabla\phi)\right] m_p \mathrm{d}\Omega \\ &- \int_{\Gamma} \left(\sigma\nabla\phi\cdot\nabla\lambda - g m_p\right) V_n \mathrm{d}\Gamma \end{aligned} \tag{6.1.15}$$

The energy bilinear form of (6.1.14) is symmetric in its arguments;

$$\int_{\Omega} \sigma\nabla\lambda\cdot\nabla\dot{\phi}\mathrm{d}\Omega = \int_{\Omega} \sigma\nabla\dot{\phi}\cdot\nabla\lambda \mathrm{d}\Omega \tag{6.1.16}$$

By using this relation and inserting (6.1.14) into (6.1.15), the sensitivity is written as;

$$\begin{aligned}\dot{G} = &\int_{\Omega} \sigma\nabla\phi\cdot\nabla(\mathbf{V}\cdot\nabla\lambda)\mathrm{d}\Omega \\ &+ \int_{\Omega} \left[\sigma\nabla\lambda\cdot\nabla(\mathbf{V}\cdot\nabla\phi) - g_{\phi}(\mathbf{V}\cdot\nabla\phi) - \mathbf{g_E}\cdot\nabla(\mathbf{V}\cdot\nabla\phi)\right] m_p \mathrm{d}\Omega \\ &- \int_{\Gamma} \left(\sigma\nabla\phi\cdot\nabla\lambda - g m_p\right) V_n \mathrm{d}\Gamma \end{aligned} \tag{6.1.17}$$

where all terms are expressed with the velocity field $\mathbf{V}$. Once the state variable ϕ and the adjoint variable λ are determined to be the solutions to (2.412) and (6.1.13), respectively, this design sensitivity is obtained.

6.1.4 Boundary Expression of Shape Sensitivity

The domain integrals of the sensitivity (6.1.17) are transformed into boundary integrals by using two variational identities for the state and the adjoint equations. First, the variational identity for the state equation was given as (2.4.7) in Sect. 2.4;

$$\int_{\Omega} \sigma\nabla\phi \cdot \nabla\bar{\phi}\mathrm{d}\Omega = \int_{\Gamma} \sigma\frac{\partial\phi}{\partial n}\bar{\phi}\mathrm{d}\Gamma \quad \forall\bar{\phi} \in \Phi \tag{6.1.18}$$

The other variational identity for the adjoint equation can be derived from the differential adjoint equation, which is obtained from (6.1.16) by the same procedure in Sect. 3.2. The variational adjoint equation (3.2.19) is equivalent to the differential adjoint equation of (3.2.21). The obtained differential adjoint equation is written as

$$-\nabla \cdot \sigma\nabla\lambda = \left(g_{\phi} - \nabla \cdot \mathbf{g}_{\mathrm{E}}\right)m_p \tag{6.1.19}$$

with the boundary condition:

$$\lambda = 0 \quad \text{on} \quad \Gamma^0 \quad \text{(homogeneous Dirichlet boundary condition)} \tag{6.1.20}$$

$$\frac{\partial\lambda}{\partial n} = 0 \quad \text{on} \quad \Gamma^1 \quad \text{(homogeneous Neumann boundary condition)} \tag{6.1.21}$$

Since the adjoint sources $g_{\phi}m_p$ and $\mathbf{g}_{\mathrm{E}}m_p$ exist only in the domain Ω, there is no adjoint surface source equivalent to Dirichlet boundary condition. While the original state equation may have surface sources equivalent to the Dirichlet boundary condition, the adjoint equation has no surface source equivalent to its Dirichlet boundary condition. Hence, the Dirichlet boundary condition of the adjoint equation is zero. In addition, since the gradient of the adjoint variable has no normal component on the conductor surface, the homogeneous Neumann condition is imposed on Γ^1 of the adjoint system.

Just as the variational identity (2.4.7) for the state equation is obtained from the differential state equation Eq. (2.4.4), the variational identity for the adjoint equation is obtained;

$$\int_{\Omega} \left(\sigma\nabla\lambda \cdot \nabla\bar{\lambda} - g_{\phi}\bar{\lambda}m_p - \mathbf{g}_{\mathrm{E}} \cdot \nabla\bar{\lambda}m_p\right)\mathrm{d}\Omega = \int_{\Gamma} \sigma\frac{\partial\lambda}{\partial n}\bar{\lambda}\mathrm{d}\Gamma \quad \forall\bar{\lambda} \in \Phi \tag{6.1.22}$$

Note that imposing the boundary conditions (6.1.20) and (6.1.21) provides the variational adjoint equation (6.1.13).

The variational identities of (6.1.18) and (6.1.22) are used to express the domain integrals of (6.1.17) as boundary integrals. Choosing $\bar{\phi} = \mathbf{V} \cdot \nabla\lambda$ in (6.1.18) yields

$$\int_{\Omega} \sigma\nabla\phi \cdot \nabla(\mathbf{V} \cdot \nabla\lambda)\mathrm{d}\Omega = \int_{\Gamma} \sigma\frac{\partial\phi}{\partial n}(\mathbf{V} \cdot \nabla\lambda)\mathrm{d}\Gamma \tag{6.1.23}$$

and choosing $\bar{\lambda} = \mathbf{V} \cdot \nabla\phi$ in (6.1.22) yields

$$\int_{\Omega} \left[\sigma\nabla\lambda \cdot \nabla(\mathbf{V} \cdot \nabla\phi) - g_{\phi}(\mathbf{V} \cdot \nabla\phi)m_p - \mathbf{g_E} \cdot \nabla(\mathbf{V} \cdot \nabla\phi)m_p\right] \mathrm{d}\Omega$$
$$= \int_{\Gamma} \sigma \frac{\partial\lambda}{\partial n}(\mathbf{V} \cdot \nabla\phi)\mathrm{d}\Gamma \qquad (6.1.24)$$

By inserting (6.1.23) and (6.1.24) into (6.1.17), the domain integrals in (6.1.17) become a boundary integral;

$$\dot{G} = \int_{\Gamma} \left[\sigma \frac{\partial\phi}{\partial n}(\mathbf{V} \cdot \nabla\lambda) + \sigma \frac{\partial\lambda}{\partial n}(\mathbf{V} \cdot \nabla\phi)\right] \mathrm{d}\Gamma - \int_{\Gamma} (\sigma\nabla\phi \cdot \nabla\lambda - gm_p) V_n \mathrm{d}\Gamma \qquad (6.1.25)$$

, which is the desired expression.

This sensitivity formula for the outer boundary variation becomes simpler by using the relations (3.2.32)–(3.2.34) in Sect. 3.2;

$$\dot{G} = \int_{\Gamma} \left(\sigma \frac{\partial\phi}{\partial n}\frac{\partial\lambda}{\partial n} - \sigma \frac{\partial\lambda}{\partial t}\frac{\partial\phi}{\partial t}\right) V_n \mathrm{d}\Gamma + \int_{\Gamma} gm_p V_n \mathrm{d}\Gamma \qquad (6.1.26)$$

The Dirichlet boundary condition in the DC current-carrying conductor is imposed on the electrode surface, where the electric field has only the normal component;

$$\frac{\partial\phi}{\partial t} = 0 \quad \text{on} \quad \Gamma^0 \qquad (6.1.27)$$

$$\frac{\partial\lambda}{\partial t} = 0 \quad \text{on} \quad \Gamma^0 \qquad (6.1.28)$$

Since the electric current has only the tangential component on the conductor surface,

$$\frac{\partial\phi}{\partial n} = 0 \quad \text{on} \quad \Gamma^1 \qquad (6.1.29)$$

$$\frac{\partial\lambda}{\partial n} = 0 \quad \text{on} \quad \Gamma^1 \qquad (6.1.30)$$

With these boundary conditions, the first integral in (6.1.26) is decomposed into the two integrals on the Dirichlet and the Neumann boundaries;

$$\dot{G} = \int_{\Gamma^0} \sigma \frac{\partial \phi}{\partial n} \frac{\partial \lambda}{\partial n} V_n \mathrm{d}\Gamma - \int_{\Gamma^1} \sigma \frac{\partial \phi}{\partial t} \frac{\partial \lambda}{\partial t} V_n \mathrm{d}\Gamma + \int_{\Gamma} g m_p V_n \mathrm{d}\Gamma \tag{6.1.31}$$

When the objective function is defined on the inner area in the domain Ω that does not intersect with the design boundary, since $m_p = 0$ on Γ, the sensitivity (6.1.31) becomes

$$\dot{G} = \int_{\Gamma^0} \sigma \frac{\partial \phi}{\partial n} \frac{\partial \lambda}{\partial n} V_n \mathrm{d}\Gamma - \int_{\Gamma^1} \sigma \frac{\partial \phi}{\partial t} \frac{\partial \lambda}{\partial t} V_n \mathrm{d}\Gamma \tag{6.1.32}$$

With (3.2.44), this sensitivity formula is expressed as

$$\dot{G} = \int_{\Gamma^0} \sigma E_{\mathrm{n}}(\phi) E_{\mathrm{n}}(\lambda) V_n \mathrm{d}\Gamma - \int_{\Gamma^1} \sigma E_t(\phi) E_t(\lambda) V_n \mathrm{d}\Gamma \tag{6.1.33}$$

6.2 Shape Sensitivity of Outer Boundary for Joule Loss Power

In this section, the shape sensitivity for the interface design is developed in the DC current-carrying conductor as in the Sect. 6.1; but the objective function is Joule loss power. The Joule loss power in the DC current-carrying conductor, which comes from the Ohm's law, is determined by the distribution of current density in the conductor domain. The shape variation of the outer boundary causes the variation in the distribution of current density. The Joule loss power is also related to the resistance of its equivalent electric circuit. The shape sensitivity can be applied to the design problems for reducing the Joule loss power, obtaining a desired resistance or reducing the leakage current in the insulator.

6.2.1 Problem Definition

A DC current-carrying conductor for the outer boundary design is given as in Fig. 6.2, where the domain Ω has a homogeneous distribution of conductivity σ and the outer boundary consists of the Dirichlet boundary Γ^0 and the Neumann boundary Γ^1. The two boundaries are taken as the design variable. The electrode surface, which is connected to an external voltage source, is the Dirichlet boundary Γ^0.

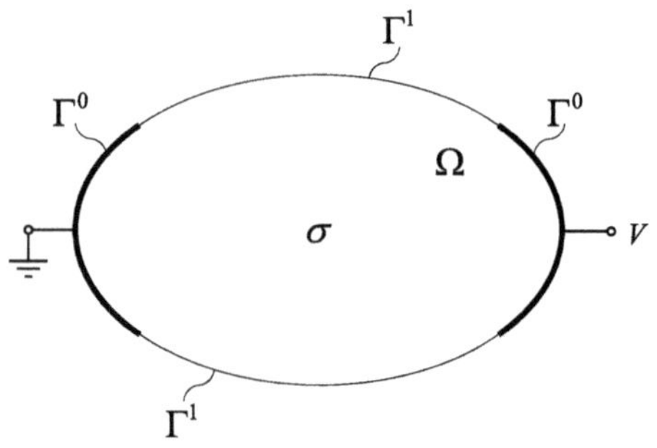

Fig. 6.2 Outer boundary design of DC conductor system for Joule loss power

The objective function P_r is the Joule loss power;

$$P_r = \int_{\Omega} \frac{1}{\sigma} J^2 \mathrm{d}\Omega \tag{6.2.1}$$

where $\frac{1}{\sigma} J^2$ is the density of Joule loss power. With the relations (2.4.2) and (2.4.3), this Joule loss power density is written as

$$\frac{1}{\sigma} J^2 = \sigma \nabla \phi \cdot \nabla \phi \tag{6.2.2}$$

This expression is inserted into (6.2.1) to provide

$$P_r = \int_{\Omega} \sigma \nabla \phi \cdot \nabla \phi \mathrm{d}\Omega \tag{6.2.3}$$

6.2.2 Boundary Expression of Shape Sensitivity

The objective function of Joule loss power can be also expressed with the stored electric field energy. When the conductor is assumed to have the dielectric constant ε, the stored electric field energy W_e inside the conductor is written as

$$W_e = \int_{\Omega} \frac{1}{2} \varepsilon \nabla \phi \cdot \nabla \phi \mathrm{d}\Omega \tag{6.2.4}$$

This field energy is inserted into the objective function (6.2.3);

$$P_r = \frac{2\sigma}{\varepsilon} \int_{\Omega} \frac{1}{2} \varepsilon \nabla \phi \cdot \nabla \phi \mathrm{d}\Omega \tag{6.2.5}$$

Comparison of (6.2.4) and (6.2.5) provides the relation;

$$P_r = \frac{2\sigma}{\varepsilon} W_e \tag{6.2.6}$$

Taking the total derivative of both sides in (6.2.6) provides the relation between the sensitivity of the Joule loss power and the sensitivity of the stored field energy:

$$\dot{P}_r = \frac{2\sigma}{\varepsilon} \dot{W}_e \tag{6.2.7}$$

The sensitivity of the stored field energy was already derived as (3.3.33) in Sect. 3.3;

$$\dot{W}_e = \int_{\Gamma^0} \frac{1}{2}\varepsilon \frac{\partial \phi}{\partial n}\frac{\partial \phi}{\partial n} V_n \mathrm{d}\Gamma - \int_{\Gamma^1} \frac{1}{2}\varepsilon \frac{\partial \phi}{\partial t}\frac{\partial \phi}{\partial t} V_n \mathrm{d}\Gamma \tag{6.2.8}$$

This sensitivity formula was derived under the condition of the external current source. But this DC current-carrying conductor was assumed to have the external voltage source on the electrode of Dirichlet boundary Γ^0. Thus, the sign of the sensitivity formula is changed;

$$\dot{W}_e = -\int_{\Gamma^0} \frac{1}{2}\varepsilon \frac{\partial \phi}{\partial n}\frac{\partial \phi}{\partial n} V_n \mathrm{d}\Gamma + \int_{\Gamma^1} \frac{1}{2}\varepsilon \frac{\partial \phi}{\partial t}\frac{\partial \phi}{\partial t} V_n \mathrm{d}\Gamma \tag{6.2.9}$$

It was explained in Sect. 3.3.5 how the sign of the energy sensitivity depends on the condition of source application. This energy sensitivity (6.2.9) is inserted into (6.2.7) to provide the sensitivity of the Joule loss power:

$$\dot{P}_r = -\int_{\Gamma^0} \sigma \frac{\partial \phi}{\partial n}\frac{\partial \phi}{\partial n} V_n \mathrm{d}\Gamma + \int_{\Gamma^1} \sigma \frac{\partial \phi}{\partial t}\frac{\partial \phi}{\partial t} V_n \mathrm{d}\Gamma \tag{6.2.10}$$

This sensitivity formula becomes simpler one under the specific condition of given problem. When the design variable is only the Dirichlet boundary of the electrode, $V_n = 0$ on Γ^1. Hence, the sensitivity formula (6.2.10) becomes

$$\dot{P}_r = -\int_{\Gamma^0} \sigma \frac{\partial \phi}{\partial n}\frac{\partial \phi}{\partial n} V_n \mathrm{d}\Gamma \tag{6.2.11}$$

On the other hand, when the design variable is only the Neumann boundary of the conductor surface, $V_n = 0$ on Γ^0. Hence, the sensitivity formula (6.2.10) becomes

$$\dot{P}_r = \int_{\Gamma^1} \sigma \frac{\partial \phi}{\partial t} \frac{\partial \phi}{\partial t} V_n \mathrm{d}\Gamma \tag{6.2.12}$$

6.2.3 Resistance Sensitivity

Here, the resistance sensitivity is derived in the electric-circuit point of view by using the sensitivity of the Joule loss power obtained in Sect. 6.2.2. For the DC current-carrying conductor in Fig. 6.2, where an external voltage source supplies the Joule power into the conductor through the electrode terminal, the relation between the current and the voltage is modeled as a resistance.

The Joule loss power supplied by the voltage source is written with the resistance;

$$P_r = \frac{V^2}{R} \tag{6.2.13}$$

The shape variation of the DC current-carrying conductor, which causes the variation of the current distribution, results in the variation of the resistance R. With a voltage V given, the sensitivity of the Joule loss power is expressed by taking the total derivative of (6.2.13);

$$\dot{P}_r = -\frac{V^2}{R^2} \dot{R} \tag{6.2.14}$$

where $\dot{R}$ is the sensitivity of the resistance.

The sensitivity of the Joule loss power (6.2.10), which was obtained in the previous section, is

$$\dot{P}_r = -\int_{\Gamma^0} \sigma \frac{\partial \phi}{\partial n} \frac{\partial \phi}{\partial n} V_n \mathrm{d}\Gamma + \int_{\Gamma^1} \sigma \frac{\partial \phi}{\partial t} \frac{\partial \phi}{\partial t} V_n \mathrm{d}\Gamma \tag{6.2.15}$$

By comparison of (6.2.14) and (6.2.15), the sensitivity of the resistance is obtained as

$$\dot{R} = \frac{R^2}{V^2} \left(-\int_{\Gamma^0} \sigma \frac{\partial \phi}{\partial n} \frac{\partial \phi}{\partial n} V_n \mathrm{d}\Gamma + \int_{\Gamma^1} \sigma \frac{\partial \phi}{\partial t} \frac{\partial \phi}{\partial t} V_n \mathrm{d}\Gamma \right) \tag{6.2.16}$$

where the resistance R can be calculated by using (6.2.13) and (6.2.1).

6.2.4 Analytical Examples

When the objective function in DC conductor is defined as the Joule loss power, the sensitivity formula was derived as (6.2.10) in Sect. 6.2.2. One-dimensional examples with the analytic field solutions are taken to show that the sensitivity formula is correct. The objective is to compare the analytical sensitivity results with the ones by the sensitivity formula to ensure that the two results are the same. The analytical examples are a cylindrical conductor and a coaxial electrode filled with conductive material, which are the one-dimensional conductor models in the cylindrical coordinate.

6.2.4.1 Cylindrical Conductor

For an example that can be analytically calculated, consider a cylindrical conductor in Fig. 6.3, where two electrodes are connected to the top and the bottom of a cylindrical conductor of radius R and conductivity σ, and a voltage ϕ_o is applied between the electrodes. An electric field, which is uniformly generated in the z-direction, causes a uniform current in the conductor. The change of the conductor radius results in the change of the resistance. The design objective is to obtain a desired Joule loss power in Ω by moving the lateral surface Γ^1. The design variable is the radius R of the cylindrical conductor, and the design sensitivity with respect to the design variable R is calculated and compared with the result by the sensitivity formula.

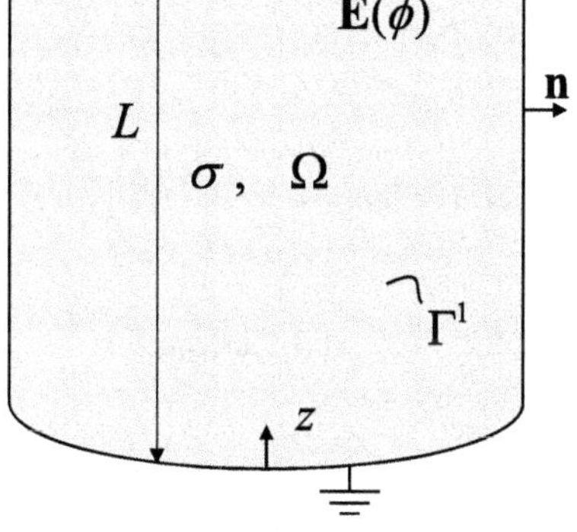

Fig. 6.3 Cylindrical conductor-outer boundary design, Joule loss power

The objective function is the Joule loss power;

$$F = P_r = \int_{\Omega} \sigma E^2(\phi) m_p \mathrm{d}\Omega \tag{6.2.17}$$

where

$$\mathbf{E}(\phi) = \frac{\phi_o}{L}\mathbf{z} \tag{6.2.18}$$

The objective function is rewritten by using the field (6.2.18):

$$F = \sigma\pi R^2 \frac{\phi_o^2}{L} \tag{6.2.19}$$

The analytical sensitivity per unit length is obtained by differentiating the objective function (6.2.19) with respect to the radius R of the conductor;

$$\frac{\mathrm{d}F}{\mathrm{d}R} = 2\pi\sigma R\left(\frac{\phi_o}{L}\right)^2 \tag{6.2.20}$$

This analytical sensitivity result is compared with the result obtained from the sensitivity formula (6.2.12):

$$\dot{G} = \int_{\Gamma^1} \sigma E_t^2(\phi) V_n \mathrm{d}\Gamma \tag{6.2.21}$$

(6.2.18) is inserted into the sensitivity formula (6.2.21);

$$\dot{G} = 2\pi\sigma R V_n \frac{\phi_o^2}{L} \tag{6.2.22}$$

Using $V_n = \frac{\mathrm{d}R}{\mathrm{d}t}$, the design sensitivity is obtained as

$$\frac{\mathrm{d}G}{\mathrm{d}R} = 2\pi\sigma R\left(\frac{\phi_o}{L}\right)^2 \tag{6.2.23}$$

which is identical to the analytical result in (6.2.20).

6.2.4.2 Coaxial Conductor

To take an example that can be analytically calculated, consider a cylindrical conductor in Fig. 6.4, where a coaxial electrode is filled with conductive material of

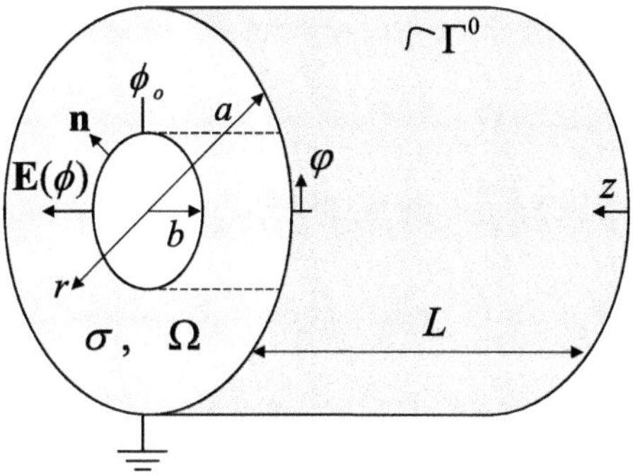

Fig. 6.4 Coaxial conductor-outer boundary design, Joule loss power

conductivity σ and a voltage ϕ_o is applied between the two electrodes. The current flows by the radial electric field $\mathbf{E}(\phi)$. The variation in the radius a of the outer electrode changes the resistance R. The objective is to obtain a desired Joule loss power V^2/R in Ω by moving the outer electrode Γ^0 . The design variable is the radius a of the outer electrode, and the design sensitivity with respect to the design variable a is calculated.

The objective function is defined as

$$F = P_r = \int_{\Omega} \sigma E^2(\phi) m_p \mathrm{d}\Omega \tag{6.2.24}$$

where

$$\mathrm{E}(\phi) = \frac{\phi_o}{r\,\ln(a/b)}\mathbf{r} \tag{6.2.25}$$

The objective function is rewritten by using the field (6.2.25):

$$F = 2\pi\sigma L \frac{\phi_o^2}{\ln(a/b)} \tag{6.2.26}$$

Differentiating the objective function (6.2.26) with respect to the radius a of the outer electrode provides the analytical sensitivity;

$$\frac{\mathrm{d}F}{\mathrm{d}a} = -2\pi\sigma \frac{\phi_o^2}{a(\ln(a/b))^2} \tag{6.2.27}$$

Alternatively, the sensitivity can be calculated by using the sensitivity formula (6.2.11):

$$\dot{G} = -\int_{\Gamma^0} \sigma E_n^2(\phi) V_n \mathrm{d}\Gamma \tag{6.2.28}$$

(6.2.25) is inserted into the sensitivity formula (6.2.28);

$$\dot{G} = -2\pi\sigma L V_n \frac{\phi_o^2}{a(\ln(a/b))^2} \tag{6.2.29}$$

Using $V_n = \frac{\mathrm{d}a}{\mathrm{d}t}$, the design sensitivity per unit length is obtained as

$$\frac{\mathrm{d}G}{\mathrm{d}a} = -2\pi\sigma \frac{\phi_o^2}{a(\ln(a/b))^2} \tag{6.2.30}$$

which is the same as the analytical result in (6.2.27).

6.2.5 *Numerical Examples*

The sensitivity formula (6.2.10) in Sect. 6.2.2 for the DC conductor system is applied to five shape optimization problems of two-dimensional design model, which does not have the analytic field solutions. If the results of the first two examples, which have the known optimal designs, are obtained as the expected optimal designs, the shape optimization using the sensitivity formula is feasible for the design of the DC conductor system. The rest three design problems, which do not have the known optimal designs, are also tested to show that the design method is useful for the design of the DC conductor system. In these two-dimensional examples, the state variable is numerically calculated by the finite element method, and its result is used to evaluate the sensitivity formula. The sensitivity evaluated is used for the optimization algorithm to evolve the shape of the DC conductor. The level set method is used to represent the shape evolution of the design model. In the level set method described in the Chap. 7, the shape evolution is expressed with the parameter t of unit s.

6.2.5.1 Plate Conductor

In Fig. 6.5, a conductor is connected with two electrodes with a potential difference ϕ_o. The electric field and the current density in the conductor are not uniform. The current path in this model is longer than the straight model, which is the optimal design with the same cross-sectional area. If the shape of the conductor becomes straight, its equivalent resistance will decrease and the Joule loss will increase. The design

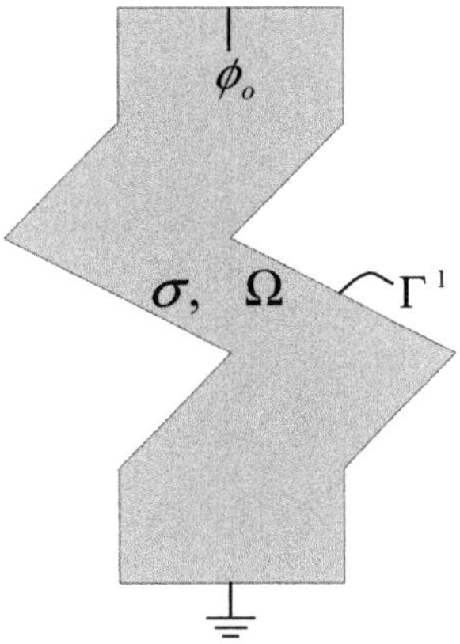

Fig. 6.5 Plate conductor-outer boundary design, Joule loss power

objective is to maximize the Joule loss power in the conductor. The design variable is the shape of the conductor, and the constraint is constant volume of the conductor [7].

The objective function is defined as the Joule loss power;

$$F = P_r = \int_{\Omega} \sigma E^2(\phi) m_p \mathrm{d}\Omega \tag{6.2.31}$$

The shape of the conductor boundary in this design problem Γ^1 is the design variable, which has a constraint of constant conductor volume;

$$\int_{\Omega} \mathrm{d}\Omega = C \tag{6.2.32}$$

where C is a constant. By using the material derivative formula (3.1.22) in Sect. 3.1.2, the material derivative of the constraint (6.2.32) is obtained as

$$\int_{\Gamma} V_n \mathrm{d}\Gamma = 0 \tag{6.2.33}$$

which is a different form of the constraint (6.2.32) expressed with the design velocity field V_n . The sensitivity formula (6.2.12) is the shape sensitivity for this outer boundary design:

$$\dot{G} = \int_{\Gamma^1} \sigma E_t^2(\phi) V_n \mathrm{d}\Gamma \tag{6.2.34}$$

For this maximization problem, the design velocity is taken as

$$V_n = \sigma E_t^2(\phi) \tag{6.2.35}$$

In order that the velocity field satisfies the constraint (6.2.33), the design velocity (6.2.35) is modified by subtracting its average V_{na} to become U_n as

$$U_n = V_n - V_{na} \tag{6.2.36}$$

where

$$V_{na} = \int_{\Gamma^1} \sigma E_t^2(\phi) \mathrm{d}\Gamma / \int_{\Gamma^1} \mathrm{d}\Gamma \tag{6.2.37}$$

The design result is shown in Fig. 6.6, where the shape of the conductor boundary becomes gradually a straight form with the increase of the iteration number as expected. The final design of the straight shape is obtained at the 400 s, when the objective function value converges to the maximum value as in Fig. 6.7.

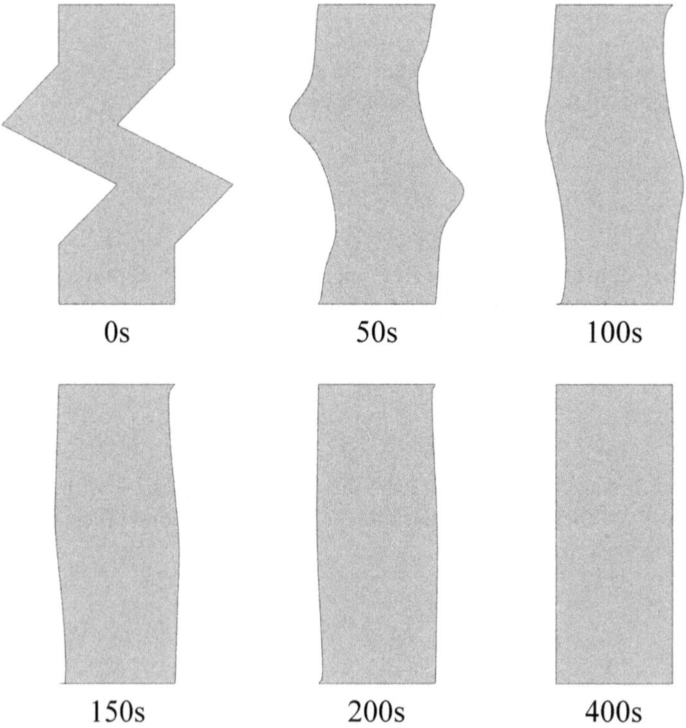

Fig. 6.6 Plate conductor: shape variation

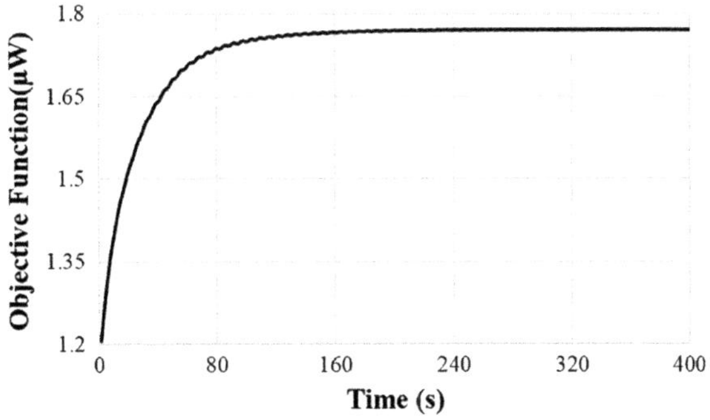

Fig. 6.7 Plate conductor: evolution of objective function

The result of this example shows the feasibility of the shape sensitivity analysis for the outer boundary in the two-dimensional DC conductor system.

6.2.5.2 Coaxial Conductor

As a second example of which the optimal design is known, consider a conductor in Fig. 6.8, where the outer electrode is circular but the inner electrode is squared. The conductor of conductivity σ is between the electrodes with a potential difference ϕ_o. The electric field and the current are formed from the inner electrode to the outer electrode. The design objective is to minimize the Joule loss power by deforming the inner electrode. Under the constraint of the constant conductor volume, it is expected that the shape of the inner electrode is changed to be a circle, which provides the maximum resistance.

The objective function to be minimized is the Joule loss power;

$$F = P_r = \int_{\Omega} \sigma E^2(\phi) m_p \mathrm{d}\Omega \tag{6.2.38}$$

Fig. 6.8 Coaxial conductor-outer boundary design, Joule loss power

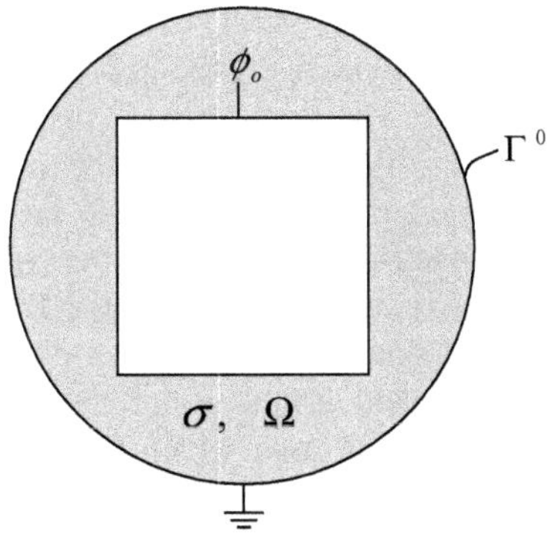

In this design problem, the design variable is the shape of the outer electrode Γ^0 and it is subject to a constraint of constant conductor volume;

$$\int_{\Omega} \mathrm{d}\Omega = C \tag{6.2.39}$$

where C is a constant. By using the material derivative formula (3.1.22) in Sect. 3.1.2, the material derivative of the constraint (6.2.39) is obtained as

$$\int_{\Gamma} V_n \mathrm{d}\Gamma = 0 \tag{6.2.40}$$

The sensitivity formula (6.2.11) is used for the shape sensitivity for this outer boundary design:

$$\dot{G} = -\int_{\Gamma^0} \sigma E_n^2(\phi) V_n \mathrm{d}\Gamma \tag{6.2.41}$$

For this minimization problem, the design velocity is taken as

$$V_n = \sigma E_n^2(\phi) \tag{6.2.42}$$

In order that the velocity field satisfies the constraint (6.2.40), the design velocity (6.2.42) is modified to be U_n by subtracting its average V_{na};

$$U_n = V_n - V_{na} \tag{6.2.43}$$

where

$$V_{na} = \int_{\Gamma^0} \sigma E_n^2(\phi) \mathrm{d}\Gamma \Big/ \int_{\Gamma^0} \mathrm{d}\Gamma \tag{6.2.44}$$

The shape evolution of the design result is shown in Fig. 6.9, where the shape of the inner electrode becomes gradually rounded and finally becomes a circle as expected. The final circular shape is obtained at the 250 s, when the objective function value converges to the minimum value as in Fig. 6.10. The results of this example show that this design method works well for the design problem of the two-dimensional DC conductor system.

The sensitivity formula (6.2.10) in Sect. 6.2.2 is also applied to three shape optimization problems of two-dimensional design model that does not have the analytic field solutions nor the known optimal designs. The design models are conductor junctions, which connect different conductors or distribute the current to other conductors. Under the constraint for constant junction volume, the shape of

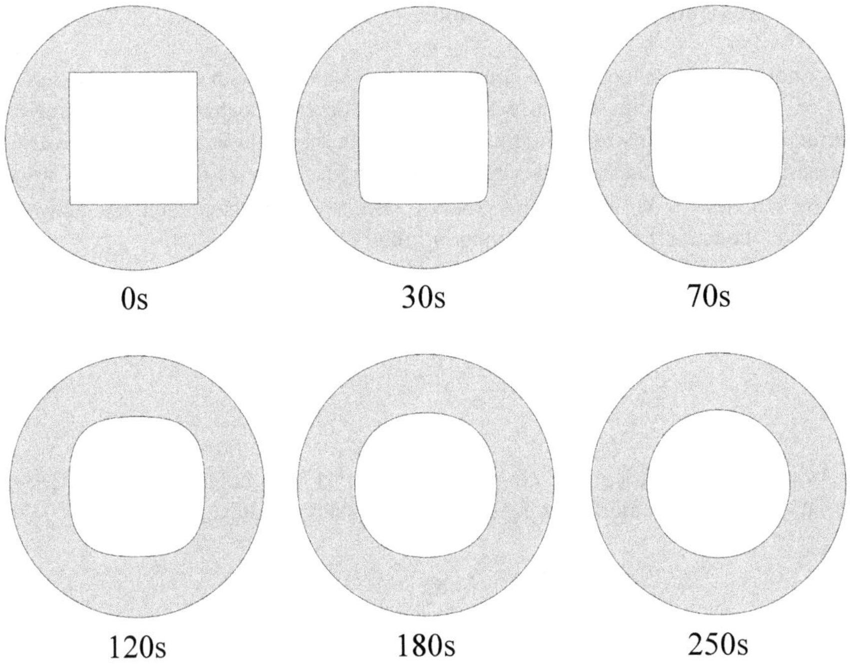

Fig. 6.9 Coaxial conductor: shape variation

Fig. 6.10 Coaxial conductor: evolution of objective function

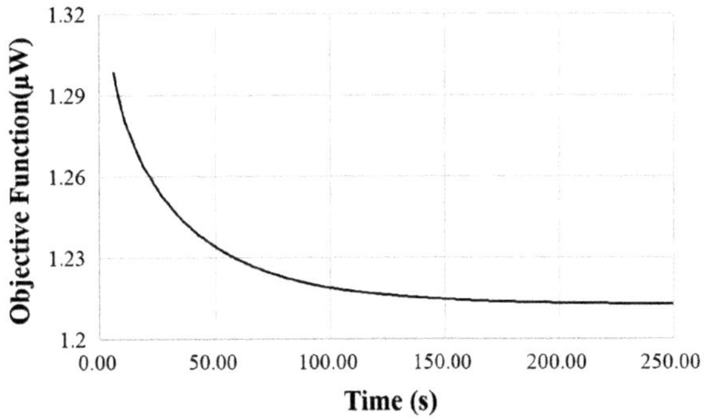

the junction is designed to maximize the Joule loss of the junction. The optimal design of the junction provides a minimized resistance. The first two examples are the shape optimization problem of the junction connecting two different conductors. In each problem, the two conductors are different in configurations and have two electrodes in contact with the junction. The last example is the shape design problem of the junction which distributes an input current to three connected conductors.

6.2.5.3 Junction of Two Conductors

The junction of conductivity σ connects the conductors with different sizes as in Fig. 6.11, where a voltage ϕ_o is applied between the two electrodes representing the surfaces of the conductors. With the voltage given, the Joule loss power increases as the resistance decreases. The design objective is to maximize the Joule loss power of the junction by deforming the junction surface of the Neumann boundary Γ^1 under the constraint of constant junction volume [7].

The objective function to be maximized is the Joule loss, which is the integration of the Joule loss power density in Ω;

$$F = P_r = \int_{\Omega} \sigma E^2(\phi) m_p \mathrm{d}\Omega \tag{6.2.45}$$

In this design problem, the Neumann boundary Γ^1 of the junction is the design variable, which is subject to a constraint of constant conductor volume;

$$\int_{\Omega} \mathrm{d}\Omega = C \tag{6.2.46}$$

where C is a constant. The material derivative of the constraint (6.2.46) is obtained as

$$\int_{\Gamma} V_n \mathrm{d}\Gamma = 0 \tag{6.2.47}$$

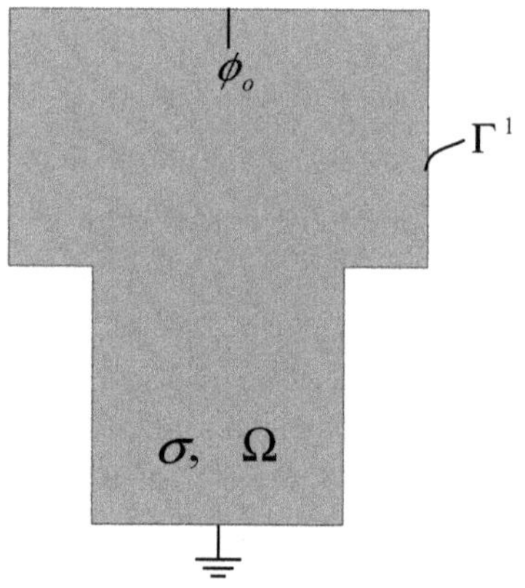

Fig. 6.11 Junction of two conductors-outer boundary design, Joule loss power

The shape sensitivity is calculated on the junction boundary by the formula;

$$\dot{G} = \int_{\Gamma^1} \sigma E_t^2(\phi) V_n \mathrm{d}\Gamma \tag{6.2.48}$$

The design velocity U_n with the constraint is taken as

$$U_n = V_n - V_{na} \tag{6.2.49}$$

where

$$V_n = \sigma E_t^2(\phi) \tag{6.2.50}$$

$$V_{na} = \int_{\Gamma^1} \sigma E_t^2(\phi) \mathrm{d}\Gamma / \int_{\Gamma^1} \mathrm{d}\Gamma \tag{6.2.51}$$

Figure 6.12 shows the shape variation during the optimization. To reduce the resistance, its length becomes shorter or its cross-sectional area wider. With the length fixed, the cross-sectional area of the junction deforms to reduce the resistance. The final design is obtained at the 39 s, when the objective function converges to the maximum value as in Fig. 6.13. The junction shape of the straight line in Fig. 6.14 is a usual design, of which the joule loss power is the dotted line in Fig. 6.13, where the Joule loss power in the optimal design is larger than the usual straight junction.

6.2.5.4 Junction of Two Misaligned Conductors

A junction of conductivity σ connects two misaligned conductors as in Fig. 6.15. A voltage ϕ_o is applied between the two electrodes on the surfaces of the conductors. To minimize the resistance, the Joule loss power of the junction is maximized by deforming the conductor surface of the Neumann boundary Γ^1 under the constraint for constant junction volume.

The objective function to be maximized is the integration of the Joule loss power density in Ω;

$$F = P_r = \int_{\Omega} \sigma E^2(\phi) m_p \mathrm{d}\Omega \tag{6.2.52}$$

In this design problem, the outer boundary Γ^1 of the junction is the design variable, which has a constraint of constant conductor volume;

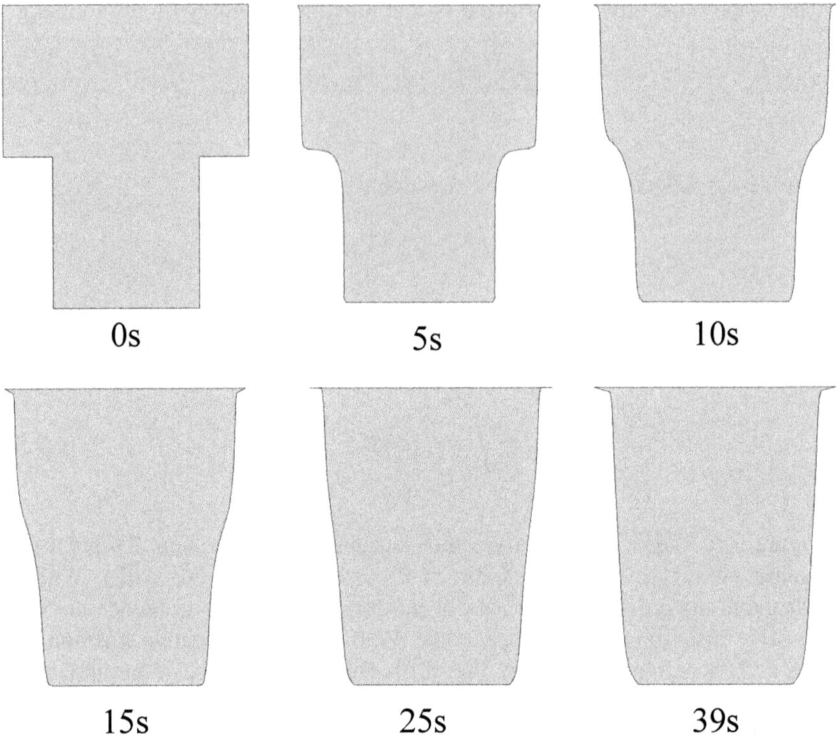

Fig. 6.12 Junction of two conductors: shape variation

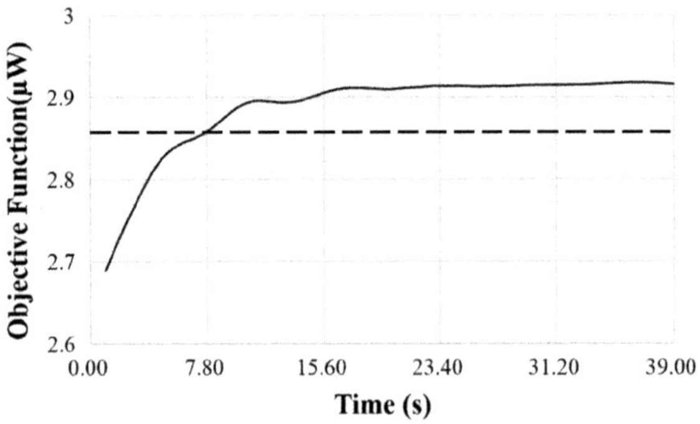

Fig. 6.13 Junction of two conductors: evolution of objective function

Fig. 6.14 Usual design of the junction

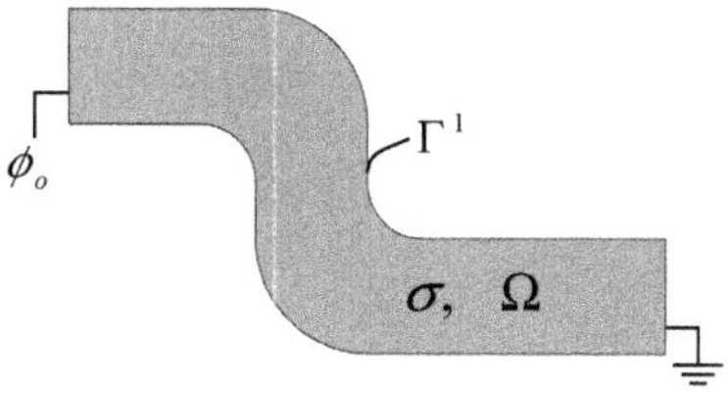

Fig. 6.15 Junction of two misaligned conductors-outer boundary design, Joule loss power

$$\int_{\Omega} \mathrm{d}\Omega = C \qquad (6.2.53)$$

where C is a constant. The material derivative of the constraint (6.2.53) is obtained as

$$\int_{\Gamma} V_n \mathrm{d}\Gamma = 0 \qquad (6.2.54)$$

The shape sensitivity can be calculated on the junction boundary by the formula.

$$\dot{G} = \int_{\Gamma^1} \sigma E_t^2(\phi) V_n \mathrm{d}\Gamma \qquad (6.2.55)$$

The design velocity U_n for the optimization is taken as for the constraint as

$$U_n = V_n - V_{na} \qquad (6.2.56)$$

where

$$V_n = \sigma E_t^2(\phi) \tag{6.2.57}$$

$$V_{na} = \int_{\Gamma^1} \sigma E_t^2(\phi) \mathrm{d}\Gamma / \int_{\Gamma^1} \mathrm{d}\Gamma \tag{6.2.58}$$

Figure 6.16 shows variation during the optimization. As the conductor shape is deformed with the iteration, the length becomes shorter and the cross-sectional area becomes wider. The final design is obtained at the 92 s, when the objective function value converges to the maximum value as in Fig. 6.17.

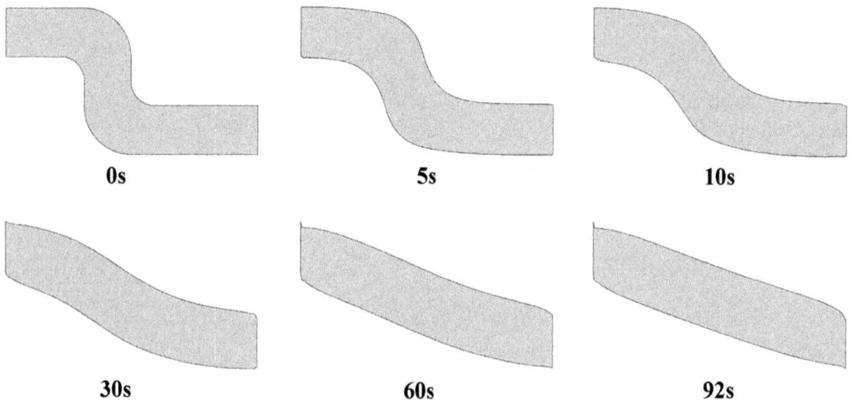

Fig. 6.16 Junction of two misaligned conductors: shape variation

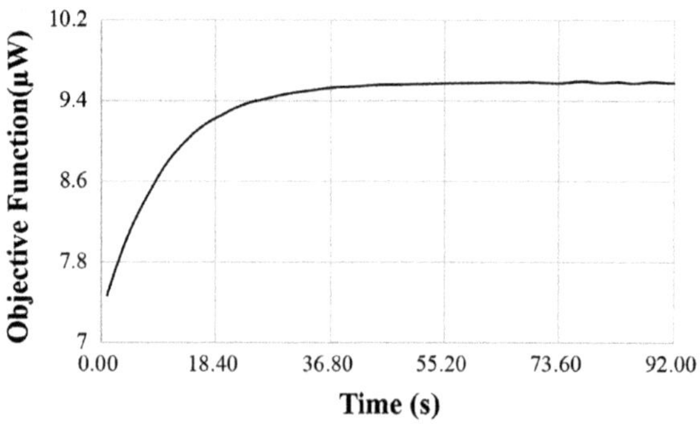

Fig. 6.17 Junction of two misaligned conductors: evolution of objective function

6.2.5.5 Junction in Multiport

As an example whose optimal design is unknown, consider a junction in Fig. 6.18, where a circular electrode is connected to other three circular electrodes. Through the junction, the input current from the left electrode is distributed to the three electrodes. Since the current path toward *gnd*3 is longer than the path toward *gnd*1, the resistance of the former is larger than the one of the latter. The design objective is to equalize and reduce the resistances between the input electrode and the three electrodes by deforming the shape of the junction under the constraint for constant junction volume.

The Joule loss power for each path is different due to the different resistances. If all three resistances are equal, the Joule loss power dissipated by each current path is equal, and vice versa. Thus, the objective function to be minimized is defined as the square of the difference between the Joule loss power of each path and a given target value. The target value is determined by using the analysis result of the initial model. To equalize and reduce the resistances, the target value should be greater than the largest of the three Joule loss powers in the initial model. If the objective function converges to zero, the three resistances become the same. Thus, the objective function is defined as

$$F = (F_{e1} - P_c)^2 + (F_{e2} - P_c)^2 + (F_{e3} - P_c)^2 \tag{6.2.59}$$

where P_c is the target value and F_{ei} is the Joule loss power for each path between ϕ_o and ith ground, and each Joule loss power F_{ei} is written as

$$F_{ei} = P_{ri} = \int_{\Omega} \sigma E_i^2(\phi) m_p \mathrm{d}\Omega \tag{6.2.60}$$

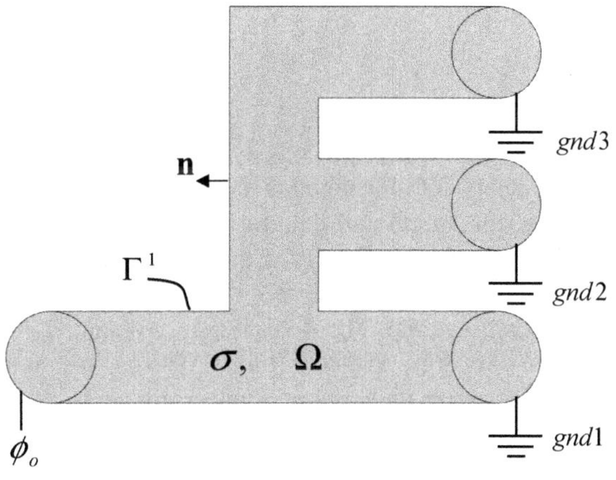

Fig. 6.18 Junction in multiport-outer boundary design

In this design problem, the outer boundary Γ^1 of the junction is the design variable, which has a constraint of constant conductor volume;

$$\int_{\Omega} \mathrm{d}\Omega = C \tag{6.2.61}$$

where C is a constant. The material derivative of the constraint (6.2.61) is obtained as

$$\int_{\Gamma} V_n \mathrm{d}\Gamma = 0 \tag{6.2.62}$$

The sensitivity of the objective function (6.2.59) is obtained as

$$\dot{F} = 2\dot{F}_{e1}(F_{e1} - P_c) + 2\dot{F}_{e2}(F_{e2} - P_c) + 2\dot{F}_{e3}(F_{e3} - P_c) \tag{6.2.63}$$

where the values in the parentheses are taken as coefficients of $\dot{F}_{ei}$ that are obtained by the sensitivity formula used in the previous examples. Each $\dot{F}_{ei}$ has its own design velocity U_{ni} . If the sign of each design velocity has the opposite sign to the corresponding coefficient, the sensitivity in (6.2.63) is always negative. The design velocity is taken with the constraint as

$$U_n = U_{n1} + U_{n2} + U_{n3} = \sum_{i=1}^{3} U_{ni} = \sum_{i=1}^{3} (V_{ni} - V_{nia}) \tag{6.2.64}$$

where

$$V_{ni} = \sigma_i E_{ti}^2(\phi)\, \sigma_i = -\sigma(F_{ei} - P_c)/|F_{ei} - P_c| \tag{6.2.65}$$

$$V_{nia} = \int_{\Gamma^1} \sigma_i E_{ti}^2(\phi) \mathrm{d}\Gamma / \int_{\Gamma^1} \mathrm{d}\Gamma \tag{6.2.66}$$

Figure 6.19 shows the shape variation of the junction during the optimization. As the shape of the junction is deformed with the increase of the number of iteration, the length and the cross-sectional area of each path become shorter and wider, respectively. Among the three paths, the path toward the third ground is most deformed, and the empty spaces between the grounded electrodes are filled with the conducting material. The final design is obtained at the 120 s, when the objective function converges to 0 as in Fig. 6.20. The variation of the resistances is shown in Fig. 6.21, where the three resistances converge to a same value.

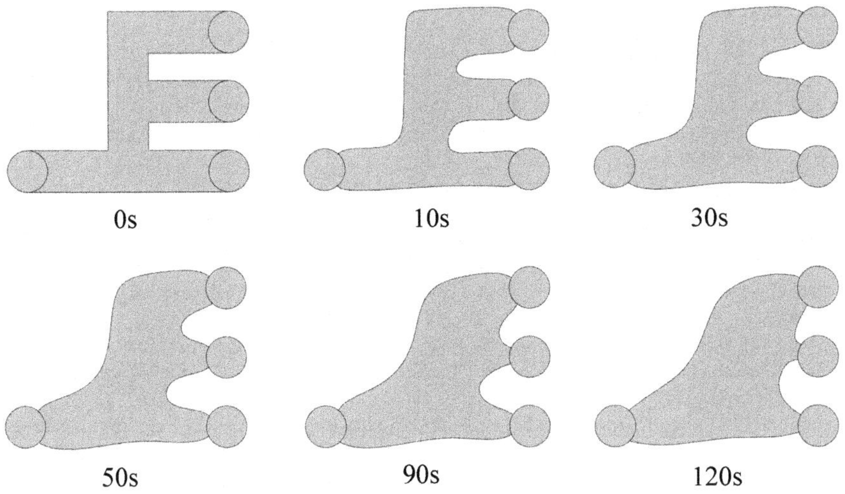

Fig. 6.19 Junction in multiport: shape variation

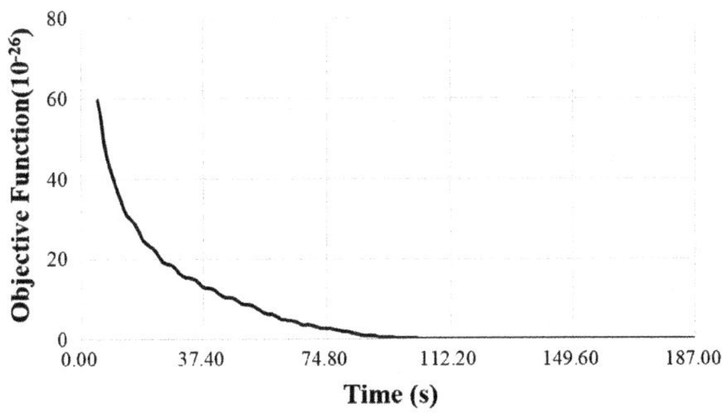

Fig. 6.20 Junction in multiport: evolution of objective function

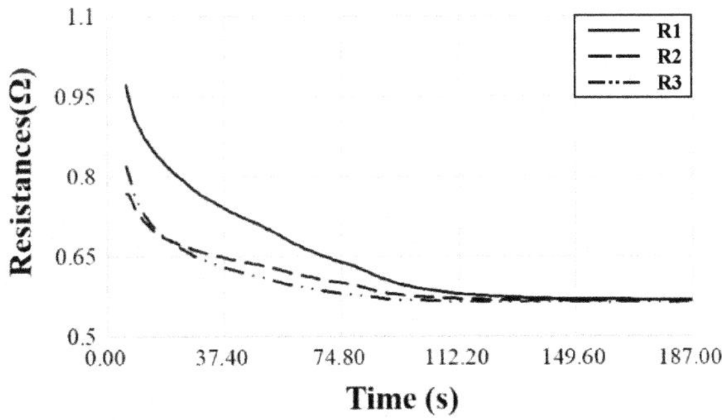

Fig. 6.21 Junction in multiport: resistance evolution for each current path

References

1. Assis, A.K.T., Hernades, J.A.: The Electric Force of a Current: Weber and the Surface Charges of Resistive Conductors Carrying Steady Currents. Apeiron, Montreal (2007)
2. Müller, R.: A semiquantitative treatment of surface charges in DC circuits. Am. J. Phys. **80**, 782–788 (2012)
3. Sommerfeld, A.J.W.: Electrodynamics. Academic Press, New York (1952)
4. Chabay, R.W., Sherwood, B.A.: Matter and interactions. In: Electric and magnetic interactions, vol. 2, John Wiley & Sons, New York (2011)
5. Borage, M., Nagesh, K.V., Bhatia, M.S., Tiwari, S.: Design of LCL-T resonant converter including the effect of transformer winding capacitance. IEEE Trans. Ind. Electron. **56**, 1420–1427 (2009)
6. Arora, N.D., Raol, K.V., Schumann, R., Richardson, L.M.: Modeling and extraction of interconnect capacitances for multilayer VLSI circuits. IEEE Trans. Comput.-Aided DesIntegr. Circuits Syst. **15**, 58–67 (1996)
7. Cheon, W.J., Lee, K.H., Seo, K.S., Park, I.H.: Shape sensitivity analysis and optimization of current-carrying conductor for current distribution control. IEEE Trans. Magn. **54**, 9401004 (2018)

Chapter 7
Level Set Method and Continuum Sensitivity

In order that the sensitivity formulas derived in the previous chapters are used for design optimization of electromagnetic systems, the optimization algorithms and the geometry modeling for evolving shapes are both required. In this chapter, the level set method is introduced for the evolving geometry modeling. While the conventional design parameterization is not only complicated but also dependent on the specific problem, the level set method can represent the design shape with a simple level set function.

The sensitivity formulas and the level set equation, both of which contain the velocity field V_n as a common term, can be coupled. This coupling enables us to transform the optimization process into a solving process of the level set equation in time domain, which does not require any additional optimization algorithm.

In Sect. 7.1, the concept of the level set method is briefly presented and the level set method for the shape variation is introduced. In Sect. 7.2, the level set equation and the sensitivity formula are coupled for the shape optimization. As an example, the continuum sensitivity formula of the electrostatic system is coupled with the level set equation for minimization or maximization problems. In Sect. 7.3, the numerical procedure for sensitivity calculation is presented, and the adaptive level set method and the artificial diffusion technique, which are used for solving the coupled level set equation with the finite element codes, are presented.

7.1 Level Set Method

Level set method, which was originally proposed by the mathematicians Osher and Sethian, has developed to be one of the most successful tools for the expression of evolving geometries, and it is being widely employed in many practical applications such as fluid mechanics, materials science, image processing and computer vision. The advantage of the level set method is that it not only provides easy and efficient schemes for shape variation but also enables handling topological changes

I. H. Park, *Design Sensitivity Analysis and Optimization of Electromagnetic Systems*, Mathematical and Analytical Techniques with Applications to Engineering, https://doi.org/10.1007/978-981-13-0230-5_7

such as merging, splitting, and even disappearing of connected components, which were hardly expressed with classical boundary parameterization [1–7].

Recently, the numerical analysis for the electromagnetic systems is being extended to complex problems such as shape optimization problems, multiphysical coupled problems, and inverse problems. These problems are all involved in geometry variation during the analysis procedure. The geometry change in an analysis system has been a laborious work for its numerical implementation since it requires the boundary parameterization techniques such as spline, Bezier curve, NURBS.

This chapter introduces the level set method for the optimization of the electromagnetic system. The level set method, which can easily express the shape variation with the velocity, matches well with the continuum shape sensitivity, which is explicitly expressed in terms of the design velocity.

7.1.1 Concept of Level Set Method

A surface in the space can be implicitly expressed as

$$\{\mathbf{x}|\phi(\mathbf{x}) = c\} \tag{7.1.1}$$

where c is a constant and $\mathbf{x} = (x,y,z)$. This surface is called the c-level set of ϕ, and the surface at $c = 0$ is the zero level set of ϕ.

For example, consider a function $\phi(x,y) = x^2 + y^2 - 1$. If $z = \phi(x,y)$ and $z = c$ $(c > -1)$, the c-level set is a circle. When $z = 0$, the zero level set of $\phi(x,y)$ is the circle of $x^2 + y^2 = 1$ as shown in Fig. 7.1.

In this figure, the level set function serves to distinguish the region Ω^- and the other region Ω^+ by the boundary of the zero level set. The level set function ϕ has negative value in the region Ω^- and positive value in the region Ω^+. The level set function, which divides an area into two regions of different materials by its sign, represents the interface with the isocontour of the zero level set. The change of the

Fig. 7.1 Level set function $\phi(x,y)$ and its zero level set

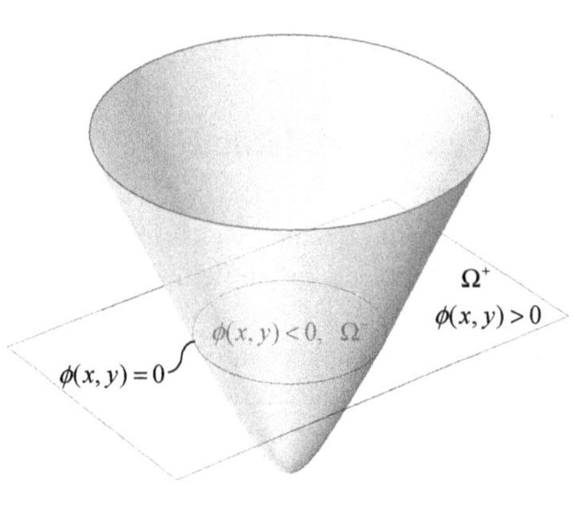

interface can be also expressed with the level set function. When the surface evolves in a temporal variable t, the zero level set is a function of t and it is written as

$$\phi(\mathbf{x}(t), t) = 0 \tag{7.1.2}$$

which is called the level set function. Since this level set function holds at any time regardless of the motion, its total derivative with respect to t is written using the Lagrange formulation as

$$\frac{\mathrm{d}\phi}{\mathrm{d}t} = 0 \tag{7.1.3}$$

Using an Eulerian formulation, it can be rewritten as

$$\frac{\partial \phi}{\partial \mathbf{x}}\frac{\partial \mathbf{x}}{\partial t} + \frac{\partial \phi}{\partial t} = 0 \tag{7.1.4}$$

where $\frac{\partial \phi}{\partial \mathbf{x}}$ is the gradient of ϕ, $\nabla\phi$, and $\frac{\partial \mathbf{x}}{\partial t}$ is the velocity on the surface, $\mathbf{V}$. Thus, (7.1.4) is expressed as

$$\nabla\phi \cdot \mathbf{V} + \frac{\partial \phi}{\partial t} = 0 \tag{7.1.5}$$

This is the level set equation for the variable ϕ, which is a kind of the first-order Hamilton–Jacobi equation. Only the normal component of the velocity $\mathbf{V}$ on the surface contributes to the shape variation; the velocity can be written as

$$\mathbf{V} = V_n\mathbf{n} \tag{7.1.6}$$

where V_n is a scalar and $\mathbf{n}$ is the out normal vector on the surface. The normal vector $\mathbf{n}$ can be expressed on the surface as

$$\mathbf{n} = \nabla\phi/|\nabla\phi| \tag{7.1.7}$$

With (7.1.6) and (7.1.7), the level set equation Eq. (7.1.5) is rewritten as

$$|\nabla\phi|V_n + \frac{\partial \phi}{\partial t} = 0 \tag{7.1.8}$$

This equation is an Eulerian formulation for capturing the interface with the implicit function ϕ. It can easily express the variation of the surface with the velocity field $\mathbf{V}$. Only if the velocity $\mathbf{V}$ is given on the surface of the zero level set, the evolution of the surface is determined. It means that with the given velocity $\mathbf{V}$, (7.1.8) is solved to provide a changed ϕ in the space, from which the zero level set of $\{\mathbf{x}|\phi(\mathbf{x}) = 0\}$ for the evolved surface is obtained. Note that the level set equation (7.1.5) has the same form as the material derivative of the state variable in Sect. 3.1.

7.2 Coupling of Continuum Sensitivity and Level Set Method

Examine the level set equation Eq. (7.1.8) and the sensitivity formulas derived in the Chaps. 3–6. In the level set method, the shape variation is determined by the velocity field V_n. The sensitivity formulas are also evaluated with the velocity field V_n. The level set equation and the sensitivity formulas have the velocity field V_n as a common term; thus, they can be coupled. This coupling makes the optimization procedure simple, providing considerable advantage in the numerical implementation of the optimal shape design [8].

The total derivative, the sensitivity, of the objective function in the electromagnetic system was derived using the material derivative of continuum mechanics and an adjoint variable technique. For example, the sensitivity formula for the outer boundary design in the electrostatic system was derived as (3.2.42) in Sect. 3.2.4:

$$\dot{F} = \int_{\Gamma^0} S(\phi, \lambda)\, V_n \mathrm{d}\Gamma \tag{7.2.1}$$

$$\text{where} \quad S(\phi, \lambda) = \varepsilon \frac{\partial \phi}{\partial n} \frac{\partial \lambda}{\partial n} \tag{7.2.2}$$

and V_n is the normal component of the velocity vector, λ the adjoint variable, ϕ the electric scalar potential, and Γ^0 the design boundary. The sensitivity formula means the variation of the objective function by the velocity field.

For the minimization problem of an objective function, the velocity field can be chosen to be

$$V_n = -S(\phi, \lambda) \tag{7.2.3}$$

By inserting (7.2.3) into (7.2.1), the sensitivity formula is expressed as

$$\dot{F} = \int_{\Gamma^0} -S^2(\phi, \lambda) V_n \mathrm{d}\Gamma \tag{7.2.4}$$

where the sign of the sensitivity is always negative-valued. That is, the shape variation by the velocity field of (7.2.3) leads to decrease the objective function. This velocity field is called gradient descent flow. For the coupling of the sensitivity and the level set equation, the velocity field of (7.2.3) is inserted into the level set equation of (7.1.5), which provides a coupled level set equation:

$$-S(\phi, \lambda)|\nabla \phi| + \frac{\partial \phi}{\partial t} = 0 \tag{7.2.5}$$

The solution ϕ of this equation provides the shape variation to decrease the objective function. As the above procedure is iterated in the optimization process, the objective function will continually decrease with the shape evolving. Finally, when the objective function arrives at a minimum value, the optimized shape is obtained.

The velocity field for the maximization problem is chosen to be

$$V_n = S(\phi, \lambda) \tag{7.2.6}$$

and the sensitivity formula is obtained as

$$\dot{F} = \int\limits_{\Gamma^0} S^2(\phi, \lambda)\, V_n \mathrm{d}\Gamma \tag{7.2.7}$$

where the sign of the sensitivity is always positive-valued. By the same procedure as the above minimization problem, when the objective function arrives at a maximum value, the optimized shape is obtained.

The usual optimization process for the shape design requires two main procedures. One is the optimization algorithm such as steepest descent method, conjugate gradient method, quasi-Newton method. The other is the repeated geometrical renewal for evolving shapes during the optimization process. These two procedures cause a laborious task. In particular, when the shape design problem has a large number of design variables and deals with complex geometries, they become more difficult and complex.

The optimization process using the coupled level set equation Eq. (7.2.5) does not require the above two processes. It is because solving the level set equation of (7.2.5) means obtaining the optimal shape design. The optimization algorithm is replaced by inserting the velocity field V_n into the level set equation Eq. (7.1.8), and the renewed geometries are automatically obtained by the zero level set, which is obtained from the solution of the level set equation. In addition, the level set method enables not only the shape design but also the topology design; it has larger design space than usual shape optimization methods. Hence, it leads to a better design and enhances the possibility of convergence to the global optimal design.

The coupling of the sensitivity and the level set equation transforms the usual iterative optimization process into the solving process of the differential equation, which is the transient analysis in the time domain. The time in the coupled equation, which is a kind of parameter for shape variation, has no physical meaning; so, it is called "pseudo-time" for optimization procedure.

7.3 Numerical Considerations

In this book, the electromagnetic equation and the coupled level set equation are solved using the finite element method. The finite element method has been well developed to be the most general method for diverse partial differential equations; many commercial codes are also available.

7.3.1 *Sensitivity Calculation*

The sensitivity is calculated as in the following manner. First, the state variable for the electromagnetic system is obtained by the finite element analysis and it is used for the evaluation of the objective function. Second, after the source of the adjoint equation is set with the obtained state variable and the objective function, the adjoint variable for the adjoint system is also obtained by the finite element analysis. Third, these two variables are used for calculating the sensitivity formulas, which are expressed with the state variable and the adjoint variable on the design boundary.

7.3.2 *Analysis of Level Set Equation*

With the sensitivity values obtained in the above process, the coupled level set equation is also solved with the finite element program. But the level set equation, which is a first-order partial differential equation, is not solved with the usual finite element code for the second-order differential equation. There are many other numerical methods for solving the equation, but most of them are based on the finite difference method. In order to solve the level set equation with the finite element method, the artificial diffusion method, which is popularly used in hydrodynamic equations for damping and smoothing effect of high-order frequencies, is employed. The artificial diffusion term of second order is added to the level set equation:

$$-S(\phi, \lambda)|\nabla\phi| + \frac{\partial\phi}{\partial t} = \alpha\nabla^2\phi \tag{7.3.1}$$

where α is the coefficient of the artificial diffusion term, which should be moderately well chosen. If α is too small, the equation is not stable to be solved. If α is needlessly large, its solution will be quite different than the original one. By adding artificial diffusion term, the level set equation becomes a second-order equation, which can be easily solved by the usual finite element code [9, 10].

For the finite element analysis for the modified level set equation of (7.3.1), a fixed mesh can be used. The zero level set, which is obtained from the solution ϕ of

(7.3.1), is the interface of two adjacent regions. The interface normally crosses in the middle of the elements in the fixed mesh. This makes it difficult to determine the material property of the elements on the interface. The smeared Heaviside function was employed to allocate smooth-distributed material to the elements near the interface. This technique, although efficient with the fixed mesh, has a serious problem caused by the smeared distribution of material property. The sensitivity, which is evaluated using the state and adjoint variables on the interface, needs the accurate value on the interface. But the smooth-distributed material causes the inaccurate sensitivity for the shape optimization.

This problem can be resolved by the adaptive level set method, where the mesh is adaptively generated to be matched with the interface of the zero level set. As a result, the regions with different materials are clearly distinguished by the interface. In an example of Figs. 7.2 and 7.3, two meshes and two distributions of the material density near the circle interface are compared.

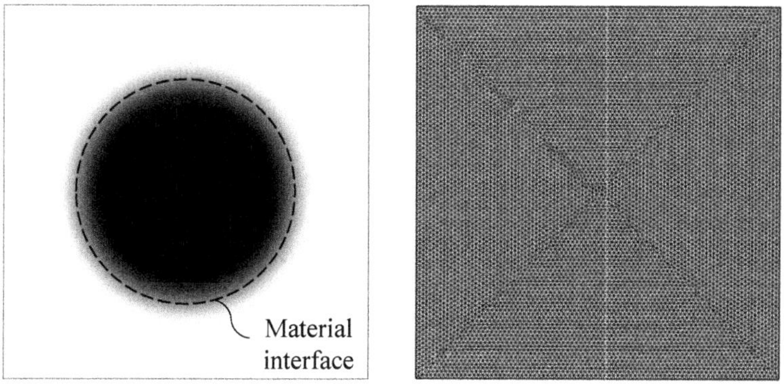

Fig. 7.2 Material and mesh distributions for smeared Heaviside function

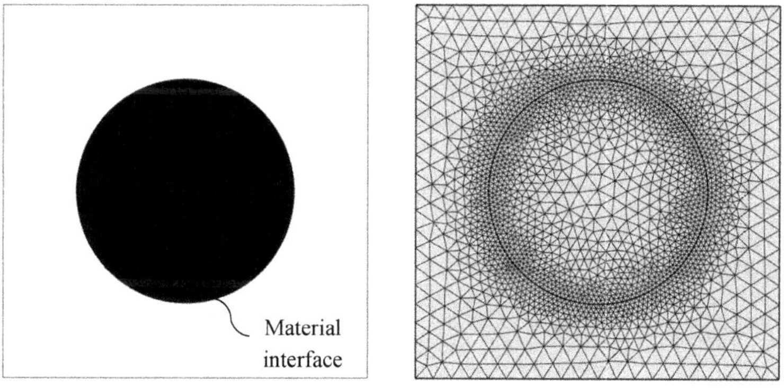

Fig. 7.3 Material and mesh distributions in adaptive level set method

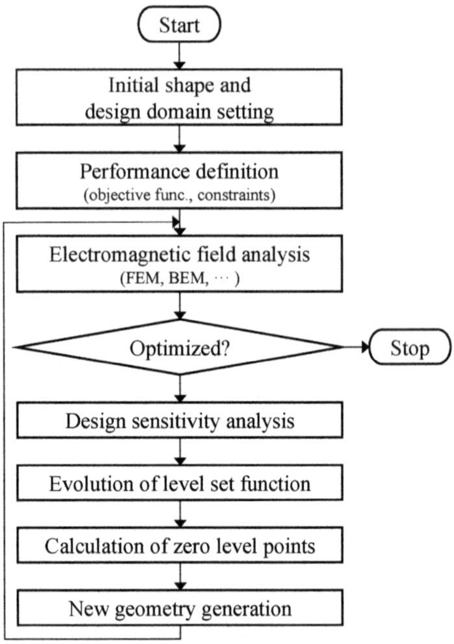

Fig. 7.4 Flowchart of optimization process in adaptive level set method

For all the numerical examples in the Chaps. 3–6 and Chap. 8, this adaptive level set method was used with a commercial finite element code. Figure 7.4 is the flowchart of the optimization process by the adaptive level set method.

References

1. Sethian, J.A.: Level Set Methods and Fast Marching Methods: Evolving Interfaces in Computational Geometry, Fluid Mechanics, Computer Vision, and Materials Science. Cambridge University Press, Cambridge (1999)
2. Osher, S., Fedkiw, R.: Level Set Methods and Dynamic Implicit Surfaces. Springer-Verlag, New York (2003)
3. Dorn, O., Miller, E.L., Rappaport, C.M.: A shape reconstruction method for electromagnetic tomography using adjoint fields and level sets. Inverse Probl. **16**, 1119–1156 (2000)
4. Allaire, G.: Shape Optimization by the Homogenization Method. Springer-Verlag, New York (2002)
5. Allaire, G., Jouve, F., Toader, A.-M.: Structural optimization using sensitivity analysis and a level-set method. J. Comput. Phys. **194**, 363–393 (2004)
6. Bendsøe, M.P., Sigmund, O.: Topology Optimization: Theory. Springer-Verlag, Berlin, Methods and Applications (2004)
7. Wang, Y., Luo, Z., Kang, Z., Zhang, N.: A multi-material level set-based topology and shape optimization method. Comput. Methods Appl. Mech. Engrg. **283**, 1570–1586 (2015)

8. Liu, Z., Korvink, J.G., Huang, R.: Structure topology optimization: fully coupled level set method via FEMLAB. Struct. Multidiscip. Optim. **29**, 407–417 (2005)
9. Kim, Y.S., Byun, J.K., Park, I.H.: A level set method for shape optimization of electromagnetic systems. IEEE Trans. Magn. **45**, 1466–1469 (2009)
10. Kim, Y.S., Park, I.H.: Topology optimization of rotor in synchronous reluctance motor using level set method and shape design sensitivity. IEEE Trans. Appl. Supercond. **20**, 1093–1096 (2010)

Chapter 8
Hole and Dot Sensitivity for Topology Optimization

The shape optimization, of which the design variable is only on the material boundary, has the design space limited to a given initial topology. It may result in a local minimum convergence of the objective function. The idea of topology optimization is, therefore, required to resolve such a shortcoming of the shape optimization. The concept of hole sensitivity can be used for the topology optimization of the electromagnetic system.

Over last two decades, the topology derivative analysis for the optimal design of mechanical structures has become a rich and fascinating investigation area for theories and numerical techniques. It has been applied to various fields, such as shape and topology optimization, inverse problems, and imaging processing. The topology design optimization began with the homogenization method and the density method [1–5]. But, this topology optimization, which is based on the initial-fixed mesh of discretized system, causes some problems in the shape representation and the convergence of the objective function.

Following this optimization method, the level set method was introduced for the topology and shape design. The level set method is well known for handling topology changes, such as breaking one component into two, merging two components into one, and forming sharp corners. It has successfully contributed to the optimal topology and shape design of electromagnetic systems. The level set method, however, has a drawback that the topology becomes gradually simpler during the design process, which results in a local minimum. Even with an initial design of high complexity in topology, the design regions only merge or disappear, but a new region is not generated [6–11].

In this chapter, the hole and dot sensitivity analyses are presented to resolve this problem. In Sect. 8.1, two hole sensitivity formulas in the dielectric material and the magnetic material are derived in the two-dimensional rectangular coordinates by using a hole sensitivity concept and the continuum sensitivity, respectively. In Sect. 8.2, the two dot sensitivity formulas, which are the dielectric dot sensitivity

I. H. Park, *Design Sensitivity Analysis and Optimization of Electromagnetic Systems*,
Mathematical and Analytical Techniques with Applications to Engineering,
https://doi.org/10.1007/978-981-13-0230-5_8

and the magnetic dot sensitivity in the vacant region, are also derived in the two-dimensional rectangular coordinates by using the dot sensitivity concept and the continuum sensitivity, respectively. At the ends of the two sections, the hole and dot sensitivity formulas are applied to design examples to be validated.

8.1 Hole Sensitivity

The hole sensitivity method has been, recently, presented to settle the problem of the conventional level set method. It can generate holes for complex topology. By using the hole sensitivity evaluated in the design region, the positions for hole generation are chosen in the region [12–15]. The hole generation and deformation are represented with the level set method. This hole sensitivity analysis is applied to the electromagnetic system in this book. To do that successfully, it needs a good understanding of the features and difference of the structural system and the electromagnetic system. In the structural system, a hole generation means that the material of a design domain is substituted with an air hole. But, inside the air hole, the stress/strain field does not exist in the structural analysis. The region of the air hole is excluded from the analysis region, and the surface of the hole is treated as a free boundary of Neumann condition. On the contrary, in the electromagnetic system, when an air hole is generated, the electromagnetic field still exists inside the air hole. The region of the air hole belongs to the analysis region. In the case of the magnetic system, the creation of an air hole means that the permeability of the magnetic material region is replaced with the permeability of the air. That is, while the hole generation in the structural problem means the generation of another outer boundary, the hole generation in the electromagnetic system means the variation of interface between two different materials. Thus, the hole sensitivity of electromagnetic problem should be derived using the shape sensitivity of the interface variation.

The analytical form of the hole sensitivity can be derived by using the continuum shape design sensitivity for the interface problem. The analytic field distribution near the circular hole is also used to derive the hole sensitivity formula in a closed form. The hole sensitivity formula is obtained as a point form, and it is expressed as a simple function of the electric, magnetic, and adjoint fields. In this section, two hole sensitivity formulas are derived for the dielectric material of electrostatic system and the magnetic material of magnetostatic system. They both are derived in the two-dimensional rectangular coordinate system.

Once the hole positions are chosen by using the hole sensitivity information in the design region, the hole shape and its variation are represented with the level set method. A zero-level set is generated for the hole to be superposed to the whole analysis region, and it is expanded or changed using the shape sensitivity formula on the zero-level set. Finally, this hole sensitivity analysis for topology design is applied to test examples to validate its feasibility.

8.1.1 Hole Sensitivity in Dielectric Material

In this section, the concept of hole sensitivity is presented and the hole sensitivity in the dielectric material is derived in the two-dimensional rectangular coordinate by using the continuum sensitivity formula for the electrostatic system.

Figure 8.1 shows a model for hole generation in a dielectric material. In the dielectric material in Fig. 8.1a, consider a small area where the electric field can be assumed to be uniform as in Fig. 8.1b. A hole is generated in the small area as in Fig. 8.1c. As the hole generation divides the material region Ω into the hole region Ω_1 of ε_1 and the dielectric region Ω_2 of ε_2, it perturbs the electric field. The perturbed electric field is shown in Fig. 8.1d, where $\mathbf{E}_h$ is the electric field inside the hole. In Fig. 8.1e, the center of the hole is at $\mathbf{x}$, and the hole radius is r_h, and $\mathbf{n}$ and $\mathbf{V}_h$ are the outward unit normal vector and the outward normal velocity vector on the interface γ_h, respectively.

The objective function for the hole sensitivity can be any function of the electric potential ϕ and the electric field $\mathbf{E}$, or it can be the system energy as in Sects. 3.4 and 3.5. The sensitivity formula for the dielectric interface in the electrostatic system was obtained as (3.4.69) in Sect. 3.4.4;

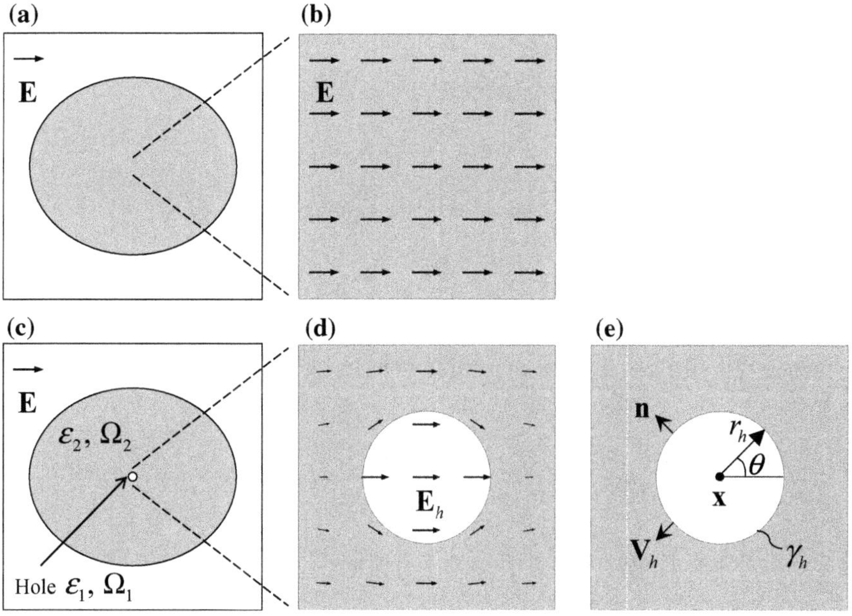

Fig. 8.1 Hole generation in dielectric material region. **a** Electrostatic system before hole generation, **b** uniform field in small area, **c** hole in dielectric material, **d** field near hole, **e** specification

$$\dot{F} = \int_{\gamma} (\varepsilon_2 - \varepsilon_1)\mathbf{E}(\phi_1) \cdot \mathbf{E}(\lambda_2) V_n \mathrm{d}\Gamma \tag{8.1.1}$$

The sensitivity formula for the system energy was also obtained as (3.5.46) in Sect. 3.5.4;

$$\dot{F} = \int_{\gamma} \frac{1}{2}(\varepsilon_2 - \varepsilon_1)\mathbf{E}(\phi_1) \cdot \mathbf{E}(\phi_2) V_n \mathrm{d}\Gamma \tag{8.1.2}$$

The sensitivity formula (8.1.1) is used for the hole sensitivity in the dielectric material. The hole sensitivity for the system energy is obtained by replacing λ in (8.1.1) with $\frac{1}{2}\phi$.

In the hole sensitivity analysis, the objective function depends on the hole position and its size;

$$F = F(\mathbf{x}, r_h) \tag{8.1.3}$$

When the hole expands by the velocity $\mathbf{V}_h$, the variation of the objective function is represented by using the sensitivity formula (8.1.1);

$$\dot{F}(\mathbf{x}, r_h) = \int_{\gamma_h} (\varepsilon_2 - \varepsilon_1)\mathbf{E}_h(\phi_1) \cdot \mathbf{E}_h(\lambda_2)(\mathbf{V}_h \cdot \mathbf{n}) \mathrm{d}\Gamma \tag{8.1.4}$$

The positions $\mathbf{x}_h$ on the hole surface is

$$\mathbf{x}_h = \mathbf{x} + r_h \mathbf{n} \tag{8.1.5}$$

The velocity $\mathbf{V}_h$ by the hole expansion is the time derivative of $\mathbf{x}_h$;

$$\mathbf{V}_h = \frac{\mathrm{d}\mathbf{x}_h}{\mathrm{d}t} = \frac{\mathrm{d}r_h}{\mathrm{d}t}\mathbf{n} \tag{8.1.6}$$

When $\frac{\mathrm{d}r_h}{\mathrm{d}t} = 1$, which is the expansion velocity of the hole radius,

$$\mathbf{V}_h \cdot \mathbf{n} = 1 \tag{8.1.7}$$

Inserting (8.1.7) into (8.1.4) yields

$$\dot{F}(\mathbf{x}, r_h) = \int_{\gamma_h} (\varepsilon_2 - \varepsilon_1)\mathbf{E}_h(\phi_1) \cdot \mathbf{E}_h(\lambda_2) \mathrm{d}\Gamma \tag{8.1.8}$$

which is calculated with the fields on both sides of the hole surface.

For a hole in a uniform electric field, the electric field near the hole is known in an analytical form. In the two-dimensional rectangular coordinates, the hole is a cylinder rod. The electric field $\mathbf{E}_h(\phi_1)$ inside the cylinder region Ω_1 is given as

$$\mathbf{E}_h(\phi_1) = \frac{2\varepsilon_2}{\varepsilon_2 + \varepsilon_1}\mathbf{E}(\phi) \tag{8.1.9}$$

where $\mathbf{E}(\phi)$ is the external uniform electric field [16]. While the electric field outside the cylinder is dependent on the position, the electric field $\mathbf{E}_h(\phi_1)$ inside the cylinder is constant and parallel to the external field $\mathbf{E}(\phi)$. The adjoint field by the adjoint state variable λ has the same form; the adjoint field $\mathbf{E}_h(\lambda_1)$ inside the cylinder is also given as

$$\mathbf{E}_h(\lambda_1) = \frac{2\varepsilon_2}{\varepsilon_2 + \varepsilon_1}\mathbf{E}(\lambda) \tag{8.1.10}$$

where $\mathbf{E}(\lambda)$ is the external uniform adjoint field. The adjoint field $\mathbf{E}_h(\lambda_1)$ inside the cylinder is constant and parallel to the external field $\mathbf{E}(\lambda)$.

To calculate the shape sensitivity (8.1.8), the adjoint field $\mathbf{E}_h(\lambda_2)$ outside the cylinder is needed, which is known in an analytical form but dependent on the position. Thus, it is referred to the inside region Ω_1 where the adjoint field is constant as in (8.1.10). On the hole surface, the adjoint field satisfies the interface conditions:

$$\varepsilon_1 E_{hn}(\lambda_1) = \varepsilon_2 E_{hn}(\lambda_2) \text{ on } \gamma_h \tag{8.1.11}$$

$$E_{ht}(\lambda_1) = E_{ht}(\lambda_2) \text{ on } \gamma_h \tag{8.1.12}$$

where the subscripts n and t mean the normal and tangential components of the field, respectively. By these interface conditions, the shape sensitivity (8.1.8) is rewritten as

$$\dot{F}(\mathbf{x}, r_h) = \int_{\gamma_h} (\varepsilon_2 - \varepsilon_1)\left(\frac{\varepsilon_1}{\varepsilon_2}E_{hn}(\phi_1)E_{hn}(\lambda_1) + E_{ht}(\phi_1)E_{ht}(\lambda_1)\right)\mathrm{d}\Gamma \tag{8.1.13}$$

The electric and adjoint fields in (8.1.13) can be expressed as

$$E_{hn} = E_{hx}\cos\theta + E_{hy}\sin\theta \text{ on } \gamma_h \tag{8.1.14}$$

$$E_{ht} = -E_{hx}\sin\theta + E_{hy}\cos\theta \text{ on } \gamma_h \tag{8.1.15}$$

where the subscripts x and y mean the x and y components of the field, respectively, and θ is shown in Fig. 8.1e. Inserting (8.1.14) and (8.1.15) into (8.1.13) yields

$$\begin{aligned}\dot{F}(\mathbf{x}, r_h) &= \int_0^{2\pi} \frac{1}{\varepsilon_2}(\varepsilon_2 - \varepsilon_1)\left[\varepsilon_1\left(E_{hx}(\phi_1)\cos\theta + E_{hy}(\phi_1)\sin\theta\right)\right.\\ &\quad \times \left(E_{hx}(\lambda_1)\cos\theta + E_{hy}(\lambda_1)\sin\theta\right) + \varepsilon_2(-E_{hx}(\phi_1)\sin\theta \\ &\quad \left. + E_{hy}(\phi_1)\cos\theta)\left(-E_{hx}(\lambda_1)\sin\theta + E_{hy}(\lambda_1)\cos\theta\right)\right] r_h \mathrm{d}\theta \\ &= \frac{\pi r_h}{\varepsilon_2}(\varepsilon_2 - \varepsilon_1)(\varepsilon_2 + \varepsilon_1)\mathbf{E}_h(\phi_1)\cdot\mathbf{E}_h(\lambda_1)\end{aligned} \tag{8.1.16}$$

By using the fields of (8.1.9) and (8.1.10), the shape sensitivity for the hole expansion is obtained as

$$\dot{F}(\mathbf{x}, r_h) = 4\pi r\varepsilon_2\left(\frac{\varepsilon_2 - \varepsilon_1}{\varepsilon_2 + \varepsilon_1}\right)\mathbf{E}(\phi)\cdot\mathbf{E}(\lambda) \tag{8.1.17}$$

Next, the concept of hole sensitivity is introduced. The hole sensitivity is defined by using the variation ΔF of the objective function before and after the hole generation as

$$S(x) \equiv \lim_{r_h \to 0}\frac{\Delta F}{\Delta A} \tag{8.1.18}$$

where $\Delta A = \pi r_h^2$ is the area of the generated hole. The variation ΔF by the hole expansion from 0 to r_h in radius is obtained by integration of the shape sensitivity (8.1.17);

$$\Delta F = F(\mathbf{x}, r_h) - F(\mathbf{x}, 0) = \int_0^{r_h} F(\mathbf{x}, r)\mathrm{d}r \tag{8.1.19}$$

The integration in (8.1.19) is obtained by using (8.1.17);

$$\Delta F = 2\pi r_h^2\varepsilon_2\left(\frac{\varepsilon_2 - \varepsilon_1}{\varepsilon_2 + \varepsilon_1}\right)\mathbf{E}(\phi)\cdot\mathbf{E}(\lambda) \tag{8.1.20}$$

Inserting this into (8.1.18) yields the desired hole sensitivity;

$$S(\mathbf{x}) = 2\varepsilon_2\left(\frac{\varepsilon_2 - \varepsilon_1}{\varepsilon_2 + \varepsilon_1}\right)\mathbf{E}(\phi)\cdot\mathbf{E}(\lambda) \tag{8.1.21}$$

When the dielectric constants of the region Ω_2 and the hole region Ω_1 are $\varepsilon_2 = \varepsilon_0\varepsilon_r$ and $\varepsilon_1 = \varepsilon_0$, respectively, the hole sensitivity is written as

$$S(\mathbf{x}) = 2\varepsilon_0\varepsilon_r\left(\frac{\varepsilon_r - 1}{\varepsilon_r + 1}\right)\mathbf{E}(\phi)\cdot\mathbf{E}(\lambda) \tag{8.1.22}$$

For the energy objective function, $\lambda = \frac{1}{2}\phi$; thus the hole sensitivity is obtained as

$$S(\mathbf{x}) = \varepsilon_0\varepsilon_r\left(\frac{\varepsilon_r - 1}{\varepsilon_r + 1}\right)\mathbf{E}^2(\phi) \tag{8.1.23}$$

As explained in Sect. 3.5.5, for the system excited by voltage source, the sign of the hole sensitivity (8.1.23) is reversed;

$$S(\mathbf{x}) = -\varepsilon_0\varepsilon_r\left(\frac{\varepsilon_r - 1}{\varepsilon_r + 1}\right)\mathbf{E}^2(\phi) \tag{8.1.24}$$

Once the state variable ϕ and the adjoint variable λ are obtained, the hole sensitivity formulas of (8.1.22) and (8.1.24), which are expressed in a point form, are straightforwardly evaluated at any point.

8.1.2 Hole Sensitivity in Magnetic Material

Here, the hole sensitivity in the magnetic material is derived in the two-dimensional rectangular coordinate by using the continuum sensitivity formula for the magnetostatic system. The derivation procedure is almost the same as that of the dielectric case in the previous Sect. 8.1.1.

A model for hole generation in a magnetic material is shown in Fig. 8.2. In the magnetic material in Fig. 8.2a, consider a small area where the magnetic field is assumed to be uniform as in Fig. 8.2b. A hole is generated in the small area as in Fig. 8.2c. As the hole generation divides the material region Ω into the hole Ω_1 of ν_1 and the magnetic region Ω_2 of ν_2, it perturbs the magnetic field. The perturbed magnetic flux near the hole is shown in Fig. 8.2d, where $\mathbf{B}_h$ is the magnetic flux density inside the hole. In Fig. 8.2e, the center of the hole is at $\mathbf{x}$, and the hole radius is r_h, and $\mathbf{n}$ and $\mathbf{V}_h$ are the outward unit normal vector and the outward normal velocity vector on the interface γ_h, respectively.

The objective function for the hole sensitivity can be any function of the magnetic vector potential $\mathbf{A}$ and the magnetic flux density $\mathbf{B}$, or it can be the system energy as in Sects. 4.1 and 4.2. The sensitivity formula for the magnetic interface in the magnetostatic system was obtained as (4.1.88) in Sect. 4.1.5;

$$\dot{F} = \int_{\gamma} (\nu_2 - \nu_1)\mathbf{B}(\mathbf{A}_1)\cdot\mathbf{B}(\boldsymbol{\lambda}_2)V_n \mathrm{d}\Gamma \tag{8.1.25}$$

The sensitivity formula for the system energy was also obtained as (4.2.54) in Sect. 4.2.5;

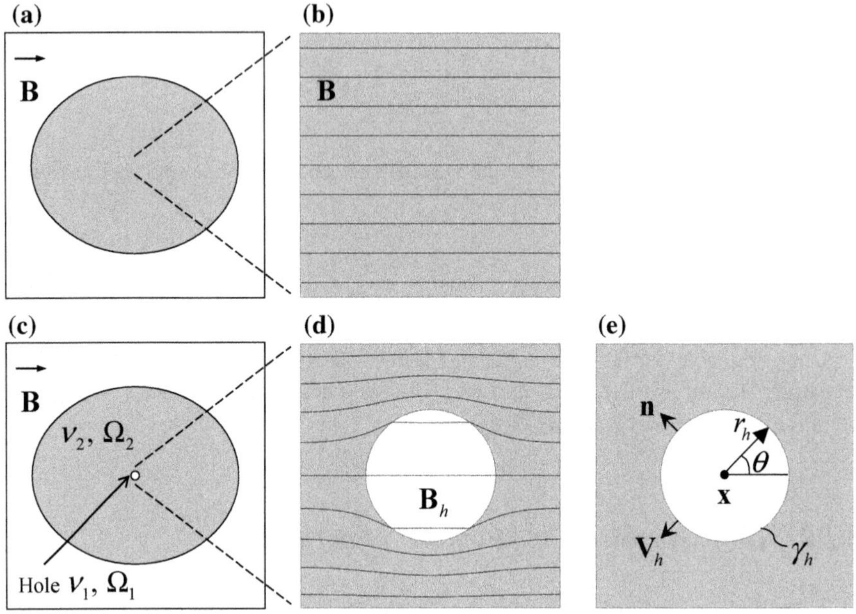

Fig. 8.2 Hole generation in magnetic material. **a** Magnetostatic system before hole generation, **b** uniform flux in small area, **c** hole in magnetic material, **d** field near hole, **e** specification

$$\dot{F} = \int_{\gamma} \frac{1}{2}(\nu_2 - \nu_1)\mathbf{B}(\mathbf{A}_1) \cdot \mathbf{B}(\mathbf{A}_2) V_n \mathrm{d}\Gamma \tag{8.1.26}$$

The sensitivity formula (8.1.25) is used for the hole sensitivity in the magnetic material. The hole sensitivity for the system energy is obtained by replacing $\boldsymbol{\lambda}$ in (8.1.25) with $\mathbf{A}$.

In the hole sensitivity analysis, the objective function depends on the hole position and its size;

$$F = F(\mathbf{x}, r_h) \tag{8.1.27}$$

When the hole expands by the velocity $\mathbf{V}_h$, the variation of the objective function is represented by using the sensitivity formula (8.1.25);

$$\dot{F}(\mathbf{x}, r_h) = \int_{\gamma_h} (\nu_2 - \nu_1)\mathbf{B}_h(\mathbf{A}_1) \cdot \mathbf{B}_h(\boldsymbol{\lambda}_2)(\mathbf{V}_h \cdot \mathbf{n}) \mathrm{d}\Gamma \tag{8.1.28}$$

The velocity vector $\mathbf{V}_h$ is set to be $\mathbf{n}$ as in the previous Sect. 8.1.1;

$$\mathbf{V}_h \cdot \mathbf{n} = 1 \tag{8.1.29}$$

Inserting (8.1.29) into (8.1.28) provides

$$\dot{F}(\mathbf{x}, r_h) = \int_{\gamma_h} (\nu_2 - \nu_1)\mathbf{B}_h(\mathbf{A}_1) \cdot \mathbf{B}_h(\boldsymbol{\lambda}_2)\mathrm{d}\Gamma \tag{8.1.30}$$

To calculate the shape sensitivity (8.1.30), the adjoint field $\mathbf{B}_h(\boldsymbol{\lambda}_2)$ outside the cylinder is needed, which is known in an analytical form but dependent on the position. Thus, it is referred to the inside region Ω_1 where the adjoint field is constant. For a hole in a uniform magnetic field, the magnetic field near the hole is known in an analytical form [16, 17]. The magnetic flux density and the adjoint field inside the hole Ω_1 are known as

$$\mathbf{B}_h(\mathbf{A}_1) = \frac{2\nu_2}{\nu_2 + \nu_1}\mathbf{B}(\mathbf{A}) \tag{8.1.31}$$

$$\mathbf{B}_h(\boldsymbol{\lambda}_1) = \frac{2\nu_2}{\nu_2 + \nu_1}\mathbf{B}(\boldsymbol{\lambda}) \tag{8.1.32}$$

The adjoint field on the hole surface satisfies the interface condition:

$$B_{hn}(\boldsymbol{\lambda}_1) = B_{hn}(\boldsymbol{\lambda}_2) \text{ on } \gamma_h \tag{8.1.33}$$

$$\nu_1 B_{ht}(\boldsymbol{\lambda}_1) = \nu_2 B_{ht}(\boldsymbol{\lambda}_2) \text{ on } \gamma_h \tag{8.1.34}$$

where the subscripts n and t mean the normal and tangential components of the field, respectively. By these interface conditions, the shape sensitivity (8.1.30) is rewritten as

$$\dot{F}(\mathbf{x}, r_h) = \int_{\gamma_h} (\nu_2 - \nu_1)\left(B_{hn}(\mathbf{A}_1)B_{hn}(\boldsymbol{\lambda}_1) + \frac{\nu_1}{\nu_2}B_{ht}(\mathbf{A}_1)B_{ht}(\boldsymbol{\lambda}_1)\right)\mathrm{d}\Gamma \tag{8.1.35}$$

The magnetic and adjoint fields in (8.1.13) can be expressed as

$$B_{hn} = B_{hx}\cos\theta + B_{hy}\sin\theta \text{ on } \gamma_h \tag{8.1.36}$$

$$B_{ht} = -B_{hx}\sin\theta + B_{hy}\cos\theta \text{ on } \gamma_h \tag{8.1.37}$$

where the subscripts x and y mean the x and y components of the field, respectively, and θ is shown in Fig. 8.2e. After inserting (8.1.36) and (8.1.37) into (8.1.35), the integral result of (8.1.35) is obtained as

$$\dot{F}(\mathbf{x}, r_h) = \frac{\pi r_h}{\nu_2}(\nu_2 - \nu_1)(\nu_2 + \nu_1)\mathbf{B}_h(\mathbf{A}_1) \cdot \mathbf{B}_h(\boldsymbol{\lambda}_1) \tag{8.1.38}$$

Inserting (8.1.31) and (8.1.32) into (8.1.38) yields

$$\dot{F}(\mathbf{x}, r_h) = 4\pi r_h \nu_2 \left(\frac{\nu_2 - \nu_1}{\nu_2 + \nu_1}\right)\mathbf{B}(\mathbf{A}) \cdot \mathbf{B}(\boldsymbol{\lambda}) \tag{8.1.39}$$

The hole sensitivity is defined in the same way as in the previous Sect. 8.1.1;

$$S(\mathbf{x}) \equiv \lim_{r_h \to 0} \frac{\Delta F}{\Delta A} \tag{8.1.40}$$

where $\Delta A = \pi r_h^2$ is the area of the generated hole. The variation ΔF by the hole expansion is obtained by integration of the shape sensitivity (8.1.39);

$$\Delta F = \int_0^{r_h} \dot{F}(\mathbf{x}, r)\mathrm{d}r = 2\pi r_h^2 \nu_2 \left(\frac{\nu_2 - \nu_1}{\nu_2 + \nu_1}\right)\mathbf{B}(\mathbf{A}) \cdot \mathbf{B}(\boldsymbol{\lambda}) \tag{8.1.41}$$

Inserting this into (8.1.40) yields the desired hole sensitivity:

$$S(\mathbf{x}) = 2\nu_2 \left(\frac{\nu_2 - \nu_1}{\nu_2 + \nu_1}\right)\mathbf{B}(\mathbf{A}) \cdot \mathbf{B}(\boldsymbol{\lambda}) \tag{8.1.42}$$

For the air Ω_1 and the linear ferromagnetic material Ω_2, the hole sensitivity is written as

$$S(\mathbf{x}) = 2\nu_0 \nu_r \left(\frac{\nu_r - 1}{\nu_r + 1}\right)\mathbf{B}(\mathbf{A}) \cdot \mathbf{B}(\boldsymbol{\lambda}) \tag{8.1.43}$$

where ν_0 is the reluctivity of the air and ν_r is the relative reluctivity of the linear ferromagnetic material.

The hole sensitivity for the energy objective function, is obtained as

$$S(\mathbf{x}) = \nu_0 \nu_r \left(\frac{\nu_r - 1}{\nu_r + 1}\right)\mathbf{B}^2(\mathbf{A}) \tag{8.1.44}$$

Once the state variable $\mathbf{A}$ and the adjoint variable $\boldsymbol{\lambda}$ are obtained, the hole sensitivity formulas of (8.1.42) and (8.1.44), which are expressed in a point form, are straightforwardly evaluated at any point.

8.1.3 Numerical Examples

The topology design method using the hole sensitivity formula, which was derived in Sects. 8.1.1 and 8.1.2, is applied to three design problems of two-dimensional model. These design problems are taken to illustrate that the hole sensitivity formula is well applied to the topology design. The design problems are a MEMS motor, a synchronous reluctance motor and a core-type transformer. While the MEMS motor is the test model for the hole sensitivity in the dielectric material region in the electrostatic system, the synchronous reluctance motor and the core-type transformer are the test models for the hole sensitivity in the magnetic material region in the magnetostatic system. The results of the three examples show that this topology optimization method is useful for the design of electromagnetic system.

In the design examples, the state and the adjoint variables, which are required to evaluate the hole sensitivity formula, are calculated by using the finite element method. After the hole sensitivity is evaluated at all points in the material region, the points with the highest sensitivity value are chosen to generate the holes. In each iteration, not only the holes are generated, but also the material boundary shapes including the hole surfaces are optimized by using the shape sensitivity method. This procedure is iterated until the objective function converges to the final design. These hole generation and shape variation are obtained by using the level set method, where the parameter t of unit s means the amount of shape change.

8.1.3.1 MEMS Motor

A four-pole MEMS motor, which was taken as a design model for the shape optimization in Sect. 3.5.7, is employed again for the topology optimization. Consider the initial design of the MEMS motor in Fig. 8.3, where the dielectric rotor is at the center and the eight electrodes are on the outer surface of the motor. The initial design of the rotor is given as a dielectric cylinder of the simplest topology. The voltage source is imposed on the electrodes to drive the dielectric rotor. As the voltage is switched in the clockwise direction, a rotating electric field is generated in the air gap. If the rotor is properly designed, the reluctance torque is generated in the rotor. The design objective is to find the topology and shape of the rotor that produces a maximum reluctance torque [18].

The reluctance torque is obtained by differentiating the system energy with respect to the rotor position; the objective function to be maximized is defined as the energy difference between the switching positions in Fig. 8.3:

$$F = W_{ea} - W_{eb} \tag{8.1.45}$$

where the subscripts a and b denote the switching positions a and b, respectively. Since this model is excited by the voltage source and the objective function is the system energy difference, the hole sensitivity of (8.1.24) is employed for this example;

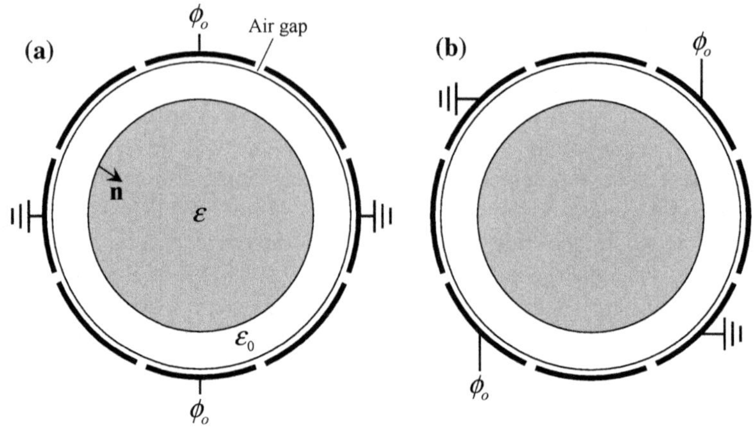

Fig. 8.3 MEMS motor-topology design (hole), torque maximization

$$S(\mathbf{x}) = -\varepsilon_0 \varepsilon_r \left(\frac{\varepsilon_r - 1}{\varepsilon_r + 1} \right) \left(\mathbf{E}_a^2(\phi) - \mathbf{E}_b^2(\phi) \right) \tag{8.1.46}$$

In the design process, the hole sensitivity formula (8.1.46) is calculated all over the dielectric region, and then the points with the highest hole sensitivity value are selected as the positions for hole generation. Depending on the circumstances, 1–8 holes are generated in each iteration step. When the hole sensitivity values become negative at all points, no more holes are generated since it does not contribute to the increase of the objective function. Even after the hole generation is finished, the shape of the dielectric rotor is still optimized until the objective function converges. For the shape optimization, the shape sensitivity formula for the objective function (8.1.45) is used;

$$\dot{G} = -\int_{\gamma} \frac{1}{2} \varepsilon_0 (\varepsilon_r - 1) (\mathbf{E}_a(\phi_1) \cdot \mathbf{E}_a(\phi_2) - \mathbf{E}_b(\phi_1) \cdot \mathbf{E}_b(\phi_2)) V_n \mathrm{d}\Gamma \tag{8.1.47}$$

The design velocity on the interface for this maximization problem is taken as

$$V_n = -\frac{1}{2} \varepsilon_0 (\varepsilon_r - 1) (\mathbf{E}_a(\phi_1) \cdot \mathbf{E}_a(\phi_2) - \mathbf{E}_b(\phi_1) \cdot \mathbf{E}_b(\phi_2)) \tag{8.1.48}$$

Figure 8.4 shows the topology and shape variation of the rotor during the optimization. In the early stage, some holes are generated in the four corners of the dielectric rotor, and then, they are combined to form the lines. After that, other holes are generated in the center region of the rotor, and then, they also become the lines. As the rotor shape touches the boundary of the air gap, the objective function increases rapidly as in Fig. 8.5 and the sliced dielectrics in the four corners

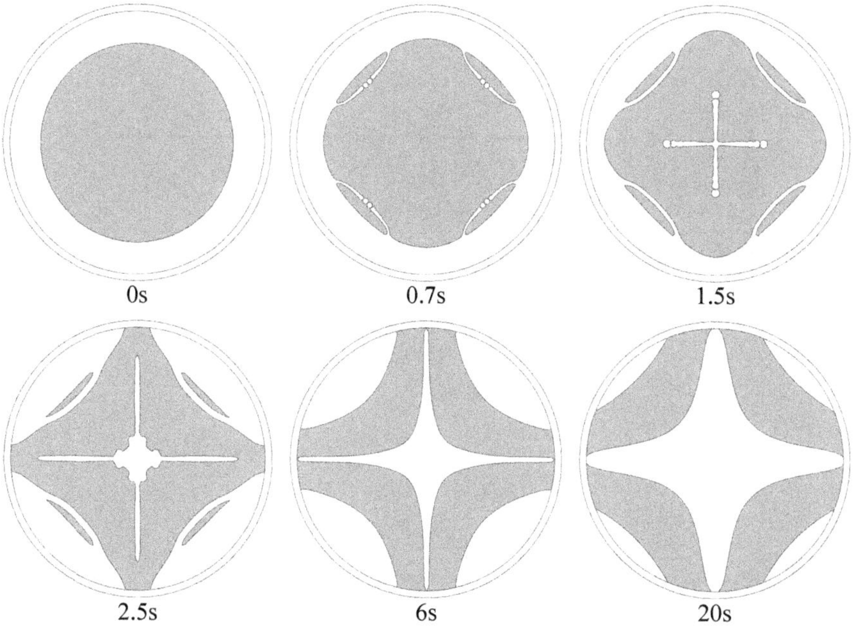

Fig. 8.4 MEMS motor-topology design (hole): variation of topology and shape

gradually disappear. After the hole generation ceases near 6 s, the rotor shape is more deformed until the objective function converges. The final design is obtained as a diamond shape with a cross void inside at 20 s in Fig. 8.4. This final result by the hole sensitivity method is compared with the result by the shape optimization at the bottom of Fig. 8.5, where the two rotor shapes are quite different. In Fig. 8.5, the variations of their objective functions are compared each other. When the final designs are obtained, the objective function by the hole sensitivity is 40% larger than the one by the shape sensitivity. Their resulting torques are also 40% different. This comparison shows that the topology design method using the hole sensitivity provides the much better design than only the shape design method.

8.1.3.2 Synchronous Reluctance Motor

A synchronous reluctance motor is taken as a numerical design model for the topology optimization. It consists of a ferromagnetic rotor and a stator excited by a poly-phase winding. The synchronous reluctance motor contains neither the permanent magnet nor the current conductor in the rotor; its structure is simple and the weight-torque ratio is high. The topology optimization method using the hole sensitivity formula is applied to the rotor design of a synchronous reluctance motor. As an example, consider a six-pole synchronous reluctance motor in Fig. 8.6, where a ferromagnetic cylinder of simple topology is initially located at the center

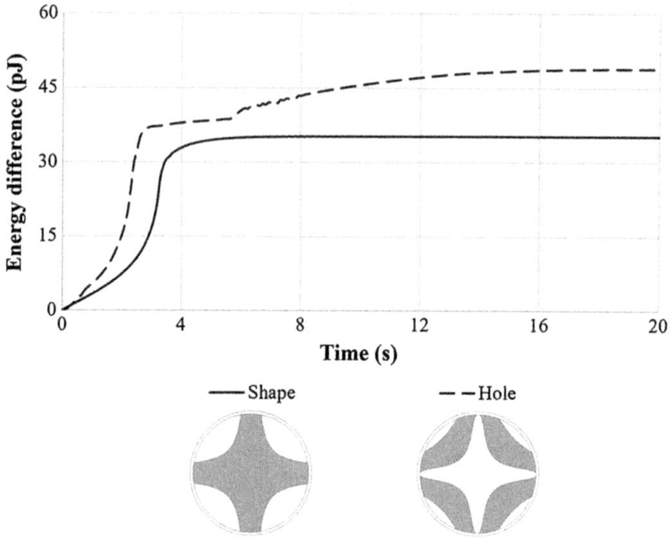

Fig. 8.5 MEMS motor-topology design (hole): evolution of objective function

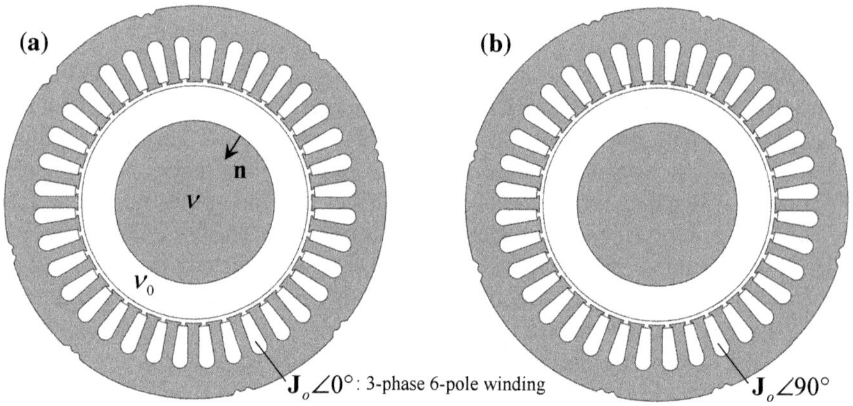

Fig. 8.6 Synchronous reluctance motor-topology design (hole), torque maximization

of the outer stator. The rotating magnetic flux is generated by the three-phase six-pole winding in the stator to drive the ferromagnetic rotor. A properly designed rotor would produce the reluctance torque by the rotating magnetic field. The design objective is to find the topology and the shape of the rotor that maximizes the reluctance torque [19].

The reluctance torque on the ferromagnetic rotor can be expressed as a derivative of the magnetic energy W_m with respect to the angular position θ;

$$T_m = \frac{\mathrm{d}W_m}{\mathrm{d}\theta} \tag{8.1.49}$$

For the six-pole stator, as the energy difference between the two positions of the rotor deviated by 90 electrical degrees increases, the reluctance torque increases. The objective function is defined as the energy difference;

$$F = W_{ma} - W_{mb} \tag{8.1.50}$$

where the subscripts a and b denote the two switching positions in Fig. 8.6, and W_{ma} and W_{mb} are the energies at the positions of a and b, respectively. The hole sensitivity (8.1.44) for the magnetic energy is used for the energy difference;

$$S(\mathbf{x}) = \nu_0 \nu_r \left(\frac{\nu_r - 1}{\nu_r + 1}\right) \left(\mathbf{B}_a^2(\mathbf{A}) - \mathbf{B}_b^2(\mathbf{A})\right) \tag{8.1.51}$$

This hole sensitivity formula is evaluated in the ferromagnetic region. The points of the highest hole sensitivity are then selected as the candidate positions for hole generation. Unless all the hole sensitivity values in the ferromagnetic rotor are negative, the holes continues to be generated. Even after the hole generation is finished, the shape of the ferromagnetic rotor is still deformed until the objective function converges. For the shape optimization of the interface between the ferromagnetic core and the air, the shape sensitivity of the objective function (8.1.50) is used;

$$\dot{G} = \int_\gamma \frac{1}{2} \nu_0 (\nu_r - 1)(\mathbf{B}_a(\mathbf{A}_1) \cdot \mathbf{B}_a(\mathbf{A}_2) - \mathbf{B}_b(\mathbf{A}_1) \cdot \mathbf{B}_b(\mathbf{A}_2)) V_n \mathrm{d}\Gamma \tag{8.1.52}$$

To maximize the objective function, the design velocity on the ferromagnetic surface is taken as

$$V_n = \frac{1}{2} \nu_0 (\nu_r - 1)(\mathbf{B}_a(\mathbf{A}_1) \cdot \mathbf{B}_a(\mathbf{A}_2) - \mathbf{B}_b(\mathbf{A}_1) \cdot \mathbf{B}_b(\mathbf{A}_2)) \tag{8.1.53}$$

Figure 8.7 shows the topology and shape variation of the rotor by the hole sensitivity analysis. In the early stage, the six rotor teeth begin to be formed and grow in the radial direction, while some holes are generated near the center of the rotor. When the tips of the rotor teeth reach to the boundary of the air gap near 10 s, the objective function rapidly increases as in Fig. 8.8. At the same time, the interior vacant region continuously expands. After the hole generation ceases at 30 s, the rotor shape is a little deformed to 50 s, when the final design is obtained. The final rotor for the maximum reluctance torque has the six salient poles with a hexagonal

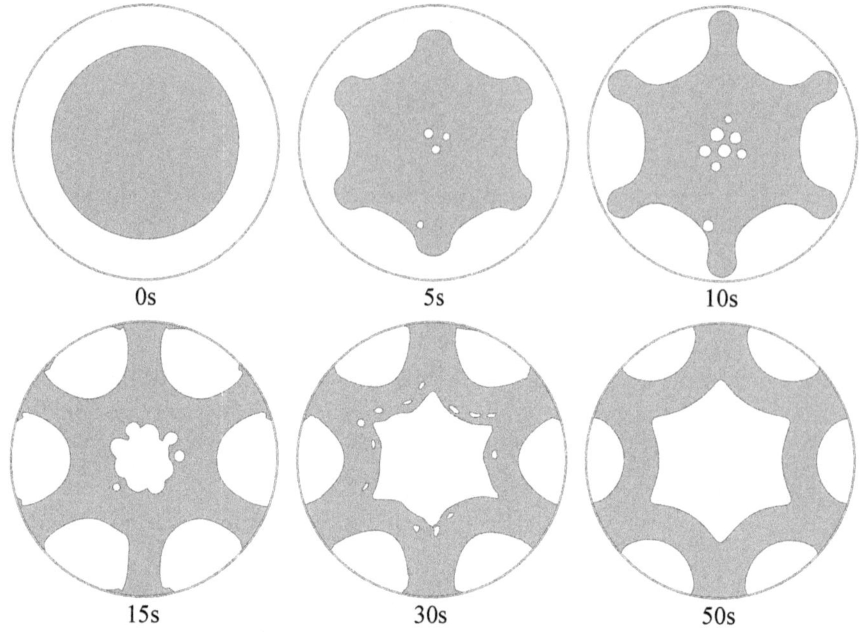

Fig. 8.7 Synchronous reluctance motor-topology design (hole): variation of rotor shape

void inside. Without the hole sensitivity design method, the void in the center of the rotor, which considerably contributes to the increase of the reluctance torque, would not have been generated.

8.1.3.3 Core-Type Transformer

A single-phase transformer is taken as a numerical design model for the topology optimization. It consists of a ferromagnetic core and two coil windings. In the magnetic circuit of the transformer, the ferromagnetic core plays the role of maximizing the mutual flux linkage between the two coil windings for efficient electric energy transfer. The rectangular core, which is obtained by the conventional magnetic circuit method, has some flux leakage near the sharp edges. In this design example, the core is designed by the topology design method using the hole sensitivity to minimize the flux leakage for the transformer efficiency. The initial state of a core-type transformer model is given as in Fig. 8.9, where the primary and the secondary coil windings are separated in parallel and the elliptical core of simple topology is filled between the two coil windings. The primary coil winding is fed by the current density $\mathbf{J}_0$, and the objective function is defined in the region $\Omega_p = \Omega_{p1} \cup \Omega_{p2}$, which is the region of the secondary coil winding.

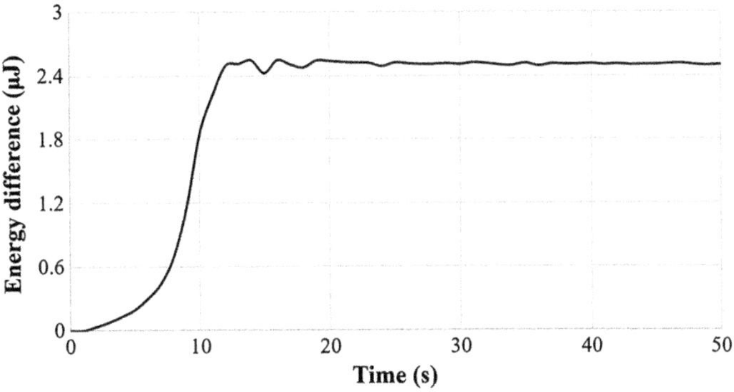

Fig. 8.8 Synchronous reluctance motor, topology design (hole): evolution of objective function

Fig. 8.9 Core-type transformer-topology design (hole), flux linkage maximization

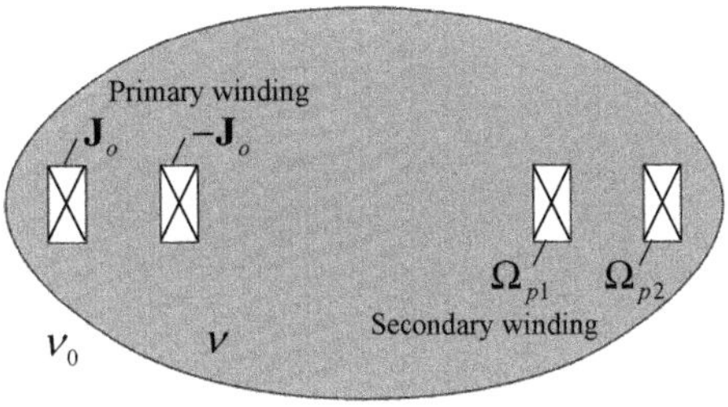

Minimization of the flux leakage results in maximization of the flux linkage; the objective function to be maximized is defined as the flux linkage in the secondary coil winding;

$$F = \frac{1}{A_w} \int_{\Omega} \mathbf{A} m_p \mathrm{d}\Omega \tag{8.1.54}$$

where

$$m_p = \begin{cases} -1 & \text{in } \Omega_{p1} \\ 1 & \text{in } \Omega_{p2} \\ 0 & \text{elsewhere} \end{cases} \tag{8.1.55}$$

and A_w is the area of the secondary coil winding. The variational adjoint equation for (8.1.54) is expressed as

$$a(\boldsymbol{\lambda}, \overline{\boldsymbol{\lambda}}) = \frac{1}{A_w} \int_{\Omega} \overline{\boldsymbol{\lambda}} m_p \mathrm{d}\Omega \quad \forall \overline{\boldsymbol{\lambda}} \in \Phi \tag{8.1.56}$$

The hole sensitivity (8.1.44) is used for the topology design by the hole sensitivity method;

$$S(\mathbf{x}) = 2\nu_0\nu_r\left(\frac{\nu_r - 1}{\nu_r + 1}\right)\mathbf{B}(\mathbf{A})\cdot\mathbf{B}(\boldsymbol{\lambda}) \tag{8.1.57}$$

At each iteration of the optimization process, the hole sensitivity formula (8.1.57) is evaluated in the ferromagnetic core region, and then, the points with the highest hole sensitivity are selected as the position for the hole generation. When the hole sensitivity values are negative at all points in the core region, the hole generation ceases since it does not contribute to the increase of the objective function. Even after the hole generation is finished, the core shape is still optimized until the objective function converges. For the shape optimization, the shape sensitivity formula for the objective function (8.1.54) is used;

$$\dot{G} = \int_{\gamma} \nu_0(\nu_r - 1)\mathbf{B}(\mathbf{A}_1)\cdot\mathbf{B}(\boldsymbol{\lambda}_2)V_n \mathrm{d}\Gamma \tag{8.1.58}$$

For this maximization problem of the flux linkage, the design velocity on the core surface is taken as

$$V_n = \nu_0(\nu_r - 1)\mathbf{B}(\mathbf{A}_1)\cdot\mathbf{B}(\boldsymbol{\lambda}_2) \tag{8.1.59}$$

Figure 8.10 shows the topology variation of the ferromagnetic material with the magnetic field distribution. In the initial design, the flux through the secondary coil winding is very small since most of the flux generated by the primary coil leaks through the ferromagnetic material filled between the two coil windings. At the

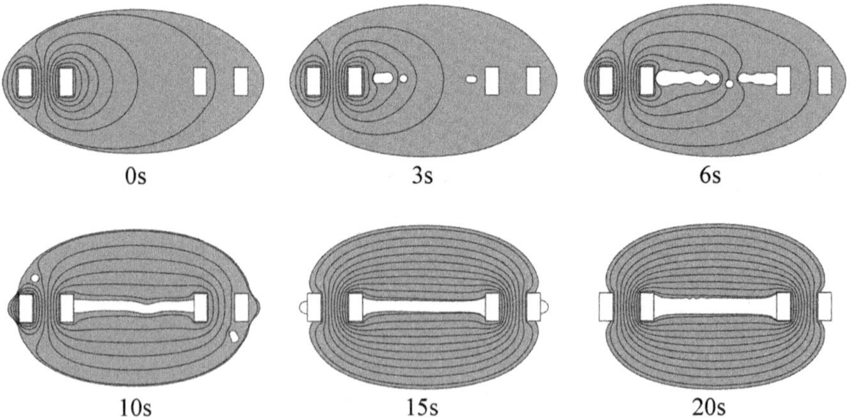

Fig. 8.10 Core-type transformer-topology design (hole): variation of core shape

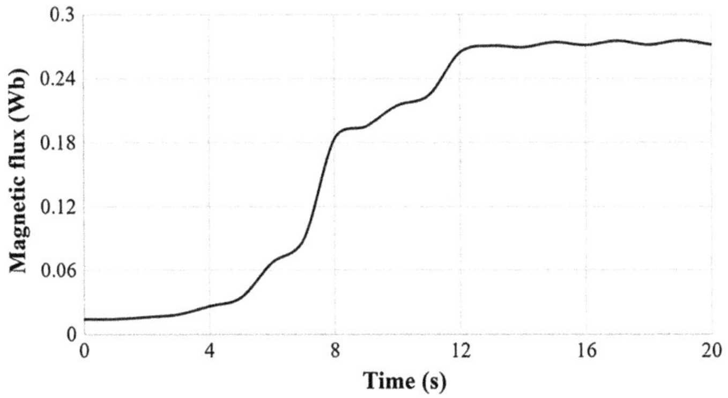

Fig. 8.11 Core-type transformer-topology design (hole): evolution of objective function

beginning of the optimization process, some holes are generated in the region between the two coil windings and then the flux linkage through the secondary winding increases gradually as in Figs. 8.10 and 8.11. When the optimization process comes to 10 s, a band of air gap is formed in the region between the two coil windings and most of the flux generated by the primary coil links the secondary winding. After the hole generation ceases near 15 s, the core shape is changed a little until the objective function converges. The final core shape is obtained at 20 s, when the core surface becomes smooth.

8.2 Dot Sensitivity

The conventional topology design method has, recently, made more progress than the boundary shape optimization. However, the topology of the final design tends to be simpler with iteration, which results in the possibility of being trapped in local minima. The hole sensitivity analysis described in the previous sect. 8.1 has partially settled the problem of the conventional shape optimization: local minimum convergence by simpler topology.

Nevertheless, the hole sensitivity analysis has still two problems. First, the hole sensitivity, although it can create the air hole in the material region, cannot generate a new material in a vacant region. As a result, the final design may converge to a local minimum. Second, it requires the designer to set up an initial topology design, which needs designer's careful effort and intuition.

A dot sensitivity analysis is presented to resolve these problems of the hole sensitivity analysis. The concept of the dot sensitivity analysis is the opposite of the hole sensitivity analysis. In the hole sensitivity analysis, unnecessary material is removed from a design region; in the dot sensitivity analysis, necessary material is added to a vacant region. The dot sensitivity analysis, which can generate the material dot in a vacant region, does not require the initial design. The dot sensitivity analysis provides a proper topology design in the vacant region to satisfy the problem requirement. Here, the analytical form of the dot sensitivity is derived by using the continuum shape design sensitivity for the interface problem. The analytic field distribution near the circular dot is also used to derive the dot sensitivity formula of a closed form. The positions for the dot generation are chosen by evaluating the dot sensitivity in a vacant region. In this section, two dot sensitivity formulas are derived for the dielectric material of the electrostatic system and the magnetic material of the magnetostatic system. They both are derived in the two-dimensional rectangular coordinate system.

8.2.1 Dot Sensitivity of Dielectric Material

In this section, the concept of dot sensitivity is presented and the dielectric dot sensitivity in the vacant region is derived in the two-dimensional rectangular coordinate by using the continuum sensitivity formula for the electrostatic system.

Figure 8.12 shows a model for a dielectric dot generation in a vacant region. In the vacant region in Fig. 8.12a, consider a small area where the electric field is assumed to be uniform as in Fig. 8.12b. When an dielectric dot is generated in the small area as in Fig. 8.12c, the region Ω is divided into the vacant region Ω_1 of ε_1 and the dielectric dot Ω_2 of ε_2. The generated dot distorts the electric field near the dot as shown in Fig. 8.12d, where $\mathbf{E}_d$ is the electric field inside the dot. In Fig. 8.12e, the center of the dot is at $\mathbf{x}$, and the dot radius is r_d, and $\mathbf{n}$ and $\mathbf{V}_d$ are the inward normal unit vector and the outward normal velocity vector on the interface γ_d, respectively.

The objective function for the dot sensitivity can be any integral of ϕ and $\mathbf{E}$ or the system energy. The sensitivity formula for the dielectric interface in the electrostatic system was obtained as (3.4.69) in Sect. 3.4.4;

$$\dot{F} = \int_{\gamma} (\varepsilon_2 - \varepsilon_1)\mathbf{E}(\phi_1) \cdot \mathbf{E}(\lambda_2) V_n \mathrm{d}\Gamma \tag{8.2.1}$$

The sensitivity formula for the system energy was derived as (3.5.46) in Sect. 3.5.4;

$$\dot{F} = \int_{\gamma} \frac{1}{2}(\varepsilon_2 - \varepsilon_1)\mathbf{E}(\phi_1) \cdot \mathbf{E}(\phi_2) V_n \mathrm{d}\Gamma \tag{8.2.2}$$

The sensitivity formula (8.2.1) is used for the dielectric dot sensitivity. The dot sensitivity for the system energy is obtained by replacing λ in (8.2.1) with $\frac{1}{2}\phi$.

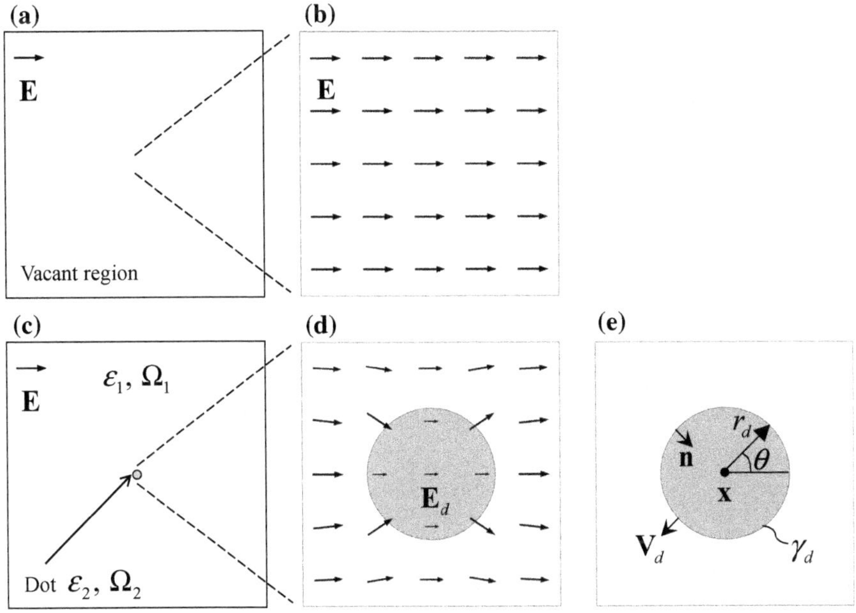

Fig. 8.12 Dielectric dot generation model in vacant region. **a** Vacant region before dot generation, **b** uniform field in small area, **c** dielectric dot in vacant region, **d** field near dot, **e** specification

In the dot sensitivity analysis, the objective function depends on the dot position and its size;

$$F = F(\mathbf{x}, r_d) \tag{8.2.3}$$

The effect of the dot expansion on the objective function is expressed by the shape sensitivity (8.2.1) on the dot surface;

$$\dot{F}(\mathbf{x}, r_d) = \int_{\gamma_d} (\varepsilon_2 - \varepsilon_1)\mathbf{E}_d(\phi_1) \cdot \mathbf{E}_d(\lambda_2)(\mathbf{V}_d \cdot \mathbf{n})\mathrm{d}\Gamma \tag{8.2.4}$$

With the positions $\mathbf{x}_d$ on the dot surface:

$$\mathbf{x}_d = \mathbf{x} - r_d\mathbf{n} \tag{8.2.5}$$

The velocity $\mathbf{V}_d$ by the dot expansion is time derivative of $\mathbf{x}_d$;

$$\mathbf{V}_d = \frac{\mathrm{d}\mathbf{x}_d}{\mathrm{d}t} = -\frac{\mathrm{d}r_d}{\mathrm{d}t}\mathbf{n} \tag{8.2.6}$$

When $\frac{\mathrm{d}r_d}{\mathrm{d}t} = 1$, which is the expansion velocity of the dot radius,

$$\mathbf{V}_d \cdot \mathbf{n} = -1 \tag{8.2.7}$$

Inserting (8.2.7) into (8.2.4) yields

$$\dot{F}(\mathbf{x}, r_d) = -\int_{\gamma_d} (\varepsilon_2 - \varepsilon_1)\mathbf{E}_d(\phi_1) \cdot \mathbf{E}_d(\lambda_2)\mathrm{d}\Gamma \tag{8.2.8}$$

For a dielectric dot in a uniform electric field, the electric field near the dot is known in an analytical form [16]. In the two-dimensional rectangular coordinate, the dot is a cylinder rod. The electric field $\mathbf{E}_d(\phi_2)$ inside the cylinder region Ω_2 is proportional to the external uniform field $\mathbf{E}(\phi)$ as

$$\mathbf{E}_d(\phi_2) = \frac{2\varepsilon_1}{\varepsilon_1 + \varepsilon_2}\mathbf{E}(\phi) \tag{8.2.9}$$

The adjoint field $\mathbf{E}_d(\lambda_2)$ inside the dot also has the same form as (8.2.9).

$$\mathbf{E}_d(\lambda_2) = \frac{2\varepsilon_1}{\varepsilon_1 + \varepsilon_2}\mathbf{E}(\lambda) \tag{8.2.10}$$

To calculate the shape sensitivity (8.2.8), the electric field $\mathbf{E}_d(\phi_1)$ outside the cylinder is needed, which is known in an analytical form but depends on the position. Thus, it is referred to the inside region Ω_2 where the electric field is constant as in (8.2.10). On the dot surface, the electric field satisfies the interface conditions:

$$\varepsilon_1 E_{dn}(\phi_1) = \varepsilon_2 E_{dn}(\phi_2) \text{ on } \gamma_d \tag{8.2.11}$$

$$E_{dt}(\phi_1) = E_{dt}(\phi_2) \text{ on } \gamma_d \tag{8.2.12}$$

where the subscripts n and t mean the normal and tangential components of the field, respectively. By these interface conditions, the shape sensitivity (8.2.8) is rewritten as

$$\dot{F}(\mathbf{x}, r_d) = \int_{\gamma_d} (\varepsilon_1 - \varepsilon_2)\left(\frac{\varepsilon_2}{\varepsilon_1} E_{dn}(\phi_2)E_{dn}(\lambda_2) + E_{dt}(\phi_2)E_{dt}(\lambda_2)\right)\mathrm{d}\Gamma \tag{8.2.13}$$

The electric and adjoint fields in (8.2.13) can be expressed as

$$E_{dn} = -E_{dx}\cos\theta - E_{dy}\sin\theta \text{ on } \gamma_d \tag{8.2.14}$$

$$E_{dt} = E_{dx} \sin\theta - E_{dy} \cos\theta \text{ on } \gamma_d \tag{8.2.15}$$

where the subscripts x and y mean the x and y components of the field, respectively, and θ is shown in Fig. 8.12e. Inserting (8.2.14) and (8.2.15) into (8.2.13) yields

$$\begin{aligned}\dot{F}(\mathbf{x}, r_d) = & \int_0^{2\pi} \frac{1}{\varepsilon_1}(\varepsilon_1 - \varepsilon_2)\left[\varepsilon_2\left(-E_{dx}(\phi_2)\cos\theta - E_{dy}(\phi_2)\sin\theta\right)\right. \\ & \times \left(-E_{dx}(\lambda_2)\cos\theta - E_{dy}(\lambda_2)\sin\theta\right) + \varepsilon_1\left(E_{dx}(\phi_2)\sin\theta\right. \\ & \left.\left.-E_{dy}(\phi_2)\cos\theta\right)\left(E_{dx}(\lambda_2)\sin\theta - E_{dy}(\lambda_2)\cos\theta\right)\right] r_d \mathrm{d}\theta \\ = & \frac{\pi r_d}{\varepsilon_1}(\varepsilon_1 - \varepsilon_2)(\varepsilon_1 + \varepsilon_2)\mathbf{E}_d(\phi_2)\cdot\mathbf{E}_d(\lambda_2)\end{aligned} \tag{8.2.16}$$

By using the fields of (8.2.9) and (8.2.10), the shape sensitivity for the dot expansion is obtained as

$$\dot{F}(\mathbf{x}, r_d) = 4\pi r_d \varepsilon_1 \frac{\varepsilon_1 - \varepsilon_2}{\varepsilon_1 + \varepsilon_2}\mathbf{E}(\phi)\cdot\mathbf{E}(\lambda) \tag{8.2.17}$$

Next, the concept of dot sensitivity is introduced. The dot sensitivity is defined by using the variation ΔF of the objective function before and after the dot generation as

$$S(\mathbf{x}) \equiv \lim_{r_d \to 0}\frac{\Delta F}{\Delta A} \tag{8.2.18}$$

where $\Delta A = \pi r_d^2$ is the area of the generated dot. The variation ΔF by the dot expansion from 0 to r_d in radius is obtained by integration of the shape sensitivity (8.2.17);

$$\Delta F = F(\mathbf{x}, r_d) - F(\mathbf{x}, 0) = \int_0^{r_d} \dot{F}(\mathbf{x}, r)\mathrm{d}r \tag{8.2.19}$$

The integration in (8.2.19) is obtained by using (8.2.17);

$$\Delta F = 2\pi r_d^2 \varepsilon_1 \frac{\varepsilon_1 - \varepsilon_2}{\varepsilon_1 + \varepsilon_2}\mathbf{E}(\phi)\cdot\mathbf{E}(\lambda) \tag{8.2.20}$$

The desired dot sensitivity is derived by inserting (8.2.20) into (8.2.18);

$$S(\mathbf{x}) = 2\varepsilon_1\left(\frac{\varepsilon_1 - \varepsilon_2}{\varepsilon_1 + \varepsilon_2}\right)\mathbf{E}(\phi)\cdot\mathbf{E}(\lambda) \tag{8.2.21}$$

When the dielectric constants of the vacant region Ω_1 and the dot Ω_2 are $\varepsilon_1 = \varepsilon_0$ and $\varepsilon_2 = \varepsilon_0\varepsilon_r$, respectively, the dot sensitivity is rewritten as

$$S(\mathbf{x}) = 2\varepsilon_0 \left(\frac{1 - \varepsilon_r}{1 + \varepsilon_r} \right) \mathbf{E}(\phi) \cdot \mathbf{E}(\lambda) \tag{8.2.22}$$

For the energy objective function, $\lambda = \frac{1}{2}\phi$; thus the dot sensitivity is obtained as

$$S(\mathbf{x}) = \varepsilon_0 \left(\frac{1 - \varepsilon_r}{1 + \varepsilon_r} \right) \mathbf{E}^2(\phi) \tag{8.2.23}$$

For the system excited by voltage source, the sign of the dot sensitivity (8.2.23) is reversed;

$$S(\mathbf{x}) = -\varepsilon_0 \left(\frac{1 - \varepsilon_r}{1 + \varepsilon_r} \right) \mathbf{E}^2(\phi) \tag{8.2.24}$$

Once the state variable ϕ and the adjoint variable λ are obtained, the dot sensitivity formulas of (8.2.22) and (8.2.24), which are expressed in a point form, are easily evaluated at any point.

8.2.2 Dot Sensitivity of Magnetic Material

The magnetic dot sensitivity in a vacant region is derived in the two-dimensional rectangular coordinate by using the continuum sensitivity formula for the magnetostatic system. The derivation procedure is almost the same as that of the dielectric dot case in the previous Sect. 8.2.1.

Figure 8.13 shows a model for a magnetic dot generation in a vacant region. In the vacant region in Fig. 8.13a, consider a small area where the magnetic field is assumed to be uniform as in Fig. 8.13b. When a magnetic dot is generated in the small area as in Fig. 8.13c, the region Ω is divided into the vacant region Ω_1 of ν_1 and the magnetic dot Ω_2 of ν_2. The dot generation distorts the magnetic field near the dot as shown in Fig. 8.12d, where $\mathbf{B}_d$ is the magnetic field inside the dot. In Fig. 8.13e, the center of the dot is at $\mathbf{x}$, and the dot radius is r_d, and $\mathbf{n}$ and $\mathbf{V}_d$ are the inward normal unit vector and the outward normal velocity vector on the interface γ_d, respectively.

The objective function for the dot sensitivity can be any integral of $\mathbf{A}$ and $\mathbf{B}$, or it can be the system energy. The sensitivity formula for the magnetic interface in the magnetostatic system was obtained as (4.1.88) in Sect. 4.1.5;

$$\dot{F} = \int_{\gamma} (\nu_2 - \nu_1) \mathbf{B}(\mathbf{A}_1) \cdot \mathbf{B}(\boldsymbol{\lambda}_2) V_n \mathrm{d}\Gamma \tag{8.2.25}$$

The sensitivity formula for the system energy was derived as (4.2.54) in Sect. 4.2.5;

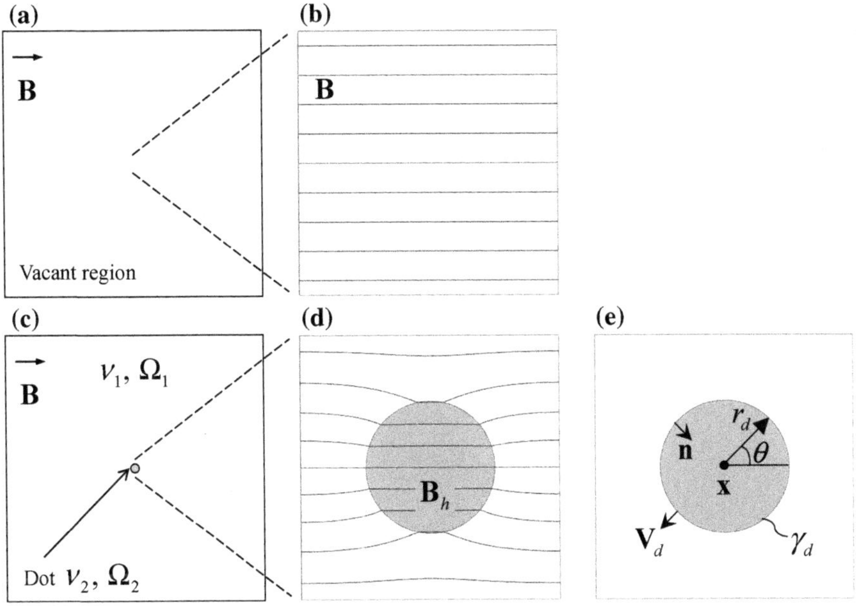

Fig. 8.13 Magnetic dot generation in vacant region. **a** Vacant region before dot generation, **b** uniform flux in small area, **c** magnetic dot in vacant region, **d** field near dot, **e** specification

$$\dot{F} = \int_{\gamma} \frac{1}{2}(\nu_2 - \nu_1)\mathbf{B}(\mathbf{A}_1) \cdot \mathbf{B}(\mathbf{A}_2)V_n \mathrm{d}\Gamma \tag{8.2.26}$$

The sensitivity formula (8.2.25) is used for the magnetic dot sensitivity. The dot sensitivity for the system energy is obtained by replacing $\boldsymbol{\lambda}$ in (8.2.25) with $\mathbf{A}$.

The objective function in the dot sensitivity analysis is a function of the center position $\mathbf{x}$ and radius r_h of the dot;

$$F = F(\mathbf{x}, r_d) \tag{8.2.27}$$

The effect of the dot expansion on the objective function is expressed by the shape sensitivity (8.2.25) on the dot surface;

$$\dot{F}(\mathbf{x}, r_d) = \int_{\gamma} (\nu_2 - \nu_1)\mathbf{B}_d(\mathbf{A}_1) \cdot \mathbf{B}_d(\boldsymbol{\lambda}_2)(\mathbf{V}_n \cdot \mathbf{n})\mathrm{d}\Gamma \tag{8.2.28}$$

The velocity vector $\mathbf{V}_d$ is set to be $-\mathbf{n}$ as in the previous Sect. 8.2.1;

$$\mathbf{V}_d \cdot \mathbf{n} = -1 \tag{8.2.29}$$

Inserting (8.2.29) into (8.2.28) yields

$$\dot{F}(\mathbf{x}, r_d) = -\int_{\gamma_d} (\nu_2 - \nu_1)\mathbf{B}_d(\mathbf{A}_1) \cdot \mathbf{B}_d(\boldsymbol{\lambda}_2)\mathrm{d}\Gamma \tag{8.2.30}$$

For a magnetic dot in a uniform magnetic field, the magnetic field near the dot is known in an analytical form. In the two-dimensional rectangular coordinate, the dot is a cylinder rod [16, 17]. The magnetic and adjoint fields $\mathbf{B}_d(\mathbf{A}_2)$ and $\mathbf{B}_d(\boldsymbol{\lambda}_2)$ inside the cylinder region Ω_2 are proportional to the external uniform field $\mathbf{B}(\mathbf{A})$ and $\mathbf{B}(\boldsymbol{\lambda})$;

$$\mathbf{B}_d(\mathbf{A}_2) = \frac{2\nu_1}{\nu_1 + \nu_2}\mathbf{B}(\mathbf{A}) \tag{8.2.31}$$

$$\mathbf{B}_d(\boldsymbol{\lambda}_2) = \frac{2\nu_1}{\nu_1 + \nu_2}\mathbf{B}(\boldsymbol{\lambda}) \tag{8.2.32}$$

To calculate the shape sensitivity (8.2.30), the magnetic field $\mathbf{B}_d(\mathbf{A}_1)$ outside the cylinder is needed, which is known in an analytical form but dependent on the position. Thus, it is referred to the inside region Ω_2 where the magnetic field is constant as in (8.2.32). On the dot surface, the magnetic field satisfies the interface conditions:

$$B_{dn}(\mathbf{A}_1) = B_{dn}(\mathbf{A}_2) \text{ on } \gamma_d \tag{8.2.33}$$

$$\nu_1 B_{dt}(\mathbf{A}_1) = \nu_2 B_{dt}(\mathbf{A}_2) \text{ on } \gamma_d \tag{8.2.34}$$

where the subscripts n and t mean the normal and tangential components of the field, respectively. By these interface conditions, the shape sensitivity (8.2.30) is rewritten as

$$\dot{F}(\mathbf{x}, r_d) = \int_{\gamma_d} (\nu_1 - \nu_2)\left(B_{dn}(\mathbf{A}_2)B_{dn}(\boldsymbol{\lambda}_2) + \frac{\nu_2}{\nu_1}B_{dt}(\mathbf{A}_2)B_{dt}(\boldsymbol{\lambda}_2)\right)\mathrm{d}\Gamma \tag{8.2.35}$$

The magnetic and adjoint fields in (8.2.25) can be expressed as

$$B_{dn} = -B_{dx}\cos\theta - B_{dy}\sin\theta \text{ on } \gamma_d \tag{8.2.36}$$

$$B_{dt} = B_{dx}\sin\theta - B_{dy}\cos\theta \text{ on } \gamma_d \tag{8.2.37}$$

where the subscripts x and y mean the x and y components of the field, respectively, and θ is shown in Fig. 8.13e. After inserting (8.2.36) and (8.2.37) into (8.2.35), the integral result of (8.2.35) is obtained as

$$\dot{F}(\mathbf{x}, r_d) = \frac{\pi r_d}{v_1}(v_1 - v_2)(v_1 + v_2)\mathbf{B}_d(\mathbf{A}_2) \cdot \mathbf{B}_d(\boldsymbol{\lambda}_2) \tag{8.2.38}$$

Inserting (8.2.31) and (8.2.32) into (8.2.38) yields

$$\dot{F}(\mathbf{x}, r_d) = 4\pi r_d v_1 \left(\frac{v_1 - v_2}{v_1 + v_2}\right)\mathbf{B}(\mathbf{A}) \cdot \mathbf{B}(\boldsymbol{\lambda}) \tag{8.2.39}$$

The dot sensitivity is defined in the same way as in the previous Sect. 8.2.1;

$$S(\mathbf{x}) \equiv \lim_{r_d \to 0} \frac{\Delta F}{\Delta A} \tag{8.2.40}$$

where $\Delta A = \pi r_d^2$ is the area of the generated dot. The variation ΔF by the dot expansion is obtained by integration of the shape sensitivity (8.2.39);

$$\Delta F = \int_0^{r_d} \dot{F}(\mathbf{x}, r)\mathrm{d}r = 2\pi r_d^2 v_1 \left(\frac{v_1 - v_2}{v_1 + v_2}\right)\mathbf{B}(\mathbf{A}) \cdot \mathbf{B}(\boldsymbol{\lambda}) \tag{8.2.41}$$

Inserting this into (8.2.40) yields the desired dot sensitivity:

$$S(\mathbf{x}) = 2v_1 \left(\frac{v_1 - v_2}{v_1 + v_2}\right)\mathbf{B}(\mathbf{A}) \cdot \mathbf{B}(\boldsymbol{\lambda}) \tag{8.2.42}$$

For the vacant region Ω_1 and a linear Ω_2 ferromagnetic material, the dot sensitivity is written as

$$S(\mathbf{x}) = 2v_0 \left(\frac{1 - v_r}{1 + v_r}\right)\mathbf{B}(\mathbf{A}) \cdot \mathbf{B}(\boldsymbol{\lambda}) \tag{8.2.43}$$

where v_0 is the reluctivity of the vacant region and v_r is the relative reluctivity of the linear ferromagnetic material.

The dot sensitivity for the energy objective function is obtained as

$$S(\mathbf{x}) = v_0 \left(\frac{1 - v_r}{1 + v_r}\right)\mathbf{B}^2(\mathbf{A}) \tag{8.2.44}$$

Once the state variable $\mathbf{A}$ and the adjoint variable $\boldsymbol{\lambda}$ are obtained, the dot sensitivity formulas of (8.2.43) and (8.2.44), which are expressed in a point form, are easily calculated at any point.

8.2.3 Numerical Examples

The topology design method using the dot sensitivity formula, which was derived in Sects. 8.2.1 and 8.2.2, is applied to three design problems of two-dimensional model. These design problems are taken to illustrate how the dot sensitivity formula is applied the topology design. The design problems are a MEMS motor, a magnetic shielding and a wireless power transfer. While the MEMS motor is the test model for the dot sensitivity in the electrostatic system, the magnetic shielding and the wireless power transfer are the test models for the dot sensitivity in the magnetostatic system. The results of the three examples show that this topology optimization method is useful for the electromagnetic system design.

In this design method the initial design is a vacant region without any material; thus dot sensitivity analysis does not require the initial design of the material. The state and the adjoint variables, which are required to evaluate the dot sensitivity formula, are calculated by the finite element method. After the dot sensitivity is evaluated at all points in the vacant region, the points with the highest sensitivity value are selected as the candidate positions to generate the dots. In each iteration, not only the dots are generated, but also the material interface including the dot surfaces are optimized by using the shape sensitivity method. This procedure is iterated until the objective function converges to the final design. The dot generation and the shape variation are represented by using the level set method, where the parameter t of unit s means the amount of shape change.

8.2.3.1 MEMS Motor

Although the four-pole MEMS motor was a design model for the topology design using the hole sensitivity in the previous Sect. 8.1.3, it is employed once more for the topology optimization using the dot sensitivity. The initial design model of the MEMS motor is given as in Fig. 8.14, where there is nothing inside the eight electrodes on the outer surface of the motor. The vacant region inside the electrodes is the design domain of this design problem. The source voltage, which is imposed on the electrodes to drive the dielectric rotor, is switched in the clockwise direction to generate the rotating electric field. A proper distribution of the dielectric material in the vacant region for the rotor will produce the reluctance torque on the rotor. The design objective is to find the topology and shape of the rotor that produces a maximum reluctance torque [20].

The reluctance torque is obtained by differentiating the system energy W_e with respect to the angular position θ of the rotor; the objective function to be maximized is defined as the energy difference between the switching positions in Fig. 8.14:

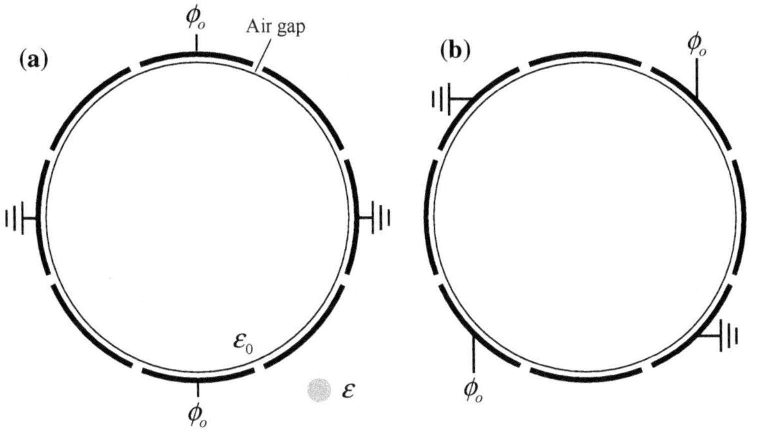

Fig. 8.14 MEMS motor-topology design (dot), torque maximization

$$F = W_{ea} - W_{eb} \tag{8.2.45}$$

where the subscripts a and b denote the two switching positions in Fig. 8.14, and W_{ea} and W_{eb} are the energies at the positions of a and b, respectively. Since this model is excited by the voltage source, the dot sensitivity of (8.2.24) is used.

$$S(\mathbf{x}) = -\varepsilon_0\left(\frac{1-\varepsilon_r}{1+\varepsilon_r}\right)\left(\mathbf{E}_a^2(\phi) - \mathbf{E}_b^2(\phi)\right) \tag{8.2.46}$$

This dot sensitivity is evaluated in the vacant region, and then the dots are generated at the points where the dot sensitivity value is the highest. One to eight dots are generated in each iteration step. When the dot sensitivity values become negative all over the vacant region, the dot generation ceases. Even after the dot generation is finished, the shape of the existing dielectric materials is optimized until the objective function converges. For the shape optimization, the shape sensitivity formula for the objective function (8.2.45) is used;

$$\dot{G} = -\int_\gamma \frac{1}{2}\varepsilon_0(\varepsilon_r - 1)(\mathbf{E}_a(\phi_1)\cdot\mathbf{E}_a(\phi_2) - \mathbf{E}_b(\phi_1)\cdot\mathbf{E}_b(\phi_2))V_n \mathrm{d}\Gamma \tag{8.2.47}$$

For this maximization problem, the design velocity on the interface between the dielectric and the air is taken as

$$V_n = -\frac{1}{2}\varepsilon_0(\varepsilon_r - 1)(\mathbf{E}_a(\phi_1)\cdot\mathbf{E}_a(\phi_2) - \mathbf{E}_b(\phi_1)\cdot\mathbf{E}_b(\phi_2)) \tag{8.2.48}$$

Figure 8.15 shows the topology and shape variation of the rotor by the dot sensitivity analysis. At the beginning of the optimization process, some dots are generated in the region near the eight edges of the electrodes, where the electric field intensity is higher than the other region. And then, they are chained to form bands of the dielectric material. After this band formation, other dots are generated on the surfaces of the electrodes and they are also chained to form other bands. While the two-layer bands are formed, the objective function increases rapidly as in Fig. 8.16. When the second band is formed, the first band becomes thicker gradually. After the dot generation ceases at 3 s, the shape variation of the dielectric material is trivial and the objective function hardly ever changes. The final shape of the dielectric rotor is obtained at 20 s. In the bottom of Fig. 8.16, this topology and shape is compared with the ones by the shape optimization and the hole sensitivity, which were already obtained in Sect. 3.5.7 and Sect.8.1.3. The rotor shapes obtained by the three design methods are topologically quite different. The variations of the objective functions by the three design methods are also compared in Fig. 8.16, where the design by the dot sensitivity method is 49 and 7% better than the ones by the shape optimization and the hole sensitivity method, respectively.

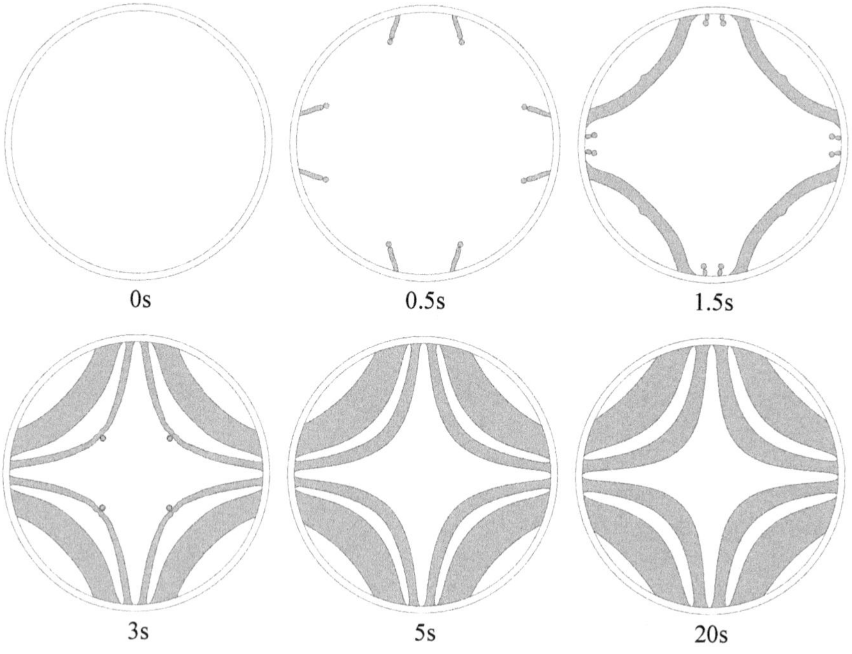

Fig. 8.15 MEMS motor-topology design (dot): variation of rotor topology and shape

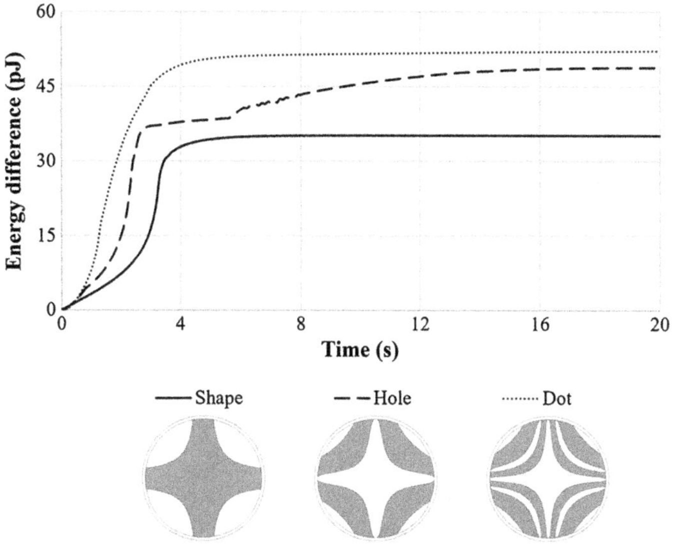

Fig. 8.16 MEMS motor, topology design (dot): evolution of objective function

8.2.3.2 Magnetic Shielding

The magnetic shielding in the low-frequency systems is one of the difficult problems in the electromagnetic design. A careless design may cause the increase of the leakage field rather than its shielding. In this example, the topology design method using the dot sensitivity is applied to a design problem of magnetic material for the magnetic field shielding. The initial state of a magnetic shielding model is shown in Fig. 8.17, where there is nothing except a current source. The vacant region near the currents is the design domain of this design problem. When a current of density $\mathbf{J}_o$ flows in the parallel conductors, the magnetic field produced by the current is distributed in the outside region Ω_p as well as in the design domain. The design objective is to minimize the magnetic field in the outside domain Ω_p by optimally distributing the magnetic material in the design domain [21].

The objective function to be minimized is defined as the magnetic leakage field in the outside domain;

$$F = \int_{\Omega} \mathbf{B}^2(\mathbf{A}) m_p \mathrm{d}\Omega \tag{8.2.49}$$

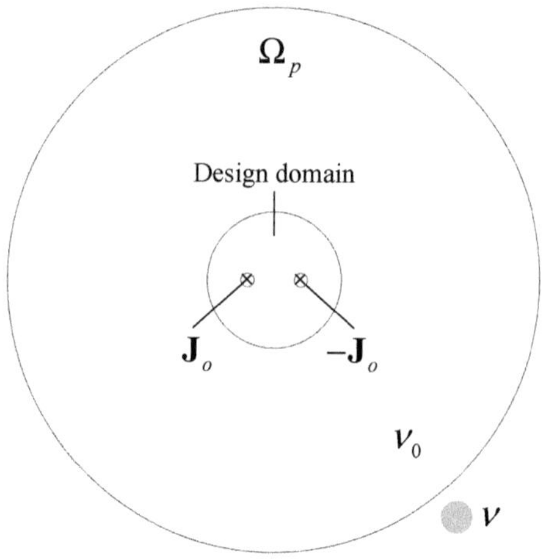

Fig. 8.17 Magnetic shielding-topology design (dot), field minimization in Ω_p

The adjoint equation for this objective function is obtained as

$$a(\boldsymbol{\lambda}, \overline{\boldsymbol{\lambda}}) = \int_{\Omega} 2\mathbf{B}(\mathbf{A}) \cdot \mathbf{B}(\overline{\boldsymbol{\lambda}}) m_p \mathrm{d}\Omega \quad \forall \overline{\boldsymbol{\lambda}} \in \Phi \tag{8.2.50}$$

The dot sensitivity formula for this minimization problem is given as

$$S(\mathbf{x}) = -2\nu_0 \left(\frac{1 - \nu_r}{1 + \nu_r} \right) \mathbf{B}(\mathbf{A}) \cdot \mathbf{B}(\boldsymbol{\lambda}) \tag{8.2.51}$$

In the design process, the dot sensitivity formula (8.2.51) is calculated in the vacant design domain. Then, the points with the highest dot sensitivity value are chosen to be the candidate positions for the dot generation. Unless the dot sensitivity values at all points in the vacant region are negative in each iteration step, the dots are generated. Even after the dot generation, the shape of the ferromagnetic material is optimized with the shape sensitivity. For the shape optimization, the shape sensitivity formula for the objective function (8.2.49) is used;

$$\dot{G} = \int_{\gamma} \nu_0(\nu_r - 1) \mathbf{B}(\mathbf{A}_1) \cdot \mathbf{B}(\boldsymbol{\lambda}_2) V_n \mathrm{d}\Gamma \tag{8.2.52}$$

The design velocity on the ferromagnetic surface for this minimization problem is taken as

$$V_n = -\nu_0(\nu_r - 1) \mathbf{B}(\mathbf{A}_1) \cdot \mathbf{B}(\boldsymbol{\lambda}_2) \tag{8.2.53}$$

The dot sensitivity analysis yields the evolving topology and shape of the ferromagnetic material as shown in Fig. 8.18. The ferromagnetic dots begin to be

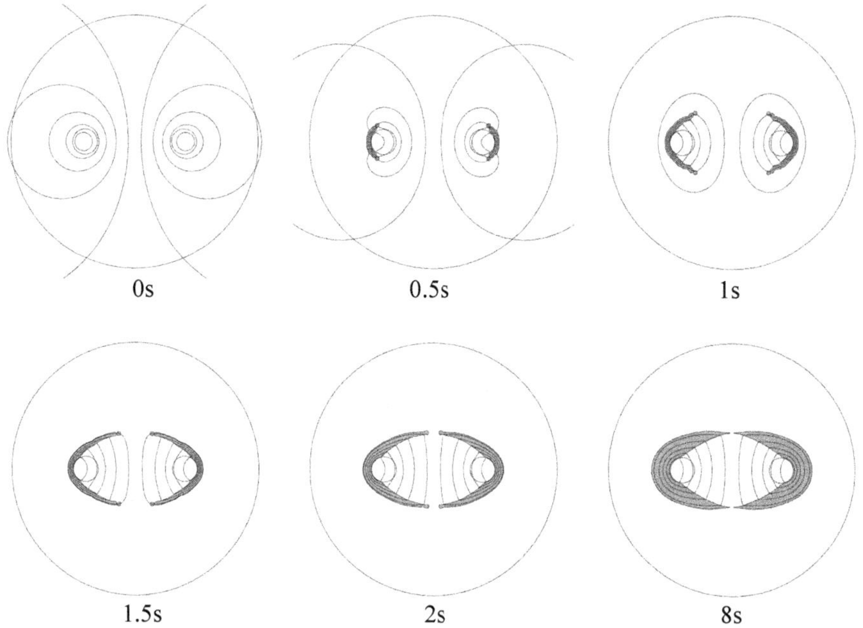

Fig. 8.18 Magnetic shielding-topology design (dot): variation of shielding core

generated at the outer sides of the current coils. Then, the ferromagnetic material almost encloses the current coils to prevent the field leakage. The dot generation ceases when the ferromagnetic materials reaches near the center line at 2 s, and then, the shape of the ferromagnetic material becomes thicker until the final design is obtained at 8 s. In the final design, most of the magnetic field is confined to the inside of the enclosing ferromagnetic material and the objective function converges to almost zero as in Fig. 8.19.

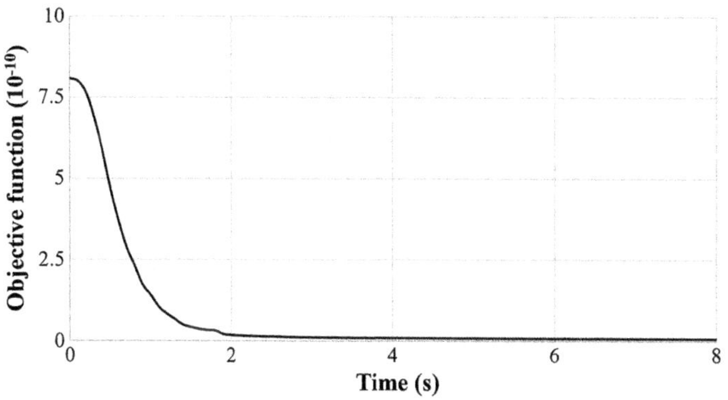

Fig. 8.19 Magnetic shielding-topology design (dot): evolution of objective function

8.2.3.3 Inductive Power Transfer

The topology design method using the dot sensitivity formula is applied to the design problem of an inductive power transfer system. Consider a wireless power transfer model in Fig. 8.20, where the transmitter and the receiver windings are placed in parallel in the vacant region. The transmitter winding is fed by the current of density $\mathbf{J}_0$. The region of the receiver winding is set as the objective function region $\Omega_p = \Omega_{p1} \cup \Omega_{p2}$. The design objective is to maximize the mutual inductance between the two windings by distributing the magnetic material in the design domain. The design domain is divided into two regions, which are separated by the air gap d [21].

The mutual inductance is maximized by maximizing the flux linkage of receiver winding. Thus, the objective function in this example is defined as the flux linkage of receiver winding;

$$F = \frac{1}{A_w} \int_{\Omega} \mathbf{A} m_p \mathrm{d}\Omega \tag{8.2.54}$$

where

$$m_p = \begin{cases} -1 & \text{in } \Omega_{p1} \\ 1 & \text{in } \Omega_{p2} \\ 0 & \text{elsewhere} \end{cases} \tag{8.2.55}$$

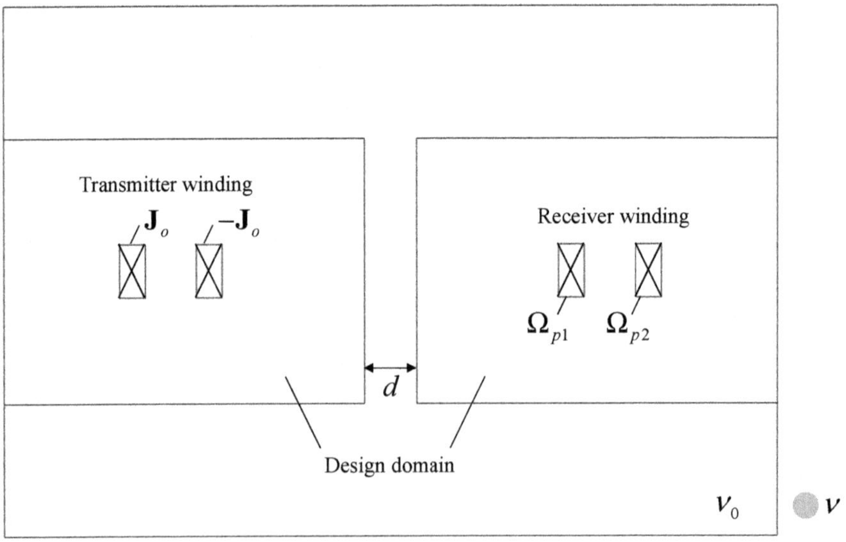

Fig. 8.20 Inductive power transfer-topology design (dot), mutual inductance

and A_w is the receiver winding area. The variational adjoint equation for the objective function (8.2.54) is obtained as

$$a(\boldsymbol{\lambda}, \overline{\boldsymbol{\lambda}}) = \int_{\Omega} \frac{1}{A_w} \overline{\boldsymbol{\lambda}} m_p \mathrm{d}\Omega \quad \forall \overline{\boldsymbol{\lambda}} \in \Phi \tag{8.2.56}$$

The dot sensitivity formula for the dot generation is given as

$$S(\mathbf{x}) = 2\nu_0 \left(\frac{1 - \nu_r}{1 + \nu_r} \right) \mathbf{B}(\mathbf{A}) \cdot \mathbf{B}(\boldsymbol{\lambda}) \tag{8.2.57}$$

This dot sensitivity formula is evaluated in the vacant design domain, and then, the points with the highest dot sensitivity are selected as the candidate positions for the dot generation. When the dot sensitivity values become negative at all points in the vacant domain, the dot generation is finished. Even after the dot generation, the shape of the existing ferromagnetic material is optimized until the objective function converges. For the shape optimization, the shape sensitivity formula for the objective function is used;

$$\dot{G} = \int_{\gamma} \nu_0 (\nu_r - 1) \mathbf{B}(\mathbf{A}_1) \cdot \mathbf{B}(\boldsymbol{\lambda}_2) V_n \mathrm{d}\Gamma \tag{8.2.58}$$

The design velocity on the ferromagnetic material for this maximization problem is taken as

$$V_n = \nu_0 (\nu_r - 1) \mathbf{B}(\mathbf{A}_1) \cdot \mathbf{B}(\boldsymbol{\lambda}_2) \tag{8.2.59}$$

The topology and the shape variation of the ferromagnetic material is shown in Fig. 8.21. In the initial state, the magnetic field generated by the transmitter winding hardly links the receiver winding. At the beginning of the optimization process, some magnetic dots are generated inside each coil winding, where the magnetic field intensity is higher than the other regions. As the generated dots are chained to form two C-type cores, the flux linkage in the receiver winding gradually increases. When the chained-cores reach the air-gap boundary, the flux linkage begins to rapidly increase. After the dot generation is finished at 1.5 s, the cores become thicker and wider and the objection function converges at 12 s as in Fig. 8.22.

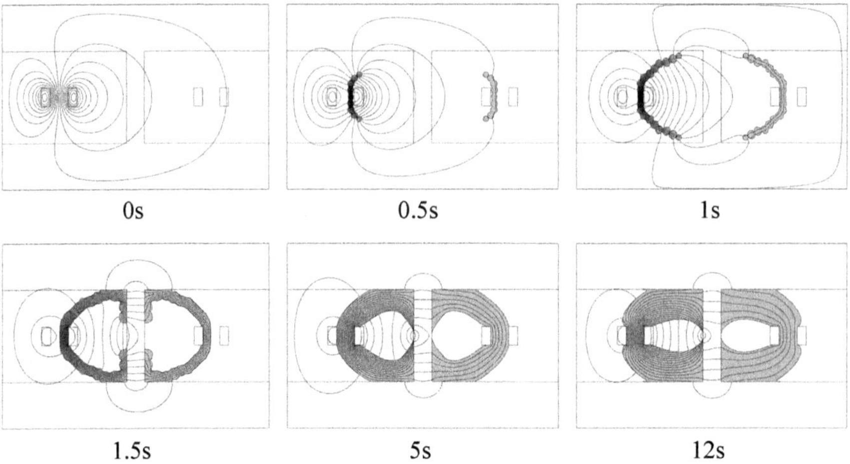

Fig. 8.21 Inductive power transfer-topology design (dot): variation of core shape

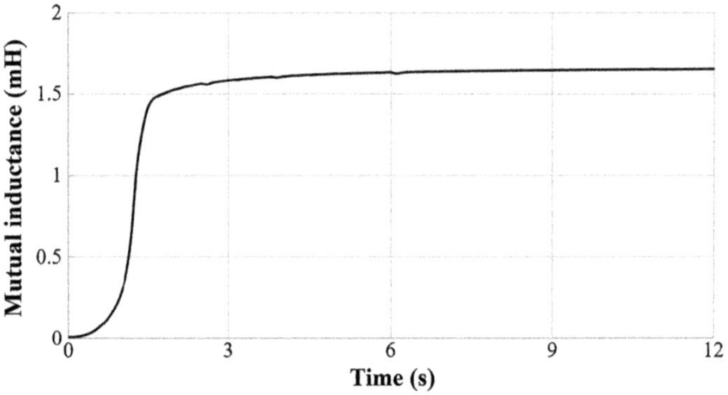

Fig. 8.22 Inductive power transfer-topology design (dot): Evolution of objective function

References

1. Bendsoe, M.P., Soares, C.A.M.: Topology Design of Structures, NATO ASI Series. Kluwer Academic Publishers, Boston (1993)
2. Yoo, J., Kikuchi, N., Volakis, J.L.: Structural optimization in magnetic devices by the homogenization design method. IEEE Trans. Magn. **36**, 574–580 (2000)
3. Byun, J.K., Park, I.H., Hahn, S.Y.: Topology optimization of electrostatic actuator using design sensitivity. IEEE Trans. Magn. **38**, 1053–1056 (2002)
4. Byun, J.K., Hahn, S.Y., Park, I.H.: Topology optimization of electrical devices using mutual energy and sensitivity. IEEE Trans. Magn. **35**, 3718–3720 (1999)
5. Byun, J.K., Park, I.H., Nah, W., Lee, J.H., Kang, J.: Comparison of shape and topology optimization methods for HTS solenoid design. IEEE Trans. Appl. Supercond. **14**, 1842–1845 (2004)

6. Peng, D., Merriman, B., Osher, S., Zhao, H., Kang, M.: A PDE-Based Fast Local Level Set Method. J. Comput. Phys. **155**, 410–438 (1999)
7. Wang, M.Y., Wang, X., Guo, D.: A level set method for structural topology optimization. Comput. Methods Appl. Mech. Eng. **192**, 227–246 (2003)
8. Lee, K.H., Hong, S.G., Baek, M.K., Choi, H.S., Kim, Y.S., Park, I.H.: Adaptive level set method for accurate boundary shape in optimization of electromagnetic systems. COMPEL **33**, 809–820 (2014)
9. Kim, Y.S., Baek, M.K., Park, I.H.: Design sensitivity and LSM for topology and shape optimization in electromagnetic system. COMPEL **31**, 803–815 (2012)
10. Lee, K.H., Hong, S.G., Baek, M.K., Choi, H.S., Kim, Y.S., Park, I.H.: Alleviation of electric field intensity in high-voltage system by topology and shape optimization of dielectric material using continuum design sensitivity and level set method. IEEE Trans. Magn. **51**, 9400404 (2015)
11. Kim, Y.S., Park, I.H.: Topology optimization of rotor in synchronous reluctance motor using level set method and shape design sensitivity. IEEE Trans. Appl. Supercond. **20**, 1093–1096 (2010)
12. Céa, J., Garreau, S., Guillaume, P., Masmoudi, M.: The shape and topological optimizations connection. Comput. Methods Appl. Mech. Eng. **188**, 713–726 (2000)
13. Bendsøe, M.P., Olhoff, N., Sigmund, O.: IUTAM Symposium on Topological Design Optimization of Structures, Machines and Materials. Springer, Dordrecht
14. He, L., Kao, C.Y., Osher, S.: Incorporating topological derivatives into shape derivatives based level set methods. J. Comput. Phys. **225**, 891–909 (2007)
15. Novotny, A.A., Sokołowski, J.: Topological Derivatives in Shape Optimization. Springer, Berlin (2012)
16. Nayfeh, M.H., Brussel, M.K.: Electricity and Magnetism. Wiley, New York (1985)
17. Stafl, M.: Electrodynamics of Electrical Machines. Academia, Prague (1967)
18. Lee, K.H., Hong, S.G., Baek, M.K., Park, I.H.: Hole sensitivity analysis for topology optimization in electrostatic system using virtual hole concept and shape sensitivity. IEEE Trans. Magn. **52**, 9401304 (2016)
19. Hong, S.G., Lee, K.H., Park, I.H.: Derivation of hole sensitivity formula for topology optimization in magnetostatic system using virtual hole concept and shape sensitivity. IEEE Trans. Magn. **51**, 9400304 (2015)
20. Lee, K.H., Hong, S.G., Park, I.H.: Dot sensitivity analysis for topology optimization of dielectric material in electrostatics system. IEEE Trans. Magn. **53**, 9401404 (2017)
21. Hong, S.G., Lee, K.H., Park, I.H.: Dot sensitivity analysis for topology optimization of ferromagnetic material in magnetostatic system. In: IEEE Optimization and Inverse Problems in Electromagnetism 2016 (2016)

Appendix A
More Examples of Electrostatic System

A.1 Outer Boundary Design

In Sect. 3.2.4, the sensitivity formula for the objective function defined on the inner area was derived as (3.2.45). To show that the sensitivity formula is correct, one-dimensional analytical examples, which have the analytic field solutions, are taken. The sensitivity results, which are analytically calculated in the examples, are compared with the results of the sensitivity formula, ensuring that they give the same result. The analytical examples are a parallel plate capacitor and a spherical capacitor, which are the one-dimensional capacitor models in the rectangular and the spherical coordinates, respectively.

A.1.1 Parallel Plate Capacitor

As an example that can be analytically calculated, consider a parallel plate capacitor shown in Fig. A.1, where two parallel electrodes of a surface area S are separated by a dielectric of thickness a and dielectric constant ε, and a voltage ϕ_o is applied between the two electrodes. The design objective is to obtain a target electric field $\mathbf{E}_o$ in Ω_p by moving the upper electrode Γ^{0d}. In this example, the design variable is the height of the upper electrode a. The design sensitivity with respect to the design variable a is analytically calculated, and then its result is compared with the result by the sensitivity formula.

The objective function is defined as

$$F = \int_{\Omega} (\mathbf{E}(\phi) - \mathbf{E}_o)^2 m_p \mathrm{d}\Omega \tag{A.1.1}$$

I. H. Park, *Design Sensitivity Analysis and Optimization of Electromagnetic Systems*, Mathematical and Analytical Techniques with Applications to Engineering,
https://doi.org/10.1007/978-981-13-0230-5

Fig. A.1 Parallel plate capacitor-outer boundary design

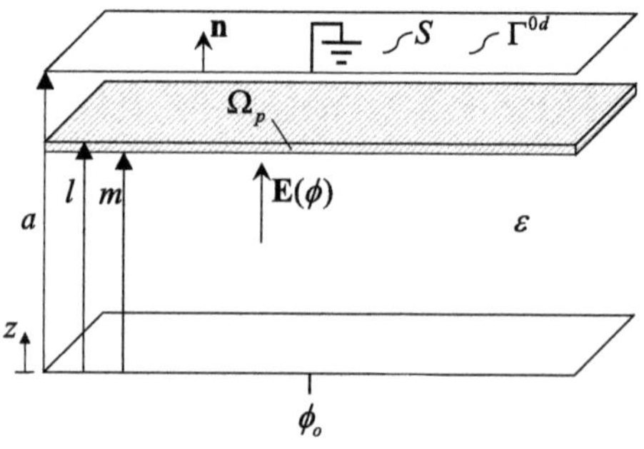

where

$$\mathbf{E}(\phi) = \frac{\phi_o}{a}\mathbf{z} \tag{A.1.2}$$

$$\mathbf{E}_o = E_o\mathbf{z} \tag{A.1.3}$$

The objective function is rewritten by using the fields (A.1.2) and (A.1.3);

$$F = (l - m)\left(\frac{\phi_o}{a} - E_o\right)^2 S \tag{A.1.4}$$

The analytical sensitivity per unit area is obtained by differentiating the objective function (A.1.4) with respect to a;

$$\frac{\mathrm{d}F}{\mathrm{d}a} = -2\phi_o\frac{l - m}{a^2}\left(\frac{\phi_o}{a} - E_o\right) \tag{A.1.5}$$

Alternatively, the sensitivity can be calculated by using the sensitivity formula (3.2.45) in Sect. 3.2.4:

$$\dot{G} = \int_{\Gamma^{0d}} \varepsilon E_n(\phi)E_n(\lambda)V_n\mathrm{d}\Gamma \tag{A.1.6}$$

This formula requires the adjoint solution, which can be obtained in the adjoint variable system in Fig. A.2.

The differential adjoint equation is given as

$$-\nabla\cdot\varepsilon\nabla\lambda = -\nabla\cdot\mathbf{g}_{\mathbf{E}}m_p \tag{A.1.7}$$

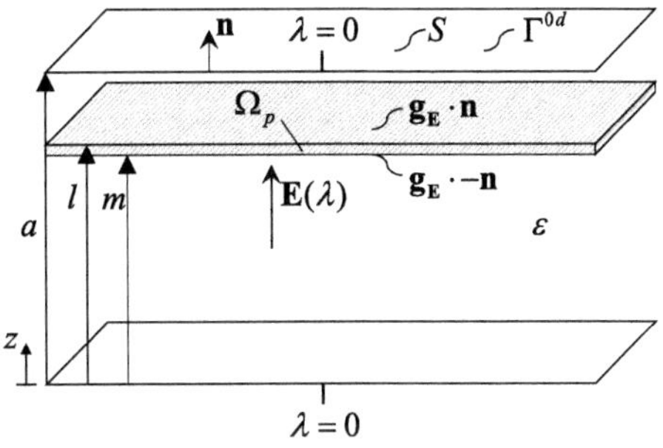

Fig. A.2 Parallel plate capacitor-outer boundary design: adjoint variable system

Inserting the electric fields (A.1.2) and (A.1.3) into (A.1.6) provides

$$\varepsilon \frac{\mathrm{d}^2 \lambda}{\mathrm{d}z^2} = -2\left(\frac{\phi_o}{a} - E_o\right)(\delta(z-m) - \delta(z-l)) \tag{A.1.8}$$

where $\delta(z)$ is the Dirac delta function. Integrating (A.1.8) yields the solution of the adjoint field as

$$E(\lambda) = -\frac{\mathrm{d}\lambda}{\mathrm{d}z} = \frac{2}{\varepsilon}\left(\frac{\phi_o}{a} - E_o\right)(H(z-m) - H(z-l)) - k \tag{A.1.9}$$

where $H(z)$ is the Heaviside function and k, the integral constant, is obtained by integrating (A.1.9) and applying the boundary condition in Fig. A.2;

$$k = \frac{2}{\varepsilon}\frac{l-m}{a}\left(\frac{\phi_o}{a} - E_o\right) \tag{A.1.10}$$

Inserting (A.1.10) into (A.1.9) provides the adjoint field;

$$\mathbf{E}(\lambda) = \frac{2}{\varepsilon}\left(\frac{\phi_o}{a} - E_o\right)\left(H(z-m) - H(z-l) - \frac{l-m}{a}\right)\mathbf{z} \tag{A.1.11}$$

(A.1.2) and (A.1.11) are inserted into the sensitivity formula (A.1.6);

$$\dot{G} = -2\phi_o \frac{l-m}{a^2}\left(\frac{\phi_o}{a} - E_o\right)V_n S \tag{A.1.12}$$

Using $V_n = \frac{\mathrm{d}a}{\mathrm{d}t}$, the design sensitivity per unit area is obtained as

$$\frac{\mathrm{d}G}{\mathrm{d}a} = -2\phi_o \frac{l-m}{a^2}\left(\frac{\phi_o}{a} - E_o\right) \tag{A.1.13}$$

which is the correct result when compared with the analytical result in (A.1.5).

A.1.2 Spherical Capacitor

To take an example that can be analytically calculated, consider a spherical capacitor in Fig. A.3, where two spherical electrodes are separated by a dielectric of permittivity ε, and a voltage ϕ_o is applied between the two electrodes. The design objective is to obtain a target electric field $\mathbf{E}_o$ in Ω_p by moving the outer electrode Γ^{0d}. The design variable is the radius a of the outer electrode. The design sensitivity with respect to the design variable a is analytically calculated, and then its result is compared with the one by the sensitivity formula.

The objective function is defined as

$$F = \int_{\Omega} (\mathbf{E}(\phi) - \mathbf{E}_o)^2 m_p \mathrm{d}\Omega \tag{A.1.14}$$

where

$$\mathbf{E}(\phi) = \frac{\phi_o}{\frac{1}{b} - \frac{1}{a}} \frac{1}{r^2} \mathbf{r} \tag{A.1.15}$$

$$\mathbf{E}_o = \frac{E_o}{r^2} \mathbf{r} \tag{A.1.16}$$

The objective function is rewritten by using the fields (A.1.15) and (A.1.16);

$$F = 4\pi \left(\frac{1}{m} - \frac{1}{l} \right) \left(\frac{\phi_o}{\frac{1}{b} - \frac{1}{a}} - E_o \right)^2 \tag{A.1.17}$$

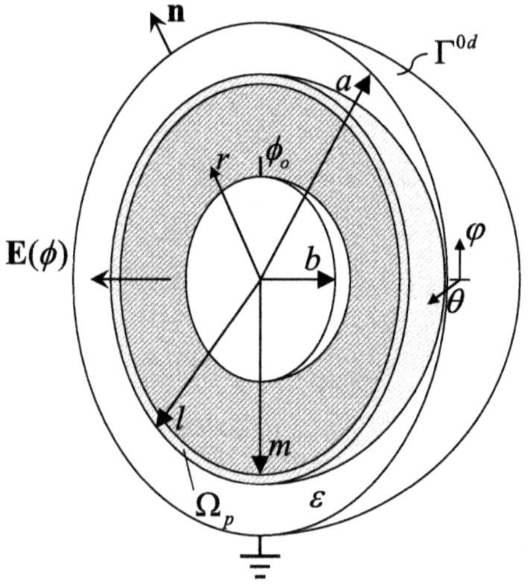

Fig. A.3 Spherical capacitor-outer boundary design

The analytical sensitivity is obtained by differentiating the objective function (A.1.17) with respect to the radius a of the outer electrode;

$$\frac{\mathrm{d}F}{\mathrm{d}a} = -8\pi\phi_o \frac{\frac{1}{m} - \frac{1}{l}}{a^2\left(\frac{1}{b} - \frac{1}{a}\right)^2}\left(\frac{\phi_o}{\frac{1}{b} - \frac{1}{a}} - E_o\right) \tag{A.1.18}$$

This analytical sensitivity result is compared with the result of the sensitivity formula (3.2.45):

$$\dot{G} = \int_{\Gamma^{0d}} \varepsilon E_n(\phi) E_n(\lambda) V_n \mathrm{d}\Gamma \tag{A.1.19}$$

The adjoint variable system shown in Fig. A.4 is analyzed prior to the evaluation of (A.1.19).

The differential adjoint equation in the spherical coordinate is obtained by using the electric fields (A.1.15) and (A.1.16);

$$\varepsilon \frac{1}{r^2}\frac{\mathrm{d}}{\mathrm{d}r}\left(r^2 \frac{\mathrm{d}\lambda}{\mathrm{d}r}\right) = -2\left(\frac{\phi_o}{\frac{1}{b} - \frac{1}{a}}\frac{1}{r^2} - \frac{E_o}{r^2}\right)(\delta(r-m) - \delta(r-l)) \tag{A.1.20}$$

Integrating (A.1.20) yields the solution of the adjoint field as

$$E(\lambda) = -\frac{\mathrm{d}\lambda}{\mathrm{d}r} = \frac{2}{\varepsilon}\left(\frac{\phi_o}{\frac{1}{b} - \frac{1}{a}} - E_o\right)(H(r-m) - H(r-l))\frac{1}{r^2} - \frac{k}{r^2} \tag{A.1.21}$$

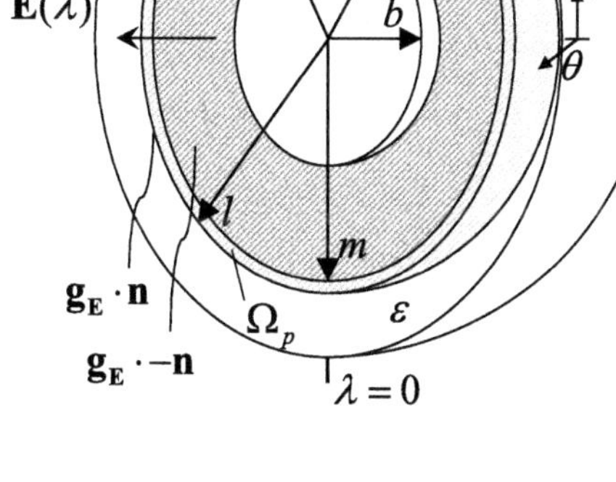

Fig. A.4 Spherical capacitor-outer boundary design: adjoint variable system

where k, the integral constant, is obtained by integrating (A.1.21) and applying the boundary condition in Fig. A.4;

$$k = \frac{2}{\varepsilon} \frac{\frac{1}{m} - \frac{1}{l}}{\frac{1}{b} - \frac{1}{a}} \left(\frac{\phi_o}{\frac{1}{b} - \frac{1}{a}} - E_o \right) \tag{A.1.22}$$

Inserting (A.1.22) into (A.1.21) yields the adjoint field;

$$\mathbf{E}(\lambda) = \frac{2}{\varepsilon} \left(\frac{\phi_o}{\frac{1}{b} - \frac{1}{a}} - E_o \right) \left(H(r - m) - H(r - l) - \frac{\frac{1}{m} - \frac{1}{l}}{\frac{1}{b} - \frac{1}{a}} \right) \frac{1}{r^2} \mathbf{r} \tag{A.1.23}$$

(A.1.15) and (A.1.23) are inserted into the sensitivity formula (A.1.19);

$$\dot{G} = -8\pi\phi_o \frac{\frac{1}{m} - \frac{1}{l}}{a^2 \left(\frac{1}{b} - \frac{1}{a} \right)^2} \left(\frac{\phi_o}{\frac{1}{b} - \frac{1}{a}} - E_o \right) V_n \tag{A.1.24}$$

Using $V_n = \frac{\mathrm{d}a}{\mathrm{d}t}$, the design sensitivity is obtained as

$$\frac{\mathrm{d}G}{\mathrm{d}a} = -8\pi\phi_o \frac{\frac{1}{m} - \frac{1}{l}}{a^2 \left(\frac{1}{b} - \frac{1}{a} \right)^2} \left(\frac{\phi_o}{\frac{1}{b} - \frac{1}{a}} - E_o \right) \tag{A.1.25}$$

which is identical to the analytical result in (A.1.18).

Next, the sensitivity formula (3.2.45) derived in Sect. 3.2.4 is applied to a shape optimization problem of two-dimensional design model, of which the analytic field solution is not given, but the optimal design is known. This design model is taken to illustrate how the sensitivity formula is applied to the shape design problem of the outer boundary in the electrostatic system. The numerical example is a circular capacitor, of which the optimal design is known. If the result of the example is obtained as the expected optimal design, it can be said that the shape optimization using the sensitivity formula is feasible for the design of the outer boundary. In this example, the sensitivity formula requires the state and the adjoint variables, which are numerically calculated by the finite element method. The sensitivity obtained is used for the optimization algorithm to evolve the electrode shape. For the optimization algorithm, the level set method is used to represent the shape evolution of the design model.

A.1.3 Circular Capacitor

As an example with a known optimal design, consider a circular capacitor in Fig. A.5, where the dielectric of permittivity ε is between two electrodes: the inner circular electrode and the outer elliptical electrode. When a voltage ϕ_o is applied between the two electrodes, the distribution of the electric field is not uniform along the circular direction. If the shape of the outer electrode is changed to be a circle, the capacitor becomes a coaxial capacitor and the electric field between the two electrodes becomes uniform along the circular direction.

The design objective is to obtain a uniform field $\mathbf{E}_o$ in the region Ω_p in Fig. A.5, which is analytically given in the coaxial capacitor. The design variable is the shape of the outer electrode Γ^{0d}, the optimal shape of which is a circle. The objective function to be minimized is defined as the integral of the field difference in Ω_p;

$$F = \int_{\Omega} (\mathbf{E}(\phi) - \mathbf{E}_o)^2 m_p \mathrm{d}\Omega \tag{A.1.26}$$

where

$$\mathbf{E}_o = \frac{E_o}{r}\mathbf{r} \tag{A.1.27}$$

The variational adjoint equation for (A.1.26) is obtained as

$$a(\lambda, \bar{\lambda}) = \int_{\Omega} 2(\mathbf{E}(\phi) - \mathbf{E}_o) \cdot \mathbf{E}(\bar{\lambda}) m_p \mathrm{d}\Omega \quad \forall\, \bar{\lambda} \in \Phi \tag{A.1.28}$$

The shape sensitivity for this outer boundary design is the sensitivity formula (3.2.45):

$$\dot{G} = \int_{\Gamma^{0d}} \varepsilon E_n(\phi) E_n(\lambda) V_n \mathrm{d}\Gamma \tag{A.1.29}$$

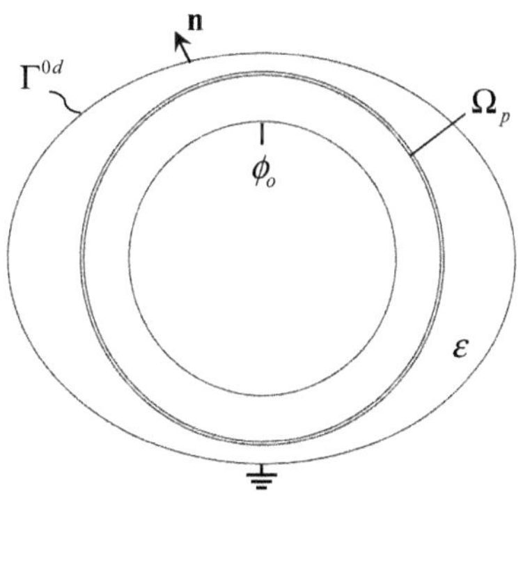

Fig. A.5 Circular capacitor-outer boundary design

For this minimization problem, the design velocity is taken as

$$V_n = -\varepsilon E_n(\phi)E_n(\lambda) \tag{A.1.30}$$

The design result is shown in Fig. A.6, where the shape of the outer electrode becomes gradually a circle with the increase of the iteration number as expected. The final design of the circular shape is obtained at the 8 s, when the objective function converges to 0 as in Fig. A.7. The result of this example shows that the shape sensitivity method is feasible for the shape design of the outer boundary in the two-dimensional electrostatic system.

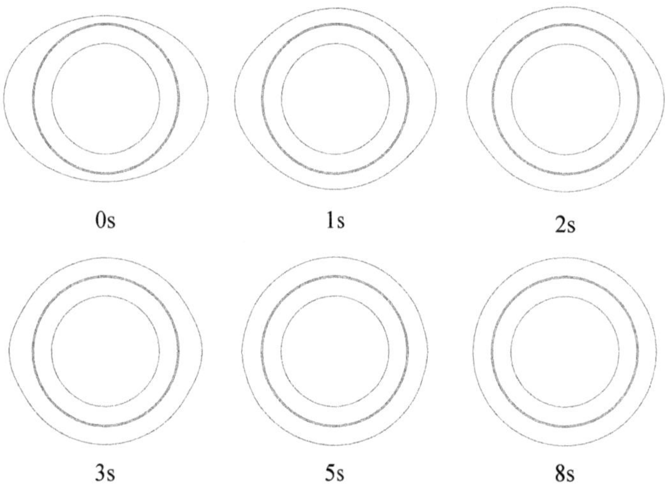

Fig. A.6 Circular capacitor-outer boundary design: shape variation

Fig. A.7 Coaxial capacitor-outer boundary design: evolution of objective function

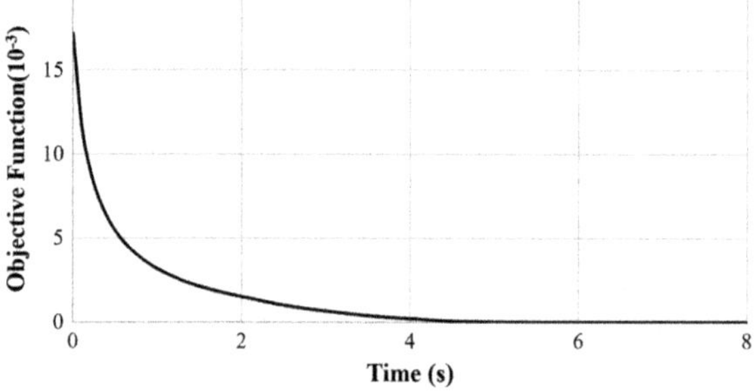

A.2 Outer Boundary Design for System Energy

When the objective function is defined as the system energy, the sensitivity formula was derived as (3.3.42) in Sect. 3.3.5. Two one-dimensional examples with the analytic field solutions are taken to show that the sensitivity formula is correct. The objective is to compare the analytical sensitivity results with the ones by the sensitivity formula to ensure that the two results are the same. The analytical example is a parallel plate capacitor and a spherical capacitor, which are the one-dimensional capacitor model in the rectangular and the spherical coordinates, respectively.

A.2.1 Parallel Plate Capacitor

For an example that can be analytically calculated, consider a parallel plate capacitor in Fig. A.8, where two parallel electrodes of the surface area S are separated by a dielectric of thickness a and dielectric constant ε, and a voltage ϕ_o is applied between the two electrodes. The design objective is to obtain a desired system energy by moving the upper electrode Γ^{0d}; so the design variable is the height a of the upper electrode. The design sensitivity is the derivative of the system energy with respect to the design variable a, and then this sensitivity is compared with the result by the sensitivity formula.

The objective function is the system energy;

$$F = W_e = \int_{\Omega} \frac{1}{2} \varepsilon \mathbf{E}^2(\phi) d\Omega \tag{A.2.1}$$

where

$$\mathbf{E}(\phi) = \frac{\phi_o}{a} \mathbf{z} \tag{A.2.2}$$

The objective function is rewritten with the field in (A.2.2);

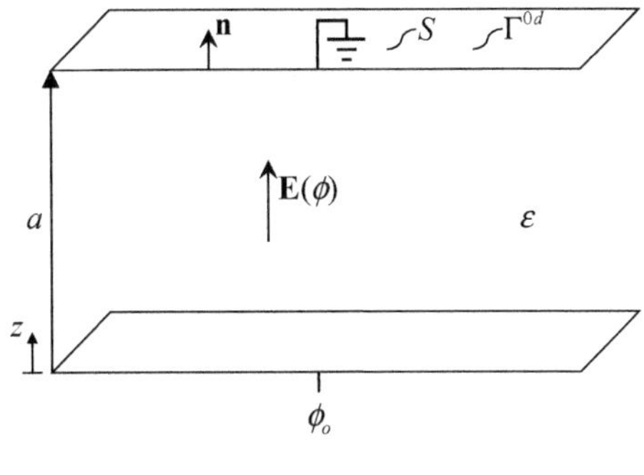

Fig. A.8 Parallel plate capacitor-outer boundary design, system energy

$$F = \frac{1}{2}\varepsilon\frac{\phi_o^2}{a}S \tag{A.2.3}$$

The analytical sensitivity per unit area is the derivative of the objective function (A.2.3) with respect to a;

$$\frac{\mathrm{d}F}{\mathrm{d}a} = -\frac{1}{2}\varepsilon\left(\frac{\phi_o}{a}\right)^2 \tag{A.2.4}$$

This analytical sensitivity is compared with the result obtained from the sensitivity formula (3.3.42) in Sect. 3.3.5:

$$\dot{G} = -\int_{\Gamma^{0d}} \frac{1}{2}\varepsilon E_n^2(\phi)V_n \mathrm{d}\Gamma \tag{A.2.5}$$

which does not require the adjoint variable. The sensitivity formula (A.2.5) is calculated by inserting (A.2.2) into (A.2.5);

$$\dot{G} = -\frac{1}{2}\varepsilon\left(\frac{\phi_o}{a}\right)^2 V_n S \tag{A.2.6}$$

Using $V_n = \frac{\mathrm{d}a}{\mathrm{d}_t}$, the design sensitivity per unit area is obtained as

$$\frac{\mathrm{d}G}{\mathrm{d}a} = -\frac{1}{2}\varepsilon\left(\frac{\phi_o}{a}\right)^2 \tag{A.2.7}$$

This is the same as the analytical one in (A.2.4).

A.2.2 Spherical Capacitor

For an example that can be analytically calculated, consider a spherical capacitor in Fig. A.9, where two spherical conducting electrodes are separated by a dielectric of permittivity ε, and a voltage ϕ_o is applied between the two electrodes. The design objective is to obtain a system energy by moving the outer electrode Γ^{0d}; thus, the design variable is the radius a of the outer electrode. The design sensitivity is calculated by differentiating the objective function to the design variable a, and then its result is compared with the result by the sensitivity formula.

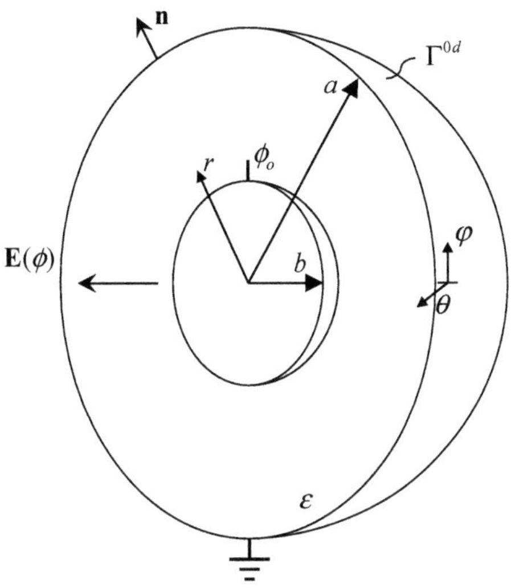

Fig. A.9 Spherical capacitor-outer boundary design, system energy problem

The objective function is the system energy;

$$F = W_e = \int_{\Omega} \frac{1}{2} \varepsilon \mathbf{E}^2(\phi) \mathrm{d}\Omega \tag{A.2.8}$$

where

$$\mathbf{E}(\phi) = \frac{\phi_o}{\frac{1}{b} - \frac{1}{a}} \frac{1}{r^2} \mathbf{r} \tag{A.2.9}$$

The objective function is rewritten by using the field (A.2.9);

$$F = 2\pi\varepsilon \frac{\phi_o^2}{\frac{1}{b} - \frac{1}{a}} \tag{A.2.10}$$

Differentiating this objective function with respect to the radius a of the outer electrode provides the analytical sensitivity;

$$\frac{\mathrm{d}F}{\mathrm{d}a} = -2\pi \frac{\varepsilon}{a^2} \left(\frac{\phi_o}{\frac{1}{b} - \frac{1}{a}} \right)^2 \tag{A.2.11}$$

By using the sensitivity formula (3.3.42), the shape sensitivity is also obtained:

$$\dot{G} = -\int_{\Gamma^{0d}} \frac{1}{2} \varepsilon E_n^2(\phi) V_n \mathrm{d}\Gamma \tag{A.2.12}$$

With (A.2.9), the sensitivity formula (A.2.12) is rewritten;

$$\dot{G} = -2\pi \frac{\varepsilon}{a^2} \left(\frac{\phi_o}{\frac{1}{b} - \frac{1}{a}} \right)^2 V_n \tag{A.2.13}$$

The design sensitivity is obtained by using $V_n = \frac{\mathrm{d}a}{\mathrm{d}t}$;

$$\frac{\mathrm{d}G}{\mathrm{d}a} = -2\pi \frac{\varepsilon}{a^2} \left(\frac{\phi_o}{\frac{1}{b} - \frac{1}{a}} \right)^2 \tag{A.2.14}$$

which is the same as the analytical sensitivity in (A.2.11).

Next, the sensitivity formula (3.3.42) in Sect. 3.3.5 is applied to a shape optimization problem of two-dimensional design model without the analytic field solution. This design model shows that the sensitivity formula is well applied to the shape design of the outer boundary for the system energy in the electrostatic system. In this two-dimensional example, the state variable, which is numerically calculated by the finite element method, is required to evaluate the sensitivity formula. The sensitivity evaluated is used for the optimization algorithm to evolve the electrode shape. The level set method is used to represent the shape evolution of the design model.

A.2.3 Circular Capacitor

For an example of which the optimal design is known, consider a capacitor in Fig. A.10, where the inner electrode is circular but the outer electrode is elliptical in the form. When a voltage ϕ_o is applied between the two electrodes, the distribution of the electric field is not uniform along the circular direction. When the shape of

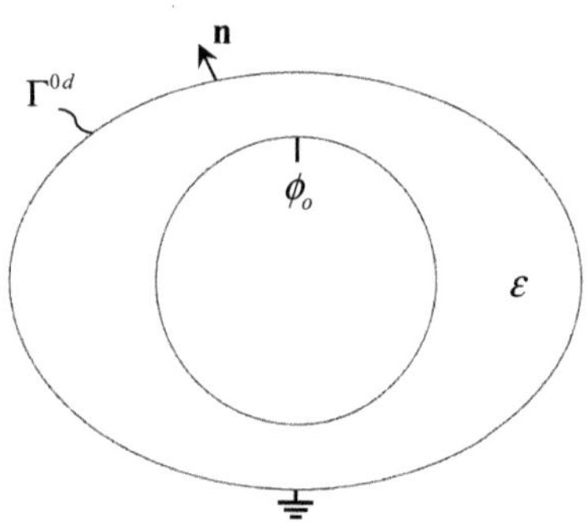

Fig. A.10 Coaxial capacitor-outer boundary design, system energy minimization

the outer electrode is changed to be a circle, the stored system energy becomes the minimum under the constraint of constant dielectric volume.

The objective function to be minimized is the system energy;

$$F = W_e = \int_{\Omega} \frac{1}{2} \varepsilon \mathbf{E}^2(\phi) d\Omega \tag{A.2.15}$$

In this design problem, the shape of the outer electrode Γ^{0d} is the design variable, which has a constraint of constant dielectric volume;

$$\int_{\Omega} d\Omega = C \tag{A.2.16}$$

which is the volume of the outer electrode per unit length. By using the material derivative formula (3.1.22) in Sect. 3.1.2, the material derivative of the constraint (A.2.16) is obtained as

$$\dot{C} = \int_{\Omega} 1' d\Omega + \int_{\Gamma} V_n d\Gamma = \int_{\Gamma^{0d}} V_n d\Gamma = 0 \tag{A.2.17}$$

which is a different form of the constraint.

The sensitivity formula (3.3.42) is the shape sensitivity for this outer boundary design:

$$\dot{G} = -\int_{\Gamma^{0d}} \frac{1}{2} \varepsilon E_n^2(\phi) V_n d\Gamma \tag{A.2.18}$$

For this minimization problem, the design velocity is taken as

$$V_n = \frac{1}{2} \varepsilon E_n^2(\phi) \tag{A.2.19}$$

In order that the velocity field satisfies the constraint (A.2.17), the design velocity (A.2.19) is modified by subtracting its average V_n to become U_n as

$$U_n = V_n - V_{na} \tag{A.2.20}$$

where

$$V_{na} = \int_{\Gamma^{0d}} \frac{1}{2} \varepsilon E_n^2(\phi) \mathrm{d}\Gamma \bigg/ \int_{\Gamma^{0d}} \mathrm{d}\Gamma \tag{A.2.21}$$

The design result is shown in Fig. A.11, where the shape of the outer electrode becomes gradually a circle as expected. The final design of the circular shape is obtained at the 10 s, when the system energy converges to the minimum as in Fig. A.12. The result of this example shows that the shape sensitivity analysis for the system energy is well applied to the two-dimensional electrostatic system.

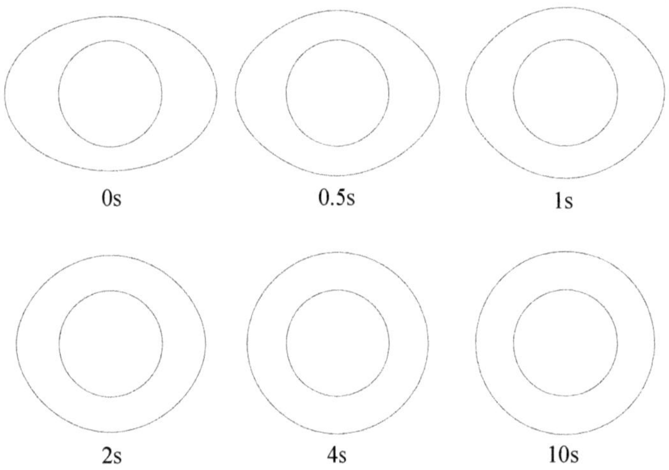

Fig. A.11 Circular capacitor: shape variation

Fig. A.12 Circular capacitor: evolution of objective function

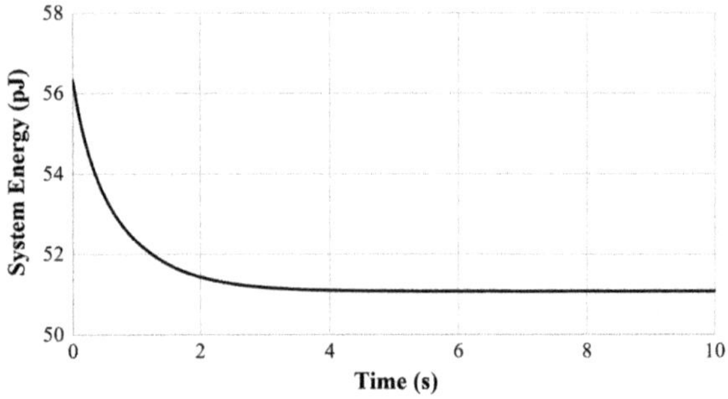

A.3 Interface Design

For the interface design problem in Sect. 3.4.4, the sensitivity formula, which is for the objective function on the inner area, was derived as (3.4.69). One-dimensional examples, which have the analytic field solutions, are taken to show that the sensitivity formula is correct. The sensitivity results analytically calculated in the examples are compared with the ones of the sensitivity formula. The analytical examples are a parallel plate capacitor and a spherical capacitor, which are the one-dimensional capacitor models in the rectangular and the spherical coordinates, respectively.

A.3.1 Parallel Plate Capacitor

For an analytical example, a parallel plate capacitor is given in Fig. A.13, where two parallel electrodes of a surface area S are separated by two dielectrics, and a voltage ϕ_o is applied between the two electrodes. The upper dielectric of dielectric constants ε_1 is $a - b$ thick, and the lower dielectric of dielectric constants ε_2 is b thick. The design objective is to obtain a target electric field $\mathbf{E}_o$ in Ω_p of the upper dielectric by moving the interface, where the two dielectrics meet; the design variable is the thickness b of the lower dielectric. The design sensitivity with respect to the design variable b is analytically calculated.

The objective function is defined as

$$F = \int_{\Omega} (\mathbf{E}(\phi) - \mathbf{E}_o)^2 m_p \mathrm{d}\Omega \tag{A.3.1}$$

where

$$\mathbf{E}(\phi_1) = \frac{\phi_o}{\frac{\varepsilon_1}{\varepsilon_2} a + \left(1 - \frac{\varepsilon_1}{\varepsilon_2}\right) b} \mathbf{z} \quad \text{for } 0 \le z \le b \tag{A.3.2}$$

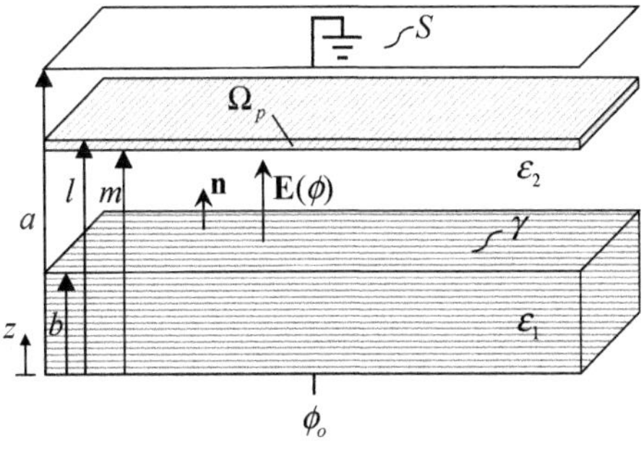

Fig. A.13 Parallel plate capacitor-interface design

$$\mathbf{E}(\phi_2) = \frac{\phi_o}{a + \left(\frac{\varepsilon_2}{\varepsilon_1} - 1\right)b}\mathbf{z} \quad \text{for } b \le z \le a \tag{A.3.3}$$

$$\mathbf{E}_o = E_o\mathbf{z} \tag{A.3.4}$$

The objective function is rewritten by using the electric field (A.3.3) and (A.3.4);

$$F = (l - m)\left[\frac{\phi_o}{a + \left(\frac{\varepsilon_2}{\varepsilon_1} - 1\right)b} - E_o\right]^2 S \tag{A.3.5}$$

The analytical sensitivity per unit area is obtained by differentiating the objective function (A.3.5) with respect to b;

$$\frac{\mathrm{d}F}{\mathrm{d}b} = -2\phi_o\frac{\left(\frac{\varepsilon_2}{\varepsilon_1} - 1\right)(l - m)}{\left[a + \left(\frac{\varepsilon_2}{\varepsilon_1} - 1\right)b\right]^2}\left[\frac{\phi_o}{a + \left(\frac{\varepsilon_2}{\varepsilon_1} - 1\right)b} - E_o\right] \tag{A.3.6}$$

This analytical sensitivity result is compared with the result obtained from the sensitivity formula (3.4.69) in Sect. 3.4.4:

$$\dot{G} = \int_{\gamma} (\varepsilon_2 - \varepsilon_1)\mathbf{E}(\phi_1) \cdot \mathbf{E}(\lambda_2) V_n \mathrm{d}\Gamma \tag{A.3.7}$$

This formula requires the adjoint solution, which can be obtained in the adjoint variable system in Fig. A.14.

The differential adjoint equations are given as

$$-\nabla \cdot \varepsilon_1 \nabla \lambda_1 = 0 \quad \text{in } \Omega_1 \tag{A.3.8}$$

$$-\nabla \cdot \varepsilon_2 \nabla \lambda_2 = -\nabla \cdot \mathbf{g}_{\mathbf{E}_2} m_p \quad \text{in } \Omega_2 \tag{A.3.9}$$

Inserting the electric fields (A.3.3) and (A.3.4) provides

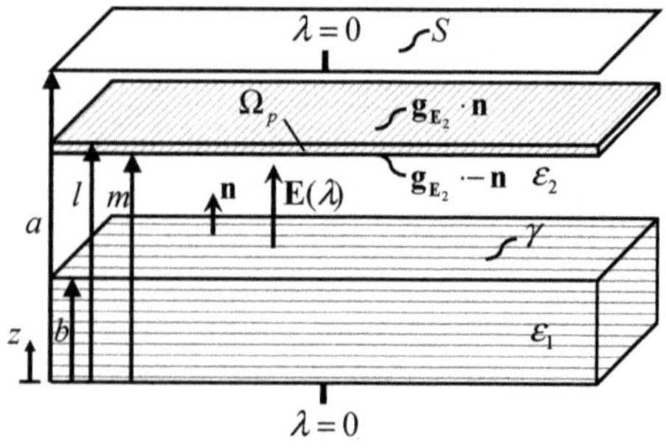

Fig. A.14 Parallel plate capacitor: adjoint variable system

$$\varepsilon_1 \frac{d^2\lambda_1}{dz^2} = 0 \quad \text{for } 0 \le z \le b \tag{A.3.10}$$

$$\varepsilon_2 \frac{d^2\lambda_2}{dz^2} = -2\left[\frac{\phi_o}{a + \left(\frac{\varepsilon_2}{\varepsilon_1} - 1\right)b} - E_o\right](\delta(z-m) - \delta(z-l)) \quad \text{for } b \le z \le a \tag{A.3.11}$$

where $\delta(z)$ is the Dirac delta function. Integrating (A.3.10) and (A.3.11) yields the solution of the adjoint fields as

$$E(\lambda_1) = -\frac{d\lambda_1}{dz} = -k_1 \quad \text{for } 0 \le z \le b \tag{A.3.12}$$

$$E(\lambda_2) = -\frac{d\lambda_2}{dz} = \frac{2}{\varepsilon_2}\left[\frac{\phi_o}{a + \left(\frac{\varepsilon_2}{\varepsilon_1} - 1\right)b} - E_o\right](H(z-m) - H(z-l)) - k_2 \quad \text{for } b \le z \le a \tag{A.3.13}$$

where $H(z)$ is the Heaviside function, and the integral constants k_1 and k_2 are obtained by integrating (A.3.12), (A.3.13) and applying the boundary and interface conditions in Fig. A.14;

$$k_1 = \frac{2}{\varepsilon_1} \frac{l-m}{a + \left(\frac{\varepsilon_2}{\varepsilon_1} - 1\right)b}\left[\frac{\phi_o}{a + \left(\frac{\varepsilon_2}{\varepsilon_1} - 1\right)b} - E_o\right] \tag{A.3.14}$$

$$k_2 = \frac{2}{\varepsilon_2} \frac{l-m}{a + \left(\frac{\varepsilon_2}{\varepsilon_1} - 1\right)b}\left[\frac{\phi_o}{a + \left(\frac{\varepsilon_2}{\varepsilon_1} - 1\right)b} - E_o\right] \tag{A.3.15}$$

Inserting (A.3.14), (A.3.15) into (A.3.12), (A.3.13) provides the adjoint fields;

$$\mathbf{E}(\lambda_1) = -\frac{2}{\varepsilon_1} \frac{l-m}{a + \left(\frac{\varepsilon_2}{\varepsilon_1} - 1\right)b}\left[\frac{\phi_o}{a + \left(\frac{\varepsilon_2}{\varepsilon_1} - 1\right)b} - E_o\right]\mathbf{z} \quad \text{for } 0 \le z \le b \tag{A.3.16}$$

$$\mathbf{E}(\lambda_2) = \frac{2}{\varepsilon_2}\left[\frac{\phi_o}{a + \left(\frac{\varepsilon_2}{\varepsilon_1} - 1\right)b} - E_o\right]\left[H(z-m) - H(z-l) - \frac{l-m}{a + \left(\frac{\varepsilon_2}{\varepsilon_1} - 1\right)b}\right]\mathbf{z} \quad \text{for } b \le z \le a \tag{A.3.17}$$

(A.3.12) and (A.3.17) are inserted into the sensitivity formula (A.3.7);

$$\dot{G} = -2\phi_o \frac{\left(\frac{\varepsilon_2}{\varepsilon_1} - 1\right)(l-m)}{\left[a + \left(\frac{\varepsilon_2}{\varepsilon_1} - 1\right)b\right]^2}\left[\frac{\phi_o}{a + \left(\frac{\varepsilon_2}{\varepsilon_1} - 1\right)b} - E_o\right]V_n S \tag{A.3.18}$$

Using $V_n = \frac{\mathrm{d}_b}{\mathrm{d}_t}$, the design sensitivity per unit area is obtained as

$$\frac{\mathrm{d}G}{\mathrm{d}b} = -2\phi_o \frac{\left(\frac{\varepsilon_2}{\varepsilon_1} - 1\right)(l - m)}{\left[a + \left(\frac{\varepsilon_2}{\varepsilon_1} - 1\right)b\right]^2} \left[\frac{\phi_o}{a + \left(\frac{\varepsilon_2}{\varepsilon_1} - 1\right)b} - E_o\right] \tag{A.3.19}$$

which is the correct result when compared with the analytical result (A.3.6).

A.3.2 Spherical Capacitor

For an analytical example, a spherical capacitor is taken as in Fig. A.15, where two spherical electrodes are separated by two dielectrics of permittivity ε_1 and ε_2, and a voltage ϕ_o is applied between the two electrodes. The design objective is to obtain a target electric field $\mathbf{E}_o$ in Ω_p by moving the interface γ. The design variable is the radius b of the interface. The design sensitivity with respect to the design variable b is analytically calculated, and then its result is compared with the result by the sensitivity formula.

The objective function to be minimized is defined as

$$F = \int_{\Omega} (\mathbf{E}(\phi) - \mathbf{E}_o)^2 m_p \mathrm{d}\Omega \tag{A.3.20}$$

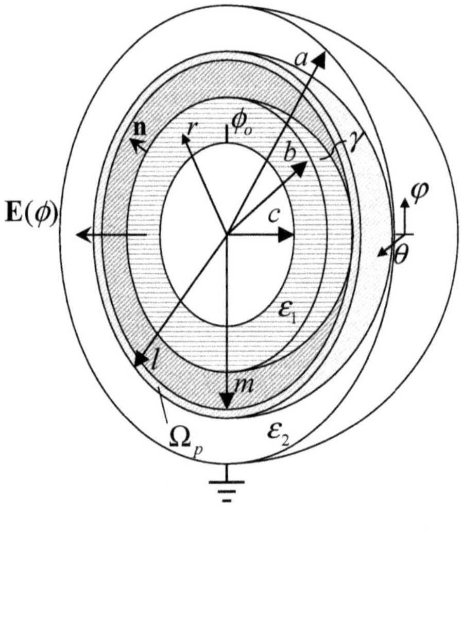

Fig. A.15 Spherical capacitor-interface design

where

$$\mathbf{E}(\phi_1) = \frac{\phi_o}{\frac{\varepsilon_1}{\varepsilon_2}\left(\frac{1}{b}-\frac{1}{a}\right)+\left(\frac{1}{c}-\frac{1}{b}\right)}\frac{1}{r^2}\mathbf{r} \quad \text{for } c \le r \le b \tag{A.3.21}$$

$$\mathbf{E}(\phi_2) = \frac{\phi_o}{\left(\frac{1}{b}-\frac{1}{a}\right)+\frac{\varepsilon_2}{\varepsilon_1}\left(\frac{1}{c}-\frac{1}{b}\right)}\frac{1}{r^2}\mathbf{r} \quad \text{for } b \le r \le a \tag{A.3.22}$$

$$\mathbf{E}_o = \frac{E_o}{r^2}\mathbf{r} \tag{A.3.23}$$

The objective function is rewritten by using the fields (A.3.22) and (A.3.23);

$$F = 4\pi\left(\frac{1}{m}-\frac{1}{l}\right)\left[\frac{\phi_o}{\left(\frac{1}{b}-\frac{1}{a}\right)+\frac{\varepsilon_2}{\varepsilon_1}\left(\frac{1}{c}-\frac{1}{b}\right)} - E_o\right]^2 \tag{A.3.24}$$

The analytical sensitivity is obtained by differentiating the objective function (A.3.24) with respect to the radius b of the interface;

$$\frac{\mathrm{d}F}{\mathrm{d}b} = -8\pi\phi_o\frac{\left(\frac{\varepsilon_2}{\varepsilon_1}-1\right)\left(\frac{1}{m}-\frac{1}{l}\right)}{b^2\left[\left(\frac{1}{b}-\frac{1}{a}\right)+\frac{\varepsilon_2}{\varepsilon_1}\left(\frac{1}{c}-\frac{1}{b}\right)\right]^2}\left[\frac{\phi_o}{\left(\frac{1}{b}-\frac{1}{a}\right)+\frac{\varepsilon_2}{\varepsilon_1}\left(\frac{1}{c}-\frac{1}{b}\right)} - E_o\right] \tag{A.3.25}$$

The shape sensitivity can also be calculated by using the sensitivity formula (3.4.69):

$$\dot{G} = \int_\gamma (\varepsilon_2 - \varepsilon_1)\mathbf{E}(\phi_1)\cdot\mathbf{E}(\lambda_2)V_n\mathrm{d}\Gamma \tag{A.3.26}$$

This sensitivity formula requires an adjoint variable λ, which can be obtained in the adjoint variable system in Fig. A.16.

The differential adjoint equation in the spherical coordinate is obtained by using the electric fields (A.3.22) and (A.3.23);

$$\varepsilon_1\frac{1}{r^2}\frac{\mathrm{d}}{\mathrm{d}r}\left(r^2\frac{\mathrm{d}\lambda_1}{\mathrm{d}r}\right) = 0 \quad \text{for } c \le r \le b \tag{A.3.27}$$

$$\varepsilon_2\frac{1}{r^2}\frac{\mathrm{d}}{\mathrm{d}r}\left(r^2\frac{\mathrm{d}\lambda_2}{\mathrm{d}r}\right) = -2\left[\frac{\phi_o}{\left(\frac{1}{b}-\frac{1}{a}\right)+\frac{\varepsilon_2}{\varepsilon_1}\left(\frac{1}{c}-\frac{1}{b}\right)}\frac{1}{r^2} - \frac{E_o}{r^2}\right](\delta(r-m)-\delta(r-l))$$
$$\text{for } b \le r \le a \tag{A.3.28}$$

Integrating (A.3.27) and (A.3.28) yields the solution of the adjoint fields as

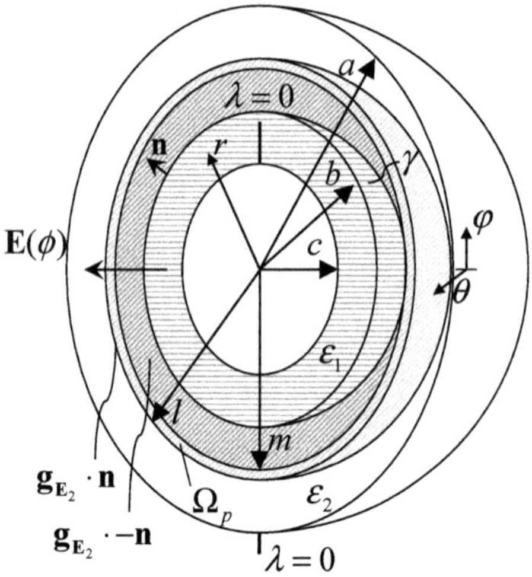

Fig. A.16 Spherical capacitor: adjoint variable system

$$E(\lambda_1) = -\frac{\mathrm{d}\lambda_1}{\mathrm{d}r} = -\frac{k_1}{r^2} \quad \text{for } c \leq r \leq b \tag{A.3.29}$$

$$E(\lambda_2) = -\frac{\mathrm{d}\lambda_2}{\mathrm{d}r} = \frac{2}{\varepsilon_2}\left[\frac{\phi_o}{\left(\frac{1}{b}-\frac{1}{a}\right)+\frac{\varepsilon_2}{\varepsilon_1}\left(\frac{1}{c}-\frac{1}{b}\right)} - E_o\right](H(r-m)-H(r-l))\frac{1}{r^2} - \frac{k_2}{r^2}$$
$$\text{for } b \leq r \leq a \tag{A.3.30}$$

where the integral constants k_1 and k_2 are obtained by integrating (A.3.29), (A.3.30) and applying the boundary and interface conditions in Fig. A.16;

$$k_1 = \frac{2}{\varepsilon_1}\frac{\frac{1}{m}-\frac{1}{l}}{\left(\frac{1}{b}-\frac{1}{a}\right)+\frac{\varepsilon_2}{\varepsilon_1}\left(\frac{1}{c}-\frac{1}{b}\right)}\left[\frac{\phi_o}{\left(\frac{1}{b}-\frac{1}{a}\right)+\frac{\varepsilon_2}{\varepsilon_1}\left(\frac{1}{c}-\frac{1}{b}\right)} - E_o\right] \tag{A.3.31}$$

$$k_2 = \frac{2}{\varepsilon_2}\frac{\frac{1}{m}-\frac{1}{l}}{\left(\frac{1}{b}-\frac{1}{a}\right)+\frac{\varepsilon_2}{\varepsilon_1}\left(\frac{1}{c}-\frac{1}{b}\right)}\left[\frac{\phi_o}{\left(\frac{1}{b}-\frac{1}{a}\right)+\frac{\varepsilon_2}{\varepsilon_1}\left(\frac{1}{c}-\frac{1}{b}\right)} - E_o\right] \tag{A.3.32}$$

Inserting (A.3.31) and (A.3.32) into (A.3.29) and (A.3.30) yields the adjoint fields, respectively;

$$\mathbf{E}(\lambda_1) = -\frac{2}{\varepsilon_1}\frac{\frac{1}{m}-\frac{1}{l}}{\left(\frac{1}{b}-\frac{1}{a}\right)+\frac{\varepsilon_2}{\varepsilon_1}\left(\frac{1}{c}-\frac{1}{b}\right)}\left[\frac{\phi_o}{\left(\frac{1}{b}-\frac{1}{a}\right)+\frac{\varepsilon_2}{\varepsilon_1}\left(\frac{1}{c}-\frac{1}{b}\right)} - E_o\right]\frac{1}{r^2}\mathbf{r} \quad \text{for } c \leq r \leq b \tag{A.3.33}$$

$$\mathbf{E}(\lambda_2) = \frac{2}{\varepsilon_2}\left[\frac{\phi_o}{\left(\frac{1}{b}-\frac{1}{a}\right)+\frac{\varepsilon_2}{\varepsilon_1}\left(\frac{1}{c}-\frac{1}{b}\right)} - E_o\right]\left[H(r-m)-H(r-l)-\frac{\frac{1}{m}-\frac{1}{l}}{\left(\frac{1}{b}-\frac{1}{a}\right)+\frac{\varepsilon_2}{\varepsilon_1}\left(\frac{1}{c}-\frac{1}{b}\right)}\right]\frac{1}{r^2}\mathbf{r} \quad \text{for } b \leq r \leq a \tag{A.3.34}$$

(A.3.21) and (A.3.34) are inserted into the sensitivity formula (A.3.26);

$$\dot{G} = -8\pi\phi_o \frac{\left(\frac{\varepsilon_2}{\varepsilon_1}-1\right)\left(\frac{1}{m}-\frac{1}{l}\right)}{b^2\left[\left(\frac{1}{b}-\frac{1}{a}\right)+\frac{\varepsilon_2}{\varepsilon_1}\left(\frac{1}{c}-\frac{1}{b}\right)\right]^2}\left[\frac{\phi_o}{\left(\frac{1}{b}-\frac{1}{a}\right)+\frac{\varepsilon_2}{\varepsilon_1}\left(\frac{1}{c}-\frac{1}{b}\right)} - E_o\right]V_n \tag{A.3.35}$$

Using $V_n = \frac{\mathrm{d}b}{\mathrm{d}t}$, the design sensitivity is obtained as

$$\frac{\mathrm{d}G}{\mathrm{d}b} = -8\pi\phi_o \frac{\left(\frac{\varepsilon_2}{\varepsilon_1}-1\right)\left(\frac{1}{m}-\frac{1}{l}\right)}{b^2\left[\left(\frac{1}{b}-\frac{1}{a}\right)+\frac{\varepsilon_2}{\varepsilon_1}\left(\frac{1}{c}-\frac{1}{b}\right)\right]^2}\left[\frac{\phi_o}{\left(\frac{1}{b}-\frac{1}{a}\right)+\frac{\varepsilon_2}{\varepsilon_1}\left(\frac{1}{c}-\frac{1}{b}\right)} - E_o\right] \tag{A.3.36}$$

which is identical to the analytical result in (A.3.25).

Next, the sensitivity formula (3.4.69) derived in Sect. 3.4.4 is applied to a shape optimization problem of two-dimensional model, of which the analytic field solution is not given. This design model shows that the sensitivity formula is well applied to the shape design problem of the interface in the electrostatic system. The numerical example is a circular capacitor, whose optimal design is known. If the result of the example is obtained as the expected optimal design, the shape optimization using the sensitivity formula can be said to be feasible for the shape design of the interface. In this example, the sensitivity formula requires the state and the adjoint variables, which are numerically calculated by the finite element method. The sensitivity obtained is used for the optimization algorithm to provide the evolution of the electrode shape. For the optimization algorithm, the level set method is used to represent the shape evolution of the design model.

A.3.3 Circular Capacitor

For an example with a known optimal design, consider a circular capacitor in Fig. A.17, where the inner electrode and the outer electrode are both circular, but the interface between the two dielectrics is elliptical. When a voltage is applied between the two electrodes, the distribution of the electric field is not uniform along the circular direction. If the shape of the interface is changed to be a circle, the electric field between the two electrodes becomes uniform along the circular direction.

The design objective is to obtain a uniform field $\mathbf{E}_o$ in region Ω_p, which is analytically given in the coaxial capacitor. The design variable is the shape of the

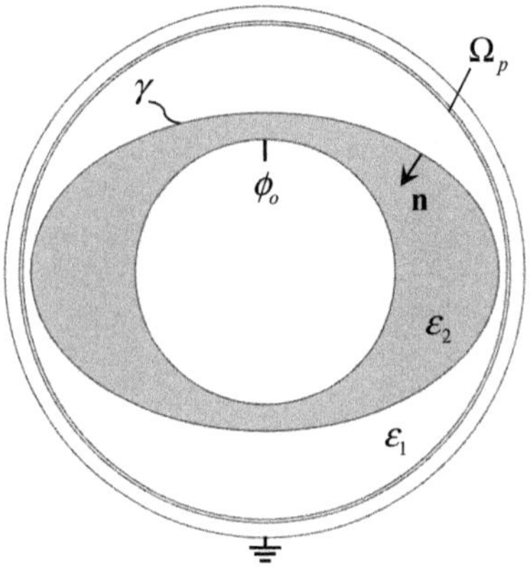

Fig. A.17 Circular capacitor-interface design

interface γ, of which the optimal shape is a circle. The objective function to be minimized is defined as the integral of the field difference in Ω_p.

$$F = \int_{\Omega} (\mathbf{E}(\phi) - \mathbf{E}_o)^2 m_p \mathrm{d}\Omega \tag{A.3.37}$$

where

$$\mathbf{E}_o = \frac{E_o}{r}\mathbf{r} \tag{A.3.38}$$

The variational adjoint equation for (A.3.37) is obtained as

$$a(\lambda, \bar{\lambda}) = \int_{\Omega} 2(\mathbf{E}(\phi) - \mathbf{E}_o) \cdot \mathbf{E}(\bar{\lambda}) m_p \mathrm{d}\Omega \quad \forall\, \bar{\lambda} \in \Phi \tag{A.3.39}$$

The sensitivity formula (3.4.69) is applied to this interface design:

$$\dot{G} = \int_{\gamma} (\varepsilon_2 - \varepsilon_1)\mathbf{E}(\phi_1) \cdot \mathbf{E}(\lambda_2) V_n \mathrm{d}\Gamma \tag{A.3.40}$$

For this minimization problem, the design velocity is taken as

$$V_n = -(\varepsilon_2 - \varepsilon_1)\mathbf{E}(\phi_1) \cdot \mathbf{E}(\lambda_2) \tag{A.3.41}$$

The design result is shown in Fig. A.18, where the shape of the interface becomes gradually a circle as expected. The final design of the circular shape is obtained at the 20 s, when the objective function converges to zero as in Fig. A.19.

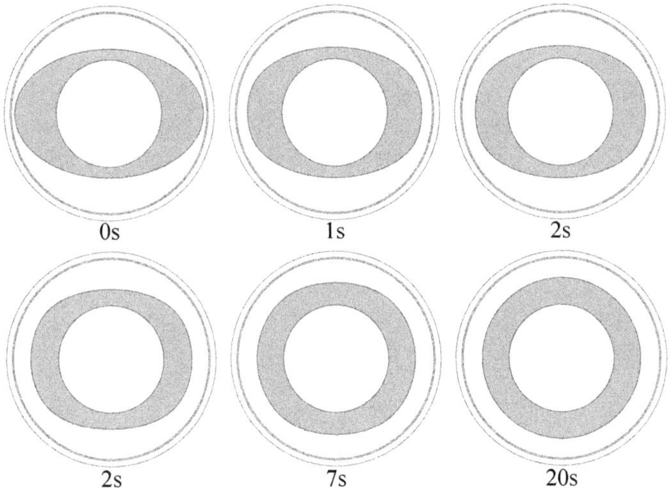

Fig. A.18 Circular capacitor: shape variation

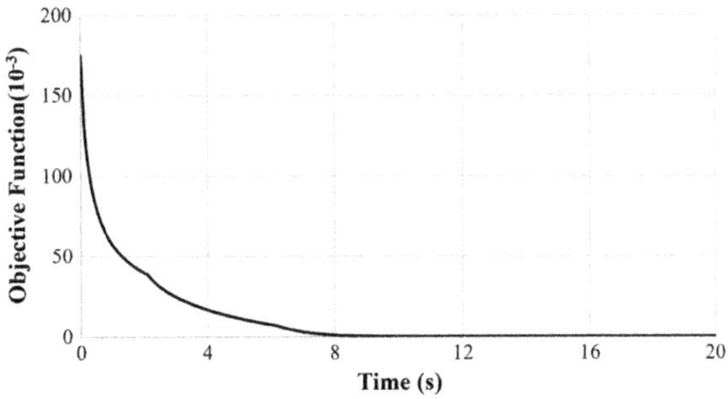

Fig. A.19 Circular capacitor: evolution of objective function

The optimal design by the shape sensitivity is well applied to the interface design problem in the two-dimensional electrostatic system.

A.4 Interface Design for System Energy

For the objective function of the system energy, the sensitivity formula for the interface variation was derived as (3.5.50) in Sect. 3.5.5. One-dimensional examples with the analytic field solutions are taken to show that the sensitivity formula is correct. Comparison of the analytical sensitivity results and the ones by the sensitivity formula shows that the two results are the same. The analytical examples are a parallel capacitor and a spherical capacitor, which are the one-dimensional capacitor models in the rectangular and the spherical coordinates, respectively.

A.4.1 Parallel Plate Capacitor

As an analytical example, a parallel plate capacitor is given as in Fig. A.20, where two parallel conducting electrodes of surface area S are separated by two dielectrics of thickness $a - b$ and b, and their dielectric constants ε_1 and ε_2, respectively, and a voltage ϕ_o is applied between the two electrodes. The design objective is to obtain a desired system energy by moving the interface γ. In this example, the design variable is the thickness of the below dielectric b. The design sensitivity is the derivative of the system energy with respect to the design variable b, and then this sensitivity is compared with the result by the sensitivity formula.

The objective function is the system energy;

$$F = W_e = \int_{\Omega} \frac{1}{2} \varepsilon \mathbf{E}^2(\phi) d\Omega \tag{A.4.1}$$

where

$$\mathbf{E}(\phi_1) = \frac{\phi_o}{\frac{\varepsilon_1}{\varepsilon_2} a + \left(1 - \frac{\varepsilon_1}{\varepsilon_2}\right) b} \mathbf{z} \quad \text{for } 0 \le z \le b \tag{A.4.2}$$

$$\mathbf{E}(\phi_2) = \frac{\phi_o}{a + \left(\frac{\varepsilon_2}{\varepsilon_1} - 1\right) b} \mathbf{z} \quad \text{for } b \le z \le a \tag{A.4.3}$$

The objective function is rewritten by using the electric fields (A.4.2) and (A.4.3);

$$F = \frac{1}{2} \varepsilon_2 \frac{\phi_o^2}{a + \left(\frac{\varepsilon_2}{\varepsilon_1} - 1\right) b} S \tag{A.4.4}$$

The analytical sensitivity per unit area is obtained by differentiating the objective function (A.4.4) with respect to b;

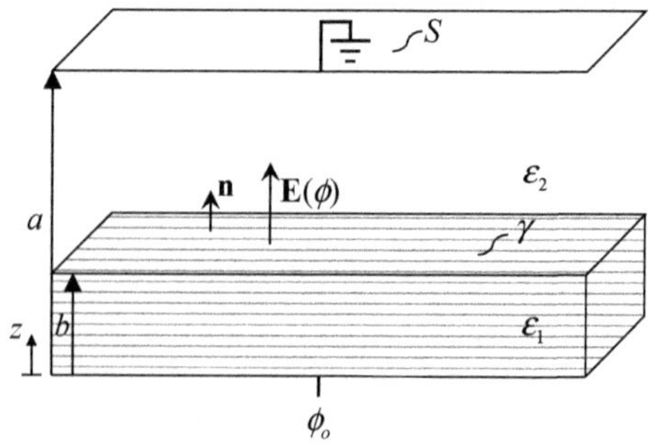

Fig. A.20 Parallel plate capacitor-interface design, system energy

$$\frac{\mathrm{d}F}{\mathrm{d}b} = -\frac{1}{2}\varepsilon_2\left(\frac{\varepsilon_2}{\varepsilon_1} - 1\right)\left[\frac{\phi_o}{a + \left(\frac{\varepsilon_2}{\varepsilon_1} - 1\right)b}\right]^2 \tag{A.4.5}$$

This analytical sensitivity result is compared with the result obtained from the sensitivity formula (3.5.50) in Sect. 3.5.5:

$$\dot{G} = -\int_{\gamma} \frac{1}{2}(\varepsilon_2 - \varepsilon_1)\mathbf{E}(\phi_1) \cdot \mathbf{E}(\phi_2)V_n \mathrm{d}\Gamma \tag{A.4.6}$$

(A.4.2) and (A.4.3) are inserted into the sensitivity formula (A.4.6);

$$\dot{G} = -\frac{1}{2}\varepsilon_2\left(\frac{\varepsilon_2}{\varepsilon_1} - 1\right)\left[\frac{\phi_o}{a + \left(\frac{\varepsilon_2}{\varepsilon_1} - 1\right)b}\right]^2 V_n S \tag{A.4.7}$$

Using $V_n = \frac{\mathrm{d}_b}{\mathrm{d}_t}$, the design sensitivity per unit area is obtained as

$$\frac{\mathrm{d}G}{\mathrm{d}b} = -\frac{1}{2}\varepsilon_2\left(\frac{\varepsilon_2}{\varepsilon_1} - 1\right)\left[\frac{\phi_o}{a + \left(\frac{\varepsilon_2}{\varepsilon_1} - 1\right)b}\right]^2 \tag{A.4.8}$$

Note that this is the same as the analytical result in (A.4.5).

A.4.2 Spherical Capacitor

To take an example which has a known optimal design, consider a spherical capacitor in Fig. A.21, where two spherical electrodes are separated by two dielectrics of permittivity ε_1 and ε_2, and a voltage ϕ_o is applied between the two

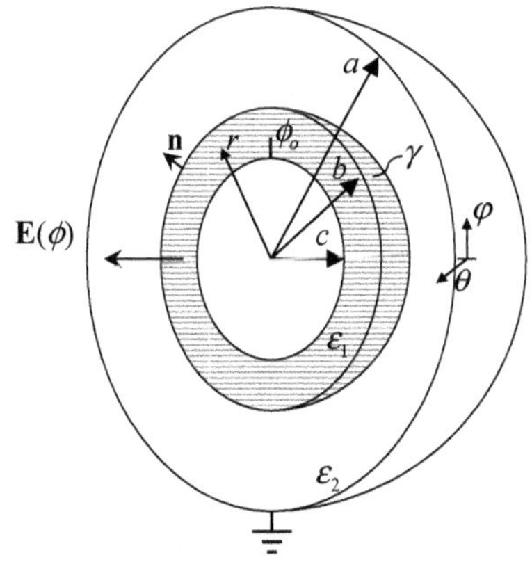

Fig. A.21 Spherical capacitor-interface design, system energy

electrodes. The design objective is to obtain a desired system energy by moving the interface γ; thus, the design variable is the radius b of the interface. The design sensitivity is calculated by differentiating the objective function to the design variable b, and then its result is compared with the result by the sensitivity formula.

The objective function is the system energy;

$$F = W_e = \int_{\Omega} \frac{1}{2} \varepsilon \mathbf{E}^2(\phi) \mathrm{d}\Omega \tag{A.4.9}$$

where

$$\mathbf{E}(\phi_1) = \frac{\phi_o}{\frac{\varepsilon_1}{\varepsilon_2}\left(\frac{1}{b} - \frac{1}{a}\right) + \left(\frac{1}{c} - \frac{1}{b}\right)} \frac{1}{r^2} \mathbf{r} \quad \text{for } c \leq r \leq b \tag{A.4.10}$$

$$\mathbf{E}(\phi_2) = \frac{\phi_o}{\left(\frac{1}{b} - \frac{1}{a}\right) + \frac{\varepsilon_2}{\varepsilon_1}\left(\frac{1}{c} - \frac{1}{b}\right)} \frac{1}{r^2} \mathbf{r} \quad \text{for } b \leq r \leq a \tag{A.4.11}$$

The objective function is rewritten by using the fields (A.4.10) and (A.4.11);

$$F = 2\pi\varepsilon_2 \frac{\phi_o^2}{\left(\frac{1}{b} - \frac{1}{a}\right) + \frac{\varepsilon_2}{\varepsilon_1}\left(\frac{1}{c} - \frac{1}{b}\right)} \tag{A.4.12}$$

Differentiating the objective function (A.4.12) with respect to the radius b of the interface provides the analytical sensitivity;

$$\frac{\mathrm{d}F}{\mathrm{d}b} = -2\pi \frac{\varepsilon_2}{b^2}\left(\frac{\varepsilon_2}{\varepsilon_1} - 1\right)\left[\frac{\phi_o}{\left(\frac{1}{b} - \frac{1}{a}\right) + \frac{\varepsilon_2}{\varepsilon_1}\left(\frac{1}{c} - \frac{1}{b}\right)}\right]^2 \tag{A.4.13}$$

The shape sensitivity can also be calculated by using the sensitivity formula (3.5.50):

$$\dot{G} = -\int_{\gamma} \frac{1}{2}(\varepsilon_2 - \varepsilon_1)\mathbf{E}(\phi_1) \cdot \mathbf{E}(\phi_2) V_n \mathrm{d}\Gamma \tag{A.4.14}$$

(A.4.10) and (A.4.11) are inserted into the sensitivity formula (A.4.14);

$$\dot{G} = -2\pi \frac{\varepsilon_2}{b^2}\left(\frac{\varepsilon_2}{\varepsilon_1} - 1\right)\left[\frac{\phi_o}{\left(\frac{1}{b} - \frac{1}{a}\right) + \frac{\varepsilon_2}{\varepsilon_1}\left(\frac{1}{c} - \frac{1}{b}\right)}\right]^2 V_n \tag{A.4.15}$$

Using $V_n = \frac{\mathrm{d}b}{\mathrm{d}t}$, the design sensitivity is obtained as

$$\frac{\mathrm{d}G}{\mathrm{d}b} = -2\pi\frac{\varepsilon_2}{b^2}\left(\frac{\varepsilon_2}{\varepsilon_1} - 1\right)\left[\frac{\phi_o}{\left(\frac{1}{b} - \frac{1}{a}\right) + \frac{\varepsilon_2}{\varepsilon_1}\left(\frac{1}{c} - \frac{1}{b}\right)}\right]^2 \quad \text{(A.4.16)}$$

which is the correct result when compared with the analytical result in (A.4.13).

Next, the sensitivity formula (3.5.50) derived in Sect. 3.5.5 is applied to a shape optimization problem of two-dimensional model, of which the analytic field solution is not given. This design model shows that the sensitivity formula is well applied to the shape design problem of the interface for the system energy in the electrostatic system. The numerical example is a circular capacitor which has the known optimal design. If the result of the example is obtained as the expected optimal design, the shape optimization using the sensitivity formula is feasible for the shape design of the interface. In this example, the evaluation of the sensitivity formula needs the state variable, which is numerically calculated by the finite element method. The sensitivity evaluated is used for the optimization algorithm, which provides the evolution of the electrode shape. The level set method is used as an optimization algorithm to provide the shape evolution of the design model.

A.4.3 Circular Capacitor

To take an example with a known optimal design, consider a circular capacitor in Fig. A.22, where the inner electrode and the outer electrode are both circular, but the interface between the two dielectrics is elliptical. When a voltage ϕ_o is applied between the two electrodes, the distribution of the electric field is not uniform along the circular direction. If the shape of the interface is changed to be a circle, the

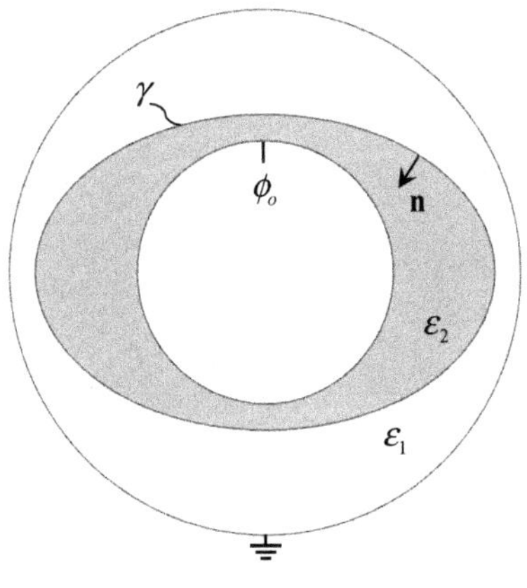

Fig. A.22 Circular capacitor-interface design, system energy

stored system energy becomes the minimum under the constraint of constant dielectric volume.

The objective function to be minimized is the system energy;

$$F = W_e = \int_{\Omega} \frac{1}{2} \varepsilon \mathbf{E}^2(\phi) \mathrm{d}\Omega \tag{A.4.17}$$

and the design variable is the shape of the interface γ. The shape sensitivity for this interface design is the sensitivity formula (3.5.50):

$$\dot{G} = -\int_{\gamma} \frac{1}{2} (\varepsilon_2 - \varepsilon_1) \mathbf{E}(\phi_1) \cdot \mathbf{E}(\phi_2) V_n \mathrm{d}\Gamma \tag{A.4.18}$$

For this minimization problem, the design velocity is taken as

$$V_n = \frac{1}{2} (\varepsilon_2 - \varepsilon_1) \mathbf{E}(\phi_1) \cdot \mathbf{E}(\phi_2) \tag{A.4.19}$$

This problem is subject to the constraint of constant dielectric volume;

$$\int_{\Omega_1} \mathrm{d}\Omega = C \tag{A.4.20}$$

where the constant C is a given dielectric volume. The constant volume (A.4.20) is equivalent to the zero sum of the design velocity on the interface, which is obtained by differentiating (A.4.20);

$$\dot{C} = \int_{\Omega_1} 1' \mathrm{d}\Omega + \int_{\gamma} V_n \mathrm{d}\Gamma = \int_{\gamma} V_n \mathrm{d}\Gamma = 0 \tag{A.4.21}$$

For the constraint of the constant volume, the modified design velocity U_n is obtained by subtracting its averaged V_{na};

$$U_n = V_n - V_{na} \tag{A.4.22}$$

where

$$V_{na} = \int_{\gamma} \frac{1}{2} (\varepsilon_2 - \varepsilon_1) \mathbf{E}(\phi_1) \cdot \mathbf{E}(\phi_2) \mathrm{d}\Gamma \bigg/ \int_{\gamma} \mathrm{d}\Gamma \tag{A.4.23}$$

The design result is shown in Fig. A.23, where the shape of the interface becomes gradually a circle as expected. The final design of the circular shape is

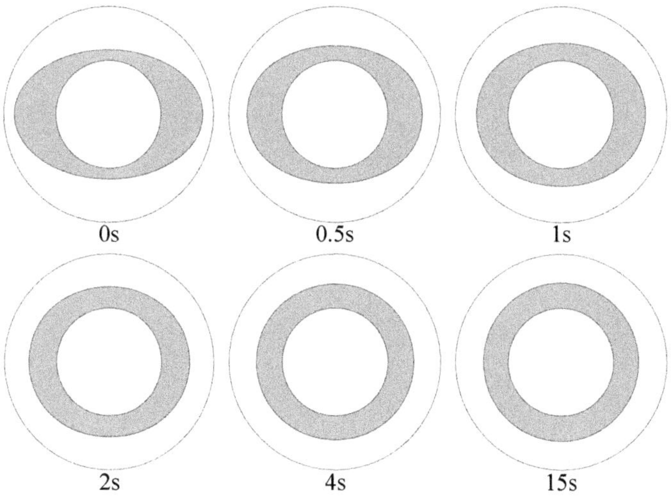

Fig. A.23 Circular capacitor: shape variation

Fig. A.24 Circular capacitor: evolution of objective function

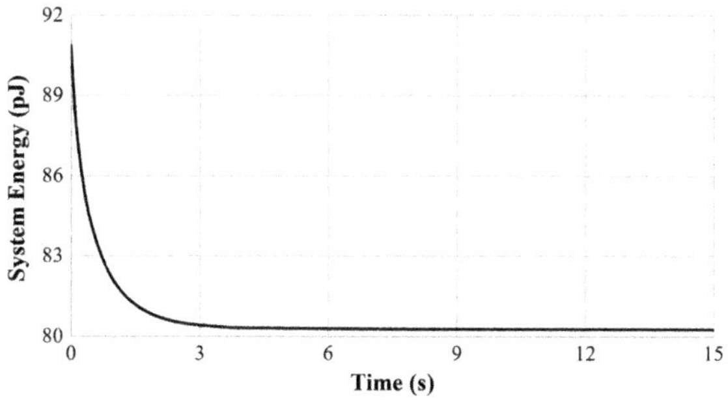

obtained at the 15, when the system energy converges to the minimum as in Fig. A.24. The result of this example shows that the shape sensitivity analysis is feasible for the interface design problem with the system energy objective function in the two-dimensional electrostatic system.

Appendix B
More Examples of Magnetostatic System

B.1 Interface Design

For the interface design problem in Sect. 4.1.4, the sensitivity formula was derived as (4.1.75). One-dimensional example, which has the analytic field solution, is taken to show the correctness of the derived sensitivity formula. The sensitivity result analytically calculated in the example is compared with the one of the sensitivity formula. The analytical example is two infinite parallel plates carrying opposite currents.

B.1.1 Parallel Current-Carrying Plates

For an analytical example, a model of two parallel current plates is given in Fig. B.1a, where the same currents of density J_z in the parallel plates flow in the opposite direction. The analysis model has both plates of thickness $b - a$ and surface area of $S = \mathrm{LD}$. The design objective is to obtain a target magnetic flux density $\mathbf{B}_o$ in region Ω_p by moving the interface γ, which is the outer surface of the plates. The design variable is the height b, which is the distance from the center to the outer surface. The design sensitivity is calculated with respect to the design variable b. Using Dirichlet boundary condition, this model is simplified to the model shown in Fig. B.1b.

The objective function is defined as

$$F = \int_{\Omega} (\mathbf{B}(\mathbf{A}) - \mathbf{B}_o)^2 m_p \mathrm{d}\Omega \tag{B.1.1}$$

I. H. Park, *Design Sensitivity Analysis and Optimization of Electromagnetic Systems*,
Mathematical and Analytical Techniques with Applications to Engineering,
https://doi.org/10.1007/978-981-13-0230-5

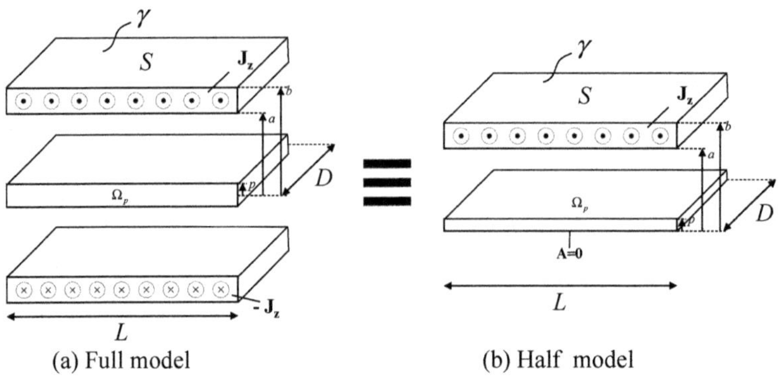

Fig. B.1 Parallel current-carrying plates model

where

$$\mathbf{B}(\mathbf{A}) = \mu_0 J_z (b - a)\mathbf{x} \quad \text{for } 0 \leq s < a \tag{B.1.2}$$

$$\mathbf{B}_o = B_o \mathbf{x} \tag{B.1.3}$$

The objective function is rewritten by using the field (B.1.2) and (B.1.3);

$$F = p[\mu_0 J_z (b - a) - B_o]^2 S \tag{B.1.4}$$

The analytical sensitivity per unit area is obtained by differentiating the objective function (B.1.4) with respect to b;

$$\frac{\mathrm{d}F}{\mathrm{d}b} = 2p\mu_0 J_z [\mu_0 J_z (b - a) - B_o] \tag{B.1.5}$$

Alternatively, the sensitivity is calculated by using the sensitivity formula (4.1.75) in Sect. 4.1.4:

$$\dot{G} = \int_{\gamma} J_z \lambda_z V_n \mathrm{d}\Gamma \tag{B.1.6}$$

This formula requires the adjoint variable, which can be obtained in the adjoint variable system in Fig. B.2.

The differential adjoint equation is given as

$$\nabla \times \nu(\nabla \times \boldsymbol{\lambda}) = g_{\mathbf{A}}(\mathbf{A}) m_p + (\nabla \times \mathbf{g}_{\mathbf{B}}(\mathbf{A})) m_p \tag{B.1.7}$$

The adjoint sources $g_{\mathbf{A}}(\mathbf{A})$ and $\mathbf{g}_{\mathbf{B}}(\mathbf{A})$ are obtained from the objective function (B.1.4).

Fig. B.2 Parallel current-carrying plates model: adjoint variable system

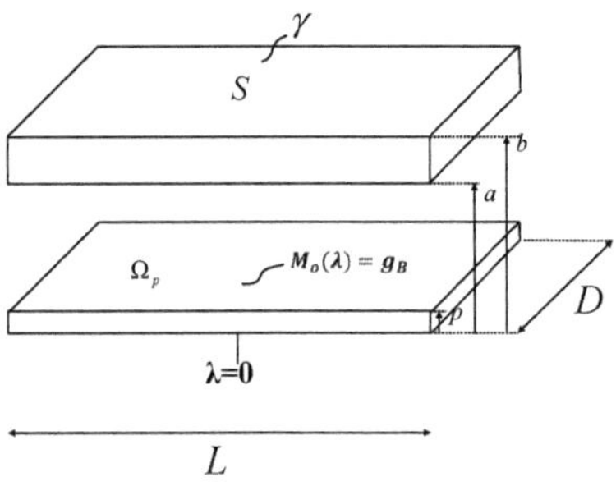

$$g_{\mathbf{A}}(\mathbf{A}) = 0 \tag{B.1.8}$$

$$\mathbf{g_B}(\mathbf{A}) = 2(\mathbf{B}(\mathbf{A}) - \mathbf{B}_o)m_p = 2[\mu_0 J_z(b-a) - B_o]m_p\mathbf{x} \tag{B.1.9}$$

$\mathbf{B}(\lambda_z)$ in this adjoint system is obtained as

$$\mathbf{B}(\lambda_z) = \mu_0 g_B(A_z) = 2\mu_0[\mu_0 J_z(b-a) - B_o]m_p\mathbf{x} \tag{B.1.10}$$

By using the notation $B(\lambda_z) = \nabla \times \lambda_z$, λ_z on the γ is obtained as

$$\lambda_z = \int_0^p B(\lambda_z)\mathrm{d}l = 2p\mu_0[\mu_0 J_z(b-a) - B_o] \tag{B.1.11}$$

(B.1.11) is inserted into the sensitivity formula (B.1.6);

$$\dot{G} = 2p\mu_0 J_z[\mu_0 J_z(b-a) - B_o]V_n S \tag{B.1.12}$$

Using $V_n = \frac{\mathrm{d}b}{\mathrm{d}t}$, the design sensitivity per unit area is obtained as

$$\frac{\mathrm{d}G}{\mathrm{d}b} = 2p\mu_0 J_z[\mu_0 J_z(b-a) - B_o] \tag{B.1.13}$$

which is the correct result when compared with the analytical result in (B.1.5).

Next, the sensitivity formula (4.1.75) derived in Sect. 4.1.4 is applied to two shape optimization problems of two-dimensional design model, of which the analytic field solutions are not given. These design models are taken to show that the sensitivity formula is well applied to the shape design problem of the material interface in the magnetostatic system. The optimal designs for the two examples are known. If the results of the examples are obtained as the expected optimal designs, it can be said that the shape optimization using the sensitivity formula is feasible for the design of the material interface. In these examples, the sensitivity formula requires the state and the adjoint variables, which are numerically calculated by the finite element method. The sensitivity information obtained is used for the optimization algorithm to evolve the material interface shape. For the optimization algorithm, the level set method is used to represent the shape evolution of the design model.

B.1.2 Coaxial Cable—Ferromagnetic Interface Design

As an example with a known optimal design, consider a coaxial cable in Fig. B.3, where the air and the ferromagnetic material of permeability μ are between two cables and the initial shape of the ferromagnetic material is given as a cross-shaped one. In the two cables, the same currents flow in the opposite direction. In this initial shape, the distribution of the magnetic flux density is not uniform along the circular direction. If the shape of the ferromagnetic material is changed to be a circle, the magnetic flux density between the two cables becomes uniform along the circular direction.

The design objective is to obtain a uniform field $\mathbf{B}_o$ in the region Ω_p in Fig. B.3, which is analytically given in the coaxial cable. The design variable is the shape of the interface γ between the ferromagnetic region and the air, the optimal shape of which is a circle. The objective function to be minimized is defined as the integral of the field difference in Ω_p;

$$F = \int_{\Omega} (\mathbf{B}(\mathbf{A}) - \mathbf{B}_o)^2 m_p \mathrm{d}\Omega \tag{B.1.14}$$

where

$$\mathbf{B}_o = \frac{B_o}{r}\mathbf{r} \tag{B.1.15}$$

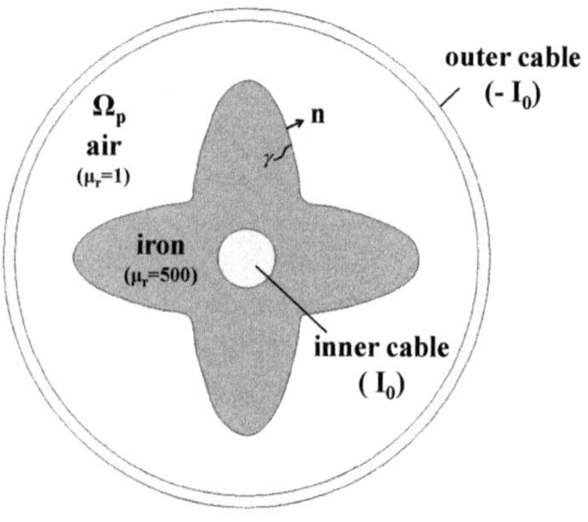

Fig. B.3 Coaxial cable-ferromagnetic material design

The variational adjoint equation for (B.1.14) is obtained as

$$a(\mathbf{A}, \bar{\mathbf{A}}) = \int_{\Omega} 2(\mathbf{B}(\mathbf{A}) - \mathbf{B}_o) \cdot \mathbf{B}(\bar{\mathbf{A}}) m_p \mathrm{d}\Omega \quad \forall\, \bar{\mathbf{A}} \in \Phi \tag{B.1.16}$$

The shape sensitivity for the ferromagnetic material interface is the sensitivity formula (4.1.75):

$$\dot{G} = \int_{\gamma} [\nu_0(1 - \nu_r)\mathbf{B}(\mathbf{A}_1) \cdot \mathbf{B}(\boldsymbol{\lambda}_2)] V_n \mathrm{d}\Gamma \tag{B.1.17}$$

For this minimization problem, the design velocity is taken as

$$V_n = -\nu_0(1 - \nu_r)\mathbf{B}(\mathbf{A}_1) \cdot \mathbf{B}(\boldsymbol{\lambda}_2) \tag{B.1.18}$$

This problem is subject to the constraint of the constant volume;

$$\int_{\Omega_1} \mathrm{d}\Omega = C \tag{B.1.19}$$

where C is a given volume of Ω_1. The constant volume (B.1.19) is equivalent to the zero sum of the design velocity on the ferromagnetic interface, which is obtained by differentiating (B.1.19);

$$\int_{\gamma} V_n \mathrm{d}\Gamma = 0 \tag{B.1.20}$$

For this minimization problem with the volume constraint, the design velocity is taken as

$$U_n = V_n - V_{na} \tag{B.1.21}$$

where

$$V_{na} = -\int_{\gamma} [\nu_0(1 - \nu_r)\mathbf{B}(\mathbf{A}_1) \cdot \mathbf{B}(\boldsymbol{\lambda}_2)] \mathrm{d}\Gamma \Big/ \int_{\gamma} \mathrm{d}\Gamma \tag{B.1.22}$$

is the average of the velocity (B.1.18) over the interface.

The design result is shown in Fig. B.4, where the shape of the ferromagnetic material becomes gradually a circle as expected. The final design of the circular shape is obtained at the 60 s, when the objective function converges to zero as in Fig. B.5. The result of this example shows the feasibility of the shape sensitivity

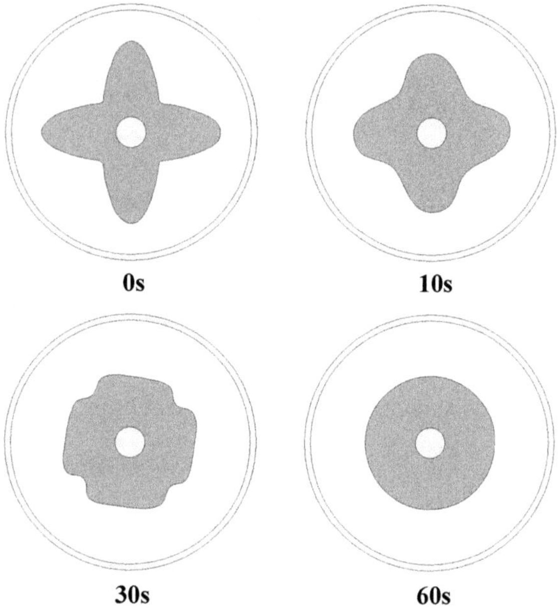

Fig. B.4 Coaxial cable: shape variation

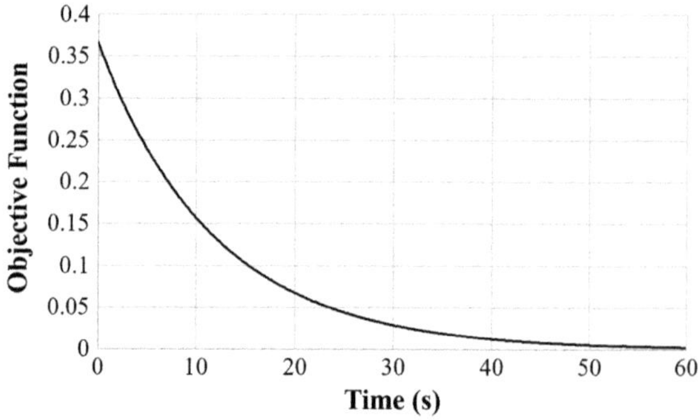

Fig. B.5 Coaxial cable: evolution of objective function

analysis for the ferromagnetic interface in the two-dimensional magnetostatic system.

B.1.3 Coaxial Cable—Current Region Interface Design

For an example with a known optimal design, a coaxial cable is given as in Fig. B.6, where the outer cable is circular and the inner cable is elliptical. In these two cables, the same currents flow in the opposite direction. In this initial shape, the distribution of the magnetic flux density is not uniform in the circular direction. If the shape of the inner cable is changed to be a circle, this model becomes a coaxial

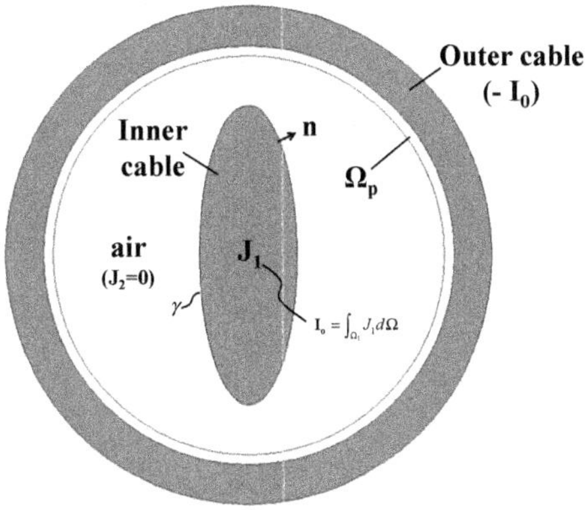

Fig. B.6 Coaxial cable-current region design

cable and the magnetic flux density between the two cables becomes uniform in the circular direction.

The design objective is to obtain a uniform field $\mathbf{B}_o$ in the region Ω_p in Fig. B.6, which is analytically given in the 1D coaxial cable. The design variable is the shape of the interface γ between the inner cable and the air, the optimal shape of which is a circle. The objective function to be minimized is defined as the integral of the field difference in Ω_p;

$$F = \int_{\Omega} (\mathbf{B}(\mathbf{A}) - \mathbf{B}_o)^2 m_p \mathrm{d}\Omega \tag{B.1.23}$$

where

$$\mathbf{B}_0 = \frac{B_o}{r}\mathbf{r} \tag{B.1.24}$$

The variational adjoint equation for (B.1.23) is obtained as

$$a(\mathbf{A}, \bar{\mathbf{A}}) = \int_{\Omega} 2(\mathbf{B}(\mathbf{A}) - \mathbf{B}_o) \cdot \mathbf{B}(\bar{\mathbf{A}}) m_p \mathrm{d}\Omega \quad \forall\, \bar{\mathbf{A}} \in \Phi \tag{B.1.25}$$

The shape sensitivity for the current region is the sensitivity formula (4.1.75):

$$\dot{G} = \int_{\gamma} \mathbf{J}_1 \cdot \lambda_2 V_n \mathrm{d}\Gamma \tag{B.1.26}$$

This problem has a constraint of constant volume. The minimization problem under the volume constraint, the design velocity is expressed as

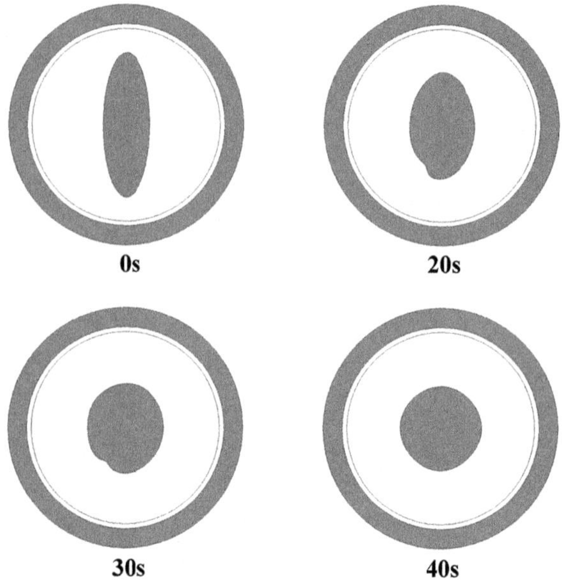

Fig. B.7 Coaxial cable: shape variation

$$U_n = V_n - V_{na} \tag{B.1.27}$$

where

$$V_n = -\mathbf{J}_1 \cdot \boldsymbol{\lambda}_2 \tag{B.1.28}$$

$$V_{na} = -\int_{\gamma} \mathbf{J}_1 \cdot \boldsymbol{\lambda}_2 \mathrm{d}\Gamma \Big/ \int_{\gamma} \mathrm{d}\Gamma \tag{B.1.29}$$

The design result is shown in Fig. B.7, where the shape of the current region becomes gradually a circle as expected. The final design of the circular shape is obtained at the 40 s, when the objective function converges to zero as in Fig. B.8. The optimal design by the shape sensitivity is well applied to the current interface design problem in the two-dimensional magnetostatic system.

B.2 Interface Design for System Energy

For the energy objective problem in Sect. 4.2.4, the sensitivity formula was derived as (4.2.40). To show that the sensitivity formula is correct, one-dimensional analytical examples, which have the analytic field solutions, are taken. The sensitivity results analytically calculated in the examples are compared with the ones of the sensitivity formula. For this purpose, an analytical example is employed to compare the sensitivities for the ferromagnetic interface.

Fig. B.8 Coaxial cable: evolution of objective function

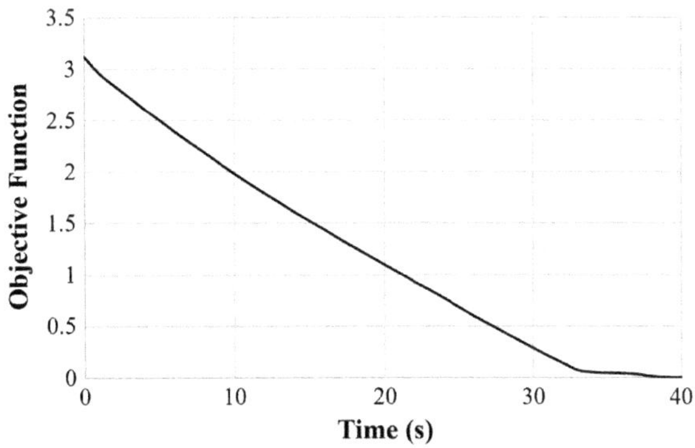

B.2.1 Coaxial Cable—Ferromagnetic Material Interface

For an analytical example, a coaxial cable is given in Fig. B.9, where two ferromagnetic materials are between two cables and the same currents I_o flow in the opposite direction through the inner and the outer cable. The inner ferromagnetic material has thickness $b - a$ and permeability μ_1, and the outer one has thickness $c - b$ and permeability μ_2. The design objective is to obtain a desired system energy by moving interface γ where the two ferromagnetic materials meet. The design variable is the radius b of the interface γ. The design sensitivity for the system energy is analytically calculated with respect to the design variable b.

The objective function is the system energy;

$$F = W_m = \frac{1}{2}\int_{\Omega} \frac{1}{\mu} B^2 \mathrm{d}\Omega \tag{B.2.1}$$

where

$$\mathbf{B}_1 = \frac{\mu_1 I_o}{2\pi s}\boldsymbol{\phi} \quad \text{for} \quad a \leq s < b \tag{B.2.2}$$

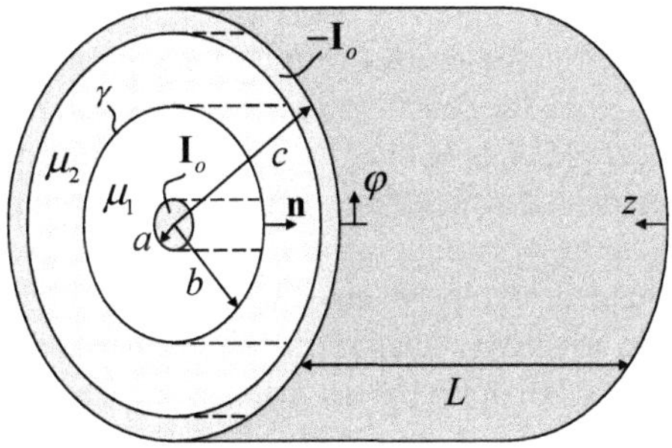

Fig. B.9 Coaxial cable-ferromagnetic material design

$$\mathbf{B}_2 = \frac{\mu_2 I_o}{2\pi s}\boldsymbol{\Phi} \quad \text{for} \quad b<s<c \tag{B.2.3}$$

$$\mathbf{B} = 0 \quad \text{for} \quad s<a,\ c<s \tag{B.2.4}$$

The objective function is rewritten by using the field (B.2.2), (B.2.3), and (B.2.4);

$$\begin{aligned} F &= \frac{1}{2}\int_0^{2\pi} \mathrm{d}\phi \int_0^L \mathrm{d}z \left(\int_a^b \frac{1}{\mu_1}\left(\frac{\mu_1 I_o}{2\pi r}\right)^2 r\mathrm{d}r + \frac{1}{2}\int_b^c \frac{1}{\mu_2}\left(\frac{\mu_2 I_o}{2\pi r}\right)^2 r\mathrm{d}r \right) \\ &= \frac{I_o^2 L}{4\pi}\left(\mu_1 \ln\frac{b}{a} + \ln\frac{c}{b}\right) \end{aligned} \tag{B.2.5}$$

The analytical sensitivity per unit length is obtained by differentiating the objective function (B.2.5) with respect to b;

$$\frac{\mathrm{d}F}{\mathrm{d}b} = \frac{I_o^2}{4\pi b}(\mu_1 - \mu_2) \tag{B.2.6}$$

Alternatively, the sensitivity can be calculated by using the sensitivity formula (4.2.40) in Sect. 4.2.4:

$$\dot{G} = \frac{1}{2}\int_\gamma \left(\frac{1}{\mu_2} - \frac{1}{\mu_1}\right)\mathbf{B}_1 \cdot \mathbf{B}_2 V_n \mathrm{d}\Gamma \tag{B.2.7}$$

(B.2.2) and (B.2.3) is inserted into the sensitivity formula (B.2.7);

$$\dot{G} = \frac{I_o^2}{4\pi b}(\mu_1 - \mu_2) V_n L \tag{B.2.8}$$

Using $V_n = \frac{\mathrm{d}_b}{\mathrm{d}_t}$, the design sensitivity per unit length is obtained as

$$\frac{\mathrm{d}G}{\mathrm{d}b} = \frac{I_o^2}{4\pi b}(\mu_1 - \mu_2) \tag{B.2.9}$$

which is the correct result when compared with the analytical result in (B.2.6).

Next, the sensitivity formula (4.2.40) for the energy objective function derived in Sect. 4.2.4 is applied to a shape optimization problem of two-dimensional design model, of which the analytic field solution is not given. This design model shows that the sensitivity formula for the energy objective function is well applied to the shape design problem of the interface in the magnetostatic system. In this example, the sensitivity analysis using the derived formula for energy objective function is applied to optimize the material interface shape in a simple numerical model, which has the analytical design. In this example, the sensitivity formula requires the state and the adjoint variables, which are numerically calculated by the finite element method. The sensitivity obtained is used for the optimization algorithm to evolve the material interface shape.

B.2.2 Infinite Solenoid—Ferromagnetic Interface Design

As an example with a known optimal design, consider a solenoid in Fig. B.10, where the length in the direction z is infinite and the ferromagnetic material of permeability μ is inside the solenoid coil. The initial shape of the ferromagnetic material is curved and an electric current flows in the solenoid coil. When the shape of the ferromagnetic material is changed to be a cylinder, the stored system energy becomes the maximum under the constraint of constant ferromagnetic material volume.

The objective function to be maximized is the magnetic system energy;

$$F = W_m = \int_{\Omega} \frac{1}{2} \nu \mathbf{B}(\mathbf{A}) \cdot \mathbf{B}(\mathbf{A}) \mathrm{d}\Omega \tag{B.2.10}$$

The shape sensitivity for this ferromagnetic material interface is the sensitivity formula (4.2.40):

$$\dot{G} = \int_{\gamma} \frac{\nu_0}{2} (1 - \nu_r) \mathbf{B}_1 \cdot \mathbf{B}_2 V_n \mathrm{d}\Gamma \tag{B.2.11}$$

For this maximization problem, the design velocity is taken as

$$V_n = \frac{\nu_0}{2} (1 - \nu_r) \mathbf{B}_1 \cdot \mathbf{B}_2 \tag{B.2.12}$$

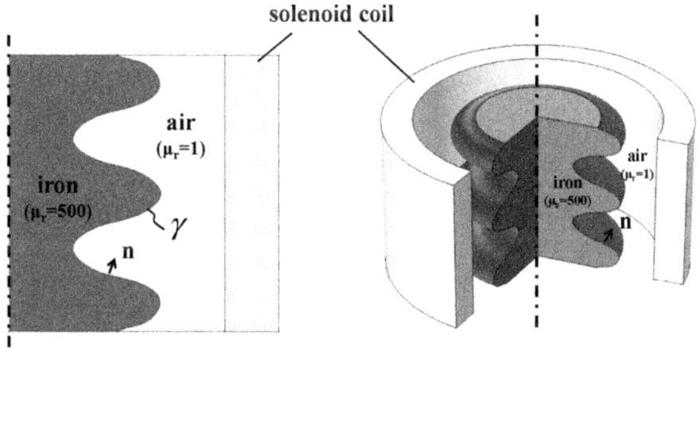

Fig. B.10 Coaxial cable-ferromagnetic material design, system energy

This problem has a constraint of constant volume;

$$\int_{\Omega_1} \mathrm{d}\Omega = C \tag{B.2.13}$$

where C is a constant. The constant volume (B.2.13) is equivalent to the zero sum of the design velocity over the ferromagnetic interface, which is obtained by differentiating (B.2.13).

$$\int_{\gamma} V_n \mathrm{d}\Gamma = 0 \tag{B.2.14}$$

For this maximization problem with the volume constraint, the design velocity is taken as

$$U_n = V_n - V_{na} \tag{B.2.15}$$

where

$$V_{na} = \int_{\gamma} \frac{v_0}{2}(1 - v_r)\mathbf{B}_1 \cdot \mathbf{B}_2 \mathrm{d}\Gamma \Big/ \int_{\gamma} \mathrm{d}\Gamma \tag{B.2.16}$$

is the averaged the velocity on the interface.

The design result is shown in Fig. B.11, where the shape of the ferromagnetic material becomes gradually a cylinder as expected. The final design of the cylindrical shape is obtained at the 40 s, when the objective function converges to the

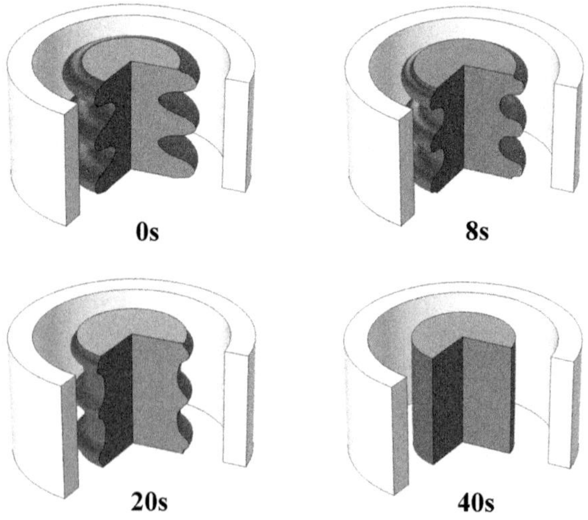

Fig. B.11 Coaxial cable: shape variation

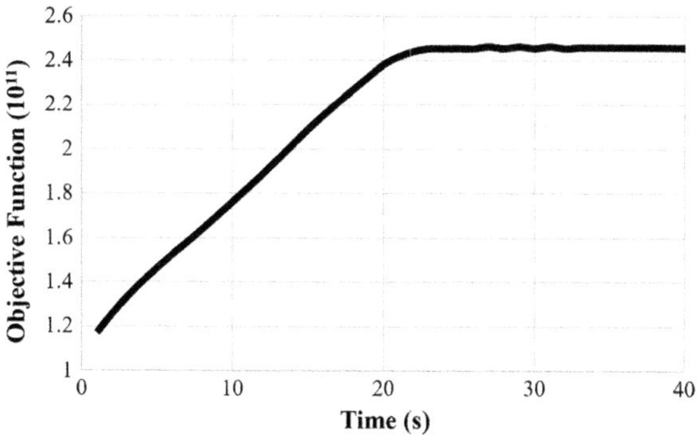

Fig. B.12 Coaxial cable: evolution of objective function

maximum as in Fig. B.12. The result of this example shows the feasibility of the shape sensitivity analysis for the interface of ferromagnetic regions in the two-dimensional axi-symmetric magnetostatic system.

Appendix C
More Examples of Eddy Current System

C.1 Interface Design for System Power

The sensitivity formula (5.2.35) in Sect. 5.2.3 is applied to three shape optimization problems of two-dimensional design model, where the analytic field solutions are unknown. These design models show that the sensitivity formula is well applied to the shape design for the Joule current loss by the eddy current. The numerical examples are the design problems of three interfaces: conductor–air, current region–air and ferromagnetic material–air. These examples have the known optimal designs. If the results of the examples are obtained as the expected optimal designs, it can be said that the shape optimization using the sensitivity formula for the Joule loss is feasible in the design of the eddy current system. In these examples, the state variable, which is required to evaluate the sensitivity formula, is numerically calculated by the finite element method. The sensitivity obtained is used for the optimization algorithm to evolve the interface shape. The level set method is used to represent the shape evolution of the design model.

C.1.1 Conductor–Air Interface Design

As an example that have a known optimal design, consider a two-dimensional eddy current system consisting of a conductor, a sinusoidal current source, and the air as shown in Fig. C.1. The eddy current is induced in the conductor by the time-varying magnetic field by the source current. If the shape of the conductor is changed to a rectangle, the system has the minimum eddy loss under the constraint of the constant conductor volume.

The design objective is to obtain the minimum Joule loss power of the eddy current system. The objective function F to be minimized is defined as

I. H. Park, *Design Sensitivity Analysis and Optimization of Electromagnetic Systems*,
Mathematical and Analytical Techniques with Applications to Engineering,
https://doi.org/10.1007/978-981-13-0230-5

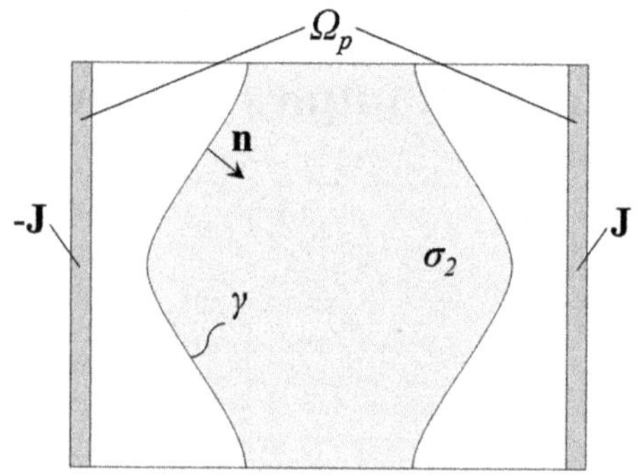

Fig. C.1 Conductor-air interface design

$$F = \mathrm{Re}\left(\int_{\Omega} \mathbf{J} \cdot \mathbf{E}^* m_p \mathrm{d}\Omega \right) \tag{C.1.1}$$

The design variable is the shape of the conductor–air interface γ as shown in Fig C.1. The shape sensitivity of this interface for the Joule loss power is obtained from the sensitivity formula (5.2.35) in Sect. 5.2.3;

$$\dot{P}_r = \mathrm{Re}\left(\int_{\gamma} \omega^2 \sigma_2 \mathbf{A}_1 \cdot \mathbf{A}_2 V_n \mathrm{d}\Gamma \right) \tag{C.1.2}$$

The velocity field V_n is taken for the negative shape sensitivity as

$$V_n = -\mathrm{Re}\left(\omega^2 \sigma_2 \mathbf{A}_1 \cdot \mathbf{A}_2\right) \tag{C.1.3}$$

This design problem has a constraint of constant conductor volume;

$$\int_{\Omega_c} \mathrm{d}\Omega = C \tag{C.1.4}$$

where Ω_c is the conductor region and C is a given value of the constraint. The material derivative of the constraint (C.1.4) is obtained as

$$\int_{\Gamma} V_n \mathrm{d}\Gamma = 0 \tag{C.1.5}$$

In order to satisfy the constraint (C.1.5), the design velocity (C.1.3) is modified by subtracting its average V_{na} to become U_n as

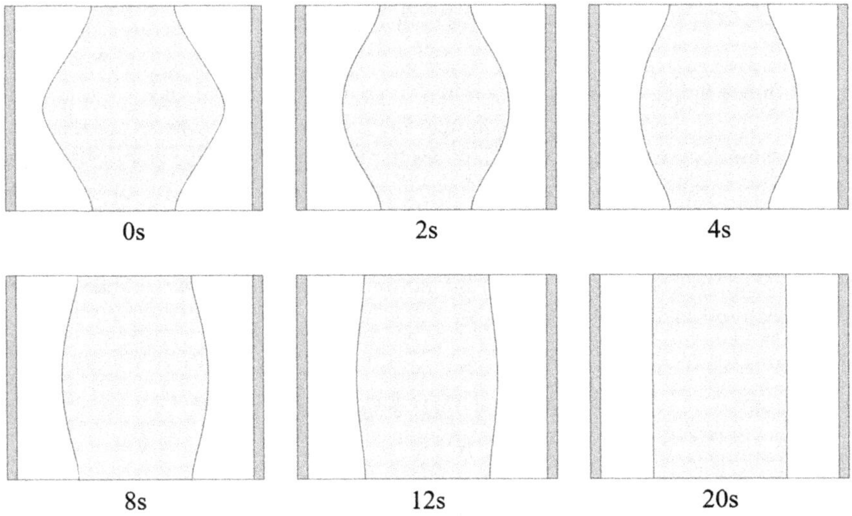

Fig. C.2 Conductor-air interface design: shape variation

$$U_n = V_n - V_{na} \tag{C.1.6}$$

where

$$V_{na} = -\mathrm{Re}\left(\omega^2 \sigma_2 \mathbf{A}_1 \cdot \mathbf{A}_2\right) \bigg/ \int_\gamma \mathrm{d}\Gamma \tag{C.1.7}$$

The design result is shown in Fig. C.2, where the shape of the conductor becomes a rectangle as expected. The final design of the rectangular shape is obtained at the 250 ns, when the objective function value converges to 136 kW/m^3 as in Fig. C.3. The result of this example shows the feasibility of the shape sensitivity for conductor–air interface in the two-dimensional eddy current system.

Fig. C.3 Conductor-ir interface design: evolution of objective function

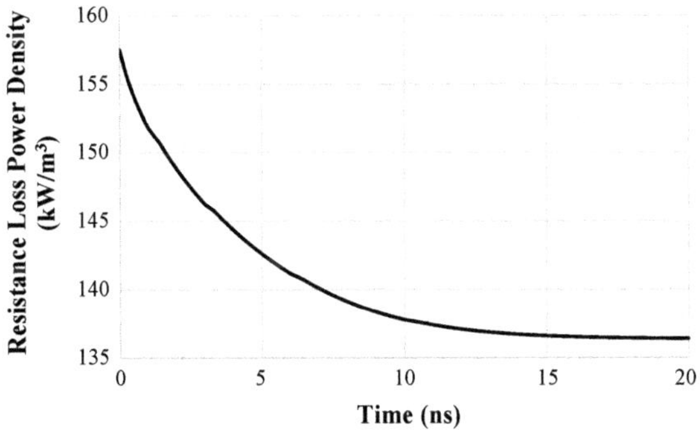

C.1.2 Current Region–Air Interface Design

A two-dimensional eddy current problem consists of a conductor, a sinusoidal current source, and the air as shown in Fig. C.4. The current produces a time-varying magnetic field, which induces the eddy current in the conductor. If the shape of the current region is changed to be a rectangle, the eddy loss of the system becomes the minimum value under the constant volume constraint of the current region.

The design objective is to obtain the minimum eddy loss power of the system. The objective function F is defined as

$$F = \mathrm{Re}\left(\int_{\Omega} \mathbf{J} \cdot \mathbf{E}^{*} m_{p} \mathrm{d}\Omega\right) \tag{C.1.8}$$

The design variable is the shape of the current region–air interface γ as shown in Fig C.4. The shape sensitivity of this interface for the eddy loss power is obtained from (5.2.35);

$$\dot{P}_{r} = \mathrm{Re}\left(\int_{\gamma} -2j\omega \mathbf{J}_{1} \cdot \mathbf{A}_{2} V_{n} \mathrm{d}\Gamma\right) \tag{C.1.9}$$

The velocity field V_n is taken for the negative shape sensitivity as

$$V_{n} = -\mathrm{Re}(-2j\omega \mathbf{J}_{1} \cdot \mathbf{A}_{2}) \tag{C.1.10}$$

In order to satisfy the constraint, the design velocity is modified by subtracting its average V_{na} to become U_n as

$$U_{n} = V_{n} - V_{na} \tag{C.1.11}$$

where

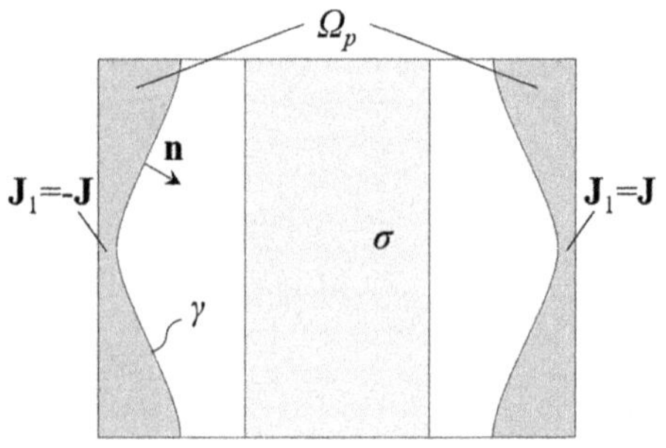

Fig. C.4 Current region-air interface design

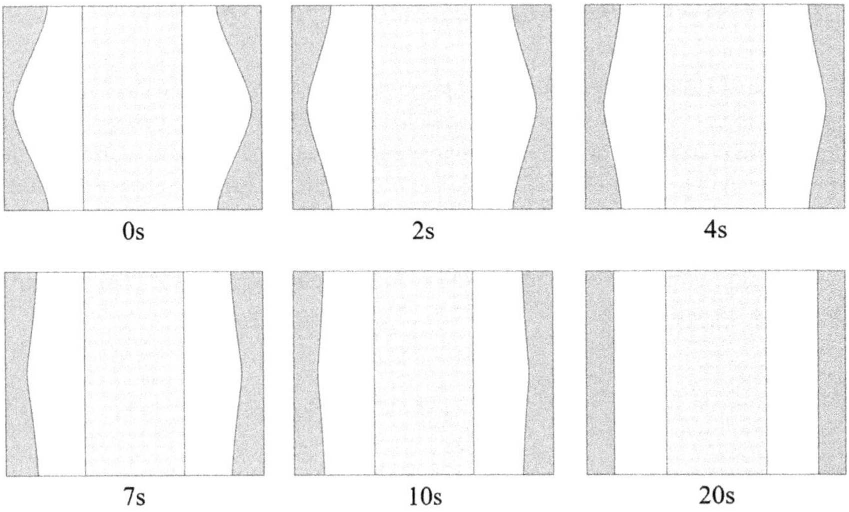

Fig. C.5 Current region-air interface design: shape variation

Fig. C.6 Current region-air interface design: evolution of objective function

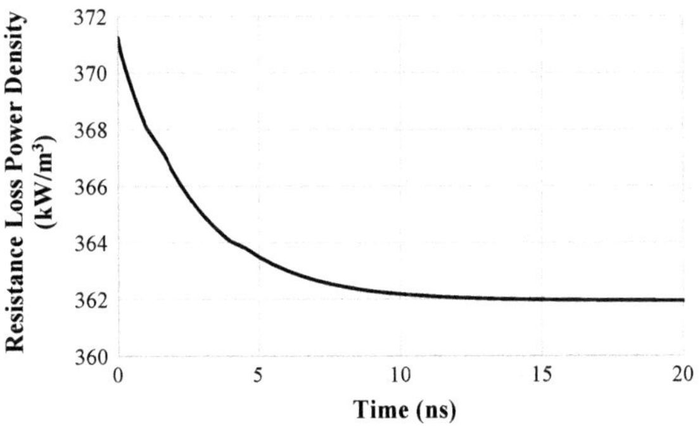

$$V_{na} = -\text{Re}(-2j\omega\mathbf{J}_1 \cdot \mathbf{A}_2) \Big/ \int_\gamma \mathrm{d}\Gamma \tag{C.1.12}$$

The design result is shown in Fig. C.5, where the shape of the current region becomes a rectangle as expected. The final design of the rectangular shape is obtained at the 200 ns, when the objective function value converges to 362 kW/m^3 as in Fig. C.6. The result of this example shows the feasibility of the shape sensitivity for the current region–air interface in the two-dimensional eddy current system.

C.1.3 Ferromagnetic Material–Air Interface Design

A two-dimensional eddy current problem consists of a conductor, a sinusoidal a current source, a ferromagnetic material, and the air as shown in Fig. C.7. The time-varying magnetic field by the source current induces the eddy current in the conductor. If the shape of the ferromagnetic material is changed to be a rectangle, the eddy loss of the system becomes the minimum value under the constant volume constraint of the ferromagnetic material.

The design objective is to obtain the minimum eddy loss power. The objective function F is defined as

$$F = \mathrm{Re}\left(\int_{\Omega} \mathbf{J} \cdot \mathbf{E}^* m_p \mathrm{d}\Omega \right) \tag{C.1.13}$$

The design variable is the shape of the ferromagnetic material–air interface γ as shown in Fig C.7. The shape sensitivity of this interface for the eddy loss power is obtained from (5.2.35);

$$\dot{P}_r = \mathrm{Re}\left(\int_{\gamma} -j\omega(v_2 - v_1)\mathbf{B}_1 \cdot \mathbf{B}_2 V_n \mathrm{d}\Gamma \right) \tag{C.1.14}$$

The velocity field V_n for the negative shape sensitivity is taken as

$$V_n = -\mathrm{Re}[-j\omega(v_2 - v_1)\mathbf{B}_1 \cdot \mathbf{B}_2] \tag{C.1.15}$$

In this problem, the modified velocity U_n for the volume constraint of the ferromagnetic material is taken as

$$U_n = V_n - V_{na} \tag{C.1.16}$$

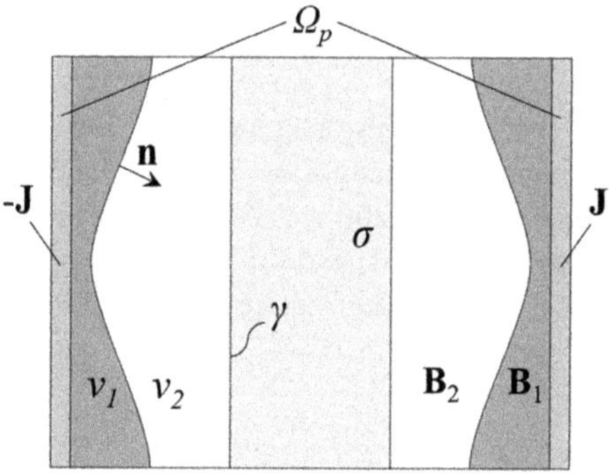

Fig. C.7 Ferromagnetic material-air interface design

where

$$V_{na} = -\mathrm{Re}[-j\omega(v_2 - v_1)\mathbf{B}_1 \cdot \mathbf{B}_2] \bigg/ \int_\gamma \mathrm{d}\Gamma \qquad \text{(C.1.17)}$$

The design result is shown in Fig. C.8, where the shape of the ferromagnetic material becomes a rectangle as expected. The final design of the rectangular shape is obtained at the 250 ns, when the objective function value converges to 17 W/m^3 as in Fig. C.9. The result of this example shows the feasibility of the shape sensitivity for ferromagnetic material–air interface in the two-dimensional eddy current system.

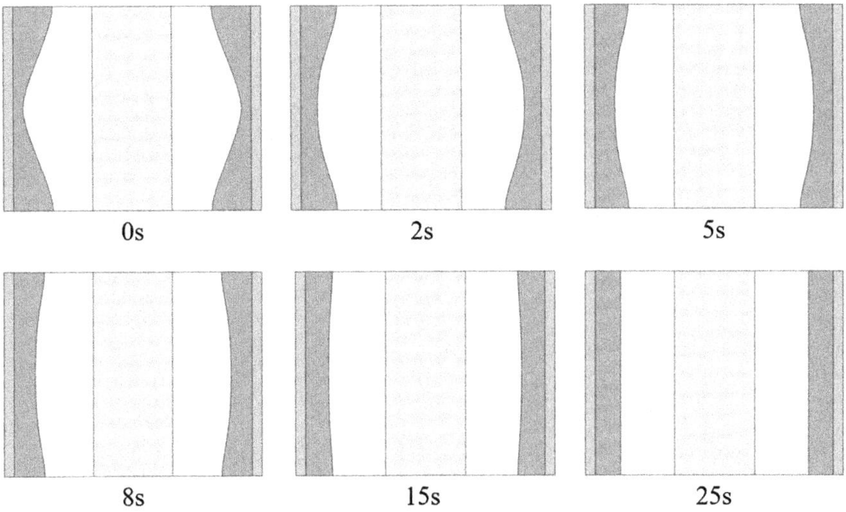

Fig. C.8 Ferromagnetic material-air interface design: shape variation

Fig. C.9 Ferromagnetic material-air interface design: evolution of objective function

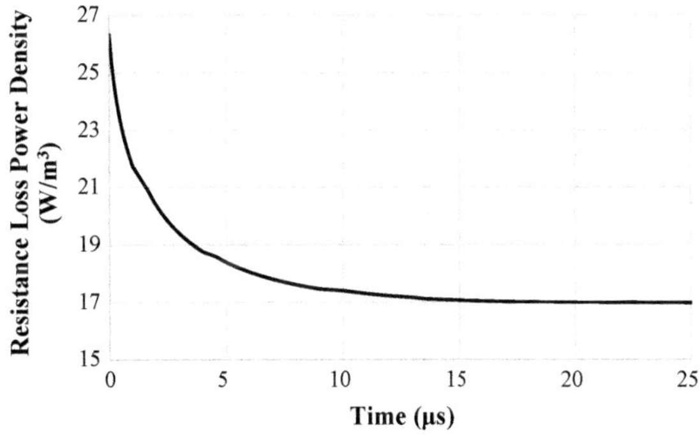

Appendix D
More Examples of DC Conductor System

D.1 Outer Boundary Design for Joule Loss Power

When the objective function in DC conductor system is defined as the Joule loss power, the sensitivity formula was derived as (6.2.10) in Sect. 6.2.2. A one-dimensional example with the analytic field solution is taken to show that the sensitivity formula is correct. The objective is to compare the analytical sensitivity result with the result by the sensitivity formula to ensure that their two results are the same. The analytical example is a rectangular conductor, which is the one-dimensional conductor models in the rectangular coordinate.

D.1.1 Rectangular Conductor

As an example that can be analytically calculated, consider a rectangular conductor in Fig. D.1, where two electrodes are connected to both sides of a conductor of length l, width w, thickness L, and conductivity σ. An electric field, which is uniformly generated in the x-direction, causes a uniform current in the conductor. The change of the conductor size results in the change of the resistance. The design objective is to obtain a desired Joule loss power V^2/R in Ω by moving the conductor boundaries Γ^0 and Γ^1 in each case. The design variables are the width w and the length l of the conductor and the design sensitivity with respect to each design variable is analytically calculated to be compared with the result by the sensitivity formula.

The objective function is the Joule loss power;

$$F = P_r = \int_{\Omega} \sigma E^2(\phi) m_p \mathrm{d}\Omega \tag{D.1.1}$$

I. H. Park, *Design Sensitivity Analysis and Optimization of Electromagnetic Systems*, Mathematical and Analytical Techniques with Applications to Engineering,
https://doi.org/10.1007/978-981-13-0230-5

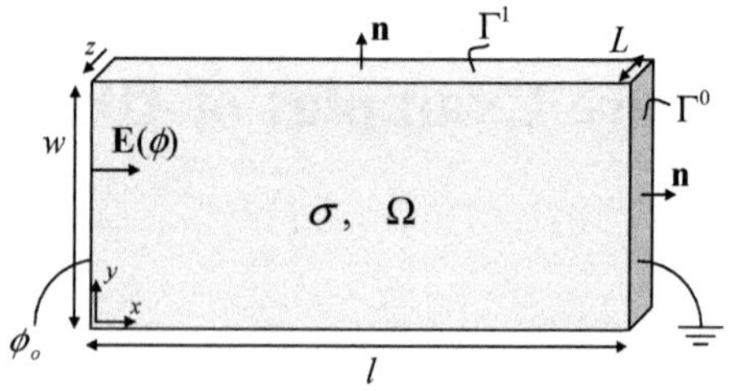

Fig. D.1 Rectangular conductor-outer boundary design, Joule loss power

where

$$\mathbf{E}(\phi) = \frac{\phi_o}{l}\mathbf{x} \tag{D.1.2}$$

The objective function is rewritten by using the field (D.1.2);

$$F = \sigma w L \frac{\phi_o^2}{l} \tag{D.1.3}$$

(1) The sensitivity for the electrode of Dirichlet boundary

The analytical sensitivity per unit length is obtained by differentiating the objective function (D.1.3) with respect to the length l of the conductor;

$$\frac{\mathrm{d}F}{\mathrm{d}l} = -\sigma w \left(\frac{\phi_o}{l}\right)^2 \tag{D.1.4}$$

This analytical sensitivity result is compared with the result obtained from the sensitivity formula (6.2.11):

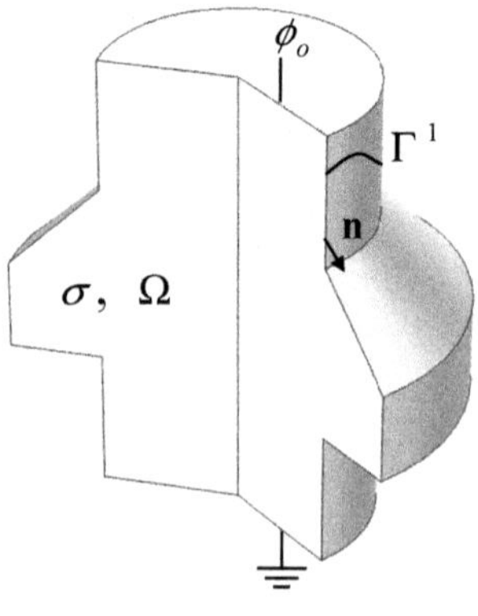

Fig. D.2 Axi-symmetry cylindrical conductor-outer boundary design, Joule loss power

$$\dot{G} = -\int_{\Gamma^0} \sigma E_n^2(\phi) V_n \mathrm{d}\Gamma \tag{D.1.5}$$

(D.1.2) is inserted into the sensitivity formula (D.1.5);

$$\dot{G} = -\sigma w L \left(\frac{\phi_o}{l}\right)^2 V_n \tag{D.1.6}$$

Using $V_n = \frac{\mathrm{d}l}{\mathrm{d}t}$, the design sensitivity per unit length is obtained as

$$\frac{\mathrm{d}G}{\mathrm{d}l} = -\sigma w \left(\frac{\phi_o}{l}\right)^2 \tag{D.1.7}$$

which is identical to the analytical result in (D.1.4).

(2) The sensitivity for the conductor surface of the Neumann boundary

The analytical sensitivity per unit length is obtained by differentiating the objective function (D.1.3) with respect to the width w;

$$\frac{\mathrm{d}F}{\mathrm{d}w} = \sigma \frac{\phi_o^2}{l} \tag{D.1.8}$$

This analytical sensitivity result is compared with the result obtained from the sensitivity formula (6.2.12):

$$\dot{G} = \int_{\Gamma^1} \sigma E_t^2 V_n \mathrm{d}\Gamma \tag{D.1.9}$$

(D.1.2) is inserted into the sensitivity formula (D.1.9);

$$\dot{G} = \sigma L \frac{\phi_o^2}{l} V_n \tag{D.1.10}$$

Using $V_n = \frac{\mathrm{d}w}{\mathrm{d}t}$, the design sensitivity per unit length is obtained as

$$\frac{\mathrm{d}G}{\mathrm{d}w} = \sigma \frac{\phi_o^2}{l} \tag{D.1.11}$$

which is the same result as the analytical one in (D.1.8).

Next, the sensitivity formula (6.2.10) in Sect. 6.2.2 for the DC conductor system is applied to three shape optimization problems of two-dimensional design model, which do not have the analytic field solutions. If the result of the first example, which have the known optimal design, is obtained as the expected optimal design,

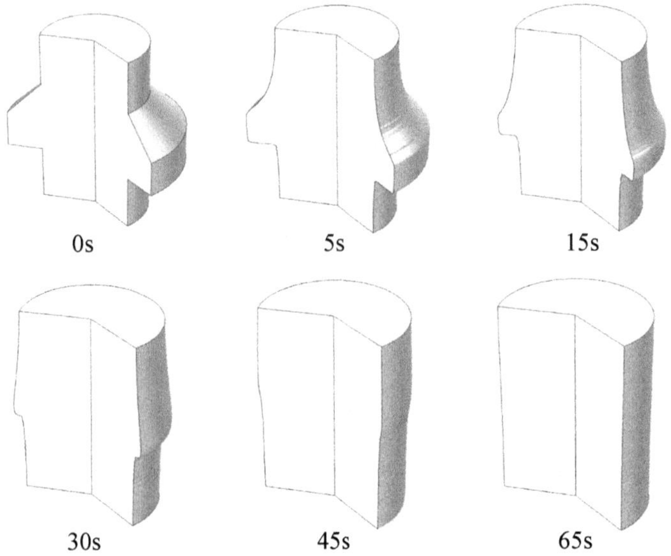

Fig. D.3 Axi-symmetry cylindrical conductor: shape variation

Fig. D.4 Axi-symmetry cylindrical conductor: evolution of objective function

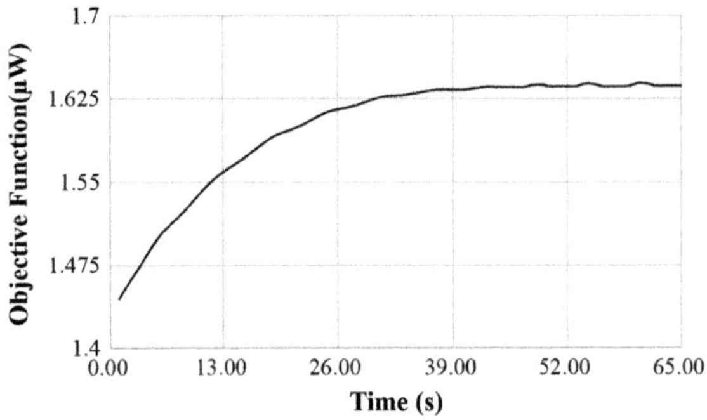

Fig D.5 Junction of two conductors with a 90° arrangement-outer boundary design, Joule loss power

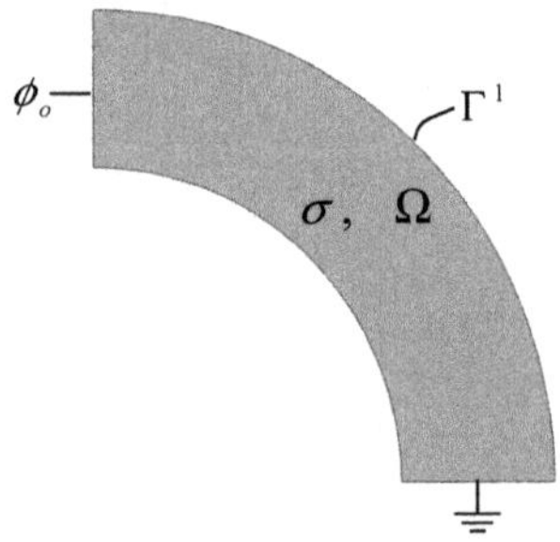

the shape optimization using the sensitivity formula is feasible for the design of the DC conductor system. The rest two design problems, which do not have the known optimal designs, are also tested to show that the design method is useful for the design of the DC conductor system. In the two-dimensional examples, the state variable is numerically calculated by the finite element method, and its result is used to evaluate the sensitivity formula. The sensitivity obtained is used for the optimization algorithm to evolve the shape of the DC conductor. The level set method is used to represent the shape evolution of the design model.

D.1.2. Axi-Symmetric Cylindrical Conductor

For an example with a known optimal design, consider a conductor of conductivity σ in Fig. D.2, where its lateral surface is not cylindrical. A potential difference ϕ_o is applied between two electrodes. The electric field and the current are not uniform in the conductor. The design objective is to maximize the Joule loss by deforming the conductor surface of Neumann boundary Γ^1. Under the constraint of the constant conductor volume, it is expected that the shape of the conductor is changed to be a cylinder, which provides the minimum resistance.

The objective function to be maximized is the Joule loss power;

$$F = P_r = \int_\Omega \sigma E^2(\phi) m_p \mathrm{d}\Omega \tag{D.1.12}$$

In this design problem, the shape of the conductor surface Γ^1 is the design variable, which has a constraint of constant conductor volume;

$$\int_\Omega \mathrm{d}\Omega = C \tag{D.1.13}$$

where C is a constant. By using the material derivative formula (3.1.22) in Sect. 3.1.2, the material derivative of the constraint (D.1.13) is obtained as

$$\int_\Gamma V_n \mathrm{d}\Gamma = 0 \tag{D.1.14}$$

The sensitivity formula (6.2.12) in Sect. 6.2.2 is the shape sensitivity for this outer boundary design:

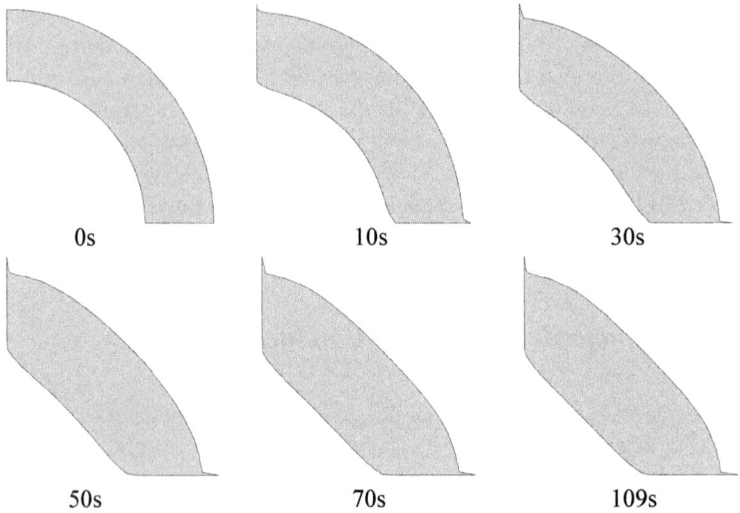

Fig. D.6 Junction of two conductors with a 90° arrangement: shape variation

Fig. D.7 Junction of two conductors with a 90° arrangement: evolution of objective function

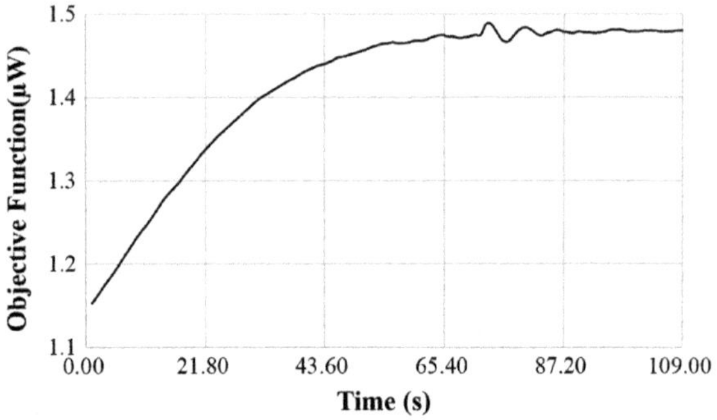

$$\dot{G} = \int_{\Gamma^1} \sigma E_t^2(\phi) V_n \mathrm{d}\Gamma \tag{D.1.15}$$

For this maximization problem, the design velocity is taken as

$$V_n = \sigma E_t^2(\phi) \tag{D.1.16}$$

In order that the velocity field satisfies the constraint (D.1.14), the design velocity (D.1.16) is modified by subtracting its average V_{na} to become U_n as

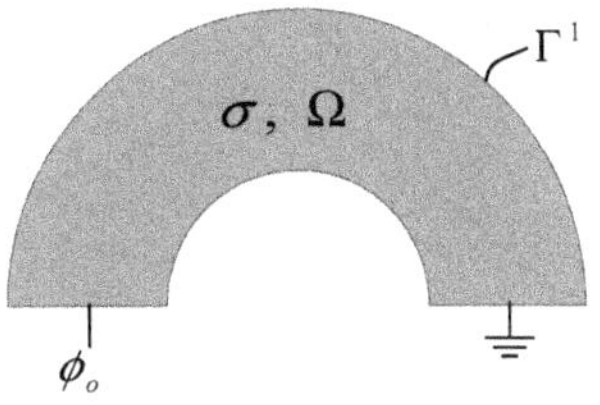

Fig D.8 Junction of two conductors with a 180° arrangement-outer boundary design, Joule loss power

$$U_n = V_n - V_{na} \tag{D.1.17}$$

where

$$V_{na} = \int_{\Gamma^1} \sigma E_t^2(\phi) \mathrm{d}\Gamma / \int_{\Gamma^1} \mathrm{d}\Gamma \tag{D.1.18}$$

The design result is shown in Fig. D.3, where the shape of the conductor boundary becomes flat with the iteration as expected. The final design of a cylindrical shape is obtained at the 65 s, when the objective function converges to the maximum value as in Fig. D.4. The result of this example shows the feasibility of the shape sensitivity analysis for the outer boundary in the axi-symmetry two-dimensional DC conductor system.

The sensitivity formula (6.2.10) in Sect. 6.2.2 is applied to two shape optimization problems of two-dimensional design model, which has neither the analytic field solutions nor the known optimal designs. The design models are conductor junctions which connect two conductors. Under the constraint for constant junction volume, the shape of the junction is designed to maximize the Joule loss of the junction. The optimal design of the junction provides a minimized resistance.

D.1.3. Junction of Two Conductors with 90° Arrangement

A junction of conductivity σ connects two conductors with 90° arrangement, as in Fig D.5, where a voltage ϕ_o is applied between the two electrodes. The change of the conductor shape results in the change of the resistance. With a voltage given, the Joule loss power increases as the resistance decreases. The design objective is to maximize the Joule loss power of the junction by deforming the junction surface of the Neumann boundary Γ^1 under the constraint of constant junction volume.

The objective function to be maximized is the Joule loss power, which is the integration of the Joule loss power density in Ω;

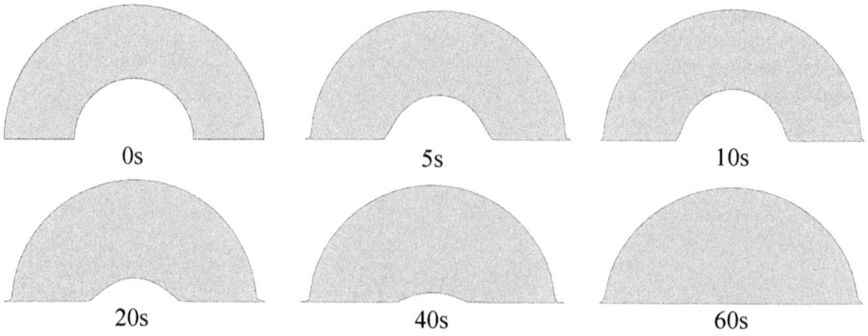

Fig. D.9 Junction of two conductors with a 180° arrangement: shape variation

Fig. D.10 Junction of two conductors with a 180° arrangement: evolution of objective function

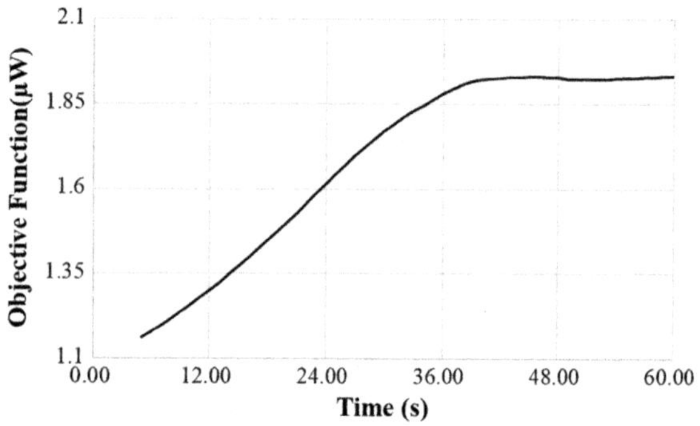

$$F = P_r = \int_{\Omega} \sigma E^2(\phi) m_p \mathrm{d}\Omega \tag{D.1.19}$$

In this design problem, the Neumann boundary Γ^1 of the junction is the design variable, which is subject to a constraint of constant conductor volume;

$$\int_{\Omega} \mathrm{d}\Omega = C \tag{D.1.20}$$

where C is a constant. The material derivative of the constraint (D.1.20) is obtained as

$$\int_{\Gamma} V_n \mathrm{d}\Gamma = 0 \tag{D.1.21}$$

The shape sensitivity is calculated on the junction boundary by the formula (6.2.12);

$$\dot{G} = \int_{\Gamma^1} \sigma E_t^2(\phi) V_n \mathrm{d}\Gamma \tag{D.1.22}$$

The design velocity U_n for the constraint is taken as

$$U_n = V_n - V_{na} \tag{D.1.23}$$

where

$$V_n = \sigma E_t^2(\phi) \tag{D.1.24}$$

$$V_{na} = \int_{\Gamma^1} \sigma E_t^2(\phi) \mathrm{d}\Gamma \Big/ \int_{\Gamma^1} \mathrm{d}\Gamma \tag{D.1.25}$$

Fig D.6 shows the shape variation of the junction during the optimization. To reduce the resistance, its length becomes shorter and its cross-sectional area becomes wider. The final design is obtained at the 109 s, when the objective function converges to the maximum value as in Fig. D.7.

D.1.4 Junction of Two Conductors with 180° Arrangement

A junction of conductivity σ connects two conductors with 180° arrangement as in Fig D.8. A voltage ϕ_o is applied between the two electrodes on the surfaces of the conductors. To minimize the resistance, the Joule loss power of the junction is maximized by deforming the conductor surface of the Neumann boundary Γ^1 under the constraint for constant junction volume.

The objective function to be maximized is the integration of the Joule loss power density in Ω;

$$F = P_r = \int_{\Omega} \sigma E^2(\phi) m_p \mathrm{d}\Omega \tag{D.1.26}$$

In this design problem, the outer boundary Γ^1 of the junction is the design variable, which has a constraint of constant conductor volume;

$$\int_{\Omega} \mathrm{d}\Omega = C \tag{D.1.27}$$

where C is a constant. The material derivative of the constraint (D.1.27) is obtained as

$$\int_{\Gamma} V_n \mathrm{d}\Gamma = 0 \tag{D.1.28}$$

The shape sensitivity is calculated on the junction boundary by the formula (6.2.12):

$$\dot{G} = \int_{\Gamma^1} \sigma E_t^2(\phi) V_n \mathrm{d}\Gamma \tag{D.1.29}$$

The design velocity U_n for the optimization algorithm is taken for the constraint as

$$U_n = V_n - V_{na} \tag{D.1.30}$$

where

$$V_n = \sigma E_t^2(\phi) \tag{D.1.31}$$

$$V_{na} = \int_{\Gamma^1} \sigma E_t^2(\phi) \mathrm{d}\Gamma \Big/ \int_{\Gamma^1} \mathrm{d}\Gamma \tag{D.1.32}$$

Fig D.9 shows the shape variation of the design variable during the optimization. As the conductor shape is deformed with the iteration, the length becomes shorter and the cross-sectional area becomes wider. The final design is obtained at the 60 s, when the objective function converges to the maximum value as in Fig. D.10.

Index

I. H. Park, *Design Sensitivity Analysis and Optimization of Electromagnetic Systems*,
Mathematical and Analytical Techniques with Applications to Engineering,
https://doi.org/10.1007/978-981-13-0230-5

Printed by Printforce, the Netherlands